# A Modern Introduction to Differential Equations

Second Edition

ELSEVIER *science & technology books*

ELSEVIER

 *Companion Web Site:*

**http://www.elsevierdirect.com/companions/9780123747464**

*A Modern Introduction to Differential Equations, Second Edition,*
*by Henry Ricardo*

## Resources for Professors:

- All figures from the book available as PowerPoint slides and as jpegs
- Links to Web sites carefully chosen to supplement the content of the textbook
- Online Student Solutions Manual is now available through separate purchase
- Also available with purchase of *A Modern Introduction to Differential Equations, 2nd Ed.*,
  password protected and activated upon registration, online Instructors' Solutions Manual

TOOLS FOR **ALL** YOUR TEACHING NEEDS
textbooks.elsevier.com

ELSEVIER

**ACADEMIC
PRESS**

# A Modern Introduction to Differential Equations

## Second Edition

Henry J. Ricardo
Medgar Evers College
City University of New York
Brooklyn, NY

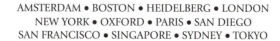

AMSTERDAM • BOSTON • HEIDELBERG • LONDON
NEW YORK • OXFORD • PARIS • SAN DIEGO
SAN FRANCISCO • SINGAPORE • SYDNEY • TOKYO

Academic Press is an imprint of Elsevier

Elsevier Academic Press
30 Corporate Drive, Suite 400, Burlington, MA 01803, USA
525 B Street, Suite 1900, San Diego, California 92101-4495, USA
84 Theobald's Road, London WC1X 8RR, UK

This book is printed on acid-free paper. ⊚

**Library of Congress Cataloging-in-Publication Data**
Ricardo, Henry.
    A modern introduction to differential equations / Henry J. Ricardo – 2nd ed.
        p. cm.
    Includes bibliographical references and index.
    ISBN 978-0-12-374746-4 (hardcover : alk. paper)

1. Differential equations.  I. Title.
    QA371.R357 2009
    515'.35–dc22

                                                                2008050079


**British Library Cataloguing in Publication Data**
A catalogue record for this book is available from the British Library.

ISBN 13: 978-0-12-374746-4

For all information on all Elsevier Academic Press publications
visit our Web site at *www.elsevierdirect.com*

Printed in Canada
09 10   9 8 7 6 5 4 3 2 1

*Typeset by*: diacriTech, India.

For Catherine, the constant function in my life, and for all the derivatives:

Henry, Marta, Tomás, and Nicholas Ricardo
Cathy, Mike, and Christopher Corcoran
Christine and Greg Gritmon

# Contents

*Denotes an optional section.

# Preface

## PHILOSOPHY

The evolution of differential equations courses I described in the Preface to the first edition of this book has progressed nicely. In particular, the quantitative, graphical, and qualitative aspects of the subject have been receiving increased attention, due in large part to the availability of technology in the classroom and at home.

As before, this text presents a solid yet highly accessible introduction to differential equations, developing the concepts from a dynamical systems perspective and employing technology to treat topics graphically, numerically, and analytically. In particular, the book acknowledges that most differential equations cannot be solved in closed form and makes extensive use of qualitative and numerical methods to analyze solutions.

The text includes discussions of several significant mathematical models, although there is no systematic attempt to teach the art of modeling. Similarly, the text introduces only the minimal amount of linear algebra necessary for an analysis of systems.

This book is intended to be the text for the one-semester ordinary differential equations course that is typically offered at the sophomore–junior level, but with some differences. The prerequisite for the course is two semesters of calculus. No prior knowledge of multivariable calculus and linear algebra is needed, because essential concepts from these subjects are developed within the text itself. This book is aimed primarily at students majoring in mathematics, the natural sciences, and engineering. However, students in economics, business, and the social sciences who have the necessary background should also benefit.

## USE OF TECHNOLOGY

This text assumes that the student has access to a computer algebra system (CAS) or perhaps some specialized software that will enable him or her to construct the required graphs (solution curves, phase planes, etc.) and numerical approximations. For example, a spreadsheet program can be used effectively to implement Euler's method of approximating solutions. Although I use *Maple*® in my own course, no specific software or hardware platform is assumed for this book. To a large extent, even a graphing calculator will suffice.

## PEDAGOGICAL FEATURES AND WRITING STYLE

This book is truly meant to be *read* by the students. The style is accessible without excessive mathematical formality and extraneous material, although it does provide a solid foundation upon which individual teachers can build according to their taste and the students' needs. (*Feedback from users of the first edition suggests that students find the book easy to read.*) Every chapter has an informal *Introduction* that sets the tone and motivates the material to come. I have tried to motivate the introduction of new concepts in various ways, including references to earlier, more elementary mathematics courses taken by the student. Each chapter concludes with a narrative *Summary* reminding the reader of the important concepts in the chapter. Within sections there are figures and tables to help students visualize or summarize concepts. There are many worked-out examples and exercises taken from biology, chemistry, and economics, as well as from traditional pure mathematics, physics, and engineering. In the text itself, I lead the student through qualitative and numerical analyses of problems that would have been difficult to handle before the ubiquitous presence of graphing calculators and computers. The exercises that appear at the end of each content section range from the routine to the challenging, the latter problems often requiring some exploration and/or theoretical justification ("proof"). Some exercises introduce students to supplementary concepts. I have provided answers to odd-numbered problems at the back of the book, with more detailed solutions to these problems in the separate *Student Solutions Manual.* Every chapter has at least one project following the *Summary.*

I have written the book the way I teach the course, using a colloquial and interactive style. The student is frequently urged to "Think about this," "Check this," or "Make sure you understand." In general there are no proofs of theorems except for those mathematical statements that can be justified by a sequence of fairly obvious calculations/algebraic manipulations. In fact, there is no general labeling of facts as theorems, although some definitions are stated formally and key results are italicized within the text or emphasized in other ways. Also, brief historical remarks related to a particular concept or result are placed throughout the text without obstructing the flow. This is not a mathematical treatise but a friendly, informative, modern introduction to tools needed by students in many disciplines. I have enjoyed teaching such a course, and I believe my students have benefited from the experience. I sincerely hope that the user of this book also gains some insight into the modern theory and applications of differential equations.

## KEY CONTENT FEATURES

**Chapters 1–3** introduce the *basic concepts* of differential equations and focus on the analytical, graphical, and numerical aspects of first-order equations, including *slope fields* and *phase lines*. In later chapters, these aspects (including the *Superposition Principle*) are generalized in natural ways to higher-order equations and systems of equations.

**Chapter 4** starts with methods of solving important *second-order homogeneous and nonhomogeneous linear equations with constant coefficients* and introduces applications to electrical circuits and spring-mass problems. The theoretical high point of the chapter is the demonstration that any higher-order differential equation is equivalent to a system of first-order equations. The student is introduced to the *qualitative* analysis of systems (*phase portraits*), the *existence and uniqueness* of solutions of systems, and the extension of *numerical methods* for first-order equations to systems of first-order equations. Among the examples treated in this chapter are *predator-prey* systems, an *arms race* illustration, and *spring-mass* systems (including one showing *resonance*).

**Chapter 5** begins with a brief introduction to the *matrix algebra* concepts needed for the systematic exposition of *two-dimensional systems of autonomous linear equations.* (This treatment is supplemented by Appendix B.) The importance of linearity is emphasized, and the *Superposition Principle* is discussed again. The *stability* of such systems is completely characterized by means of the *eigenvalues* of the matrix of coefficients. *Spring-mass* systems are discussed in terms of their eigenvalues. There is also a brief introduction to the complexities of *nonhomogeneous* systems. Finally, via $3 \times 3$ and $4 \times 4$ examples, the student is shown how the ideas previously developed can be extended to *nth-order equations* and their equivalent systems.

**Chapter 6** reveals the *Laplace transform* and its applications to the solution of differential equations and systems of differential equations. This is perhaps the most traditional topic in the book; it is included because of its usefulness in many applied areas. In particular, students can deal with *nonhomogeneous* linear equations and systems more easily and handle *discontinuous driving forces.* The Laplace transform is applied to electric circuit problems, the deflection of beams (a boundary-value problem), and spring-mass systems. However, in the spirit of the rest of the book, Section 6.6 shows the applicability of the Laplace transform to a *qualitative* analysis of linear differential equations (*transfer functions, impulse response functions*).

**Chapter 7** presents systems of *nonlinear* equations in a systematic way. The *stability* of nonlinear systems is analyzed. The important notion of a *linear approximation* to a nonlinear equation or system is developed, including the use of a qualitative result due to Poincaré and Lyapunov. Some important examples of nonlinear systems are treated in detail, including the *Lotka-Volterra equations*, the *undamped pendulum*, and the *van der Pol oscillator. Limit cycles* are discussed.

**Appendices A–C** present important prerequisite/corequisite material from *calculus* (*single-variable and multivariable*), *vector/matrix algebra*, and *complex numbers*, respectively. **Appendix D** supplements the text by introducing the *series solutions of ordinary differential equations.*

## SECOND EDITION FEATURES

- Overall, there has been a strengthening of the exposition, ranging from individual words to entire paragraphs. The result is increased clarity.

- *First-order initial-value* and *boundary-value problems* are now in a section of their own.

- The discussion of *compartment problems* has been expanded and appears in a separate section.

- To improve the flow of the exposition, some text material on *error* in numerical approximation has been removed to Section A.3.

- The treatments of *undetermined coefficients* and *variation of parameters* have been expanded, and each topic has a section of its own, with helpful tables and examples.

- *Spring-mass problems* are now in a separate section.

- There are *new examples and figures* in this edition.

- Many *new exercises* have been added (with a few culled). All exercises are now divided into A, B, and C problems, and the range of problems from drill exercises (A) to challenging problems (C) has been increased.

- One *project* has been replaced.

## SUPPLEMENTS

- **Instructor's Solutions Manual**   Contains solutions to all exercises in the text. This is available free to instructors who adopt the text.

- **Student Solutions Manual**   Provides complete solutions to every odd-numbered exercise in the text.

# Acknowledgments

The early influences on the approach and content of both editions of this book were (1) The Boston University Differential Equations Project; (2) The Consortium for Ordinary Differential Equations Experiments (C•ODE•E); (3) The Special Issue on Differential Equations: *College Mathematics Journal*, Vol. 25, No. 5 (November 1994); and (4) David Sánchez's review of ODE texts in the April 1998 issue of the *American Mathematical Monthly* (pp. 377–383).

Over the years I have enjoyed the cooperation and candor of several classes of Medgar Evers College students who learned from various versions of this text. I single out Tamara Battle, Hibourahima Camara, Lenston Elliott, Eleanor Holder, Patrice Williams, Barry Gregory, Fitzgerald Providence, and Cindy James as representatives of these patient students.

I thank my chairperson, Darius Movasseghi, for his encouragement and for his support in such crucial areas as course scheduling and ensuring the availability of technology. Our senior college laboratory technician, Ernst Gracia, also deserves thanks for his frequent interventions when technology (or my use of it) has gone awry. I am grateful to my colleague Mahendra Kawatra for his continuing encouragement and support.

At Academic Press/Elsevier, I thank my editor Lauren Schultz Yuhasz for her support of this second edition and for her encouragement in the past, Sarah Hajduk for her timely advice about manuscript preparation, and Julie Ochs for her guidance throughout the production process.

Above all, I am (still) grateful to my wife, Catherine, for her love, steadfast support, and patience during the writing of this book and at all other times. Her encouragement and practical suggestions as I hunted and pecked my way through the manuscript were invaluable.

I welcome any questions, comments, and suggestions for improvement. Please contact me at: *henry@mec.cuny.edu.*

Henry Ricardo

# CHAPTER 1

# Introduction to Differential Equations

## INTRODUCTION

What do the following situations have in common?

- An arms race between nations
- Tracking of the rate at which HIV-positive patients come to exhibit AIDS
- The dynamics of supply and demand in an economy
- The interaction between two or more species of animals on an island

The answer is that each of these areas of investigation can be modeled with differential equations. This means that the essential features of these problems can be represented using one or several differential equations, and the solutions of the mathematical problems provide insights into the future behavior of the systems being studied.

This book deals with *change*, with *flux*, with *flow*, and, in particular, with the *rate* at which change takes place. Every living thing changes. The tides fluctuate over the course of a day. Countries increase and diminish their stockpiles of weapons. The price of oil rises and falls. The proper framework of this course is **dynamics**—the study of systems that evolve over time.

The origin of dynamics (originally an area of physics) and of differential equations lies in the earliest work by the English scientist and mathematician Sir Isaac Newton (1642–1727) and the German philosopher and mathematician Gottfried Wilhelm Leibniz (1646–1716) in developing the new science of calculus in the seventeenth century. Newton in particular was concerned with determining the laws governing motion, whether of an apple falling from a tree or of the planets moving in their orbits. He was concerned with *rates of change*. However, you mustn't think that the subject of differential equations is all about physics. The same types of equations and the same kind of analysis of dynamical systems can be used to model and understand situations in biology, economics, military strategy, and chemistry, for example. Applications of this sort will be found throughout this book.

In the next section, we will introduce the language of differential equations and discuss some applications.

## 1.1 BASIC TERMINOLOGY

### 1.1.1 Ordinary and Partial Differential Equations

*Ordinary Differential Equations*

> **Definition 1.1.1**
>
> An **ordinary differential equation** (**ODE**) is an equation that involves an unknown function of a single variable, its independent variable, and one or more of its derivatives.

### ■ Example 1.1.1 An Ordinary Differential Equation

Here's a typical elementary ODE, with some of its components indicated:

unknown function, $y \downarrow$

$$3\frac{dy}{dt} = y$$

independent variable, $t \uparrow$ .

This equation describes an unknown function of $t$ that is equal to three times its own derivative. Expressed another way, the differential equation describes a function whose rate of change is proportional to its size (value) at any given time, with constant of proportionality one-third. ■

The Leibniz notation for a derivative, $\frac{d()}{d()}$, is helpful because the independent variable (the fundamental quantity whose change is causing other changes) appears in the denominator, the dependent variable in the numerator. The three equations

$$\frac{dy}{dx} + 2xy = e^{-x^2}$$

$$x''(t) - 5x'(t) + 6x(t) = 0$$

$$\frac{dx}{dt} = \frac{3t^2 + 4t + 2}{2(x-1)}$$

leave no doubt about the relationship between independent and dependent variables. But in an equation such as $(w')^2 + 2t^3 w' - 4t^2 w = 0$, we must *infer* that the unknown function $w$ is really $w(t)$, a function of the independent variable $t$.

In many dynamical applications, the independent variable is time, represented by $t$, and we may denote the function's derivative using Newton's dot notation,[1] as in the equation $\ddot{x} + 3t\dot{x} + 2x = \sin(\omega t)$. You should be able to recognize a differential equation no matter what letters are used for the independent and dependent variables and no matter what derivative notation is employed. The context will determine what the various letters mean, and it's the *form* of the equation that should be recognized. For example, you should be able to see that the two ordinary differential equations

$$\text{(A)} \quad \frac{d^2 u}{dt^2} - 3\frac{du}{dt} + 7u = 0 \quad \text{and} \quad \text{(B)} \quad \frac{d^2 y}{dx^2} = 3\frac{dy}{dx} - 7y$$

are the same—that is, they are describing the same mathematical or physical behavior. In Equation (A) the unknown function $u$ depends on $t$, whereas in Equation (B) the function $y$ is a function of the independent variable $x$, but both equations describe the same relationship that involves the unknown function, its derivatives, and the independent variable. Each equation is describing a function whose second derivative equals three times its first derivative minus seven times itself.

### The Order of an Ordinary Differential Equation

One way to classify differential equations is by their **order**.

---

**Definition 1.1.2**

An ordinary differential equation is of **order $n$**, or is an **$n$th-order equation**, if the highest derivative of the unknown function in the equation is the $n$th derivative.

---

The equations

$$\frac{dy}{dx} + 2xy = e^{-x^2}$$

$$(w')^2 + 2t^3 w' - 4t^2 w = 0$$

$$\frac{dx}{dt} = \frac{3t^2 + 4t + 2}{2(x - 1)}$$

are all first-order differential equations because the highest derivative in each equation is the first derivative. The equations

$$x''(t) - 5x'(t) + 6x(t) = 0$$

---

[1] In this notation, $\dot{x} = dx/dt$, $\ddot{x} = d^2x/dt^2$, and $\dddot{x} = d^3x/dt^3$.

and

$$\ddot{x} + 3t\dot{x} + 2x = \sin(\omega t)$$

are second-order equations, and $e^{-x}y^{(5)} + (\sin x)y''' = 3e^x$ is of order 5.

### A General Form for an Ordinary Differential Equation

If $y$ is the unknown function with a single independent variable $x$, and $y^{(k)}$ denotes the $k$th derivative of $y$, we can express an $n$th-order differential equation in a concise mathematical form as the relation

$$F\left(x, y, y', y'', y''', \ldots, y^{(n-1)}, y^{(n)}\right) = 0$$

or often as

$$y^{(n)} = G\left(x, y, y', y'', y''', \ldots, y^{(n-1)}\right).$$

The next example shows what these forms look like in practice.

### ■ Example 1.1.2 General Form for a Second-Order ODE

If $y$ is an unknown function of $x$, then the second-order ordinary differential equation $2\frac{d^2y}{dx^2} + e^x\frac{dy}{dx} = y + \sin x$ can be written as $2\frac{d^2y}{dx^2} + e^x\frac{dy}{dx} - y - \sin x = 0$ or as

$$\underbrace{2y'' + e^xy' - y - \sin x}_{F(x,y,y',y'')} = 0.$$

■

Note that $F$ denotes a mathematical expression involving the independent variable $x$, the unknown function $y$, and the first and second derivatives of $y$.

Alternatively, in this last example we could use ordinary algebra to solve the original differential equation for its highest derivative and write the equation as $y'' = \underbrace{\frac{1}{2}\sin x + \frac{1}{2}y - \frac{1}{2}e^xy'}_{G(x,y,y')}$.

### Partial Differential Equations

If we are dealing with functions of *several* variables and the derivatives involved are *partial* derivatives, then we have a **partial differential equation (PDE)**. (See Section A.7 if you are not familiar with partial derivatives.) For example, the partial differential equation $\frac{\partial^2 u}{\partial x^2} - \frac{1}{c^2}\frac{\partial^2 u}{\partial t^2} = 0$, which is called the *wave equation*, is of fundamental importance in many areas of physics and engineering. In this equation we are assuming that $u = u(x, t)$, a function of the two variables $x$ and $t$. However, in this text, when we use the term *differential equation*, we'll mean an *ordinary* differential equation. Often we'll just write *equation*, if the context makes it clear that an ordinary differential equation is intended.

### Linear and Nonlinear Ordinary Differential Equations

Another important way to categorize differential equations is in terms of whether they are *linear* or *nonlinear*.

---

**Definition 1.1.3**

If $y$ is a function of $x$, then the general form of a **linear ordinary differential equation of order $n$** is

$$a_n(x)y^{(n)} + a_{n-1}(x)y^{(n-1)} + \cdots + a_2(x)y'' + a_1(x)y' + a_0(x)y = f(x). \qquad (1.1.1)$$

---

What is important here is that each coefficient function $a_i$, as well as $f$, depends on the independent variable $x$ alone and doesn't have the dependent variable $y$ or any of its derivatives in it. In particular, Equation (1.1.1) involves no products or quotients of $y$ and/or its derivatives.

### ■ Example 1.1.3 A Second-Order Linear Equation

The equation $x'' + 3tx' + 2x = \sin(\omega t)$, where $\omega$ is a constant, is linear. We can see the form of this equation as follows:

$$\underbrace{1}_{a_2(t)} \cdot x'' + \underbrace{3t}_{a_1(t)} \cdot x' + \underbrace{2}_{a_0(t)} \cdot x = \underbrace{\sin(\omega t)}_{f(t)}.$$

The coefficients of the various derivatives of the unknown function $x$ are functions (sometimes constant) of the independent variable $t$ alone. ■

The next example shows that not all first-order equations are linear.

### ■ Example 1.1.4 A First-Order Nonlinear Equation (an HIV Infection Model)

The equation $\frac{dT}{dt} = s + rT\left(1 - \frac{T}{T_{\max}}\right) - \mu T$ models the growth and death of T cells, an important component of the immune system.[2] Here $T(t)$ is the number of T cells present at time $t$. If we rewrite the equation by removing parentheses, we get $\frac{dT}{dt} = s + rT - \left(\frac{r}{T_{\max}}\right)T^2 - \mu T$, and we see that there is a term involving the square of the unknown function. Therefore, the equation is not linear. ■

In general, there are more systematic ways to analyze linear equations than to analyze nonlinear equations, and we'll see some of these methods in Chapters 2, 5, and 6. However, nonlinear equations are important and appear throughout this book. In particular, Chapter 7 is devoted to their analysis.

---

[2] E. K. Yeargers, R. W. Shonkwiler, and J. V. Herod, *An Introduction to the Mathematics of Biology: With Computer Algebra Models* (Boston: Birkhäuser, 1996): 341.

## 1.1.2 Systems of Ordinary Differential Equations

In earlier mathematics courses, you have had to deal with *systems* of algebraic equations, such as

$$3x - 4y = -2$$
$$-5x + 2y = 7.$$

Similarly, in working with differential equations, you may find yourself confronting **systems of differential equations**, such as

$$\frac{dx}{dt} = -3x + y$$
$$\frac{dy}{dt} = x - 3y$$

or

$$\dot{x} = -sx + sy$$
$$\dot{y} = -xz + rx - y$$
$$\dot{z} = xy - bz$$

where $b, r$, and $s$ are constants. (Recall that $\dot{x} = \frac{dx}{dt}$, $\dot{y} = \frac{dy}{dt}$, and $\dot{z} = \frac{dz}{dt}$.) The last system arose in a famous study of meteorological conditions.

Note that each of these systems of differential equations has a different number of equations and that each equation in the first system is *linear*, whereas the last two equations in the second system are *nonlinear* because they contain products—$xz$ in the second equation and $xy$ in the third—of some of the unknown functions. Naturally, we'll call a system in which all equations are linear a **linear system**, and we'll refer to a system with at least one nonlinear equation as a **nonlinear system**. In Chapters 4, 5, 6, and 7, we'll see how systems of differential equations arise and learn how to analyze them. For now, just try to understand the *idea* of a system of differential equations.

## Exercises 1.1

### A

In Problems 1–12, (a) identify the independent variable and the dependent variable of each equation; (b) give the order of each differential equation; and (c) state whether the equation is linear or nonlinear. If your answer to (c) is nonlinear, explain why this is true.

**1.** $y' = y - x^2$
**2.** $xy' = 2y$

3.  $x'' + 5x = e^{-x}$

4.  $(y')^2 + x = 3y$

5.  $xy'(xy' + y) = 2y^2$

6.  $\dfrac{d^2r}{dt^2} = 3\dfrac{dr}{dt} + \sin t$

7.  $y^{(4)} + xy''' + e^x = 0$

8.  $y'' + ky'(y^2 - 1) + 3y = -2\cos t$

9.  $\ddot{x} - 2\ddot{x} + 4t\dot{x} - e^t x = t + 1$

10.  $x^{(7)} + t^2 x^{(5)} = xe^t$

11.  $e^{y'} + 3xy = 0$

12.  $t^2 R''' - 4tR'' + R' + 3R = e^t$

13.  Classify each of the following systems as linear or nonlinear:

a.  $\dfrac{dy}{dt} = x - 4xy$

$\dfrac{dx}{dt} = -3x + y$

b.  $Q' = tQ - 3t^2 R$

$R' = 3Q + 5R$

c.  $\dot{x} = x - xy + z$

$\dot{y} = -2x + y - yz$

$\dot{z} = 3x - y + z$

d.  $\dot{x} = 2x - ty + t^2 z$

$\dot{y} = -2tx + y - z$

$\dot{z} = 3x - t^3 y + z$

## B

1.  For what value(s) of the constant $a$ is the differential equation

$$\frac{d^2x}{dt^2} + (a^2 - a)x\frac{dx}{dt} = te^{(a-1)x}$$

a linear equation?

2.  Rewrite the following equations as linear equations, if possible.

a.  $\dfrac{dx}{dt} = \ln(2^x)$

b.  $x' = \begin{cases} \dfrac{x^2 - 1}{x - 1} & \text{for} \quad x \neq 1 \\ 2 & \text{for} \quad x = 1 \end{cases}$

c.  $x' = \begin{cases} \dfrac{x^4 - 1}{x^2 - 1} & \text{for} \quad x \neq 1 \\ 2 & \text{for} \quad x = 1. \end{cases}$

## 1.2 SOLUTIONS OF DIFFERENTIAL EQUATIONS

### 1.2.1 Basic Notions

In past mathematics courses, whenever you encountered an equation, you were probably asked to *solve* it, or find a *solution*. Simply put, a *solution* of a differential equation is a function that *satisfies* the equation: When you substitute this function into the differential equation, you get a true mathematical statement—an *identity*.

> ### Definition 1.2.1
>
> A **solution** of an $n$th-order differential equation $F(x, y, y', y'', y''', \ldots, y^{(n-1)}, y^{(n)}) = 0$, or $y^{(n)} = G(x, y, y', y'', y''', \ldots, y^{(n-1)})$, on an interval $(a, b)$ is a real-valued function $y = y(x)$ such that all the necessary derivatives of $y(x)$ exist on the interval and $y(x)$ satisfies the equation for every value of $x$ in the interval. **Solving** a differential equation means finding all possible solutions of a given equation.

Even before we begin learning formal solution methods in Chapter 2, we can *guess* the solutions of some simple differential equations. The next example shows how to guess intelligently.

### ■ Example 1.2.1 Guessing and Verifying a Solution to an ODE

The first-order linear differential equation $\frac{dB}{dt} = kB$, where $k$ is a given positive constant, is a simple model of a bank balance, $B(t)$, under continuous compounding $t$ years after the initial deposit. The rate of change of $B$ at any instant is proportional to the *size* of $B$ at that instant, with $k$ as the constant of proportionality. This equation expresses the fact that the larger the bank balance at any time $t$, the faster it will grow.

You can guess what kind of function describes $B(t)$ if you think about the elementary functions you know and their derivatives. What kind of function has a derivative that is a constant multiple of itself? You should be able to see why $B(t)$ must be an *exponential* function of the form $ae^{kt}$, where $a$ is any constant. By substituting $B(t) = ae^{kt}$ into the original differential equation, you can verify that you have guessed correctly. The left-hand side of the equation becomes $\frac{d(ae^{kt})}{dt}$, which equals $kae^{kt}$, and the right-hand side of the equation is $k(ae^{kt})$. The left-hand side equals the right-hand side for all values of $t$, giving us an identity.

Anticipating an idea we'll discuss later in this section, we can let $t = 0$ in our solution function to conclude that $B(0) = ae^{k(0)} = a$—that is, the constant $a$ must equal the initial deposit. Finally, we can express the solution as $B(t) = B(0)e^{kt}$.    ■

Note that in Definition 1.2.1 we say "a" solution rather than "the" solution. A differential equation, if it has a solution at all, usually has more than one solution. Also, we should pay attention to the interval on which the solution may be defined. Later in this section and in Section 2.8, we will discuss in more detail the question of the existence and uniqueness of solutions. For now, let's just learn to recognize when a function is a solution of a differential equation, as in the next example.

## ■ Example 1.2.2 Verifying a Solution of a Second-Order Equation

Suppose that someone claims that $x(t) = 5e^{3t} - 7e^{2t}$ is a solution of the second-order linear equation $x'' - 5x' + 6x = 0$ on the whole real line—that is, for all values of $t$ in the interval $(-\infty, \infty)$. You can prove that this claim is correct by calculating $x'(t) = 15e^{3t} - 14e^{2t}$ and $x''(t) = 45e^{3t} - 28e^{2t}$ and then substituting these expressions into the original equation:

$$x''(t) - 5x'(t) + 6x(t) = \overbrace{(45e^{3t} - 28e^{2t})}^{x''(t)} - 5\overbrace{(15e^{3t} - 14e^{2t})}^{x'(t)} + 6\overbrace{(5e^{3t} - 7e^{2t})}^{x(t)}$$

$$= 45e^{3t} - 28e^{2t} - 75e^{3t} + 70e^{2t} + 30e^{3t} - 42e^{2t}$$

$$= -30e^{3t} + 42e^{2t} + 30e^{3t} - 42e^{2t} = 0.$$

Because $x(t) = 5e^{3t} - 7e^{2t}$ satisfies the original equation, we see that $x(t)$ is a solution. But this is not the only solution of the given differential equation. For example, you can check that $x_2(t) = -\pi e^{3t} + \frac{2}{3}e^{2t}$ is also a solution. We'll discuss this kind of situation in more detail a little later. ■

### *Implicit Solutions*

Think back to the concept of *implicit functions* in calculus. The idea here is that sometimes functions are not defined cleanly (explicitly) by a formula in which the dependent variable (on one side) is expressed in terms of the independent variable and some constants (on the other side), as in the solution $x = x(t) = 5e^{3t} - 7e^{2t}$ of Example 1.2.2. For instance, you may be given the *relation $x^2 + y^2 = 5$*, which can be written in the form $G(x, y) = 0$, where $G(x, y) = x^2 + y^2 - 5$. The graph of this relation is a circle of radius $\sqrt{5}$ centered at the origin, and this graph does not represent a function. (*Why?*) However, this relation *does* define two functions implicitly: $y_1(x) = \sqrt{5 - x^2}$ and $y_2(x) = -\sqrt{5 - x^2}$, both having domains $[-\sqrt{5}, \sqrt{5}]$. More advanced courses in analysis discuss when a relation actually defines one or more implicit functions. For now, just remember that even if you can't untangle a relation to get an explicit formula for a function, you can use implicit differentiation to find derivatives of any differentiable functions that may be buried in the relation.

In trying to solve differential equations, often we can't find an explicit solution and must be content with a solution defined implicitly.

## ■ Example 1.2.3 Verifying an Implicit Solution

We want to show that any function $y$ that satisfies the relation $G(x, y) = x^2 + y^2 - 5 = 0$ is a solution of the differential equation $\frac{dy}{dx} = -\frac{x}{y}$.

First, we differentiate the relation implicitly, treating $y$ as $y(x)$, an implicitly defined function of the independent variable $x$:

$$(1) \quad \frac{d}{dx}G(x,y) = \frac{d}{dx}(x^2 + y^2 - 5) = \frac{d}{dx}(0) = 0$$

$$(2) \quad 2x + \overbrace{2y\frac{dy}{dx}}^{\text{Chain Rule}} - \frac{d}{dx}(5) = 0$$

$$(3) \quad 2x + 2y\frac{dy}{dx} = 0.$$

Now we solve Equation (3) for $\frac{dy}{dx}$, getting $\frac{dy}{dx} = \frac{-2x}{2y} = -\frac{x}{y}$ and proving that any function defined implicitly by the relation above is a solution of our differential equation. ■

## 1.2.2 Families of Solutions I

Next, we want to discuss how many solutions a differential equation could have. For example, the equation $(y')^2 + 1 = 0$ has *no* real-valued solution (*think about this*), whereas the equation $|y'| + |y| = 0$ has exactly one solution, the function $y \equiv 0$. (*Why?*) As we saw in Example 1.2.2, the differential equation $x'' - 5x' + 6x = 0$ has at least two solutions.

The situation gets more complicated, as the next example shows.

## ■ Example 1.2.4 An Infinite Family of Solutions

Suppose two students, Lenston and Cindy, look at the simple first-order differential equation $\frac{dy}{dx} = f(x) = x^2 - 2x + 7$. A solution of this equation is a function of $x$ whose first derivative equals $x^2 - 2x + 7$. Lenston thinks the solution is $\frac{x^3}{3} - x^2 + 7x$, and Cindy thinks the solution is $\frac{x^3}{3} - x^2 + 7x - 10$. Both answers seem to be correct.

Solving this problem is simply a matter of integrating both sides of the differential equation:

$$y = \int dy = \int \frac{dy}{dx}dx = \int x^2 - 2x + 7 \, dx.$$

Because we are using an *indefinite* integral, there is always a constant of integration that we mustn't forget. The solution to our problem is actually an *infinite family of solutions*, $y(x) = \frac{x^3}{3} - x^2 + 7x + C$, where $C$ is any real constant. Every particular value of $C$ gives us another member of the family. We have just solved our first differential equation in this course without guessing! Every time we performed an indefinite integration (found an antiderivative) in calculus class, we were solving a simple differential equation. ■

When describing the set of solutions of a first-order differential equation such as the one in the previous example, we usually refer to it as a **one-parameter family of solutions**. The *parameter* is the constant $C$. Each definite value of $C$ gives us what is called a **particular solution** of

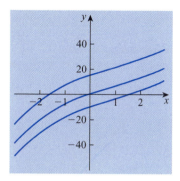

**FIGURE 1.1**
Integral curves of $\frac{dy}{dx} = x^2 - 2x + 7$ with parameters 15, 0, and $-10$

the differential equation. In the preceding example, Lenston and Cindy produced particular solutions, one with $C = 0$ and the other with $C = -10$. A particular solution is sometimes called an **integral** of the equation, and its graph is called an **integral curve** or a **solution curve**.

Figure 1.1 shows three of the integral curves of the equation $\frac{dy}{dx} = x^2 - 2x + 7$, where $C = 15, 0$, and $-10$ (from top to bottom).

The curve passing through the origin is Lenston's particular solution; the solution curve passing through the point $(0, -10)$ is Cindy's.

## Exercises 1.2

In Problems 1–10, verify that the indicated function is a solution of the given differential equation. The letters $a$, $b$, $c$, and $d$ denote constants.

**A**

1. $y'' + y = 0$; $y = \sin x$

2. $x'' - 5x' + 6x = 0$; $x = -\pi e^{3t} + \frac{2}{3}e^{2t}$

3. $\frac{1}{4}\left(\frac{dy}{dx}\right)^2 - x\frac{dy}{dx} + y = 0$; $y = x^2$

4. $t\frac{dR}{dt} - R = t^2 \sin t$; $R = t(c - \cos t)$

5. $\frac{d^4y}{dt^4} = 0$; $y = at^3 + bt^2 + ct + d$

6. $\frac{dr}{dt} = at + br$; $r = ce^{bt} - \frac{a}{b}t - \frac{a}{b^2}$

7. $xy' - 2 = 0$; $y = \ln(x^2)$

8. $y'' = a\sqrt{1 + (y')^2}$; $y = \dfrac{e^{ax} + e^{-ax}}{2a}$

9. $xy' - \sin x = 0$; $y = \int_1^x \frac{\sin t}{T} dt$ [Think of the *Fundamental Theorem of Calculus*. See Section A.4]

10. $y'' + 2xy' = 0$; $y = \int_3^x e^{-t^2} dt$ [Think of the *Fundamental Theorem of Calculus*. See Section A.4]

11. For each function, find a differential equation satisfied by that function:

   a. $y = c + \dfrac{x}{c}$, where $c$ is a constant

   b. $y = e^{ax} \sin bx$, where $a$ and $b$ are constants

   c. $y = (A + Bt)e^t$, where $A$ and $B$ are constants

   d. $y(t) = e^{-3t} + \int_1^t uy(u)du$

In each of Problems 12–15, assume that the function $y$ is defined implicitly as a function of $x$ by the given equation, where $C$ is a constant. In each case, use the technique of implicit differentiation to find a differential equation for which $y$ is a solution.

12. $xy - \ln y = C$

13. $y + \arctan y = x + \arctan x + C$

14. $y^3 - 3x + 3y = 5$

15. $1 + x^2 y + 4y = 0$

16. Is a function $y$ satisfying $x^2 + y^2 - 6x + 10y + 34 = 0$ a solution of the differential equation $\dfrac{dy}{dx} = \dfrac{3 - x}{y + 5}$? Explain your answer.

**B**

1. Verify that $y = \dfrac{x^2}{2} + \dfrac{x}{2}\sqrt{x^2 + 1} + \ln\sqrt{x + \sqrt{x^2 + 1}}$ is a solution of the equation $2y = xy' + \ln(y')$.

2. Write a paragraph explaining why $B(t)$ in Example 1.2.1—a solution of the differential equation $\dfrac{dB}{dt} = kB$—can't be a polynomial, trigonometric, or logarithmic function.

3. a. Why does the equation $(y')^2 + 1 = 0$ have no real-valued solution?

   b. Why does the equation $|y'| + |y| = 0$ have only one solution? What is the solution?

4. Explain why the equation $\dfrac{dx}{dt} = \sqrt{-|x - t|}$ has *no* real-valued solution.

5. If $c$ is a positive constant, show that the two functions $y = \sqrt{c^2 - x^2}$ and $y = -\sqrt{c^2 - x^2}$ are both solutions of the nonlinear equation $y\dfrac{dy}{dx} + x = 0$ on the interval $-c < x < c$. Explain why the solutions are not valid outside the open interval $(-c, c)$.

6. a. Verify that the function $y = \ln(|C_1 x|) + C_2$ is a solution of the differential equation $y' = \dfrac{1}{x}$ for each value of the parameters $C_1$ and $C_2$ and $x$ in the interval $(0, \infty)$.

   b. Show that there is only one genuine parameter needed for $y$. In other words, write $y = \ln(|C_1 x|) + C_2$ using only one parameter $C$.

7. Find a solution of $\dfrac{dy}{dx} + y = \sin x$ of the form $y(x) = c_1 \sin x + c_2 \cos x$, where $c_1$ and $c_2$ are constants.

8. Find a second degree polynomial $y(x)$ that is a (particular) solution of the linear differential equation $2y' - y = 3x^2 - 13x + 7$.

9. Show that the first-order nonlinear equation $(xy' - y)^2 - (y')^2 - 1 = 0$ has a one-parameter family of solutions given by $y = Cx \pm \sqrt{C^2 + 1}$, but that any function $y$ defined implicitly by the relation $x^2 + y^2 = 1$ is also a solution—one that does not correspond to a particular value of $C$ in the one-parameter solution formula.

10. Find a differential equation satisfied by the function

$$y(t) = \cos t + \int_0^t (t - u)y(u)du.$$

## C

1. Consider the equation $xy'' - (x + n)y' + ny = 0$, where $n$ is a nonnegative integer.

   a. Show that $y = e^x$ is a solution.

   b. Show that $y = 1 + x + \dfrac{x^2}{2!} + \dfrac{x^3}{3!} + \cdots + \dfrac{x^n}{n!}$ is a solution.

## 1.3 INITIAL-VALUE PROBLEMS AND BOUNDARY-VALUE PROBLEMS

Now suppose that we want to solve a first-order differential equation for $y$, a function of the independent variable $t$, and we specify that one of its integral curves must pass through a particular point $(t_0, y_0)$ in the plane. We are imposing the condition $y(t_0) = y_0$, which is called an **initial condition**, and the problem is then called an **initial-value problem (IVP)**. Note that we are trying to pin down a particular solution this way. We find this solution by choosing a specific value of the constant of integration (the parameter).

Next, we will see how to solve a simple initial-value problem.

### ■ Example 1.3.1 A First-Order Initial-Value Problem

Suppose that an object is moving along the $x$-axis in such a way that its instantaneous velocity at time $t$ is given by $v(t) = 12 - t^2$. First, we will find the *position* $x$ of the object measured from the origin at any time $t > 0$.

Because the velocity function is the derivative of the position function, we can set up the first-order differential equation $\frac{dx}{dt} = 12 - t^2$ to describe our problem.

Simple integration of both sides yields

$$x(t) = \int dx = \int \frac{dx}{dt}dt = \int 12 - t^2 dt = 12t - \frac{t^3}{3} + C.$$

This last result tells us that the position of the object at an arbitrary time $t > 0$ can be described by any member of the one-parameter family $12t - \frac{t^3}{3} + C$, which is not a very satisfactory conclusion. But if we have some additional information, we can find a definite value for $C$ and end the uncertainty.

Suppose we know, for example, that the object is located at $x = -5$ when $t = 1$. Then we can use this *initial condition* to get

$$-5 = x(1) = 12(1) - \frac{1^3}{3} + C, \text{ or } -5 = \frac{35}{3} + C.$$

This last equation implies that $C = \frac{-50}{3}$, so the position of the object at time $t$ is given by the particular function $x(t) = 12t - \frac{t^3}{3} - \frac{50}{3} = 12t - \frac{(t^3+50)}{3}$.

We selected the initial condition $x(1) = -5$ randomly. Any other choice $x(t_0) = x_0$ would have led to a definite value for $C$ and a particular solution of our problem.    ∎

### 1.3.1 An Integral Form of an IVP Solution

If a first-order equation can be written in the form $y' = f(x)$—that is, if the right-hand side is a continuous (or piecewise continuous) function of the independent variable alone—then we can always express the solution to the IVP $y' = f(x), y(x_0) = y_0$ on an interval $(a, b)$ as

$$y(x) = \int_{x_0}^{x} f(t)dt + y_0 \tag{1.3.1}$$

for $x$ in $(a, b)$. Note that we use the $x$ value of the initial condition as the lower limit of integration and the $y$ value of the initial condition as a particular constant of integration. We use $t$ as a *dummy variable*. Given Equation (1.3.1), the *Fundamental Theorem of Calculus (FTC)* (Section A.4) implies that $y' = f(x)$, and we see that $y(x_0) = \int_{x_0}^{x_0} f(t)dt + y_0 = 0 + y_0 = y_0$, which is what we want. This way of handling certain types of IVPs is common in physics and engineering texts. In Example 1.2.4, the solution of the equation with $y(-1) = 2$, for example, is

$$y(x) = \int_{-1}^{x} t^2 - 2t + 7dt + 2$$

$$= \left(\frac{t^3}{3} - t^2 + 7t\right)\Big|_{t=x} - \left(\frac{t^3}{3} - t^2 + 7t\right)\Big|_{t=-1} + 2$$

$$= \left(\frac{x^3}{3} - x^2 + 7x\right) - \left(\frac{-25}{3}\right) + 2 = \frac{x^3}{3} - x^2 + 7x + \frac{31}{3}.$$

You should also solve this problem the way we did in Example 1.3.1—that is, without using a definite integral formula.

### 1.3.2 Families of Solutions II

Although we have seen examples of first-order equations that have no solution or only one solution, in general we should expect a first-order differential equation to have an infinite set of solutions, described by a single parameter.

Extending the discussion in Section 1.2, we state that an $n$th-order differential equation may have an **$n$-parameter family of solutions**, involving $n$ arbitrary constants $C_1, C_2, C_3, \ldots, C_n$

(the parameters). For example, a solution of a second-order equation $y'' = g(t, y, y')$ may have *two* arbitrary constants. By prescribing the **initial conditions** $y(t_0) = y_0$ *and* $y'(t_0) = y_1$, we can determine specific values for these two constants and obtain a *particular* solution. Note that we use the same value, $t_0$, of the independent variable for each condition.

The next example shows how to deal with a second-order IVP.

## ■ Example 1.3.2 A Second-Order IVP

We will show in Section 4.1 that any solution of the second-order linear equation $y'' + y = 0$ has the form $y(t) = A \cos t + B \sin t$ for arbitrary constants $A$ and $B$. (You should verify that any function having the form indicated in the preceding sentence is a solution of the differential equation.) If a solution of this equation represents the *position* of a moving object relative to some fixed location, then the derivative of the solution represents the *velocity* of the particle at time $t$. If we specify, for example, the initial conditions $y(0) = 1$ and $y'(0) = 0$, we are saying that we want the position of the particle when we begin our study to be 1 unit in a positive direction from the fixed location and we want the velocity to be 0. In other words, our particle starts out at rest 1 unit (in a positive direction) from the fixed location.

We can use these initial conditions to find a particular solution of the original differential equation:

1. $y(0) = 1$ implies that $1 = y(0) = A \cos(0) + B \sin(0) = A$.

2. $y'(0) = 0$ implies that $0 = y'(0) = -A \sin(0) + B \cos(0) = B$.

Combining the results of (1) and (2), we find the particular solution $y(t) = \cos t$.  ■

---

### Definition 1.3.1

Finding the particular solution of the $n$th degree equation

$$F\left(t, y, y', y'', y''', \ldots, y^{(n-1)}, y^{(n)}\right) = 0$$

such that $y(t_0) = y_0$, $y'(t_0) = y_1$, $y''(t_0) = y_2, \ldots$, and $y^{(n-1)}(t_0) = y_{n-1}$, where $y_0, y_1, \ldots, y_{n-1}$ are arbitrary real constants, is called solving an **initial-value problem** (**IVP**). The $n$ specified values $y(t_0) = y_0$, $y'(t_0) = y_1$, $y''(t_0) = y_2, \ldots$, and $y^{(n-1)}(t_0) = y_{n-1}$ are called **initial conditions**.

---

Right now we can't be sure of the circumstances under which we can solve such an initial-value problem. We will discuss the existence and uniqueness of solutions of single equations in Section 2.8. Then in Section 4.9 we will consider IVPs for systems of differential equations.

### Boundary-Value Problems

For second- and higher-order differential equations, we can also determine a particular solution by specifying what are called **boundary conditions**. The idea here is to give conditions that must be satisfied by the solution function and/or its derivatives at *two different points* of the domain of the solution.

---

### Definition 1.3.2

A **boundary-value problem** (**BVP**) is a problem of determining a solution to a differential equation subject to conditions on the unknown function specified at *two or more* values of the independent variable. Such conditions are called **boundary conditions**.

---

The points chosen depend on the nature of the problem we are trying to solve and on the data we are given about the problem. For example, if you are analyzing the stresses on a steel girder of length $L$ whose ends are imbedded in concrete, you may want to find $y(x)$, the bend or "give" at a point $x$ units from one end if a load is placed somewhere on the beam (Figure 1.2). Note that the domain of $y$ is $[0, L]$. In this problem it is natural to specify $y(0) = 0$ and $y(L) = 0$, reasonable values at the endpoints, or *boundaries*, of the solution interval. Graphically, we are requiring the solution $y$ to pass through the points $(0, 0)$ and $(L, 0)$. (See Problem C2 in Exercises 1.3 for an applied problem of this type.)

The next example shows that, just as in the case of an initial-value problem, without further analysis we can't be sure whether there are solutions of a particular BVP or whether any solution we find is unique. In general, BVPs are harder to solve than IVPs. Although BVPs will appear in this book from time to time, we'll focus most of our attention on initial-value problems.

As the next example shows, some boundary-value problems have no solution, others have one solution, and some have (infinitely) many solutions.

### ■ Example 1.3.3 A BVP Can Have Many, One, or No Solutions

We'll use the second-order differential equation from Example 1.3.2, $y'' + y = 0$, which has the two-parameter family of solutions $y(t) = c_1 \cos t + c_2 \sin t$.

Now let's see what happens if we impose the boundary conditions $y(0) = 1$, $y(\pi) = 1$. The first condition implies that $1 = y(0) = c_1 \cos(0) + c_2 \sin(0) = c_1$, and the second condition tells us that $1 = y(\pi) = c_1 \cos(\pi) + c_2 \sin(\pi) = -c_1$. Because we can't have $c_1$ equaling 1 and $-1$ at the same time, this contradiction says that the boundary-value problem has *no solution*.

**FIGURE 1.2**
*A solution $y(x)$ satisfying the boundary conditions $y(0) = 0$ and $y(L) = 0$*

On the other hand, the boundary conditions $y(0) = 1$, $y(2\pi) = 1$ lead to a different conclusion. If we use the first condition, we get $1 = y(0) = c_1 \cos(0) + c_2 \sin(0) = c_1$. The second condition yields the result $1 = y(2\pi) = c_1 \cos(2\pi) + c_2 \sin(2\pi) = c_1$. The fact that we can't pin down the value of $c_2$ tells us that *any* value is all right. In other words, the BVP has *infinitely many solutions* of the form $y(t) = \cos t + c_2 \sin t$.

Finally, if we demand that $y(0) = 1$ and $y(\pi/4) = 1$, we find that $1 = y(0) = c_1 \cos(0) + c_2 \sin(0) = c_1$ and

$$1 = y\left(\frac{\pi}{4}\right) = c_1 \cos(\pi/4) + c_2 \sin\left(\frac{\pi}{4}\right) = c_1\left(\frac{\sqrt{2}}{2}\right) + c_2\left(\frac{\sqrt{2}}{2}\right)$$

$$= \frac{\sqrt{2}}{2} + c_2\left(\frac{\sqrt{2}}{2}\right)$$

which implies that $c_2 = \sqrt{2} - 1$. Therefore, this BVP has the *unique solution* $y(t) = \cos t + (\sqrt{2} - 1) \sin t$. ■

You should realize that for a general $n$th-order equation (or for a system of equations), there are many possible ways to specify boundary conditions, not always at the endpoints of solution intervals. The idea is to have a number of conditions that will enable us to solve for (specify) the appropriate number of arbitrary constants.

The following example shows how boundary conditions occur naturally in the solution of an interesting problem.

### ■ Example 1.3.4 A Practical BVP

The *Car and Driver* magazine website (January 23, 2005) reports that the two-passenger Ferrari F430 will go from 0 to 60 (mph) in 3.5 seconds. Assuming constant acceleration, we ask how far the car travels before it reaches 60 mph.

If $s(t)$ denotes the position of the car after $t$ seconds, then we must calculate $s(3.5) - s(0)$, the total distance covered by the car in the 3.5-second interval. We know the acceleration can be described as $a(t) = \frac{d^2 s}{dt^2}$, which in this problem equals some constant $C$; and we know that $s(0) = s'(0) = 0$—that is, our initial position is considered 0, and the velocity when we first put our foot on the gas pedal is also 0. The last bit of information we have is that $s'(3.5)$, the velocity at the end of 3.5 seconds, is 60 mph. Thus, we have a second-order differential equation $\frac{d^2 s}{dt^2} = C$, initial conditions, and some boundary conditions, and we must solve for the unknown function $s(t)$.

Now the basic rules of integral calculus tell us that when we find the antiderivative of each side of the differential equation in the last paragraph, we get

$$\int \frac{d^2 s}{dt^2} dt = \int C dt = Ct + C_1,$$

where $C_1$ is a constant of integration. But $\int \frac{d^2s}{dt^2} dt = \frac{ds}{dt}$, so $\frac{ds}{dt} = Ct + C_1$. Integrating each side of this last equation gives us

$$s(t) = \frac{Ct^2}{2} + C_1 t + C_2.$$

Thus, we have an expression for $s(t)$, but it contains three arbitrary constants. Now we can use the condition $s(0) = 0$ to write

$$0 = s(0) = \frac{C(0)^2}{2} + C_1(0) + C_2,$$

which boils down to $0 = C_2$, so we can say

$$s(t) = \frac{Ct^2}{2} + C_1 t.$$

Because $s'(0) = 0$, we can see that $0 = s'(0) = (Ct + C_1)|_{t=0} = C_1$, and thus $s(t) = \frac{Ct^2}{2}$.

We still have one unknown constant, $C$, in our formula, but we know that at the end of 3.5 seconds, the velocity is 60 miles per hour. *We have to be careful of our units here.* We don't want to mix seconds and hours. To make all our units consistent, we'll convert 3.5 seconds to $7/7200 (= 3.5/3600)$ of an hour. Then we can claim that $60 = s'(7/7200) = C \cdot (7/7200)$, so we have

$$C = \frac{60(7200)}{7} = 61{,}714.3 \text{ (miles/hr}^2\text{)},$$

$$s(t) = \frac{Ct^2}{2} = 30{,}857.2 t^2,$$

and

$$s\left(\frac{7}{7200}\right) = 30{,}857.2 \left(\frac{7}{7200}\right)^2 = 0.291667\ldots \text{ mile} \approx 154 \text{ feet}.$$

We have shown that in going from 0 to 60, the 2005 Ferrari F430 will travel approximately 154 feet.   ∎

### General Solutions

If *every* solution of an $n$th-order differential equation $F\left(x, y, y', y'', y''', \ldots, y^{(n-1)}, y^{(n)}\right) = 0$ on an interval $(a, b)$ can be obtained from an $n$-parameter family by choosing appropriate values for the $n$ constants, we say that the family is the **general solution** of the differential equation. In this case, we will need $n$ initial conditions or $n$ boundary conditions (or a combination of $n$ conditions) to determine the constants.

Sometimes, however, we can't find *every* solution somewhere among the members of an $n$-parameter family. For example, you should verify that the first-order nonlinear differential equation $2xy' + y^2 = 1$ has a one-parameter family of solutions given by $y = \frac{Cx-1}{Cx+1}$. However, for all values of $x$, the constant function $y \equiv 1$ is also a solution, but it can't be obtained from the family by choosing a particular value of the parameter $C$. Suppose we *could* find a value of $C$ such that $\frac{Cx-1}{Cx+1} = 1$. Cross-multiplication gives us $Cx - 1 = Cx + 1$, so that $-1 = 1$!

Also, $y(x) = kx^2$ is a solution of $x^2 y'' - 3xy' + 4y = 0$ for any constant $k$ and for all values of $x$, but $y(x) = x^2 \ln |x|$ is also a solution for all $x$. (*Check these claims.*) Of course, because the equation is second-order, we should realize that a one-parameter family can't be the general solution.

A solution of an $n$th-order differential equation that can't be obtained by picking particular values of the parameters in an $n$-parameter family of solutions is called a **singular solution**. We'll see in Chapter 2 that some of these singular solutions are created when we perform certain algebraic manipulations on differential equations.

## 1.3.3 Solutions of Systems of Odes

For a *system* of two equations with unknown functions $x(t)$ and $y(t)$, a solution on an interval $(a, b)$ consists of a *pair* of differentiable functions $x(t)$, $y(t)$ satisfying both equations that make up the system at all points of the interval. Initial conditions are given as $x(t_0) = x_0$ and $y(t_0) = y_0$.

## ■ Example 1.3.5 A System IVP

In Section 4.9, we will see why the only solution of the linear system

$$\frac{dx}{dt} = -3x + y$$

$$\frac{dy}{dt} = x - 3y$$

satisfying the conditions $x(0) = 0$ and $y(0) = 7$ is $\left\{x(t) = \frac{7}{2}e^{-2t} - \frac{7}{2}e^{-4t}, \; y(t) = \frac{7}{2}e^{-2t} + \frac{7}{2}e^{-4t}\right\}$.

Verify that these functions constitute a solution and accept the uniqueness as a fact for now. ■

You can think of the solution pair in the preceding example as coordinates of a point $(x(t), y(t))$ in two-dimensional space, $R^2$. As the independent variable $t$ changes, the points trace out a curve in the $x$-$y$ plane called a **trajectory**. The *positive* direction of the curve is the direction it takes as $t$ *increases*. Figure 1.3a shows the curve in the $x$-$y$ plane corresponding to the system solution in Example 1.3.5, together with arrows indicating its direction. The initial point $(x(0), y(0)) = (0, 7)$ is indicated. Looking at the solution formulas for $x(t)$ and $y(t)$, we see that $\lim_{t \to \infty} x(t) = 0 = \lim_{t \to \infty} y(t)$, so that the curve tends toward the origin as $t$ increases.

Figure 1.3b shows $x$ plotted against $t$, and Figure 1.3c shows $y$ plotted against $t$.

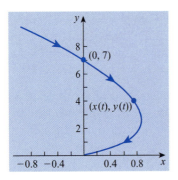

**FIGURE 1.3a**
Plot of $(x(t), y(t)) = \left( \dfrac{7}{2}e^{-2t} - \dfrac{7}{2}e^{-4t}, \dfrac{7}{2}e^{-2t} + \dfrac{7}{2}e^{-4t} \right)$ in the x-y plane, $-0.1 \le t \le 4$

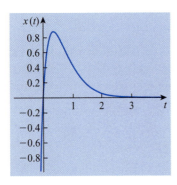

**FIGURE 1.3b**
Plot of $x(t) = \dfrac{7}{2}e^{-2t} - \dfrac{7}{2}e^{-4t}$, $-0.1 \le t \le 4$

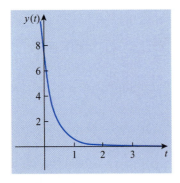

**FIGURE 1.3c**
Plot of $y(t) = \dfrac{7}{2}e^{-2t} + \dfrac{7}{2}e^{-4t}$, $-0.1 \le t \le 4$

A very important, *dynamical* way of looking at the situation in the preceding example is to think of the curve in Figure 1.3a as the path (or trajectory) of an object or quantity whose motion or change is governed by the system of differential equations. Initial conditions specify the behavior (the value, rate of change, and so on) at a single point on the path of the moving object or changing quantity. The proper graph of the solution of the system in Example 1.3.5 is a **space curve**, the set of points $(t, x(t), y(t))$. We'll see further graphical interpretations of system solutions in Chapter 4. Boundary conditions also determine certain aspects of the path of the phenomenon under study.

Similarly, each solution of the nonlinear system

$$\dot{x} = -sx + sy$$

$$\dot{y} = -xz + rx - y$$

$$\dot{z} = xy - bz$$

where $b, r$, and $s$ are constants, is an ordered triple $(x(t), y(t), z(t))$, and initial conditions have the form $x(t_0) = x_0$, $y(t_0) = y_0$, and $z(t_0) = z_0$. Boundary conditions in this situation can take various forms. The trajectory in this case is a *space curve*, a path in three-dimensional space. The true graph of the solution is the set of points $(t, x(t), y(t), z(t))$ in *four-dimensional space*. These points of view, especially the idea of a trajectory, are very useful, and we'll follow up on these concepts in Chapters 4, 5, and 7.

## Exercises 1.3

**A**

1. Consider the equation and solution in Problem A4 in Exercises 1.2. Find the particular solution that satisfies the initial condition $R(\pi) = 0$.

2. Consider the equation and solution in Problem A5 in Exercises 1.2. Find the particular solution that satisfies the initial conditions $y(0) = 1, y'(0) = 0, y''(0) = 1$, and $y'''(0) = 6$. [*Hint:* Use the initial conditions one at a time, beginning from the left.]

3. Consider the equation and solution in Problem A6 in Exercises 1.2. Find the particular solution that satisfies the initial condition $r(0) = 0$. (Your answer should involve only the constants $a$ and $b$.)

4. Consider the equation and solution in Problem A8 in Exercises 1.2. Find the particular solution that satisfies the initial conditions $y(0) = 2, y'(0) = 0$.

5. Find constants $A$, $B$, and $C$ such that $\frac{1}{8} - \frac{1}{4}x + \frac{11}{296}e^{6x} + Ax^2 + B\sin x + C\cos x$ is a solution of the IVP $y''' - 6y'' = 3 - \cos x, y(0) = 0, y'(0) = 0, y''(0) = 1$.

**B**

1. A particle moves along the $x$-axis so that its velocity at any time $t \geq 0$ is given by $v(t) = 1/(t^2 + 1)$. Assuming that the particle is at the origin initially, show that it will never get past $x = \pi/2$.

**2.** Show that the functions $y_1(x) \equiv 0$ and $y_2(x) = (x - x_0)^3$, each defined for $-\infty < x < \infty$, are both solutions of the initial-value problem

$$\frac{dy}{dx} = 3y^{2/3}, \quad y(x_0) = 0.$$

**3.** Show that $y = e^{x^2} \int_1^x e^{-t^2} \, dt$ is a solution of the IVP $y' = 1 + 2xy, \, y(1) = 0$.

**4.** The differential equation of a family of curves in the $x$-$y$ plane is given by

$$y''' = -24 \cos(\pi x / 2).$$

   **a.** Find an equation for the family and give the number of parameters involved.

   **b.** Find a member of the family that passes through the points $(0, -4)$ and $(1, 0)$ and that has a slope of 6 at the point where $x = 1$.

**5.** Is it possible for the differential equation corresponding to a three-parameter family of solutions to be of order four? Explain.

**6.** Barry leaves his home at noon and drives to his aunt's house, arriving at 3:20 P.M. He started from a parked position and steadily increased his speed in such a way that when he reached his aunt's house he was driving at 60 miles per hour. (The house had been repainted recently, and Barry didn't recognize it.) How far is it from Barry's home to his aunt's house?

**7.** A 727 jet needs to be flying 200 mph to take off. If the plane can accelerate from 0 to 200 mph in 30 seconds, how long must the runway be, assuming constant acceleration?

**8.** An automobile website reports that a 2008 Mercedes-Benz SLR McLaren will go from 0 to 62 (mph) in 3.8 seconds.

   **a.** Assuming constant acceleration, how far will the car travel before it reaches 60 mph?

   **b.** The car's "carbon ceramic" brakes are applied when the car is going 62 mph. Assuming constant deceleration, how long will it take the car to stop if it stops (according to the report) in 114 feet?

**9. a.** Show that the functions $x(t) = (A + Bt)e^{3t}$ and $y(t) = (3A + B + 3Bt)e^{3t}$ are solutions of the system

$$x' = y$$

$$y' = -9x + 6y$$

    for all values of the parameters $A$ and $B$.

   **b.** Find the solution to the system in part (a) with $x(0) = 1$ and $y(0) = 0$.

**10.** Show that the functions $x(t) = e^{-t/10} \sin t$ and $y(t) = \frac{1}{10} e^{-t/10}(-10 \cos t + \sin t)$ are solutions of the initial value problem

$$\frac{dx}{dt} = -y$$

$$\frac{dy}{dt} = (1.01)x - (0.2)y; \quad x(0) = 0, y(0) = -1.$$

**11.** A mathematical model of an idealized company consists of the equations

$$\frac{du}{dt} = kau, \quad u(0) = A$$

$$\frac{dw}{dt} = a(1-k)u, \quad w(0) = 0.$$

Here $u(t)$ represents the capital invested in the company at time $t$, $w(t)$ denotes the total dividend paid to shareholders in the period $[0, t]$, and $a$ and $k$ are constants with $0 \leq k \leq 1$.

**a.** Solve the first equation for $u(t)$. (See Example 1.2.1.)

**b.** Substitute your answer for part (a) in the differential equation for $w$ and integrate to find $w(t)$. (Distinguish between $w(t)$ for $0 < k \leq 1$ and for $k = 0$.)

## C

**1.** Let $W = W(t)$ denote your weight on day $t$ of a diet. If you eat $C$ calories per day and your body burns $EW$ calories per day, where $E$ represents calories per pound, then the equation $\frac{dW}{dt} = k(C - EW)$ models your change of weight.[3] (This equation says that your change of weight is proportional to the difference between calories eaten and calories burnt off, with constant of proportionality $k$.)

**a.** Show that $W = \frac{C}{E} + \left(W_0 - \frac{C}{E}\right)e^{-kEt}$ is a solution of the equation, where $W_0 = W(0)$, your weight at the beginning of the diet.

**b.** Given the solution in part (a), what happens to $W(t)$ as $t \to \infty$?

**c.** If $W_0 = 180$ lb, $E = 20$ cal/lb, and $k = 1/3500$ lb/cal, then how long will it take to lose 20 lb? How long for 30 lb? 35 lb? What do your answers seem to say about the process of weight loss? *take C = 2500 cal/dy.*

**2.** Solve the equation $EI\frac{d^4y}{dx^4} = -\frac{W}{L}$, with the boundary conditions $y(0) = 0, y'(0) = 0; y(L) = 0, y'(L) = 0$. (This problem arises in the analysis of the stresses on a uniform beam of length $L$ and weight $W$, both of whose ends are fixed in concrete. The solution $y$ describes the shape of the beam when a certain type of load is placed on it. Here, $E$ and $I$ are constants, and the product $EI$ is a constant called the *flexural rigidity* of the beam.) [*Hint:* Integrate successively, introducing a constant of integration at each stage. Then use the boundary conditions to evaluate these constants of integration.]

**3.** The *logistic equation* $\frac{dy}{dt} = ky(t)\left(1 - \frac{y(t)}{M}\right)$ is used to describe the growth of certain kinds of human and animal populations. Here, $k$ and $M$ denote constants describing characteristics of the population being modeled.

**a.** Show that the function $y(t) = \frac{M}{1+Ae^{-kt}}$ satisfies the logistic equation with $y(0) = \frac{M}{1+A}$.

[3] A. C. Segal, "A Linear Diet Model," *College Mathematics Journal* **18** (1987): 44–45.

**b.** A study of U.S. population data[4] indicates that the solution given in part (a) provides a good fit if $M = 387.9802$, $A = 54.0812$, and $k = 0.0270347$. Using technology, plot the graph of $y(t)$ using these values of $M$, $A$, and $k$. (Here, $t$ denotes the time in years since 1790, the year of the first U.S. census.)

**c.** In 1790, the U.S. population was 3,929,214. In 1980, the figure was 226,545,805; while for 1990, the population was 248,709,873. By evaluating the function plotted in part (b) at $t = 0, 90$, and 100, compare the values (in millions) given by $y(t)$ to the actual populations. [Check the official website of the Bureau of the Census for additional information: www.census.gov.]

**d.** According to the model with parameters as given in part (b), what happens to the population of the U.S. as $t \to \infty$?

**4.** The equations

$$\frac{dT^*}{dt} = kV_1 T_0 - \delta T^*$$

$$\frac{dV_1}{dt} = -cV_1$$

are used in modeling HIV-1 infections.[5] Here, $T^* = T^*(t)$ denotes the number of infected cells, $T_0 = T(0)$ is the number of potentially infected cells at the time therapy is begun, $V_1 = V_1(t)$ is the concentration of viral particles in plasma, $k$ is the rate of infection, $c$ is the rate constant for viral particle clearance, and $\delta$ is the rate of loss of virus-producing cells.

**a.** Imitate the analysis shown in Example 1.2.1 and solve the second equation for $V_1(t)$, expressing your solution in terms of $V_0 = V_1(0)$.

**b.** Using the solution found in part (a), show that the solution of the differential equation for $T^*$ can be written as

$$T^*(t) = T^*(0)e^{-\delta t} + \frac{kT_0 V_0}{c - \delta}(e^{-ct} - e^{-\delta t}).$$

**c.** What does the solution in part (a) say about the number of infected cells as $t \to \infty$?

**5.** Consider the linear equation $x^2 y'' + xy' - 4y = x^3$ (*). Let $y_{GR}$ be the *general* solution of the "reduced" (or "complementary") equation $x^2 y'' + xy' - 4y = 0$ and let $y_P$ be a *particular* solution of (*). Show that $y_{GR} + y_P$ is the *general* solution of (*). [For this problem, define the general solution of a second-order ODE as a solution having two arbitrary constants. A particular solution, of course, has *no* arbitrary constants.]

---

[4] E. K. Yeargers, R. W. Shonkwiler, and J. V. Herod, *An Introduction to the Mathematics of Biology: With Computer Algebra Models* (Boston: Birkhäuser, 1996): 117.

[5] A. S. Perelson, A. U. Neumann, M. Markowitz, J. M Leonard, and D. D. Ho, "HIV-1 Dynamics in Vivo: Virion Clearance Rate, Infected Cell Life-Span, and Viral Generation Time," *Science* **271** (1996): 1582–1586.

## SUMMARY

The study of differential equations is as old as the development of calculus by Newton and Leibniz in the late seventeenth century. Motivation was provided by important questions about change and motion on earth and in the heavens.

An **ordinary differential equation (ODE)** is an equation that involves an unknown function, its independent variable, and one or more of its derivatives:

$$F\left(x, y, y', y'', y''', \ldots, y^{(n-1)}, y^{(n)}\right) = 0.$$

Such an equation can be described in terms of its **order**, the order of the highest derivative of the unknown function in the equation.

Differential equations can also be classified as either **linear** or **nonlinear**. **Linear equations** can be written in the form

$$a_n(x)y^{(n)} + a_{n-1}(x)y^{(n-1)} + \cdots + a_2(x)y'' + a_1(x)y' + a_0(x)y = f(x)$$

where each coefficient function $a_i(x)$ depends on $x$ alone and doesn't involve $y$ or any of its derivatives. **Nonlinear equations** usually contain products, quotients, or more elaborate combinations of the unknown function and its derivatives.

A **solution** of an ODE is a real-valued function that, when substituted in the equation, makes the equation valid on some interval. A given $n$th-order ODE may have *no* solutions, *only one* solution, or *infinitely many solutions*. An *infinite family of solutions* may be characterized by $n$ constants (parameters). These arbitrary constants, if present, may be evaluated by imposing appropriate **initial conditions** (usually $n$ of them, involving behavior of the solution at a single point of its domain) or **boundary conditions** (at two or more points). Solving a differential equation with initial conditions is referred to as solving an **initial-value problem (IVP)**. Solving a differential equation with boundary conditions is referred to as solving a **boundary-value problem (BVP)**. In general, BVPs are harder to solve than IVPs. The result of solving either an IVP or a BVP is called a **particular solution** of the equation or an **integral** of the equation. The graph of a particular solution is called an **integral curve** or **solution curve**. In Chapters 2 and 3, we'll discuss the question of *existence and uniqueness* for IVPs: Does the equation or system have a solution that satisfies the initial conditions? If so, is there only one solution?

If *every* solution of an $n$th-order ODE on an interval can be obtained from an $n$-parameter family by choosing appropriate values for the $n$ constants, then we say that the family is the **general solution** of the differential equation. In this case we need $n$ initial conditions or $n$ boundary conditions to determine the constants. However, sometimes there are **singular solutions** that can't be found just by choosing particular values of the constants.

Just as high school or college algebra introduces systems of algebraic equations, the study of certain problems often leads to **systems of differential equations**. These, in turn, can be classified as either **linear systems** or **nonlinear systems**. We can specify initial or boundary conditions for systems. Whether we're considering single equations or systems of equations we are dealing with **dynamical** situations—situations in which objects and quantities are moving and changing. In such a dynamical situation, it is often useful to focus on a **trajectory**— for a single equation, a curve made up of points $(x(t), x'(t))$, where $x$ is a solution; for a system of two equations, the set of points $(x(t), y(t))$, where $x$ and $y$ are solutions of the system.

## PROJECT 1-1

### Draw Your Own Conclusions

Even before you learn techniques for solving differential equations, you may be able to analyze equations *qualitatively*. As an example, look at the nonlinear equation $\frac{dy}{dt} = y(1 - y)$. You are going to analyze the solutions, $y$, of this equation without actually finding them.

In what follows, picture the $t$-axis running horizontally and the $y$-axis running vertically.

**a.** For what values of $y$ is the graph of $y$ as a function of $t$ increasing? For what values of $y$ is it decreasing?

**b.** For what values of $y$ is the graph of $y$ concave up? For what values of $y$ is it concave down? (What information do you need to answer a question about concavity? Remember that $y$ is an implicit function of $t$.)

**c.** Say you are given the initial condition $y(0) = 0.5$. Use the information found in parts (a) and (b) to sketch the graph of $y$. What is the *long-term* behavior of $y(t)$? That is, what is $\lim_{t \to \infty} y(t)$?

**d.** Say you are given the initial condition $y(0) = 1.5$. Use the information found in parts (a) and (b) to sketch the graph of $y$. What is the *long-term* behavior of $y(t)$? That is, what is $\lim_{t \to \infty} y(t)$?

**e.** Sketch the graph of $y$ if $y(0) = 1$. (Look at the original equation.)

**f.** If $y(t)$ represents the population of some animal species, and if units on the $y$-axis are in thousands, interpret the results of parts (c), (d), and (e).

# CHAPTER 2

# First-Order Differential Equations

## INTRODUCTION

The various examples in the preceding chapter should have convinced you that there are different possible answers to the question of what the solution or solutions to a differential equation look like. In this chapter, we'll examine first-order differential equations from both the analytic and the qualitative point of view.

First, we'll learn *analytic* solution techniques for two important types of first-order equations. For these kinds of equations, we'll come up with explicit or implicit formulas for their solution curves. This method of solution is often referred to as *integrating* a differential equation.

Next, there is a *qualitative* way of viewing differential equations. This is a neat geometrical way of studying the behavior of solutions without actually solving the differential equation. The idea is to examine certain pictures or graphs derived from differential equations. Although we can do some of this work by hand, computer algebra systems and many graphing calculators can produce these graphs, and you'll be expected to use technology when appropriate. There are also specialized programs just for doing this sort of thing. (Follow your instructor's guidance in using technology.)

After the analytic and qualitative treatments in this chapter, Chapter 3 will focus on some numerical solution methods concerned with approximating values of solutions. As we'll see, both qualitative and numerical methods are necessary, because it is often impossible to represent the solutions of differential equations—even first-order equations—by formulas involving elementary functions.[1]

---

[1] In general, "elementary functions" are finite combinations of integer powers of the independent variable, roots, exponential functions, logarithmic functions, trigonometric functions, and inverse trigonometric functions.

## 2.1 SEPARABLE EQUATIONS

The simplest type of differential equation to solve is one in which the variables are *separable*. Formally, a first-order differential equation $\frac{dy}{dx} = F(x, y)$ is called **separable** if it can be written in the form $\frac{dy}{dx} = f(x)g(y)$, where $f$ denotes a function of the independent variable $x$ alone and $g$ denotes a function of the dependent variable $y$ alone. For example, the equation $\frac{dy}{dx} = e^x y^2$ is separable.

If $y(x)$ is a nonconstant solution of the equation $\frac{dy}{dx} = f(x)g(y)$ and $g(y(x))$ is nonzero on an interval $(a, b)$, we can divide both sides of the equation by $g(y(x))$ (a process called **separating variables**) to get

$$\frac{1}{g(y(x))} \frac{dy}{dx} = f(x).$$

If $F(x)$ is an antiderivative of $f(x)$ and $G(y)$ is an antiderivative of $\frac{1}{g(y)}$, we can integrate both sides of the preceding equation to obtain

$$\int \frac{1}{g(y(x))} \frac{dy}{dx} dx = \int \frac{1}{g(y)} dy = \int f(x) dx,$$

or $G(y(x)) = F(x) + C$, where the constants of integration associated with $G$ and $F$ have been combined in the single constant $C$ (note how we have used the *Chain Rule* in working with $G(y(x))$):

$$\frac{d}{dx} G(y(x)) = G'(y(x)) \frac{dy}{dx} = \frac{1}{g(y(x))} \frac{dy}{dx}.$$

There are three things to be careful about: (1) Not every first-order differential equation is separable; (2) even after you have separated the variables and integrated, it may not be possible to solve for one variable (say $y$) in terms of the other (say $x$); you may have to express your answer *implicitly*; (3) you may not be able to carry out the integration(s) in terms of elementary functions. We'll see examples of these situations.

Also, note that in a separable equation $y' = f(x)g(y)$, a solution of $g(y) \equiv 0$ is also a solution of the differential equation—possibly a *singular* solution (see Section 1.2). If $g(y) = 0$, then $y' = 0$, implying that $y$ is a constant. Conversely, if $y(x) = c$ is a constant solution, then $y' = 0$, which implies that $g(y) = 0$ because $f(x) = 0$ is unlikely in a physical problem. This says that the zeros of $g$ are constant solutions; and, in general, they are the only constant solutions.

The preceding analysis can be refined by considering three cases for a separable differential equation $\frac{dy}{dx} = f(x) g(y)$: (1) $g(y) \equiv 1$; (2) $f(x) \equiv 1$; (3) neither (1) nor (2). In case 1, the equation takes the simple form $\frac{dy}{dx} = f(x)$. If $f(x)$ is continuous on some interval $a < x < b$, then the IVP $\frac{dy}{dx} = f(x), y(x_0) = y_0$ has a unique solution on $(a, b)$ given by $y(x) = y_0 + \int_{x_0}^{x} f(r) dr$. [See Equation (1.2.1). The uniqueness of this solution will be discussed in Section 2.8.]

## ■ Example 2.1.1 A Separable Equation: Case 1

The initial-value problem $\frac{dy}{dx} = -x^3 + \cos x, y(1) = 1$ is a case 1 situation with $f(x) = -x^3 + \cos x$, a continuous function for all real values of $x$. We can solve the IVP by integrating both sides of the equation to obtain the general solution and then substituting the initial condition. Alternatively, we can use the simple solution formula

$$y(x) = 1 + \int_1^x (-r^3 + \cos r)\, dr = -\frac{x^4}{4} + \sin x + \left(\frac{5}{4} - \sin 1\right).$$

In case 2, we have $\frac{dy}{dx} = g(y)$. We can rewrite the equation as $\frac{1}{g(y)}\frac{dy}{dx} = 1$, or more accurately as $\frac{1}{g(y(x))}\frac{dy}{dx} = 1$. Integrating both sides with respect to $x$, we get $\int \frac{1}{g(y(x))}\frac{dy}{dx}\, dx = \int 1\, dx$. If we make the substitution $y = y(x)$, then $dy = y'(x)\, dx = \frac{dy}{dx}\, dx$, and we have $\int \frac{1}{g(y)}\, dy = \int 1\, dx = x + C$. This gives us a solution (possibly implicit) of our case 2 ordinary differential equation (ODE). Letting $G(y) = \int \frac{1}{g(y)}\, dy$, we can express the solution of the ODE in the form $G(y) = x + C$. If we are given the initial condition $y(x_0) = y_0$, we choose the constant of integration $C$ so that $G(y_0) = x_0 + C$. Finally, if $G$ has an inverse, we can write $y(x) = G^{-1}(x + C) = G^{-1}(x + G(y_0) - x_0)$.   ■

The next example—so basic, yet so important in many applications—is one we've seen before. Back in Chapter 1, we *guessed* at the solution and then verified that our guess was correct. Note that in this example we use the fact that the logarithmic and exponential functions are inverses of each other.

## ■ Example 2.1.2 Solving a Separable Equation, Case 2—Example 1.2.1 Revisited

The way the balance $B(t)$ of a bank account grows under continuous compounding demonstrates the "snowball effect": The larger the balance at a given time, the more rapid the growth—that is, the greater the rate of growth. In the language of differential equations, this becomes $\frac{dB}{dt} = rB$, where $r$, the constant of proportionality, is the annual interest rate (expressed as a decimal). If the initial balance (the principal at $t = 0$) is positive, we want to find the balance at time $t$.

Separating the variables, we can write $\frac{dB}{B} = r\, dt$, so that $\int \frac{dB}{B} = \int r\, dt$ and $\ln |B| = rt + C$. Then we exponentiate: $e^{\ln |B|} = e^{rt+C} = e^{rt}e^C$, or $|B| = Ke^{rt}$, where $K = e^C$, a positive constant. Given that the initial balance was positive, we realize that $B$ must be positive, so that we can just write $B(t) = Ke^{rt}$, with $K > 0$. Finally, we can bring in the positive initial balance to write $B(0) = Ke^0 = K$, so that our final formula is $B = B(t) = B(0)e^{rt}$. For example, if we invest \$1000.00 at 4% interest compounded continuously for 6 years, we will have $(1000)e^{(0.04)6} \approx \$1271.25$ in our account. (You should use your calculator or CAS to verify this.)   ■

Of course, there is another way to do the preceding problem, but you have to know the formula for the balance $B_n(t)$ if you invest $P$ dollars at interest rate $r$ compounded $n$ times a year for $t$ years. This formula is

$$B_n(t) = P\left(1 + \frac{r}{n}\right)^{nt}.$$

Continuous compounding involves compounding infinitely often, compounding at *every* *instant* of the year. Mathematically, we want

$$\lim_{n\to\infty} B_n(t) = \lim_{n\to\infty} P\left(1 + \frac{r}{n}\right)^{nt} = P \cdot \lim_{n\to\infty}\left(1 + \frac{r}{n}\right)^{nt} = P \cdot \left\{\lim_{n\to\infty}\left(1 + \frac{r}{n}\right)^{n}\right\}^{t}$$

$$= P \cdot \{e^r\}^t = Pe^{rt} = B(0)e^{rt}.$$

You may have seen this derivation in calculus class, as well as the fact that if you invest $1.00 for 1 year at 100% interest compounded continuously, you will have $e(\approx $2.72) at the end of the year.

Case 3—$\frac{dy}{dx} = f(x)g(y)$, where neither $f$ nor $g$ is a constant function—is the most interesting case. Now we can separate variables to write

$$\frac{1}{g(y)}\frac{dy}{dx} = f(x), \quad \int \frac{1}{g(y(x))}\frac{dy}{dx}dx = \int f(x)dx, \text{ and } \int \frac{1}{g(y)}dy = \int f(x)dx.$$

The case 3 example that follows adds the use of technology.

### ■ Example 2.1.3 A Separable Equation and the Graph of a Solution

Suppose that an insect population $P$ shows seasonal growth modeled by the differential equation $\frac{dP}{dt} = kP\cos(\omega t)$, where $k$ and $\omega$ are positive constants. (The cosine factor suggests periodic fluctuation.)

You should be able to see that the equation is separable: $\frac{dP}{dt} = f(P)g(t)$, where $f(P) = P$ and $g(t) = k\cos(\omega t)$. (We could have stuck the constant $k$ with the factor $P$, but if we think ahead, we'll realize that there's one less algebraic step if we keep the constant with the cosine term.) Separating the variables, we get $\frac{dP}{P} = k\cos(\omega t)dt$, so that $\int \frac{dP}{P} = k\int \cos(\omega t)dt$, or $\ln|P| = \frac{k}{\omega}\sin(\omega t) + C$.

Exponentiating, we see that $P(t) = Re^{\frac{k}{\omega}\sin(\omega t)}$, where $R > 0$. (This is a population problem, so $R > 0$ is a realistic assumption.) Letting $P_0 = P(0)$ denote the initial insect population, we have $P(t) = P_0 e^{\frac{k}{\omega}\sin(\omega t)}$ as the solution.

Graphing the solution curve for $P_0 = 100, k = 2$, and $\omega = \pi$ (Figure 2.1), we can see that the population varies periodically, fluctuating from a minimum value of $100e^{-\frac{2}{\pi}}$ (approximately 53) to a maximum value of $100e^{\frac{2}{\pi}}$ (approximately 189).  ■

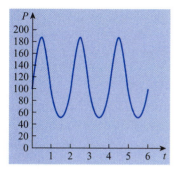

**FIGURE 2.1**
*The solution of the IVP $\frac{dP}{dt} = 2P\cos(\pi t); P(0) = 100$*

Sometimes, as the next example shows, it may not be easy to find an explicit solution for a separable differential equation.

### ■ Example 2.1.4 A Separable Equation with Implicit Solutions

The equation $\frac{dy}{dx} = \frac{x^2}{1+y^2}$ can be written as $\frac{dy}{dx} = f(x)g(y)$, where $f(x) = x^2$ and $g(y) = \frac{1}{1+y^2}$. Separating the variables, we get $(1 + y^2)dy = x^2 dx$.

Integrating both sides, we find that $y + \frac{y^3}{3} = \frac{x^3}{3} + C$, or $\frac{x^3}{3} - \left(y + \frac{y^3}{3}\right) = C$. This gives the solution *implicitly*. (See Chapter 1, right after Example 1.2.2.) To get an *explicit* solution, we must solve this last equation for $y$ in terms of $x$ or for $x$ in terms of $y$. Either way is acceptable, although solving for $x$ as a function of $y$ is easier algebraically. But even if we don't find an explicit solution, we can plot solution curves for different values of the constant $C$. (This may be a good time to find out how to graph implicit functions using your available technology.) In Figure 2.2 we use (from top to bottom) $C = -7, -5, -3, 0, 3, 5$, and $7$.

**FIGURE 2.2**
*Implicit solutions of $\frac{dy}{dx} = \frac{x^2}{1+y^2}$ : the curves $\frac{x^3}{3} - \left(y + \frac{y^3}{3}\right) = C$*
$C = -7, -5, -3, 0, 3, 5$, and $7$; $-4 \le x \le 4, -4 \le y \le 4$

The third concern we mentioned earlier is that you may not be able to integrate one or both of the sides after you have separated the variables. We will address this problem next.

### ■ Example 2.1.5 A Function with No Explicit Integral

The differential equation $\frac{dy}{dt} = e^{y^2} t$ is clearly separable—we can write $e^{-y^2} dy = t\, dt$. However, we can't carry out the integration $\int e^{-y^2}\, dy$ on the left-hand side because there is no combination of elementary functions whose derivative is $e^{-y^2}$. Consequently, we are forced to write the family of solutions as

$$\int e^{-y^2}\, dy = \frac{t^2}{2} + C, \quad \text{or} \quad 2\int e^{-y^2}\, dy = t^2 + K,$$

where $K = 2C$. ■

Integrals of the form $\int_a^b e^{-y^2}\, dy$ have many applications in mathematics and science, especially in problems dealing with probability and statistics. For instance, the **error function** $erf(x) = \frac{2}{\sqrt{\pi}} \int_0^x e^{-y^2}\, dy$ appears in many applied problems and can be evaluated easily by any CAS.

Dealing with separable equations often requires some algebraic skills and some integration intuition, although technology can help in tough situations. The next example introduces a common algebraic problem.

### ■ Example 2.1.6 Using Partial Fractions

The equation $\frac{dz}{dt} + 1 = z^2$ looks simple enough but requires some algebraic manipulation to get a neat solution. Separating variables, we get $\frac{dz}{z^2-1} = dt$. Using the method of partial fractions (see Section A.5), we can write $\frac{1}{z^2-1}$ as $\frac{1}{2}\left(\frac{1}{z-1} - \frac{1}{z+1}\right)$, so integration gives us $\int \frac{1}{2}\left(\frac{1}{z-1} - \frac{1}{z+1}\right) dz = \int 1 dt$, or $\frac{1}{2}(\ln|z-1| - \ln|z+1|) = t + C_1$. Multiplying both sides of this last equation by 2 and then simplifying the logarithmic expression, we get $\ln\left|\frac{z-1}{z+1}\right| = 2t + C_2$. Exponentiating, we find that $\frac{z-1}{z+1} = Ke^{2t}$. Finally, solving this last equation for $z$ (*a bit tricky—so do it*), we conclude that $z = \frac{1+Ke^{2t}}{1-Ke^{2t}}$, a one-parameter family of solutions.

Note that in going through the process of separating variables, we divided by $z^2 - 1$, implicitly assuming that this expression was not zero. Going back to this, we see that the constant function $z \equiv 1$ corresponds to $K = 0$ in our one-parameter family, whereas $z \equiv -1$ is a singular solution. (*Why?*). ■

As simple as separable equations may seem, they have some very important applications. The calculations and manipulations involved in the next example may seem tedious, but they should remind you of things you have seen in previous classes. The analysis at the end of the example should convince you that a graphical approach can be enlightening.

# ■ Example 2.1.7 A Model of a Bimolecular Chemical Reaction

Most chemical reactions can be viewed as interactions between two molecules that undergo a change and result in a new product. The rate of a reaction, therefore, depends on the number of interactions or collisions, which in turn depends on the concentrations (in moles per liter) of both types of molecules. Consider the simple (*bimolecular*) reaction $A + B \rightarrow X$, in which molecules of substance A collide with molecules of substance B to create substance X.

Let's designate the concentrations at time 0 of A and B by $\alpha$ and $\beta$, respectively. We'll assume that the concentration of X at the beginning is 0 and that at time $t$ it is $x = x(t)$. The concentrations of A and B at time $t$ are, correspondingly, $\alpha - x$ and $\beta - x$. Note that $\alpha - x > 0$ and $\beta - x > 0$ (*Why?*). The rate of formation (the *velocity of reaction* or *reaction rate*) is given by the differential equation $\frac{dx}{dt} = k(\alpha - x)(\beta - x)$, where $k$ is a positive number called the *velocity constant*. The product on the right-hand side of the equation reflects the interactions or collisions between the two kinds of molecules. We want to determine $x(t)$.

Separating variables and integrating, we get

$$\int \frac{dx}{(\alpha - x)(\beta - x)} = \int k \, dt.$$

To simplify the integrand $\frac{1}{(\alpha-x)(\beta-x)}$, we use the technique of partial fractions so that we can write

$$\int \frac{dx}{(\alpha - x)(\beta - x)} = \frac{1}{\beta - \alpha} \int \frac{dx}{\alpha - x} + \frac{1}{\alpha - \beta} \int \frac{dx}{\beta - x} = \int k \, dt$$

or

$$-\frac{1}{\beta - \alpha} \ln(\alpha - x) - \frac{1}{\alpha - \beta} \ln(\beta - x) = k t + C,$$

which simplifies to

$$\frac{1}{\alpha - \beta} \ln \left( \frac{\alpha - x}{\beta - x} \right) = k t + C.$$

The initial condition $x(0) = 0$ leads us to conclude that

$$C = \frac{1}{\alpha - \beta} \ln \left( \frac{\alpha}{\beta} \right).$$

Then

$$\frac{1}{\alpha - \beta} \ln \left( \frac{\alpha - x}{\beta - x} \right) = k t + \frac{1}{\alpha - \beta} \ln \left( \frac{\alpha}{\beta} \right),$$

so

$$\ln \left( \frac{\alpha - x}{\beta - x} \right) = (\alpha - \beta) k t + \ln \left( \frac{\alpha}{\beta} \right)$$

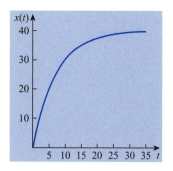

**FIGURE 2.3**

*Solution of the IVP $\frac{dx}{dt} = 0.0006(250 - x)(40 - x); x(0) = 0$*
*$0 \le t \le 35, 0 \le x \le 40$*

or

$$\frac{\alpha - x}{\beta - x} = \frac{\alpha}{\beta} e^{(\alpha - \beta)kt}.$$

A few more algebraic manipulations lead to the solution

$$x = x(t) = \frac{\alpha\beta\left(1 - e^{(\alpha - \beta)kt}\right)}{\beta - \alpha e^{(\alpha - \beta)kt}}. \tag{2.1.1}$$

Formula (2.1.1) does not seem very informative as far as understanding the nature of the chemical reaction goes, but Problem B9 of Exercises 2.1 suggests some useful ways of analyzing the formula. A CAS-generated graph of a solution (Figure 2.3) of the equation with $\alpha = 250$, $\beta = 40$ and $k = 0.0006$ is more informative, showing the steady rise in the concentration of molecule X to what is called an *equilibrium value* of 40. (We'll explore the idea of an equilibrium value in Section 2.6.) The particular solution shown corresponds to $x(0) = 0$.

Of course, as we've noted previously, the right-hand side of the original differential equation is positive, so we know ahead of time that the concentration function is increasing. Also, you can calculate $\frac{d^2x}{dt^2}$ from the original differential equation to see why the graph of $x$ is concave down. Remember that $k > 0$ and $0 \le x < \alpha, 0 \le x < \beta$. ■

## Exercises 2.1

### A

Solve the equations or IVPs in Problems 1–9 by separating variables. Be sure to describe any singular solutions where appropriate.

1. $\dfrac{dy}{dx} = \dfrac{A - 2y}{x}$, where $A$ is a constant

2. $\dfrac{dy}{dx} = \dfrac{-xy}{x + 1}$

**3.** $y' = 3\sqrt[3]{y^2}; \quad y(2) = 0$

**4.** $\dfrac{dy}{dx} = \dfrac{(y-1)(y-2)}{x}$

**5.** $(\cot x)y' + y = 2; \quad y(0) = -1$

**6.** $x' = -\dfrac{\sin t \cos^2 x}{\cos^2 t}; \quad x(0) = 0$

**7.** $x^2 y^2 y' + 1 = y$

**8.** $xy' + y = y^2; \quad y(1) = 0.5$

**9.** $z' = 10^{x+z}$

**10.** Solve the equation $y' = 1 + x + y^2 + xy^2$. [*Hint*: Factor cleverly.]

**11.** Solve the equation $(y')^2 + (x+y)y' + xy = 0$. [*Hint*: Solve this quadratic equation for $y'$ by factoring or by using the quadratic formula, and then solve the two resulting differential equations separately.]

An equation of the form $dy/dx = f(ax + by)$ can be transformed into an equation with separable variables by making the substitution $z = ax + by$ or $z = ax + by + c$, where $c$ is an arbitrary constant. For example, the equation $y' = (y-x)^2$ is not separable, but the substitution $z = y - x$ leads to the separable equation $z' + 1 = z^2$, which was solved as Example 2.1.6. Then substitute the original variables for $z$. Use this technique to solve the equations in Problems 12–14.

**12.** $y' - y = 2x - 3$

**13.** $(x + 2y)y' = 1; \quad y(0) = -1$

**14.** $y' = \sqrt{4x + 2y - 1}$

A **homogeneous** equation has the form $dy/dx = f(x, y)$, where $f(x, y)$ can be expressed in the form $g(y/x)$ or $g(x/y)$—that is, as a function of the quotient $y/x$ or the quotient $x/y$ alone. For example, by dividing numerator and denominator by $x^2$, we can write the equation $\dfrac{dy}{dx} = \dfrac{2x^2 - y^2}{3xy}$ in the form $\dfrac{dy}{dx} = \dfrac{2 - (y/x)^2}{3(y/x)} = g\left(\dfrac{y}{x}\right)$. Any such equation can be changed into a separable equation by making the substitution $z = y/x$ (or $z = x/y$). Making the substitution $z = y/x$ in our example, we have $\dfrac{dy}{dx} = \dfrac{d}{dx}(xz) \overset{\text{(Product Rule)}}{=} 1 \cdot z + x\left(\dfrac{dz}{dx}\right)$, so that our equation becomes $x\dfrac{dz}{dx} + z = \dfrac{2 - z^2}{3z}$ or, separating variables, $\left(\dfrac{3z}{2 - 4z^2}\right) dz = \dfrac{1}{x}dx$. After integrating, remember to replace $z$ by $y/x$ (or $x/y$). Use this technique to solve the equations in Problems 15–18.

**15.** $y' = \dfrac{x + y}{x - y}$

**16.** $\dot{x} = \dfrac{t - 3x}{3t + x}$

**17.** $y' = \dfrac{x}{y} + \dfrac{y}{x}$

**18.** $\dfrac{dy}{dx} = \dfrac{y^2 + 2xy - x^2}{x^2 + 2xy - y^2}$

**B**

1. Suppose that $f$ is a function such that $f(x) = \int_0^x f(t)dt$ for all real numbers $x$. Show that $f(x) \equiv 0$. [*Hint*: Use the Fundamental Theorem of Calculus to get a differential equation. Then think of an appropriate initial condition.]

2. Consider the equation $\dot{x} = \dfrac{x^2 + x}{t}$.

   a. Find a one-parameter family of solutions.
   b. Can you find a solution satisfying the initial condition $x(0) = -1$? If so, give it. If not, give a reason.
   c. Find a singular solution.

3. a. Solve the initial value problem $\dot{x} = x^2, x(1) = 1$.
   b. If the solution in part (a) is valid over an interval $I$, how large can $I$ be?
   c. Use technology to draw the graph of the solution $x(t)$ found in part (a).
   d. Solve the initial value problem $\dot{x} = x^2, x(0) = 0$.

4. The equation

$$\frac{dQ}{dP} = -\frac{cQ}{1 + cP}$$

   is one model used to estimate the cost of national health insurance,[2] where $Q(P)$ represents the quantity of health services performed at price $P$, $P$ represents the proportion of the total cost of health services that an individual pays directly ("out of pocket expenses," or coinsurance), and $c$ is a constant.

   a. Solve the equation for $Q$.
   b. If $Q(0)/Q(1)$ is approximately 2, what is the value of $c$?
   c. Using the value of $c$ found in part (b), determine $Q(0.20)$ in terms of $Q(0)$. What does your answer tell you about the effect of a 20% coinsurance (versus no coinsurance)?

5. A quantity $y$ varies in such a way that $\frac{dy}{dt} = -\frac{\ln 2}{30}(y - 20)$. If $y = 60$ when $t = 30$, find the value of $t$ for which $y = 40$.

6. In analyzing the change in the percentage of red blood cells in a hospital patient undergoing surgery, the following equation has been used[3]

$$\frac{dH}{dV_L} = -\frac{H}{EBV},$$

   where $H$ denotes the *hematocrit* (percentage of red blood cells in the total volume of blood), $V_L$ represents the volume of blood loss, and $EBV$ is the patient's estimated total blood volume.

   a. Solve the differential equation for $H$.

---

[2] A. J. Kroopnick, "Estimating the Cost of National Health Insurance Using Three Simple Models," *Math. and Comp. Ed.* **30** (1996): 267–271.
[3] M. E. Brecher and M. Rosenfeld, "Mathematical and Computer Modeling of Acute Normovolemic Hemodilution," *Transfusion* **34** (1994): 176–179. Also see "Calculus in the Operating Room" by P. Toy and S. Wagon, *Amer. Math. Monthly* **102** (1995): 101.

**b.** If the patient's total blood volume of 5 liters is maintained throughout surgery via the injection of saline solution, and the initial value of $H$ is 0.40, what is the patient's volume of red blood cells at the end of the operation?

7. The volume $V$ of water in a particular container is related to the depth $h$ of the water by the equation $\frac{dV}{dh} = 16\sqrt{4 - (h - 2)^2}$. If $V = 0$ when $h = 0$, find $V$ when $h = 4$.

8. The slope $m$ of a curve is 0 where the curve crosses the $y$-axis, and $\frac{dm}{dx} = \sqrt{1 + m^2}$. Find $m$ as a function of $x$.

9. Consider Formula (2.1.1), the solution to Example 2.1.7.

   **a.** If $\alpha > \beta$, factor $e^{(\alpha - \beta)kt}$ from the numerator and denominator and show that $x(t) \to \beta$ as $t \to \infty$.

   **b.** If $\alpha < \beta$, explain what happens to $e^{(\alpha - \beta)kt}$ as $t \to \infty$ and show that $x(t) \to \alpha$ as $t \to \infty$.

10. Solve the initial value problem $\frac{dQ}{dt} = \frac{Q^3 + 2Q}{t^2 + 3t}$, $Q(1) = 1$ explicitly for $Q(t)$ and state the interval for which the solution is valid.

## C

1. A police department forensics expert checks a gun by firing a bullet into a bale of cotton. The friction force resulting from the passage of the bullet through the cotton causes the bullet to slow down at a rate proportional to the square root of its velocity. It stopped in 0.1 second and penetrated 10 feet into the bale of cotton. How fast was the bullet going when it hit the bale?

2. The relationship between the velocity $v$ of a rifle bullet and the distance $L$ traveled by it in the barrel of the gun is established in ballistics by the equation $v = \frac{aL^n}{b + L^n}$, where $v = \frac{dL}{dt}$ and $n < 1$. Find the relationship between the time $t$ during which the bullet moves in the barrel and the distance $L$ covered.

3. In trying to determine the shape of a flexible nonstretching cable suspended between two points $A$ and $B$ of equal height, we can analyze the forces acting on the cable and get the differential equation

$$\frac{d^2 y}{dx^2} = k\left[1 + \left(\frac{dy}{dx}\right)^2\right]^{1/2},$$

where $k > 0$ is a constant.

   **a.** Use the substitution $p(x) = dy/dx$ to reduce the second-order equation to a separable first-order equation.

   **b.** Express the general solution of the equation in terms of exponential functions. (You may need a table of integrals here. Your CAS may evaluate the more difficult integral in an awkward way.)

4. When the drug Theophyllin is administered for asthma, a concentration in the blood below 5 mg/liter of blood has little effect, while undesirable side effects appear if the concentration exceeds 20 mg/liter. Suppose a dose corresponding to 14 mg/liter of blood is administered initially. The concentration satisfies the differential equation $\frac{dC}{dt} = -\frac{C}{6}$, where the time $t$ is measured in hours.

    **a.** Find the concentration at time $t$.

    **b.** Show that a second injection will need to be given after about 6 hours to prevent the concentration becoming ineffective.

    **c.** Given that the second injection also increases the concentration by 14 mg/liter, how long is it before another injection is necessary?

    **d.** What is the shortest safe time that a second injection may be given so that side effects do not occur?

    **e.** Sketch graphs of the situations in parts (b), (c), and (d).

**5.** One method of administering a drug is to feed it continuously into the bloodstream by a process called *intravenous infusion*. This process may be modeled by the separable (and linear) differential equation $\frac{dC}{dt} = -\mu C + D$, where $C$ is the concentration in the blood at time $t$, $\mu$ is a positive constant , and $D$ is also a positive constant, the rate at which the drug is administered.

    **a.** Find the equilibrium solution of the differential equation, the solution such that $\frac{dC}{dt} = 0$.

    **b.** Given $C = C_0$ when $t = 0$, find the concentration at time $t$. What limit does the concentration approach as $t \to \infty$? Compare with your answer to part (a).

    **c.** Sketch the graph of a typical solution.

**6.** Let $\frac{dP}{dt} = P(1 - P)$.

    **a.** Find all solutions by separating variables. (You will have to integrate by using partial fractions.)

    **b.** Let $P(0) = P_0$. Suppose $0 < P_0 < 1$. What happens to $P(t)$ as $t \to \infty$?

    **c.** Let $P(0) = P_0$. Suppose $P_0 > 1$. What happens to $P(t)$ as $t \to \infty$?

## 2.2 LINEAR EQUATIONS

We introduced the idea of a linear differential equation in Section 1.1. Now let's see what we can do when the order of the differential equation is 1.

---

**Definition 2.2.1**

A **linear first-order differential equation** is an equation of the form

$$a_1(x)\frac{dy}{dx} + a_0(x)y = f(x),$$

where $a_1, a_0$, and $f$ are functions of the independent variable alone.

---

After dividing through by $a_1(x)$—being careful to note where this function is zero—we can write the equation in the *standard form*

$$\frac{dy}{dx} + P(x)y = Q(x), \tag{2.2.1}$$

where $P$ and $Q$ are functions of $x$ alone. In this standard form, if the function $Q(x)$ is the zero function, we call Equation (2.2.1) **homogeneous**. Otherwise, we say that the equation is

**nonhomogeneous**. (Don't confuse this terminology with the use of the term *homogeneous* as explained before Problems A15–A18 in Exercises 2.1.) In certain applied problems, $Q(x) \neq 0$ may be referred to as the **forcing term**, the **driving term**, or the **input**, as we'll see in Example 2.2.5, for instance. The solution $y$ can be called the **output**.

For example, $\frac{dy}{dx} + \sin(x)y = e^{-x}$ is linear with $P(x) = \sin x$ and $Q(x) = e^{-x}$. The equation $x\frac{dy}{dx} + y^2 = 0$ is not linear, because even when we divide by $x$ (assuming that $x$ is nonzero), we get $\frac{dy}{dx} + \left(\frac{y}{x}\right)y = 0$. The function $Q(x)$ can be taken as $Q(x) \equiv 0$, but the coefficient of $y$, $\frac{y}{x}$, is not a function of $x$ alone.

However, even the complicated-looking equation $2tz^3 + 3t^2z^2\frac{dz}{dt} = t^5z^2$ can be made linear. Just divide by $3t^2z^2$ to get $\frac{dz}{dt} + \left(\frac{2}{3t}\right)z = \frac{1}{3}t^3$, so that $P(t) = \frac{2}{3t}$ and $Q(t) = \frac{1}{3}t^3$. Of course, we must consider the cases $t = 0$ and $z \equiv 0$ separately.

## 2.2.1  The Superposition Principle

In some applications, it is useful to think of a linear first-order equation in terms of an **operator**, or **transformation**, $L$, that changes a differentiable function $y$ into the left-hand side of Equation (2.2.1): $L(y) = \frac{dy}{dx} + P(x)y$. Then Equation (2.2.1) can be expressed simply as $L(y) = Q(x)$. For example, if the nonhomogeneous linear equation in standard form is $\frac{dy}{dx} - y = x$, then we have the operator $L$ defined as $L(y) = \frac{dy}{dx} - y$. If $y(x) = x^2$, for instance, then $L(y) = 2x - x^2$. A solution $y$ of the differential equation $\frac{dy}{dx} - y = x$ would have to satisfy $L(y) = x$. (We can see that $y = x^2$ is not a solution.)

In this general context, suppose $y_1$ is a solution of $L(y) = Q_1(x)$, $y_2$ is a solution of $L(y) = Q_2(x)$, and $c_1, c_2$ are arbitrary constants. Then $c_1y_1 + c_2y_2$ is called a **linear combination** of $y_1$ and $y_2$, and we have

$$L(c_1y_1 + c_2y_2) = \frac{d}{dx}(c_1y_1 + c_2y_2) + P(x)\left(c_1y_1 + c_2y_2\right)$$

$$= c_1\frac{d}{dx}y_1 + c_2\frac{d}{dx}y_2 + c_1P(x)y_1 + c_2P(x)y_2$$

$$= c_1\left(\frac{d}{dx}y_1 + P(x)y_1\right) + c_2\left(\frac{d}{dx}y_2 + P(x)y_2\right)$$

$$= c_1Q_1 + c_2Q_2 = c_1L(y_1) + c_2L(y_2).$$

Any operator that satisfies the condition $L(c_1y_1 + c_2y_2) = c_1L(y_1) + c_2L(y_2)$ is called a **linear operator**. Otherwise, it is called **nonlinear**. We can describe this situation by saying that adding two inputs ($Q_1$ and $Q_2$) of a linear equation gives us an output that is the sum ($y_1 + y_2$) of the individual outputs.

The general form of this last observation is called the **Superposition Principle** and, as we will see later, it applies to linear equations of any order. In particular, when $f(x) \equiv 0$, we see that *any linear combination of solutions to a homogeneous linear equation is also a solution.*

> **SUPERPOSITION PRINCIPLE FOR HOMOGENEOUS EQUATIONS**
>
> Suppose $y_1$ and $y_2$ are solutions of the homogeneous linear first-order differential equation
>
> $$a_1(x)\frac{dy}{dx} + a_0(x)y = 0$$
>
> on an interval $(a, b)$. Then the linear combination $c_1 y_1(x) + c_2 y_2(x)$, where $c_1$ and $c_2$ are arbitrary constants, is also a solution on this interval.

The next two examples should clarify the difference between linear and nonlinear operators.

### ■ Example 2.2.1 A Linear Operator

We can check that $y_1 = e^{-x}$ is a solution of the homogeneous linear equation $L(y) = y' + y = 0 = Q_1$ and that $y_2 = \sin x$ is a solution of $L(y) = y' + y = \cos x + \sin x = Q_2$. (*Note:* The same left-hand side but different right-hand sides.) You should see that $y_1 + y_2 = e^{-x} + \sin x$ is a solution of the equation $y' + y = Q_1 + Q_2 = 0 + \cos x + \sin x = \cos x + \sin x$—that is, that $L(y_1 + y_2) = Q_1 + Q_2 = L(y_1) + L(y_2)$. ■

However, not every operator defined by a first-order equation is linear.

### ■ Example 2.2.2 A Nonlinear Operator

Now consider the operator defined as $T(y) = xy' + y^2$ and suppose that $T(y_1) = 0$ and $T(y_2) = 0$. Then

$$T(y_1 + y_2) = x(y_1 + y_2)' + (y_1 + y_2)^2$$

$$= xy_1' + xy_2' + y_1^2 + y_2^2 + 2y_1 y_2$$

$$= \overbrace{(xy_1' + y_1^2)}^{T(y_1)} + \overbrace{(xy_2' + y_2^2)}^{T(y_2)} + 2y_1 y_2$$

$$= 0 + 0 + 2y_1 y_2 = 2y_1 y_2 \neq T(y_1) + T(y_2).$$

The equation $T(y) = xy' + y^2 = 0$ is nonlinear, and the operator $T$ is not a linear operator. ■

### 2.2.2 The Integrating Factor

Note that if the equation $\frac{dy}{dx} + P(x)y = Q(x)$ is homogeneous, then the equation is separable. (*Do you see why?*) Clearly, the more interesting problems are those for which $Q(x)$ is not the zero function. In addition to their applicability to significant problems, linear first-order equations are nice because you can always solve them explicitly and find the general solution. This is done by a clever technique, the use of something called an **integrating factor**—a special multiplier function that has been used to solve first-order linear equations since the late 1600s.

We will demonstrate the method in the next example, at the same time explaining its effectiveness.

### ■ Example 2.2.3 Using an Integrating Factor

Suppose we want to solve the linear nonhomogeneous equation $y' + xy = 2x$. One way would be to reach into our sleeve and pluck out the magic function $\mu(x) = e^{x^2/2}$, an *integrating factor* for this equation. Now we get an equivalent differential equation by multiplying each side of the original equation by $\mu(x)$:

$$e^{x^2/2}y' + xe^{x^2/2}y = 2xe^{x^2/2}.$$

"Why would we do such a crazy thing?" you're probably asking yourself. Well, just notice that if we assume that $y = y(x)$, an implicit function of $x$, the Product Rule gives us $\left(e^{x^2/2}y\right)' = xe^{x^2/2}y + e^{x^2/2}y'$, the left-hand side of our new differential equation. This observation tells us that the left side is an *exact derivative* and enables us to write the differential equation in a more compact form: $\left(e^{x^2/2}y\right)' = 2xe^{x^2/2}$. (*Be sure that you see this.*) Now we can integrate each side with respect to $x$ to get $e^{x^2/2}y = \int 2xe^{x^2/2}dx = 2e^{x^2/2} + C$. Solving for $y$ by multiplying each side of this last equation by $e^{-x^2/2}$, we get $y(x) = 2 + Ce^{-x^2/2}$, valid for $-\infty < x < \infty$.

We see from the closed-form solution that all solutions approach 2 as $x \to \pm\infty$: If $y(x)$ is any solution of the differential equation $y' + xy = 2x$, then $\lim_{x\to\infty} y(x) = 2 = \lim_{x\to-\infty} y(x)$. Figure 2.4 shows five solutions of this linear equation.

From top to bottom, the five particular solutions plotted correspond to $C = 4, 2, 0, -2$, and $-4$, respectively. The choice $C = 0$ gives us the asymptotic solution $y \equiv 2$.

Finally, you may have recognized that our original equation is actually a separable equation. You should solve by separating variables and then compare your solution to the one we have given.

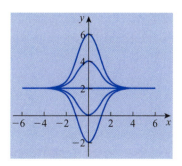

**FIGURE 2.4**

*Solutions of the IVP $y' + xy = 2x$;   $y(0) = -2, 0, 2, 4$, and 6*
*$-6 \le x \le 6, -2 \le y \le 6$*

## Rationale

Now let's step back and look at this integrating-factor technique in more generality. Suppose we have written a linear first-order differential equation in the standard form $\frac{dy}{dx} + P(x)y = Q(x)$. Let's take $P(x)$, the coefficient of $y$ in the equation, and form the new function $\mu(x) = e^{\int P(x)dx}$. (Note that in Example 2.2.3, $P(x) = x$ and $e^{\int P(x)dx} = e^{\int x\,dx} = e^{x^2/2+K}$, where we chose $K = 0$ for convenience.) The Chain Rule and the Fundamental Theorem of Calculus tell us that

$$\frac{d}{dx}\left(e^{\int P(x)dx}\right) = e^{\int P(x)dx} \cdot \frac{d}{dx}\left(\int P(x)dx\right) = e^{\int P(x)dx}P(x).$$

Then, if we multiply each side of the standard form equation by $\mu(x) = e^{\int P(x)dx}$, we have

$$e^{\int P(x)dx}\frac{dy}{dx} + e^{\int P(x)dx}P(x)y = e^{\int P(x)dx}Q(x)$$

and we can rewrite the last line as

$$\frac{d}{dx}\left(e^{\int P(x)dx}y\right) = e^{\int P(x)dx}Q(x).$$

If we integrate both sides of this equation, we get

$$e^{\int P(x)dx}y = \int e^{\int P(x)dx}Q(x)dx + C$$

so we can multiply each side by $e^{-\int P(x)dx}$ to find that

$$y = e^{-\int P(x)dx} \cdot \int e^{\int P(x)dx}Q(x)dx + Ce^{-\int P(x)dx}. \tag{2.2.2}$$

This is an explicit formula for the general solution of any first-order linear differential equation in standard form. Even if the integrals involved can't be evaluated in closed form, they can still be approximated by numerical methods usually learned in a calculus course. (Try out the formula on Example 2.2.3.) *Do not bother memorizing this formula.* Just remember that *any linear first-order equation has an explicit general solution* and understand how to find the appropriate integrating factor.

Technically, there is an infinite family of integrating factors for a given linear equation (so we should say *an* integrating factor, rather than *the* integrating factor), but we can always take the family member with $K = 0$: If $R(x)$ is an antiderivative of $P(x)$ (so that $R'(x) = P(x)$), then $\mu(x) = e^{\int P(x)dx} = e^{R(x)+K} = e^{R(x)}e^K$, and $\mu(x)\frac{dy}{dx} + \mu(x)P(x)y = \mu(x)Q(x)$ has the form $e^K\left\{e^{R(x)}\frac{dy}{dx} + e^{R(x)}P(x)y\right\} = e^K\left\{e^{R(x)}Q(x)\right\}$. Because $e^K$ is always positive, we can cancel it out in the last equation to get $\frac{d}{dx}\left(e^{R(x)}y\right) = e^{R(x)}Q(x)$, and we can continue to the solution as before. The constant $K$ disappears and plays no part in the final stage of the solution, so we can take its value to be 0 from the very beginning.

Now that we know how to choose an integrating factor and find an explicit solution, let's practice.

### ■ Example 2.2.4 Using an Integrating Factor

Let's try the integrating-factor technique on the linear equation $x\frac{dy}{dx} - 2y = x^3 e^{-2x}$. The standard form of the equation is $\frac{dy}{dx} - \left(\frac{2}{x}\right)y = x^2 e^{-2x}$. Our integrating factor is $\mu(x) = e^{\int -\frac{2}{x}dx} = e^{-2\ln|x|} = x^{-2}$.

Multiplying both sides of the equation in standard form by this factor, we get $x^{-2}\frac{dy}{dx} - 2x^{-3}y = e^{-2x}$. Recognizing the left side as the derivative of the product $\mu(x)y = x^{-2}y$, we can write the differential equation as $\frac{d}{dx}\left(x^{-2}y\right) = e^{-2x}$.

Integrating both sides, we find that $x^{-2}y = \int e^{-2x}dx = -\frac{1}{2}e^{-2x} + C$. Now we can solve for $y$ and see that $y = -\frac{1}{2}x^2 e^{-2x} + Cx^2$.  ■

The next example, an important application of linear differential equations to electrical network theory, shows that the details of using an integrating factor may get messy.

### ■ Example 2.2.5 A Circuit Problem

As a consequence of one of *Kirchhoff's laws* in physics, suppose we know that the current $I$ flowing in a particular electrical circuit satisfies the first-order linear differential equation $L\frac{dI}{dt} + RI = v_0 \sin(\omega t)$, where $L, R, v_0$, and $\omega$ are positive constants that give information about the circuit. (See Problems B9–B11 of Exercises 2.2 for related problems.) Let's try to find the current $I(t)$ at time $t$, for $t > 0$, given that $I(0) = 0$. This initial condition says that at the beginning of our analysis ($t = 0$), there is no current flowing in the circuit.

First, we divide both sides of the differential equation by $L$ to get our equation in standard form: $\frac{dI}{dt} + \left(\frac{R}{L}\right)I = \left(\frac{v_0}{L}\right)\sin(\omega t)$. Now, in terms of the standard form [Equation (2.2.1)], we make the identifications $P(t) \equiv R/L$, a constant function, and $Q(t) = (v_0/L)\sin(\omega t)$. In this problem, the forcing term $Q(t)$ represents an (alternating) electromotive force supplied by a generator. Next, we compute the integrating factor

$$\mu(t) = e^{\int P(t)dt} = e^{\int \frac{R}{L}dt} = e^{\frac{R}{L}t}.$$

Multiplying each side of the equation in standard form by $\mu(t)$, we get

$$e^{\frac{R}{L}t}\frac{dI}{dt} + e^{\frac{R}{L}t}\left(\frac{R}{L}\right)I = \left(\frac{v_0}{L}\right)e^{\frac{R}{L}t}\sin(\omega t), \quad \text{or} \quad \frac{d}{dt}\left(e^{\frac{R}{L}t}I\right) = \left(\frac{v_0}{L}\right)e^{\frac{R}{L}t}\sin(\omega t).$$

Integrating each side yields $e^{\frac{R}{L}t}I = \left(\frac{v_0}{L}\right)\int e^{\frac{R}{L}t}\sin(\omega t)\,dt.$

To evaluate this last integral, we have three choices: (1) integrate by parts twice, (2) use a table of integrals, or (3) submit the integral to a computer algebra system capable of integration. In any case, we get

$$e^{\frac{R}{L}t}I = \left(\frac{v_0}{L}\right)\left[\frac{Re^{\frac{R}{L}t}\sin(\omega t)}{L\left(\frac{R^2}{L^2}+\omega^2\right)} - \frac{\omega e^{\frac{R}{L}t}\cos(\omega t)}{\left(\frac{R^2}{L^2}+\omega^2\right)}\right] + C$$

$$= \left(\frac{v_0}{L}\right)\frac{e^{\frac{R}{L}t}}{\left(\frac{R^2}{L^2}+\omega^2\right)} \cdot \left[\frac{R}{L}\sin(\omega t) - \omega\cos(\omega t)\right] + C.$$

To find the general solution, we multiply each side of this last equation by $e^{-\frac{R}{L}t}$ to get

$$I(t) = \frac{\left(\frac{v_0}{L}\right)}{\left(\frac{R^2}{L^2}+\omega^2\right)} \cdot \left[\frac{R}{L}\sin(\omega t) - \omega\cos(\omega t)\right] + Ce^{-\frac{R}{L}t}.$$

Now we use the initial condition $I(0) = 0$:

$$0 = I(0) = \frac{\left(\frac{v_0}{L}\right)}{\left(\frac{R^2}{L^2}+\omega^2\right)} \cdot \left[\frac{R}{L}\sin(\omega \cdot 0) - \omega\cos(\omega \cdot 0)\right] + Ce0$$

$$= \frac{-\omega\left(\frac{v_0}{L}\right)}{\left(\frac{R^2}{L^2}+\omega^2\right)} + C$$

so that $C = \dfrac{\omega\left(\frac{v_0}{L}\right)}{\left(\frac{R^2}{L^2}+\omega^2\right)}$, and we have (finally!)

$$I(t) = \frac{\left(\frac{v_0}{L}\right)}{\left(\frac{R^2}{L^2}+\omega^2\right)} \cdot \left[\frac{R}{L}\sin(\omega t) - \omega\cos(\omega t)\right] + \frac{\omega\left(\frac{v_0}{L}\right)}{\left(\frac{R^2}{L^2}+\omega^2\right)}e^{-\frac{R}{L}t}$$

$$= \frac{\left(\frac{v_0}{L}\right)}{\left(\frac{R^2}{L^2}+\omega^2\right)} \cdot \left[\frac{R}{L}\sin(\omega t) - \omega\cos(\omega t) + \omega e^{-\frac{R}{L}t}\right].$$

In the preceding example, we call the term $\omega e^{-\frac{R}{L}t}$ (or its constant multiple) in the formula for $I(t)$ a **transient** term because it eventually goes to 0. "Eventually" means as $t \to \infty$. The trigonometric terms make up the **steady-state** part of the solution and have the same period as the original forcing term. (*Can you see that this last claim is true?*)

## Exercises 2.2

### A

For Problems 1–18, solve each equation or initial value problem.

1. $y' + 2y = 4x$

2. $y' + 2xy = xe^{-x^2}$

3. $\dot{x} + 2tx = t^3$

4. $y' + y = \cos x$

5. $ty' = -3y + t^3 - t^2$

6. $\dfrac{dx}{ds} = \dfrac{x}{s} - s^2$

7. $y = x(y' - x\cos x)$

8. $(1 + x^2)y' - 2xy = (1 + x^2)^2$

9. $t(x' - x) = (1 + t^2)e^t$

10. $Q' - (\tan t)Q = \sec t; \quad Q(0) = 0$

11. $xy' + y - e^x = 0; \quad y(a) = b$ [$a$ and $b$ are constants.]

12. $(xy' - 1)\ln x = 2y$

13. $y' + ay = e^{mx}$ [Consider two cases: $m \neq -a$ and $m = -a$.]

14. $y' + \left(\dfrac{1 - 2x}{x^2}\right)y = 1$    $y = x^2\left(1 + Ce^{1/x}\right):$ gen. solution.

15. $tx' - \left(\dfrac{x}{t + 1}\right) = t; x(1) = 0$

16. $y = (2x + y^3)y'$ [*Hint:* Think of $y$ as the independent variable, $x$ as the dependent variable, and rewrite the equation in terms of $dx/dy$.]

17. $x(e^y - y') = 2$ [Use the hint from Problem 16.]

18. $y(x) = \int_0^x y(t)dt + x + 1$ [Use the Fundamental Theorem of Calculus to get an ODE.]

### B

An equation of the form $y' + a(x)y = b(x)y^n$ is called a **Bernoulli equation** (named for Jakob Bernoulli, 1654–1705, one of a family of noted Swiss scientists/mathematicians). Notice that if $n = 0$ or $n = 1$, we just have a linear equation. Now if $n$ is not equal to 0 or 1 and we divide both sides of the equation by $y^n$, we can let $z = y^{1-n}$ and get a linear equation in the variable $z$. We solve the linear equation for $z$ in terms of $x$ and then return to the original variables $x$ and $y$. This substitution method was found by Leibniz in 1696. For example, $y' - y = xy^2$ is a Bernoulli equation with $a(x) \equiv -1$, $b(x) = x$, and $n = 2$. Divide by $y^2$ and the equation becomes $y^{-2}y' - y^{-1} = x$. Letting $z = y^{-1}$, we get the linear equation $-z' - z = x$, or $z' + z = -x$. Solving for $z$, we find that $z = 1 - x + ce^{-x}$. Since $z = y^{-1}$, we conclude that $y = (1 - x + ce^{-x})^{-1}$. Note that we divided by $y^2$ and that $y \equiv 0$ is a singular solution. Find all solutions of each Bernoulli equation in Problems 1–6.

1. $y' = \dfrac{4}{t}y - 6ty^2$

2. $\dot{x} = \dfrac{1}{t}x + \sqrt{x}$

3. $\dfrac{dy}{dx} + y = xy^3$

4. $y' + xy = \sqrt{y}$

5. $y' = 2ty + ty^2$

6. $y' = x^3y^2 + xy$ [*Hint*: Consider $x$ as the dependent variable.]

7. In trying to regulate fishing in the oceans, international commissions have been set up to implement controls. To understand the effect of such controls, mathematical models of fish populations have been constructed. One stage in this modeling effort involves predicting the growth of an individual fish. The *von Bertalanffy growth model* is reflected in the Bernoulli equation (see above):

$$\frac{dW}{dt} = \alpha W^{2/3} - \beta W,$$

where $W = W(t)$ denotes the weight of a fish and $\alpha$, $\beta$ are positive constants.

   a. Find the general solution of the equation.
   b. Calculate $W_\infty = \lim\limits_{t \to \infty} W(t)$, the limiting weight of the fish.
   c. Using the answer to part (b) and the initial condition $W(0) = 0$, write the formula for $W(t)$ free of any arbitrary constants.
   d. Sketch a graph of $W$ against $t$.

8. Show that if a linear first-order differential equation is homogeneous, then the equation is separable.

9. When a switch is closed in a circuit containing a resistance $R$, an inductance $L$, and a battery that supplies a constant voltage $E$, the current $I$ builds up at a rate described by the equation $L\frac{dI}{dt} + RI = E$. [In Example 2.2.5, the electromotive force on the right-hand side of the equation is not constant. Instead of a battery, there is a generator supplying an alternating voltage equal to $(v_0/L)\sin(\omega t)$.]

   a. Find the current $I$ as a function of time.
   b. Evaluate $\lim\limits_{t \to \infty} I(t)$.
   c. How long will $I$ take to reach one-half its "final" value?
   d. Find $I$ if $I_0 = I(0) = E/R$.

10. In an electrical circuit, when a capacitor of capacitance $C$ is being charged through a resistance $R$ by a battery which supplies a constant voltage $E$, the instantaneous charge $Q$ on the capacitor

satisfies the differential equation

$$R\frac{dQ}{dt} + \frac{Q}{C} = E.$$

**a.** Find $Q$ as a function of time if the capacitor is initially uncharged—that is, if $Q_0 = Q(0) = 0$.

**b.** How long will it be before the charge on the capacitor is one-half its "final" value?

11. In Problem 10, determine $Q$ if $Q_0 = 0$ and if the battery is replaced by a generator that supplies an alternating voltage equal to $E_0 \sin(\omega t)$.

12. In analyzing the effect of advertising on the sales of a product, we can extract the following model from work done by the economists Vidale and Wolf[4]:

$$\frac{dS}{dt} + \left(\frac{rA}{M} + \lambda\right)S = rA.$$

Here, $S = S(t)$ denotes sales, $A = A(t)$ indicates the amount of advertising, $M$ is the saturation level of the product (the practical limit of sales that can be generated), and $r$ and $\lambda$ are positive constants. Clearly, the solution of this linear equation depends on the form of the advertising function $A$.

**a.** Solve the equation if $A$ is constant over a particular time interval and zero after this:

$$A(t) = \begin{cases} \overline{A} & \text{for} \quad 0 < t < T \\ 0 & \text{for} \quad t > T. \end{cases}$$

(You really have to solve two equations and then combine the solutions appropriately.)

**b.** Sketch a typical graph of $S$ against $t$. (Choose reasonable values for any constants in your solution.)

13. In the study of population genetics, biological units called *genes* determine what characteristics living things inherit from their parents. Suppose we look at a gene with two "flavors" $A$ and $a$ that occur in the proportions $p(t)$ and $q(t) = 1 - p(t)$, respectively, at time $t$ in a particular population. Suppose that we have the relation $\frac{dp}{dt} = v - (\mu + v)p$, where $\mu$ is a constant describing a "forward mutation rate" and $v$ is another constant representing the "backward mutation rate."

**a.** Determine $p(t)$ and $q(t)$ in terms of $p(0), q(0), \mu$, and $v$.

**b.** Show that $\lim_{t\to\infty} p(t) = v/(\mu + v)$ and $\lim_{t\to\infty} q(t) = \mu/(\mu + v)$. These are called the *equilibrium gene frequencies*.

14. If $V = V(t)$ represents the value of a bond at time $t$, $r(t)$ is the interest rate, and $K(t)$ is the coupon payment, then $\frac{dV}{dt} + K(t) = r(t)V$ describes the value of the bond at a time before maturity.

---

[4] M. L. Vidale and H. B. Wolfe, "Response of Sales to Advertising," in *Mathematical Models in Marketing*, ed. Robert G. Murdick (Scranton, PA: Intext Educational Publishers, 1971): 249–256.

**a.** If $T$ is the time of the bond's maturity and $V(T) = Z$, show that

$$V(t) = e^{-\int_t^T r(x)dx} \left( Z + \int_t^T K(u)e^{\int_u^T r(x)dx} du \right).$$

**b.** What does $V(t)$ look like if you have a zero-coupon bond—that is, if $K(t) \equiv 0$?

## C

1. Prove that any Bernoulli differential equation $y' + a(x)y = b(x)y^n$, where $n \neq 0, 1$, can be converted into a linear equation by the special substitution $y = u^{1/(1-n)}$. (See the paragraph at the beginning of the B exercises.)

2. A disease has spread throughout a community. The number of infected individuals $I$ in the community at any time $t > 0$ is given by $\frac{dI}{dt} - k(P_0 + rt)I = -kI^2$, where $I(0) = I_0$, $P_0$ is the population of the community at $t = 0$, $r$ is the constant growth rate of the community, and $k$ is a constant. Show that

$$I = e^{kP_0t + (1/2)krt^2} \left[ \frac{1}{I_0} + k \int_0^t e^{kP_0u + (1/2)kru^2} du \right]^{-1}.$$

[*Hint*: The differential equation is a Bernoulli equation, as discussed in Problem C1.]

3. Suppose you have a linear first-order differential equation in the standard form $\frac{dy}{dx} + P(x)y = Q(x)$, where $Q(x)$ is not the zero function.

    **a.** Looking at the general solution given by Equation (2.2.2), show that the term $Ce^{-\int P(x)dx}$ is the general solution, $y_{GH}$, of the homogeneous equation you get by setting $Q(x) \equiv 0$.

    **b.** Show that the term $e^{-\int P(x)dx} \cdot \int e^{\int P(x)dx} Q(x)dx$ is a particular solution, $y_{PNH}$, of the original nonhomogeneous equation. (Thus, we can express the general solution, $y_{GNH}$, of the nonhomogeneous equation as follows: $y_{GNH} = y_{GH} + y_{PNH}$.) (See Problem C5 of Exercises 1.3.)

    **c..** Examine the result of part (b) in light of the Superposition Principle.

## 2.3 COMPARTMENT PROBLEMS

In analyzing certain systems in biology and chemical engineering, researchers encounter a class of problems called **mixing problems** or **compartment problems**.

Suppose we have a single container, or compartment, containing some substance. Now think of some other substance entering the compartment at a certain rate, and imagine that a mixture of the two substances leaves the compartment at another rate. For example, we could be talking about a tank of water into which some chemical is introduced via a pipe. What emerges from the tank through another pipe will be a mixture of the water and the chemical. In biology and physiology, the compartment may be the bloodstream or a particular organ, such as the kidneys. In fact, mathematical analyses in these research fields often regard the organism under study as a whole collection of individual components (compartments). In the human body, these could be different organs or groups of cells, for example.

**FIGURE 2.5**

*A one-compartment model*

To get our bearings, we can start with a simple one-compartment model (Figure 2.5).

We have a single tank with a certain amount of material in it. The amount of substance (or the concentration of the substance) that is added to the tank is called the **inflow**, and the amount (or concentration) of substance leaving the tank is called the **outflow**. We assume that there is a thorough mixing process taking place in the tank—an almost instantaneous uniform blending of the two substances. To model this process using a differential equation, it is important to focus on three different rates associated with this situation: (1) the rate of inflow, (2) the net rate at which some aspect of the mixture in the tank is changing, and (3) the rate of outflow of the mixture.

The basic principle we will use in solving compartment problems is the **Balance Law** (or **Balance Equation**):

$$\boxed{\textbf{Net Rate of Change} = \textbf{Rate of Inflow} - \textbf{Rate of Outflow.}} \qquad (2.3.1)$$

As our first example, let's look at a simple model of medicine in the bloodstream.

### ■ Example 2.3.1 Medicine in the Bloodstream

*Intravenous infusion* is the process of administering a substance into the veins at a steady rate. (See Problem C5 of Exercises 2.1.) Suppose a patient in a hospital is receiving medication through an intravenous tube that drips the substance into the bloodstream at a constant rate of *I* milligrams (mg) per minute. Also suppose that the medication is dispersed through the body and eliminated at a rate proportional to the concentration of the medication at the time. In this problem, concentration is defined to be

$$\frac{\text{Quantity of medication}}{\text{Volume of blood plus medication}},$$

where we assume that the volume *V* of blood plus medication remains constant. The problem is to find the concentration of the medication in the body at any time *t*. To do this, we can consider the bloodstream as a single compartment and examine a differential equation that models the process.

If we let $C = C(t)$ denote the concentration of the medication at time $t$ (in mg/cm$^3$), then the Balance Law (2.3.1) leads us to the relation

$$V\frac{dC}{dt} = I - kC,$$

where $k$ is a positive constant of proportionality that depends on the specific medication and the physiological characteristics of the patient.

Note that the left-hand side of the differential equation is in units of cm$^3 \times \frac{\frac{mg}{cm^3}}{min} = \frac{mg}{min}$ and that the right-hand term $I$ is also in mg/min. Because $C$ is expressed in units of $\frac{mg}{cm^3}$, we see that the appropriate unit for $k$, representing a removal rate, must be cm$^3$/min. (Be sure that you understand this "dimensional analysis.")

This is a linear equation that we can write in the standard form

$$\frac{dC}{dt} + \left(\frac{k}{V}\right)C = \frac{I}{V}.$$

An integrating factor for this equation is $\mu = e^{\int \frac{k}{V}dt} = e^{\frac{kt}{V}}$. Multiplying each side of this last differential equation by $\mu$ gives us

$$e^{\frac{kt}{V}}\frac{dC}{dt} + \frac{k}{V}e^{\frac{kt}{V}}C = \left(\frac{I}{V}\right)e^{\frac{kt}{V}}$$

or

$$\frac{d}{dt}\left(e^{\frac{kt}{V}}C\right) = \left(\frac{I}{V}\right)e^{\frac{kt}{V}},$$

so that integrating each side gives us

$$e^{\frac{kt}{V}}C = \int\left(\frac{I}{V}\right)e^{\frac{kt}{V}}dt$$

and

$$C(t) = e^{-\frac{kt}{V}}\int\left(\frac{I}{V}\right)e^{\frac{kt}{V}}dt = e^{-\frac{kt}{V}}\left[\frac{V}{k}\left(\frac{I}{V}\right)e^{\frac{kt}{V}} + \alpha\right] = \frac{I}{k} + \alpha e^{-\frac{kt}{V}}.$$

Using the implied initial condition $C(0) = 0$, we find that $\alpha = -\frac{I}{k}$, so that we can write our solution as

$$C(t) = \frac{I}{k} - \frac{I}{k}e^{-\frac{kt}{V}} = \frac{I}{k}\left(1 - e^{-\frac{kt}{V}}\right).$$

Note what happens as time goes by. Analytically, $\lim_{t\to\infty} C(t) = \frac{I}{k}$. This says that the concentration of medication in the patient's body reaches a *threshold*, or *saturation level*, of $\frac{I}{k}$. Figure 2.6

**FIGURE 2.6**

$C(t) = 20\left(1 - e^{-0.2t}\right), 0 \le t \le 30, 0 \le C \le 20$

is a graph of the concentration when $I = 4$, $V = 1$, and $k = 0.2$, showing a saturation level of 20 mg/cm$^3$. ■

In compartment model problems, it is often important to determine *how long* it may take for a certain result to occur.

### ■ Example 2.3.2 Air Pollution

By 10:00 P.M. on a lively Friday night, a club of dimensions 30 feet by 50 feet by 10 feet is full of customers. Sadly, many of these customers are smokers, so cigarette smoke containing 4% carbon monoxide is introduced into the room at a rate of 0.15 cubic foot per minute. Suppose that this rate does not vary significantly during the evening. Before 10:00 there is no trace of carbon monoxide in the club; and, fortunately, this club is equipped with good ventilators. These ventilators allow the formation of a uniform smoke-air mixture in the room, and they provide for the ejection of this mixture to the outside at the rate of 1.5 cubic feet per minute—that is, at a rate 10 times greater than that of the arrival of pollutants.

You want to dance and socialize, but you also want to preserve your health. A prolonged exposure to a concentration of carbon monoxide greater than or equal to 0.012% is considered dangerous by the Health Department. Knowing that the club closes its doors at 3 A.M., will you allow yourself to stay until the end? To be more precise, you want to find the time when the concentration of carbon monoxide reaches the critical concentration of 0.012%.

The key to this type of single-compartment problem is the fundamental relation we saw used in the preceding example:

**Net rate of change = rate of inflow − rate of outflow.**

Let $C(t)$ be the concentration of carbon monoxide in the club (the grams of carbon monoxide per cubic foot of air, abbreviated g/ft$^3$) at any time $t$, where $t = 0$ represents 10 P.M. Then $Q(t)$, the *amount* of pollutant in the room at time $t$, is described by the equation

$Q(t) = $ (volume of room) $\times C(t)$. Because the room is $30 \times 50 \times 10 = 15,000$ cubic feet, this expression for the amount of monoxide in the room at time $t$ becomes $Q(t) = 15,000\, C(t)$.

Now the *rate* at which carbon monoxide is *entering* the room is given by

$$\left(0.15 \frac{\text{ft}^3}{\text{min}}\right)\left(0.04 \frac{\text{g}}{\text{ft}^3}\right) = 0.006 \frac{\text{g}}{\text{min}}.$$

Similarly, the *rate* at which carbon monoxide is *leaving* the room (via the ventilators) is $\left(1.5 \frac{\text{ft}^3}{\text{min}}\right) \cdot C(t)$.

The Relationship (2.3.1) tells us that the rate of change of the amount of carbon monoxide in the room is equal to the rate at which the pollutant is introduced minus the rate at which it leaves:

$$\frac{dQ(t)}{dt} = \frac{d}{dt}\{15,000\, C(t)\} = \text{rate of inflow} - \text{rate of outflow}$$

$$= \left(0.15 \frac{\text{ft}^3}{\text{min}}\right)\left(0.04 \frac{\text{g}}{\text{ft}^3}\right) - \left(1.5 \frac{\text{ft}^3}{\text{min}}\right) C(t)$$

$$= 0.006 - 1.5 C(t)\, \text{g}/\text{min}$$

so we have the differential equation

$$15,000 \frac{d}{dt} C(t) = 0.006 - 1.5\, C(t).$$

This is a linear equation, and we can write it in the form

$$\frac{d}{dt} C(t) + (0.0001)C(t) = \left(4 \times 10^{-7}\right).$$

An integrating factor is $\mu(t) = e^{\int (0.0001)\, dt} = e^{0.0001t}$, so the last equation has the form

$$\frac{d}{dt}\left\{e^{0.0001t} C(t)\right\} = \left(4 \times 10^{-7}\right) e^{0.0001t}.$$

Integrating, we find that

$$C(t) = \left(4 \times 10^{-7}\right) e^{-0.0001t} \int e^{0.0001t}\, dt$$

$$= \left(4 \times 10^{-7}\right) e^{-0.0001t} \left(\frac{e^{0.0001t}}{0.0001} + k\right)$$

$$= 0.004 + \alpha \cdot e^{-0.0001t}$$

where $\alpha = \left(4 \times 10^{-7}\right) k$.

Because we are told that $C(0) = 0$, we have $0 = C(0) = 0.004 + \alpha$, which gives us the information that $\alpha = -0.004$. Therefore, we can write the solution of our differential equation as

$$C(t) = 0.004 \left(1 - e^{-0.0001t}\right).$$

Because we want to know the time $t$ at which the concentration equals 0.012%, we must solve the equation $C(t) = 0.00012$ for $t$. Hence, we must have

$$0.00012 = 0.004 \left(1 - e^{-0.0001t}\right)$$

$$0.03 = 1 - e^{-0.0001t}$$

$$e^{-0.0001t} = 1 - 0.03 = 0.97$$

$$-0.0001t = \ln(0.97)$$

$$t = \frac{\ln(0.97)}{(-0.0001)}$$

so $t = 304.59$ minutes $\approx 5.08$ hours $\approx 5$ hours, 5 minutes. Therefore, the critical concentration of carbon monoxide is reached at 3:05 A.M. *That's cutting it too close!* ∎

The next example shows a different sort of compartment and alerts us to the fact that not all compartments have constant "volumes."

### ∎ Example 2.3.3 Fairness in Employment

Suppose that a government agency has a current staff of 6000, of whom 25% are women. Employees are quitting randomly at the rate of 100 per week. If we know that replacements are being hired at the rate of 50 per week, with the requirement that half be women, what is the size of the agency staff in 40 weeks, and what percentage is then female?

This is a compartment problem, with the agency as the compartment. We note that, in contrast to the previous examples, our compartment size (agency staff size) varies with time. Let $W(t)$ be the number of women at time $t$, with $W(0) = 25\%$ of $6000 = 1500$. Now the net change in total staff is $50 - 100 = -50$ people/week, so that the staff size at time $t$ is $6000 - 50t$ people. Summarizing this information, we have

$$\underbrace{\frac{dW}{dt}}_{\substack{\text{rate of change} \\ \text{in no. of women}}} = \underbrace{25 \text{ women/week}}_{\substack{\text{rate of inflow of women} \\ = 50\% \text{ of all replacements}}} - \underbrace{\left(\underbrace{100 \text{ people/week}}_{\text{rate of people leaving}} \cdot \underbrace{\frac{W(t)}{6000 - 50t}\text{women/people}}_{\substack{\text{proportion of women on staff} \\ \text{at time of leaving}}}\right)}_{\text{rate of women leaving}}$$

or

$$\frac{dW}{dt} + \left(\frac{100}{6000 - 50\,t}\right) W = 25.$$

The integrating factor is $\mu(t) = (6000 - 50\,t)^{-2}$, so we get $W(t) = \frac{1}{2}(6000 - 50\,t) + C(6000 - 50\,t)^2$. Because $W(0) = 1500$, we find that $C = -1/24,000$. Thus, $W(t) = \frac{1}{2}(6000 - 50\,t) - \frac{1}{24,000}(6000 - 50\,t)^2$; and when $t = 40$ we get a staff total equal to $6000 - 50(40) = 4000$ and $W(40) = 2000 - 2000/3$, so that the staff is about $(2000 - 2000/3)/4000 = 1/3$ or $33\frac{1}{3}\%$ female.   ∎

After we've treated systems of equations in Chapter 5, we'll be able to solve *multicompartment* problems.

## Exercises 2.3

### A

1. Suppose a population has a constant per capita birth rate $b > 0$ and a constant per capita death rate $d > 0$. Using the balance Equation (2.3.1), write a differential equation for the population $p(t)$ at time $t$. (Do not solve this equation.)

2. In Problem 1, suppose the per capita death rate $d$ is not constant, but is instead proportional to the population $p(t)$. Write (but do not solve) a differential equation for the population $p(t)$.

3. In Problem 1, suppose the per capita birth rate $b$ is not constant, but is instead proportional to the population $p(t)$. Write (but do not solve) a differential equation for the population $p(t)$.

4. Suppose a country with constant per capita birth and death rates $b$ and $d$, respectively, has an influx of immigrants at a constant rate $I$ (not a per capita rate, but a constant rate). Write a differential equation for the size of the population $p(t)$.

### B

1. A study of the population of Botswana from 1975 to 1990 leads to the following model for the country's growth rate: $\frac{dP}{dt} = kP - \alpha t$, where $t$ denotes time in years with 1990 corresponding to $t = 0$, $P(0) = 1.285$ (million), $k = 0.0355$, and $\alpha = 1.60625 \times 10^{-3}$. (The term $kP$ reflects births and immigration, while the term $\alpha t$ captures deaths and emigration.)

   a. Find a formula for $P(t)$.
   b. Estimate Botswana's population in the year 2010.

2. A tank with a capacity of 100 gallons is half full of fresh water. A pipe is opened which lets treated sewage enter the tank at the rate of 4 gal/min. At the same time, a drain is opened to allow 3 gal/min of the mixture to leave the tank. If the treated sewage contains 10 grams per gallon of usable potassium, what is the *concentration* of potassium in the tank when it is full? (Be careful of your units!)

3. A tank having a capacity of 100 gallons is initially full of water. Pure water is allowed to run into the tank at the rate of 1 gallon per minute. At the same time, brine (a mixture of salt and water)

containing 1/4 pound of salt per gallon flows into the tank at the rate of 1 gallon per minute. (Assume that there is perfect mixing.) The mixture flows out at the rate of 2 gallons per minute. Find the *amount* of salt in the tank after $t$ minutes.

4. Suppose you have a 200-gallon tank full of fresh water. A drain is opened that removes 3 gal/sec from the tank and, at the same moment, a valve is opened that lets in a 1% solution (a 1% concentration) of chlorine at 2 gal/sec.

   a. When is the tank half full and what is the concentration of chlorine then?
   b. If the drain is closed when the tank is half full and the tank is allowed to fill, what will be the final concentration of chlorine in the tank?

5. A tank contains 50 gallons of fresh water. Brine (see Problem 3) at a concentration of 2 lbs/gal (that is, 2 lbs of salt per gallon) starts to run into the tank at 3 gal/min. At the same time, the mixture of fresh water and brine runs out at 2 gal/min.

   a. How much liquid is there in the tank after 50 minutes?
   b. How many pounds of dissolved salt is in the tank then?

6. In a large tank are 100 gallons of brine containing 75 pounds of dissolved salt. Water runs into the tank at the rate of 3 gal/min, and the mixture runs out at the rate of 2 gal/min. The concentration is kept uniform by stirring. How much salt is there in the solution after 1.5 hours?

7. A swimming pool holds 10,000 gallons of water. It can be filled at the rate of 100 gal/min and emptied at the same rate. Right now, the pool is filled, but there are 20 pounds of an impurity dissolved in the water. For the safety of the swimmers, this must be reduced to less than 1 pound. It would take 200 minutes to empty the pool completely and refill it, but during part of this time the pool could not be used. How long will it take to restore the pool to a safe condition if at all times the pool must be at least half full?

8. Assuming the information in Example 2.3.3, what would have been the percentage of female staff members after 40 weeks if it had been required that *all* new employees were women?

## C

1. Suppose that the maximum concentration of a drug present in a given organ of constant volume $V$ must be $c_{max}$. Assuming that the organ does not contain the drug initially, that the liquid carrying the drug into the organ has constant concentration $c > c_{max}$, and that the inflow and outflow rates are both equal to $r$, show that the liquid must not be allowed to enter for a time longer than

$$\frac{V}{r} \ln \left( \frac{c}{c - c_{max}} \right).$$

2. A tank that holds 100 gallons is half full of a brine solution with a concentration of $\frac{1}{3}$ pound of salt per gallon. Two pipes lead into it at the top, one supplying a brine solution of $\frac{1}{2}$ lb/gal and the other pure water. Each pipe has a flow of 4 gal/min. One pipe leads out at the bottom and removes the mixture at 3 gal/min. What is the *concentration* of the mixture that first flows out the overflow pipe at the top? Assume uniform mixing.

## 2.4 SLOPE FIELDS

Now that we have become familiar with the basic concepts of ordinary differential equations (ODEs) and have learned how to solve separable and linear equations, we can consider a *qualitative* approach to understanding solutions of first-order equations. This is a graphical approach to an equation that provides insights into the behavior of solutions, even when we may not know the techniques for solving the equation.

Let's look at first-order equations of the general form

$$\frac{dy}{dx} = y' = f(x, y).$$

Many equations can be written in this way, with the derivative isolated. For example, we could have $\frac{dy}{dx} = f(x, y) = 3y - 4x$, $y' = g(x, y) = \sqrt{xy}$, $y' = F(x) = 2x^3 - 1$, or $y' = G(y) = 2 - y^2$. Now remember what a first derivative tells us. One interpretation of a derivative is as the slope of the tangent line drawn to a curve at a particular point. The equation $y' = f(x, y)$ means that at the point $(x, y)$ of any solution curve of the differential equation, the slope of the tangent line is given by the value of the function $f$ at that point—that is, the slope is given by $f(x, y)$. Remember that there may be a whole family of solution curves as well as singular solutions.

For a first-order differential equation, a set of possible tangent line segments (sometimes called **lineal elements**), whose slopes at $(x, y)$ are given by $f(x, y)$, is called a **slope field** (or **direction field**) of the equation. Visually, this establishes a flow pattern for solutions of the equation. A slope field includes tangent line segments for many solutions of the equation, but the general shapes of the integral curves should be clear. You can think of these outlines as the "ghosts" of solution curves, and they may reveal certain *qualitative* aspects of the solutions, even if a closed form solution is difficult or impossible to find.

Our first example will indicate how to generate a slope field.

### ■ Example 2.4.1 A Slope Field

To get a feeling for these ideas, let's get a piece of graph paper and plot some tangent line segments for the first-order linear equation $y' - y = x$, which we can write as $y' = f(x, y) = x + y$. To make things a bit easier, we can construct a table (Table 2.1).

We've made things even easier for ourselves by choosing points at which the slopes are 0, 1, and $-1$. Now we can draw some tangent line segments corresponding to these slopes (Figure 2.7a).

Note that we have drawn the little tangent line segments so that the midpoint of each segment is the point $(x, y)$. We have used portions of the slope field given by $f(x, y) = 0$ and $f(x, y) = \pm 1$. Figure 2.7b is a computer-drawn direction field for the same ODE, with some solution curves superimposed on the slope field.

Note that as $x \to -\infty$, the solution curves seem to be approaching a straight line as an asymptote. The solution curves seem to be veering *away* from this line as $x \to +\infty$. If you

**Table 2.1** Slopes at Points $(x, y)$ for $y' = x + y$

| Point | | | $y' = x + y$ |
|---|---|---|---|
| $x$ | $y$ | $(x, y)$ | Slope at $(x, y)$ |
| $-3$ | $3$ | $(-3, 3)$ | $0$ |
| $1$ | $-1$ | $(1, -1)$ | $0$ |
| $0$ | $0$ | $(0, 0)$ | $0$ |
| $0$ | $1$ | $(0, 1)$ | $1$ |
| $1$ | $0$ | $(1, 0)$ | $1$ |
| $2$ | $-1$ | $(2, -1)$ | $1$ |
| $-1$ | $2$ | $(-1, 2)$ | $1$ |
| $0$ | $-1$ | $(0, -1)$ | $-1$ |
| $-1$ | $0$ | $(-1, 0)$ | $-1$ |
| $2$ | $-3$ | $(2, -3)$ | $-1$ |
| $\vdots$ | $\vdots$ | $\vdots$ | $\vdots$ |

look very closely, you may be able to guess that the straight line is $x + y = -1$, or $y = -1 - x$. In Section 2.2 we learned how to find the general solution, $y = -x - 1 + Ce^x$, for this linear equation. The straight line $y = -1 - x$ is the particular solution of the ODE corresponding to $C = 0$, a solution of the initial-value problem (IVP) $y' - y = x$, $y(0) = -1$. Also note that if $y$ is the general solution and $C \neq 0$, then $\lim_{x \to +\infty} y(x) = \infty$ if $C > 0$ and $\lim_{x \to +\infty} y(x) = -\infty$ if $C < 0$. ∎

Although the slope field suggests some features of the solution curves, we have to be careful not to read too much into it. In the preceding example, without the analytic form of the general solution or some sound numerical evidence, we can't be sure that $y$ doesn't have vertical asymptotes, so that $y \to \pm\infty$ as $x$ approaches some *finite* value $x_0$.

Note that in Example 2.4.1 we used portions of the slope field given by $f(x, y) = 0$ and $(x, y) = \pm 1$. For any first-order equation $y' = f(x, y)$, if we look at the set of points $(x, y)$ such that $f(x, y) = C$, a constant, we get an **isocline**—a curve along which the slopes of the tangent lines are all the same. (The word *isocline* is made up of parts that mean "equal" and "inclination" or "slope.") Isoclines are used to simplify the construction of a slope field because once you draw the isoclines, you can quickly and easily draw, for each $C$, a series of parallel line segments of slope $C$, all having their midpoints on the curve $f(x, y) = C$. In Example 2.4.1 the isoclines are the curves $x + y = C$, which are straight lines through $(0, C)$ and $(C, 0)$ with slope $-1$.

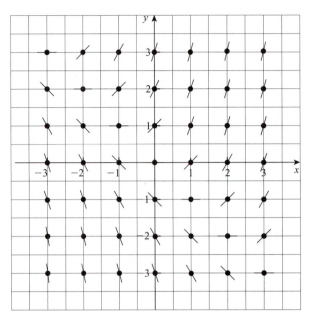

**FIGURE 2.7a**

*Some lineal elements for $y' = x + y$*

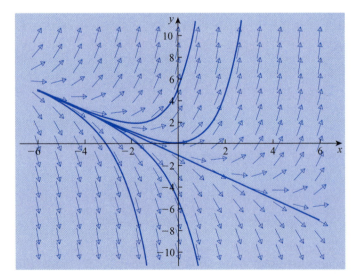

**FIGURE 2.7b**

*Slope field for $y' = x + y$, $-6 \leq x \leq 6$, $-10 \leq y \leq 10$ and five computer-generated solution curves*

It is important to realize that *an isocline is usually not a solution curve* but that through any point on an isocline, a solution to the differential equation passes with slope C. However, as we'll see in Section 2.5, isoclines corresponding to $C = 0$—called **nullclines**—turn out to be important solutions (*equilibrium solutions*) of equations in which the independent variable does not appear explicitly—that is, equations of the form $y' = f(y)$.

The next example has something important to say about the difference between equations of the general form $y' = f(x, y)$ and equations of the special form $y' = f(y)$.

### ■ Example 2.4.2 A Special Slope Field

The slope field (Figure 2.8) corresponding to the equation $x' = f(x) = -2x$ reveals something interesting about certain kinds of equations and their corresponding slope fields. (*Don't be confused by the labeling of the axes. Here, we are assuming that t is the independent variable and x is the dependent variable: $x = x(t)$.*) First of all, note that algebraically we can write the equation in the form $F(x, x') = x' + 2x = 0$, or $x' = f(x) = -2x$. In other words, we have a first-order equation in which the independent variable $t$ does not appear explicitly. This says that the slopes of the tangent line segments making up the slope field of this equation depend only on the values of $x$.

In the slope field plot given in Figure 2.8, if you fix the value of $x$ by drawing a horizontal line $x = C$ for any constant $C$, you will see that all the tangent line segments along this line have the same slope, no matter what the value of $t$. Another way to look at this is to realize

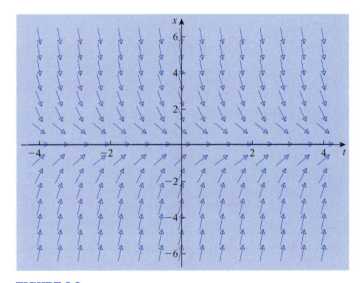

**FIGURE 2.8**
*Slope field for $x' = -2x$, $-4 \leq t \leq 4$, $-6 \leq x \leq 6$*

that you can generate infinitely many solutions by taking any one solution and translating (shifting) its graph left or right. (See Problem C1 of Exercises 2.4.)    ∎

### 2.4.1 Autonomous and Nonautonomous Equations

A differential equation, such as the one in the preceding example, in which the independent variable does not appear explicitly is called an **autonomous** equation. If the independent variable does appear, the equation is called **nonautonomous**. This definition is valid for an equation of any order.

For example, $y' = y^2 - t^2$ is nonautonomous because the independent variable $t$ appears explicitly, whereas $y' = 3y^4 + 2\sin(y)$ is autonomous because the independent variable $(t, x,$ or whatever) is missing. Note that the independent variable is always present *implicitly* (in the background), but if you don't see it "up front," the equation is autonomous. Example 2.3.1 discusses a nonautonomous equation. If we look carefully at its slope field (Figure 2.7b), we see that the slopes change as we move along any horizontal line.

Autonomous equations arise frequently in physical problems because the physical components generally depend on the *state* of the system, but not on the actual time. We can define the **state** of a system loosely as the set of values of the dependent variables in the system. For example, according to *Newton's Second Law of Motion* (to be discussed and applied to spring-mass problems in Chapter 4), an object of mass $m$ falling under the influence of gravity satisfies the autonomous equation $\ddot{x} = -g$, where $x(t)$ is the position of the mass measured from the earth's surface and $g$ is the acceleration due to gravity. Gravity is considered time-independent because the mass follows the same path no matter when the mass is dropped.

Now let's see how to recognize the correspondence between first-order differential equations and their slope fields.

### ∎ Example 2.4.3 Matching Equations and Slope Fields

$$(A)\frac{dx}{dt} = x^2 - t^2 \quad \text{and} \quad (B)\frac{dx}{dt} = x^2 - 1$$

Looking at the two differential equations and the accompanying slope fields 1–4, let's try to match each equation with exactly one of the slope fields (Figures 2.9a–d).

We can start with Equation (A) and note that it is a *nonautonomous* equation. This tells us that we should not expect equal slopes along horizontal lines. As we move horizontally—that is, if we fix the value of $x$ and vary the value of $t$—the value of the slope changes according to the formula $x^2 - t^2$. This analysis eliminates slope fields 2 and 3 because the inclinations of the tangent line segments clearly remain constant along horizontal lines. Now if we write Equation (A) in factored form, $\frac{dx}{dt} = (x + t)(x - t)$, we can see that the tangent line segments must be horizontal where $x = t$ or $x = -t$ because that's where the slope $\frac{dx}{dt}$ equals 0. (These

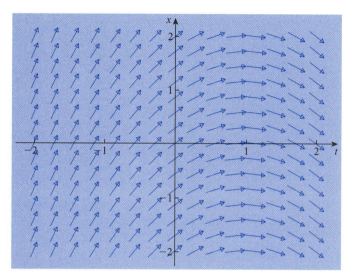

**FIGURE 2.9a**

*Slope field 1*

**FIGURE 2.9b**

*Slope field 2*

are the *nullclines*—isoclines corresponding to $C = 0$.) Looking carefully at slope fields 1 and 4, we see that field 4 exhibits a series of horizontal "steps" forming an X through the origin. If we look closely, it seems that these horizontal line segments lie on the lines $x = t$ and $x = -t$, so we conclude that Equation (A) corresponds to slope field 4.

**FIGURE 2.9c**

*Slope field 3*

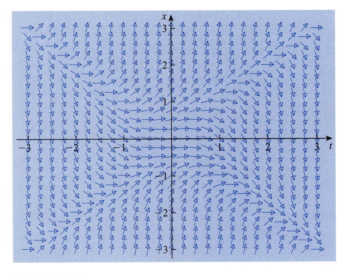

**FIGURE 2.9d**

*Slope field 4*

Equation (B) is *autonomous* because the independent variable $t$ does not appear explicitly. The corresponding slope field must show equal slopes along any horizontal line. Only fields 2 and 3 exhibit this behavior. What else can we look for? Well, if we factor Equation (B) to get $\frac{dx}{dt} = (x + 1)(x - 1)$, we realize that the slope field must show horizontal line segments when

$\frac{dx}{dt}$ equals 0—that is, where $x = 1$ or $x = -1$. Slope field 2 has horizontal tangents at $x = 1$ but doesn't have them at $x = -1$. Only slope field 3 shows zero slopes along both horizontal lines $x = 1$ and $x = -1$, so we conclude that Equation (B) must match up with slope field 3. ∎

The next example shows an advantage—and a possible drawback—of using slope fields.

### ∎ Example 2.4.4 A Slope Field for an Autonomous Equation

The first-order nonlinear autonomous equation $y' = y^4 + 1$ looks innocent, but (*surprise!*) it has the one-parameter family of implicit solutions:

$$\frac{\sqrt{2}}{8} \ln \left( \frac{y^2 + \sqrt{2}y + 1}{y^2 - \sqrt{2}y + 1} \right) + \frac{\sqrt{2}}{4} \left\{ \arctan \left( y\sqrt{2} + 1 \right) + \arctan \left( y\sqrt{2} - 1 \right) \right\} = t + C.$$

(The equation is separable, but the integration required to solve it is tricky. Use your CAS to evaluate the integral, but don't be surprised if your answer doesn't look exactly like the one given here.) Without looking at the solution formula, you can see immediately that the differential equation has no constant function as a solution: If $y$ is constant, then $y' = 0$; but the right-hand side of the differential equation, $y^4 + 1$, can never be zero. In fact, this simple analysis shows that any solution $y$ must be an *increasing* function. (*Why?*)

The fearsome formula describing a family of implicit solutions gives little useful information. However, let's take a look at the equation's slope field (Figure 2.10). First of all,

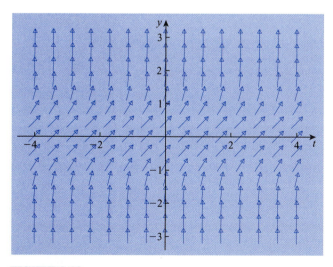

**FIGURE 2.10**
*Slope field for $y' = y^4 + 1$, $-4 \le t \le 4$, $-3 \le y \le 3$*

the autonomous nature of the equation is clear from the fact that along any horizontal line, the inclinations of the tangent line segments are equal.

Furthermore, it should be evident that any solution curve is increasing. In fact, any solution curve has vertical asymptotes holding it in on the left and on the right. Of course, we can't tell whether this last statement is true by merely looking at the slope field. A purely graphical analysis can't reveal this. But the slope field does give us an idea of what to expect when we try to solve the equation analytically or to approximate a solution numerically. ∎

As we will see in later chapters, a type of slope field can help us analyze certain *systems* of differential equations as well.

## Exercises 2.4

### A

In Problems 1–15, first sketch the slope field for the given equation by hand, and then try using a computer or graphing calculator to generate the slope field. Sketch several possible solution curves for each equation. (Your graphing calculator or CAS may have trouble with Problems 5 and 10.)

1. $y' = x$
2. $\dfrac{dx}{dt} = t$
3. $\dfrac{dr}{dt} = t - 2r$
4. $\dfrac{dx}{dt} = 1 - 0.01x$
5. $Q' = |Q|$
6. $y' = y - x$
7. $r\dfrac{dr}{dt} = -t$
8. $\dfrac{dy}{dx} = \dfrac{1}{y}$
9. $\dfrac{dy}{dx} = \dfrac{2y}{x}$
10. $y' = \max(x, y)$, the larger of the two values $x$ and $y$
11. $y' = x^2 + y^2$
12. $x' = 1 - tx$
13. $\dfrac{dy}{dt} = \dfrac{ty}{t^2 - 1}$
14. $\dfrac{dP}{dt} = 2P(1 - P)$
15. Use technology to determine the slope field for the equation $y' = \frac{\cos x}{\cos y}$. Describe the nullclines of the equation.
16. In any way your instructor tells you, manually or using technology, sketch the slope field for each of the following equations and then sketch the solution curve that passes through the given point $(x_0, y_0)$.

**a.** $\frac{dy}{dx} = x^2; (x_0, y_0) = (0, -2)$

**b.** $\frac{dy}{dx} = -xy; (x_0, y_0) = (0, 3)$

**17.** The German physiologist Gustav Fechner (1801–1887) devised the model expressed as $\frac{dR}{dS} = \frac{k}{S}$, where $k$ is a constant, to describe the response, $R$, to a stimulus, $S$. Use technology to sketch the slope field for $k = 0.1$.

**18.** Describe the isoclines of the equation $\frac{dy}{dt} = \frac{y+t}{y-t}$.

**19.** Which of the equations in Problems 1–15 are autonomous? If you have done some of these problems, look at their slope fields to confirm your answers.

**B**

**1.** A bimolecular chemical reaction is one in which molecules of substance A collide with molecules of substance B to create substance X. The rate of formation (the *velocity of reaction*) is given by a differential equation of the form $\frac{dx}{dt} = k(\alpha - x)(\beta - x)$, where $\alpha$ and $\beta$ represent the initial amounts of substances A and B, respectively, and $x(t)$ denotes the amount of substance X present at time $t$. (See Example 2.1.7.)

    **a.** Use technology to plot the slope field when $\alpha = 250$, $\beta = 40$, and $k = 0.0006$.

    **b.** If $x(0) = 0$, what seems to be the behavior of $x$ as $t \to \infty$?

**2.** The one-parameter family $y = \frac{c}{t}$ represents a solution of $\frac{dy}{dt} = f(t, y)$. Sketch (by hand) the slope field of the differential equation.

**3.** Describe the isoclines of the equation $y' = \frac{1}{\sqrt{1+t^2+y^2}}$.

**4.** Describe the nullclines of the equation $xy\frac{dy}{dx} = y^2 - x^2$.

**5.** Describe the nullclines of the equation in Problem B1.

**6.** In your own words, explain the important differences in the slope fields for the following forms of first-order differential equations:

    **a.** $y' = f(t, y)$
    **b.** $y' = f(t)$
    **c.** $y' = f(y)$

**7.** Match each of the Equations (1)–(3) with one of the accompanying slope fields.

    **a.** $\frac{dy}{dt} = y + 1$

    **b.** $\frac{dy}{dt} = y - t$

    **c.** $\frac{dy}{dt} = t + 1$

Slope field 1

Slope field 2

Slope field 3

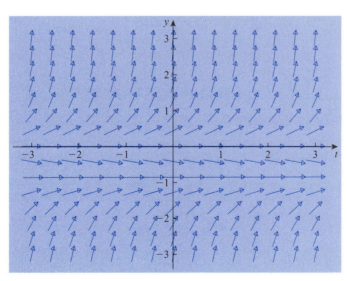

Slope field 4

8. By looking at the slope field for each of the following equations, describe the behavior of solutions of each equation as $t \to \infty$. How do your answers seem to depend on initial conditions in each case?

   a. $y' = 3y$
   b. $\dfrac{dP}{dt} = P(1 - P)$
   c. $y' = e^{-t} + y$
   d. $y' = 3 \sin t + 1 + y$

9. Examine the slope field for the first-order nonlinear equation $\dfrac{dy}{dx} = e^{-2xy}$. Based on this examination, what can you say about the solutions to this equation? (You may want to look at parts of various quadrants more closely.) In particular, what can you say about the behavior of solutions as $x \to \pm\infty$? (Be careful: Some solutions become infinite as $x$ approaches finite values.)

10. Consider the autonomous equation $\dfrac{dx}{dt} = x(1 - x)(x + 1)$.

    a. Use technology to determine the slope field of this equation.
    b. Describe the behavior of a solution satisfying $x(0) = 0.7$.
    c. Describe the behavior of a solution satisfying $x(0) = 1.3$.
    d. Describe the behavior of a solution satisfying $x(0) = -0.7$.
    e. Describe the behavior of a solution satisfying $x(0) = -1.3$.

**C**

1. If $\varphi(t)$ is a solution of an autonomous differential equation $x' = f(x)$ and $k$ is any real number, show that $\varphi(t + k)$ is also a solution. [*Hint*: Use the Chain Rule.]
2. Use the result of Problem C1 to show that if $\sin t$ is a solution of an autonomous differential equation $x' = f(x)$, then $\cos t$ is also a solution. [*Hint*: How are the graphs of sine and cosine related?]

## 2.5 PHASE LINES AND PHASE PORTRAITS

### 2.5.1 The Logistic Equation

When we are dealing with an *autonomous* first-order equation, a qualitative analysis can be used quite effectively to provide useful information about solution curves.

We'll begin to examine this new analysis technique by using an important population growth model, first studied by the Belgian mathematician Pierre Verhulst in 1838 and later rediscovered independently by the American scientists Raymond Pearl and Lowell Reed in the 1920s.

### ■ Example 2.5.1 The Qualitative Analysis of the Logistic Equation

The autonomous differential equation $\dfrac{dP}{dt} = P(1 - P)$, a particular example of something called a **logistic equation**, is useful, for instance, in analyzing such phenomena as epidemics. (We dealt with this equation in another way in part (b) of Problem B8 in Exercises 2.4.) In an epidemiological situation, $P$ could represent the infected population (or the *percentage* of the total population that is infected) as a function of time. We'll work more with this kind of model later, but for now let's ignore the fact that this is a separable equation that we can solve explicitly and see what basic calculus can tell us.

First of all, the right-hand side represents a derivative, the instantaneous rate at which $P$ is changing with respect to time. From calculus, we know that if the derivative is *positive*, then $P$ is *increasing*, and if the derivative is *negative*, then $P$ is *decreasing*. Now when is $dP/dt$ positive? The answer is when $P(1 - P)$ is greater than zero. Similarly, $dP/dt$ is negative when $P(1 - P)$ is less than zero. Finally, we see that $dP/dt = 0$ when $P(1 - P) = 0$—that is, when $P = 0$ or $P = 1$. These two critical points split the $P$-axis into three pieces (Figure 2.11): $-\infty < P < 0$, $0 < P < 1$, and $1 < P < \infty$.

What is the sign of $dP/dt$ when $P$ satisfies $-\infty < P < 0$? Well, for these values of $P$, $P$ is negative and $1 - P$ is positive, making the product $dP/dt = P(1 - P)$ *negative*. This means that $P$ is *decreasing*. When $P$ is between 0 and 1, we see that $P$ is positive and $1 - P$ is positive, so $dP/dt$ is *positive* and $P$ is *increasing*. Finally, when $P$ is greater than 1, we see that $P$ is positive and $1 - P$ is negative, so $dP/dt$ is *negative* and $P$ is *decreasing*.

We can redraw Figure 2.11 with arrows indicating whether $P$ is increasing or decreasing on a particular interval for $P$. The direction of any arrow shows the algebraic sign of $dP/dt$ in a subinterval and so indicates whether $P$ is increasing or decreasing: $\rightarrow$ means "positive derivative/increasing $P$" and $\leftarrow$ means "negative derivative/decreasing $P$." Figure 2.12 is called the **(one-dimensional) phase portrait** of the differential equation $\frac{dP}{dt} = P(1 - P)$. The horizontal line itself is called the **phase line**.

We can actually do a little more in this situation. If we differentiate each side of our original differential equation with respect to $t$, we get

$$\frac{d^2P}{dt^2} = \frac{dP}{dt} \cdot (1 - 2P) = P(1 - P)(1 - 2P),$$

where we have replaced $dP/dt$ by the right-hand side of the original differential equation. (*Check all this!*) Remember that the second derivative of a function tells us about the *concavity* of the function: $P$ is concave *up* when $d^2P/dt^2 > 0$ and $P$ is concave *down* when $d^2P/dt^2 < 0$.

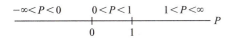

**FIGURE 2.11**

*P-axis divided by critical points*

**FIGURE 2.12**

*Phase portrait of $\frac{dP}{dt} = P(1 - P)$*

**Table 2.2** Table of Signs

| P Interval | P | $1-P$ | $1-2P$ | $P'' = P(1-P)(1-2P)$ | Concavity |
|---|---|---|---|---|---|
| $(-\infty, 0)$ | $-$ | $+$ | $+$ | $-$ | Down |
| $(0, \frac{1}{2})$ | $+$ | $+$ | $+$ | $+$ | Up |
| $(\frac{1}{2}, 1)$ | $+$ | $+$ | $-$ | $-$ | Down |
| $(1, \infty)$ | $+$ | $-$ | $-$ | $+$ | Up |

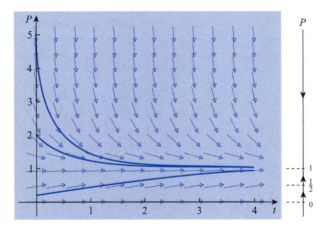

**FIGURE 2.13**

*Sketch of three solutions of $\frac{dP}{dt} = P(1-P)$, based on the phase portrait and concavity*
*Initial conditions are $P(0) = 0.2, 2,$ and $5; 0 \leq t \leq 4, 0 \leq P \leq 5$*

Using the critical points $0$, $\frac{1}{2}$, and $1$ of $\frac{d^2P}{dt^2}$ as a guide, we can construct the table of signs (Table 2.2).

We have to remember that $t$ is the independent variable in this problem and $P$ is the dependent variable. It's easy to lose sight of this because the (autonomous) form of this differential equation makes us focus on $P$ alone.

On the basis of our analysis of $\frac{dP}{dt}$ and $\frac{d^2P}{dt^2}$, let's take a look at what the graph of $P$ could look like in the $t-P$ coordinate plane (Figure 2.13). We'll focus on the first quadrant because $t \geq 0$ and $P \geq 0$ are realistic assumptions when one is dealing with a population growth model. Note that the phase line (representing the $P$-axis) is now drawn vertically and placed next to the graph and that we've marked the important values from our previous investigation of $\frac{dP}{dt}$ and $\frac{d^2P}{dt^2}$.

The graph indicates the change of concavity at $P = \frac{1}{2}$. Notice how the three solutions we have sketched seem to approach $P = 1$ as an asymptote as $t$ increases. In terms of a realistic scenario, this says that if the initial population is below 1 (the unit could be thousands or millions), the population will increase to 1 asymptotically. On the other hand, any population starting above 1 will eventually decrease toward 1. If we had drawn the rest of the phase line (for $P < 0$) and solutions in the fourth quadrant ($t \geq 0, P < 0$), we would have seen these solutions moving away from the $t$-axis. We'll say more about this phenomenon in the next section. ∎

The **logistic equation**, which is commonly used to model population growth when resources (such as food) are limited, is usually written as $\frac{dP}{dt} = rP(1 - \frac{P}{k})$, where $r$ is a per capita growth rate balancing births and deaths and $k$ represents the theoretical maximum population that a given environment (forest, petri dish, etc.) can sustain. The value $k$ is called the **carrying capacity**. This model will reappear from time to time in this text.

Sometimes an autonomous differential equation will contain a parameter whose possible values affect the behavior of solutions.

### ■ Example 2.5.2 An Equation with a Parameter

Consider the equation $x' = x^2 - ax = x(x - a)$, where $a$ is a constant. There are two apparent critical points: $x = 0$ and $x = a$. The first challenge is to position the critical points properly on the $x$-axis.

First, let's assume that $a > 0$. Then the phase line is

If $x < 0$, then $x - a < 0$ and $x' = x(x - a) > 0$. If $0 < x < a$, then $x > 0$ and $x - a < 0$, so $x(x - a) < 0$. Finally, if $x > a$, we have $x > 0$ and $x - a > 0$, so $x(x - a) > 0$. Thus, if $a > 0$, the phase portrait is as shown in Figure 2.14.

Now suppose $a < 0$. The phase line is

If $x < a$, then $x < 0$ and $x - a < 0$, so $x' = x(x - a) > 0$. When $a < x < 0$, we have $x < 0$ and $x - a > 0$, so $x(x - a) < 0$. If $x > 0$, then $x - a > 0$, so $x(x - a) > 0$. For $a < 0$, the phase portrait is as shown in Figure 2.15.

**FIGURE 2.14**
*Phase portrait of $x' = x^2 - ax, a > 0$*

**FIGURE 2.15**
*Phase portrait of $x' = x^2 - ax$, $a < 0$*

**FIGURE 2.16**
*Phase portrait of $x' = x^2 - ax$, $a = 0$*

There is one last case, $a = 0$. Now there is only one critical point, and it is easy to see that the phase portrait is as shown in Figure 2.16.

In summary, we see that the presence of parameters in a differential equation may affect the behavior of its solutions. Sketches of typical solution curves corresponding to $a > 0$, $a < 0$, and $a = 0$ are left as an exercise (see Problem A16 of Exercises 2.5). ■

## Exercises 2.5

### A

Draw phase portraits for each of the equations in Problems 1–13.

1. $\dfrac{dy}{dt} = y^2 - 1$
2. $y' = y^2(1 - y)^2$
3. $x' = (x + 1)(x - 3)$
4. $\dot{x} = \cos x$
5. $y' = e^y - 1$
6. $y' = y(1 - y)(2 - y)$
7. $\dot{y} = \sin y$
8. $x' = 1 - \dfrac{x}{1 + x}$
9. $y' = ye^{y-1}$
10. $\dot{y} = \sin y \cos y$
11. $x' = -3x(1 - x)(3 - x)$
12. $y' = y(y^2 - 4)$
13. $x' = x(1 - e^x)$
14. The equation in Problem 11 could represent a model of a population that can become extinct if it drops below a particular critical value. What is this critical value?
15. Consider the equation $\dot{x} = 1 + \frac{1}{2}\cos x$.

    a. Draw the phase portrait of this equation.
    b. What does your qualitative analysis tell you about solutions of this equation?

16. Consider the equation $x' = x^2 - ax$ of Example 2.5.2. For the cases $a > 0$, $a < 0$, and $a = 0$, sketch solution curves that illustrate Figure 2.14, Figure 2.15, and Figure 2.16, respectively.

**B**

1. Using one of *Kirchhoff's laws* in physics, we find that the current, $I$, flowing in a particular electric circuit satisfies the equation $0.5\frac{dI}{dt} + 10I = 12$. (The resistance is 10 ohms, the inductance is 0.5 henry, and there is a 12-volt battery.)

   **a.** Sketch the phase portrait of the equation.
   **b.** If the initial current, $I(0)$, is 3 amps, use part (a) to describe the behavior of $I$ for large values of $t$.

2. Example 2.1.7 and Problem B1 of Exercises 2.4 indicated that a type of chemical reaction can be modeled by the equation $\frac{dx}{dt} = k(\alpha - x)(\beta - x)$.

   **a.** If $\alpha = 250$, $\beta = 40$, and $k$ is a positive constant, produce the phase portrait of the equation.
   **b.** If $x(0) = 0$, how does $x$ behave as $t \to \infty$?

3. A skydiver's velocity $v \geq 0$ satisfies the equation $\frac{dv}{dt} = 9.8 - c \cdot v^2$, where $c > 0$ is the (per unit mass) coefficient of friction.

   **a.** Draw the phase portrait for $v \geq 0$.
   **b.** Find and classify the positive equilibrium solution.

4. Consider the equation $\dfrac{dy}{dt} = (1 + y)^2$.

   **a.** What happens to solutions with initial conditions $y(0) > -1$ as $t$ increases?
   **b.** Describe the behavior of solutions with initial conditions $y(0) < -1$ as $t$ increases.

5. Consider the equation $\frac{dP}{dt} = \left(1 - \frac{P}{15}\right)^3 \left(\frac{P}{7} - 1\right) P^5$, with $P(0) = 3$.

   **a.** Use the phase portrait for this equation to give a rough sketch of the solution $P(t)$.
   **b.** What happens to $P(t)$ as $t$ becomes very large?

6. For each of the phase portraits shown below, write down a corresponding first-order equation of the form $x' = f(x)$.

   **a.**

   **b.**

   **c.**

7. Given the following phase portrait for $\frac{dy}{dt} = f(y)$, make a rough sketch of the graph of $f(y)$, assuming that $y = 0$ is in the center of the phase line.

8. Given the following phase portrait, find a first-order ODE that is consistent with this phase portrait.

**C**

1.  Consider the equation $x' = (2\alpha - 1)x + 1$, where $\alpha$ is a parameter. Describe the behavior of solutions $x(t)$ of this equation as $t \to \infty$, noting how this behavior depends on the value of $\alpha$.

2.  Consider the equation $x' = (\alpha^2 - 1)x + 1 + \alpha$, where $\alpha$ is a parameter. Describe the behavior of solutions $x(t)$ of this equation as $t \to \infty$, noting how this behavior depends on the value of $\alpha$.

3.  The *Landau equation* arises in the analysis of the dynamics of fluid flow. It is $\frac{dx}{dt} = ax - bx^3$, where $a$ and $b$ are positive real constants.

    a.  Draw the phase portrait of Landau's equation.
    b.  What happens to $x$ as $t$ increases if $x(0) = \sqrt{\frac{a}{b}} + \varepsilon$, where $\varepsilon$ is a small positive quantity?
    c.  What happens to $x$ as $t$ increases if $x(0) = 0$?
    d.  How does $x$ behave as $t$ increases if $x(0) = \sqrt{\frac{a}{b}} - \varepsilon$?

## 2.6 EQUILIBRIUM POINTS: SINKS, SOURCES, AND NODES

Let's take another look at Figure 2.13 and focus on the **critical points**—the places where $\frac{dP}{dt} = 0$. Geometrically, these are the horizontal lines $P = 0$ and $P = 1$, which represent the functions $P(t) \equiv 0$ and $P(t) \equiv 1$, constant solutions of the differential equation $\frac{dP}{dt} = P(1 - P)$. These values of $P$ are called **equilibrium points** or **stationary points** of the autonomous differential equation. We also say that $P(t) \equiv 0$ and $P(t) \equiv 1$ are **equilibrium solutions** or **stationary solutions** of the equation. Assuming that the solutions of an autonomous differential equation describe some physical, economic, or biological system, we can conclude that if the system actually reaches an equilibrium point $P$, it must always have been at $P$—and will always remain at $P$. (*Think about this. You have $\frac{dP}{dt} = 0$ at an equilibrium point.*)

We can go further in this analysis and classify equilibrium points for autonomous first-order differential equations. It turns out that there are only three basic kinds of equilibrium points: **sinks**, **sources**, and **nodes**.

If we look at the equilibrium point $P = 1$ in Example 2.5.1, we see from Figure 2.13 that the solution curves near the line $P = 1$ seem to swarm into (or converge to) the horizontal line. We call $P = 1$ a **sink**. A little more accurately, an equilibrium solution $P \equiv k$ is a **sink** if solutions with initial conditions sufficiently close to $P \equiv k$ are asymptotic to $P \equiv k$ as $t \to \infty$. This idea of being "sufficiently close" can be made mathematically precise, but we will just consider the situation intuitively. Sinks are also called **attractors** or **asymptotically stable solutions**. The term *sink* is meant to suggest the drain of a bathroom or kitchen sink: Along the sides, water that is close enough will flow into the drain.

On the other hand, we see different behavior near $P = 0$. As we look along solution curves from left to right, they seem to be moving *away* from the line $P = 0$. The equilibrium point $P = 0$ is called a **source**. In other words, an equilibrium solution $P \equiv k$ is a **source** if solutions with initial conditions sufficiently close to $P \equiv k$ are asymptotic to $P \equiv k$ as $t \to -\infty$—that is, as we go backward in time. A source is also called a **repeller** or an **unstable equilibrium**

**solution**. Here, we can think of a faucet discharging water or a hand-held hair dryer putting out streams of hot air.

If we had another equilibrium point such that nearby solutions showed any other kind of behavior—perhaps somewhat like a sink and somewhat like a source at the same time— we would call that equilibrium point a **node**. (See Example 2.6.2 and Figure 2.16.) More technically, we can refer to a node as a **semistable equilibrium solution**.

### 2.6.1  A Test for Equilibrium Points

We can test equilibrium points/solutions for autonomous first-order equations using a criterion that should remind you of the *First* or *Second Derivative Test* from calculus:

### 2.6.2  Derivative Test

If $x^*$ is an equilibrium point for the autonomous equation $\frac{dx}{dt} = f(x)$, it is true that (1) if $f'(x^*) > 0$, then $x^*$ is a source; (2) if $f'(x^*) < 0$, then $x^*$ is a sink; and (3) if $f'(x^*) = 0$, the test fails—that is, we can't tell what sort of equilibrium point $x^*$ may be without further investigation.

In Example 2.5.1, we saw that $P = 0$ and $P = 1$ were equilibrium points. Because $f(P) = P(1 - P)$, we have $f'(P) = 1 - 2P$, so $f'(0) = 1 > 0$ indicates that $P = 0$ is a *source* and $f'(1) = -1 < 0$ shows that $P = 1$ is a *sink*.

We can understand why the Derivative Test works by using the concept of *local linearity*: Near an equilibrium solution $x^*$, we can approximate $f(x)$ by the equation of its *tangent line* at $x^*$. (See Section A.1 if necessary.) Therefore, if $x$ is close enough to $x^*$, we can write

$$\frac{dx}{dt} = f(x) \approx f(x^*) + f'(x^*)(x - x^*) = f'(x^*)(x - x^*)$$

because $f(x^*) = 0$ when $x^*$ is an equilibrium solution. Now use Table 2.3 to compare the signs of $f'(x^*)$ and $(x - x^*)$.

The first row of signs in Table 2.3, for example, tells us that if $(x - x^*)$ is positive—so that a solution $x$ is slightly *above* the equilibrium solution—and $f'(x^*) > 0$, then $\frac{dx}{dt} > 0$, which

**Table 2.3** Signs of $\frac{dx}{dt}$

| $(x - x^*)$ | $f'(x^*)$ | $\frac{dx}{dt}$ |
|:---:|:---:|:---:|
| + | + | + |
| + | − | − |
| − | + | − |
| − | − | + |

means that the solution $x$ is moving *away* from $x^*$. This last statement says that $x^*$ must be a *source*. Similarly, the third row of signs indicates that if $x$ starts out *below* $x^*$ and $f'(x^*) > 0$, then $x$ falls away from $x^*$ as $t$ increases, so $x^*$ is a source. The remaining two rows describe a sink.

The next two examples show us the power and the limitations of the Derivative Test.

## ■ Example 2.6.1 Using the Derivative Test

If we examine the autonomous equation $\frac{dx}{dt} = x - x^3 = x(1 - x^2)$, we see that the equilibrium points are $x = 0$, $x = -1$, and $x = 1$. Can we determine what kinds of equilibrium points these are without any kind of graph?

Yes, we just apply the Derivative Test given previously. First of all, we have $\frac{dx}{dt} = f(x)$, where $f(x) = x - x^3$, so $f'(x) = 1 - 3x^2$. Because $f'(0) = 1 > 0$, we know that $x = 0$ is a *source*. The fact that $f'(-1) = -2 < 0$ tells us that $x = -1$ is a *sink*. Finally, because $f'(1) = -2 < 0$, we see that $x = 1$ is another *sink*.

The phase portrait shown in Figure 2.17 reflects this information. Finally, the slope field (Figure 2.18) confirms our analysis.

**FIGURE 2.17**
*Phase portrait of $\frac{dx}{dt} = x - x^3$*

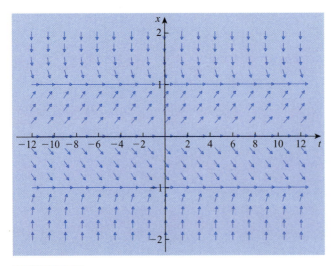

**FIGURE 2.18**
*Slope field for $\frac{dx}{dt} = x - x^3$*

### ■ Example 2.6.2 Failure of the Derivative Test

Let's look at the first-order nonlinear equation $\frac{dx}{dt} = f(x) = (1 - x)^2$. The only equilibrium solution is $x \equiv 1$, and we have $f'(x) = 2(1 - x)(-1) = 2(x - 1)$. Because $f'(1) = 0$, our test doesn't allow us to draw any conclusion. However, we can examine the behavior of $f'(x)$ *near* $x = 1$ to get an idea of what's going on.

We can see that $f'(x)$ is greater than zero for values of $x$ greater than 1, so $x \equiv 1$ looks like a source; but values of $x$ just below 1 give us *negative* values of the derivative, so $x \equiv 1$ looks like a sink. This ambivalent behavior enables us to conclude that $x \equiv 1$ is a *node*. Figure 2.19 shows the phase portrait of this equation. Figure 2.20 shows the slope field with some particular solutions superimposed.

Notice that solution curves starting *above* the line $x = 1$ seem to flow *away* from the line $x = 1$, whereas those starting *below* the equilibrium solution flow *toward* the line $x = 1$. In other words, the point $x = 1$ is neither a sink nor a source. It is a *node*. ■

In analyzing an autonomous first-order differential equation $\frac{dx}{dt} = f(x)$, it is useful to sketch the phase line using the graph of $f(x)$. First of all, equilibrium points are the zeros of $f$. Furthermore, regions where $f$ is positive and regions where $f$ is negative correspond to parts of the phase line where the arrows point to the right and to the left, respectively.

**FIGURE 2.19**

*Phase portrait of $\frac{dx}{dt} = (1 - x)^2$*

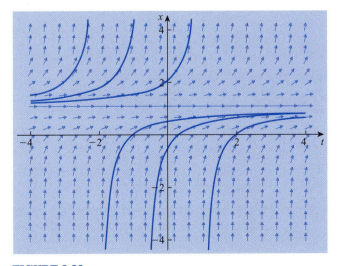

**FIGURE 2.20**

*Solutions of $\frac{dx}{dt} = (1 - x)^2 : x(0) = -\frac{1}{2}, \frac{1}{2}$, and 2*

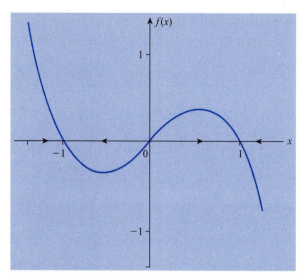

**FIGURE 2.21**
$f(x) = x - x^3$ compared to the phase line of $\frac{dx}{dt} = x - x^3$

### ■ Example 2.6.3 Using the Graph of $f(x)$ to Sketch the Phase Portrait

Let's return to the equation in Example 2.6.1, $\frac{dx}{dt} = x - x^3$, this time focusing on what the graph of $f(x) = x - x^3$ reveals. Figure 2.21 shows the graph of $f(x)$ aligned with the phase portrait of the differential equation.

Note the equivalence between equilibrium points and the zeros of $f$ and the correspondence between regions of positivity and negativity for $f(x)$ and the directions of the arrows on the phase line.                                                                     ■

Equilibrium solutions and their nature will be particularly useful when we discuss qualitative aspects of systems of linear and nonlinear equations in Chapters 4, 5, and 7.

### Exercises 2.6

**A**

For Problems 1–15, find the equilibrium point(s) of each equation and classify it as **sinks**, **sources**, or **nodes**.

**1.** $y' = y^2(1 - y)^2$

**2.** $\dot{x} = \cos x$

**3.** $y' = e^y - 1$

**4.** $y' = y^2(y^2 - 1)$

**5.** $\dot{x} = ax + bx^2, a > 0, b > 0$

**6.** $\dot{x} = x^3 - 1$

**7.** $\dot{x} = x^2 - x^3$

**8.** $\dot{y} = 10 + 3y - y^2$

**9.** $\dot{x} = x(2 - x)(4 - x)$

**10.** $\dot{x} = -x^3$

**11.** $\dot{x} = x^3$

**12.** $\dot{y} = y \ln(y + 2)$

**13.** $\dot{x} = x - \cos x$ [*Hint*: Either use technology to find the equilibrium point explicitly (via a *solve* command) or graph $y = x$ and $y = \cos x$ separately to estimate the graphs' point of intersection.]

**14.** $\dot{x} = x - e^{-x}$ (Use technology—see the preceding exercise.)

**15.** $x' = x(x + 1)(x - 0.5)^6$

**16.** A lake has two rivers flowing into it, one discharging a certain amount of water containing a concentration of pollutant and the other discharging an amount of clean water per day. Assuming that the lake volume is constant, the total amount of pollution in the lake, $Q(t)$, can be modeled by the balance equation $\frac{dQ}{dt} = D(Q^* - Q)$, where $D$ is a positive constant involving the two rates of flow into the lake and the lake's volume and $Q^*$ is a positive constant involving volume, rates of flow, and the pollutant concentration.

    **a.** What is the equilibrium solution of this equation?

    **b.** Is the solution found in part (a) stable or unstable? (For example, a *sink* would indicate that the clean river input *reduces* the long-term amount of pollution in the lake.)

**B**

**1.** The following equation has been proposed for determining the speed of a rowing boat[5]: $M\frac{du}{dt} = \frac{8P}{u} - bSu^2$, where $u(t)$ denotes the speed of the boat at time $t$; $M$, its mass; and $P$, $S$, and $b$ are positive constants describing various other aspects of the boat and the person rowing it.

    **a.** Determine the equilibrium speed of the boat.

    **b.** Determine whether the speed found in part (a) is a sink or a source.

    **c.** Interpret the result of part (b) physically.

**2.** Given $\frac{dx}{dt} = f(x)$ and the following graph of $f(x)$,

    **a.** Sketch the phase portrait of the equation.

    **b.** Identify all equilibrium points and classify each as a sink, a source, or a node.

---

[5] M. Mesterton-Gibbons, *A Concrete Approach to Mathematical Modelling* (New York: John Wiley & Sons, 1995): 32–34; 53–56; 130–132.

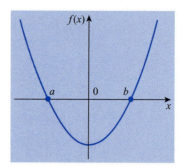

**3.** A population growth model that is fairly simple yet amazingly accurate in predicting tumor growth is described by the *Gompertz equation*, $\frac{dN}{dt} = -aN \ln(bN)$, where $N(t) > 0$ is proportional to the number of cells in the tumor and $a, b > 0$ are parameters that are determined experimentally. [Benjamin Gompertz (1779–1865) was an English mathematician/actuary.]

    **a.** Sketch the phase portrait for this equation.

    **b.** Sketch the graph of $f(N)$ against $N$.

    **c.** Find and classify all equilibrium points for this equation.

    **d.** For $0 < N \leq 1$, determine where the graph of $N(t)$ against $t$ is concave up and where it is concave down. (You may want to review Example 2.5.1.)

    **e.** Sketch $N(t)$.

**4.** Find an equation $\dot{x} = f(x)$ with the property that there are exactly 3 equilibrium points and all of them are *sinks*.

**C**

**1.** A population of animals following the logistic growth pattern (see Section 2.5) is harvested at a constant rate—that is, as long as the population size, $P$, is positive, a fixed number, $h$, of animals is removed per unit of time. The equation modeling the dynamics of this situation is $\frac{dP}{dt} = rP(1 - \frac{P}{k}) - h$ for $P > 0$.

    **a.** Show that if $h < \frac{rk}{4}$, there are two nonzero equilibrium solutions.

    **b.** Show that the smaller of the equilibrium solutions in part (a) is a source, whereas the larger of the two is a sink.

**2.** Consider the equation $\dot{x} = -x^3 + (1 + \alpha)x^2 - \alpha x$, where $\alpha$ is a constant.

    **a.** If $\alpha < 0$, find all equilibrium solutions of this equation and classify them.

    **b.** If $0 < \alpha < 1$, find all equilibrium solutions and classify them.

    **c.** If $\alpha > 1$, find all equilibrium solutions and classify them.

    **d.** Describe the equilibrium solutions if $\alpha = 0$.

    **e.** Describe the equilibrium solutions if $\alpha = 1$.

## *2.7 BIFURCATIONS

### 2.7.1 Basic Concepts

To get an idea of what this topic is all about, let's go back to elementary algebra and look at the quadratic function $f(x) = x^2 + x + c$, where $c$ is a constant. We should realize that the zeros of this function depend on the parameter $c$. To see this, let's write

$$x^2 + x + c = \left(x + \frac{1}{2}\right)^2 + \left(c - \frac{1}{4}\right). \qquad (2.7.1)$$

Clearly, the term $\left(x + \frac{1}{2}\right)^2$ is always nonnegative, so that if $c > \frac{1}{4}$, the expression (2.7.1) is always *strictly greater than zero*, and the quadratic equation $x^2 + x + c = 0$ has *no* real solutions. If $c = \frac{1}{4}$, then the equation has $x = -\frac{1}{2}$ as its *only* root, a repeated root. Finally, if $c < \frac{1}{4}$, we have *two* solutions, $x = -\frac{1}{2} \pm \sqrt{\frac{1}{4} - c}$. *Verify all the assertions in this paragraph.* Figure 2.22 shows the graph of $y = x^2 + x + c$ for three values of $c$.

The important point in this example is that $\frac{1}{4}$ is the value of the parameter $c$ at which the nature of the solutions of the quadratic equation changes. We say that $c = \frac{1}{4}$ is a **bifurcation point** because as $c$ decreases through $\frac{1}{4}$, the solution $x = 0$ splits into two solutions. (The word *bifurcation* refers to a splitting or branching.)

We can see the effect of the bifurcation most clearly by plotting the solution $x$ against the parameter $c$—in our example, showing the graph of the relationship $x = -\frac{1}{2} \pm \sqrt{\frac{1}{4} - c}$ (Figure 2.23). This graph showing the dependence of a solution on a parameter is called the **bifurcation diagram** for the equation $x^2 + x + c = 0$. Be sure you understand what this diagram tells you. Note, in particular, what happens as $c$ passes through the value $\frac{1}{4}$.

### 2.7.2 Application to Differential Equations

This sort of qualitative change caused by a change of parameter value is particularly interesting when we observe it in an autonomous differential equation. What changes for such an ODE at a bifurcation point is the number and/or nature of the equilibrium solutions.

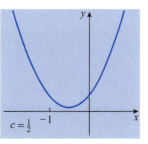

**FIGURE 2.22**

*Graphs of $y = x^2 + x + c$*

---

* Denotes an optional section.

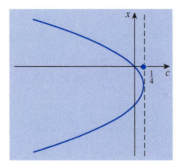

**FIGURE 2.23**
*Bifurcation diagram for $x^2 + x + c = 0$*

---

**Definition 2.7.1**

Given a family of autonomous differential equations $\frac{dx}{dt} = f(x, \lambda)$ containing a parameter $\lambda$, a **bifurcation point** (or **bifurcation value**) is a value $\lambda_0$ of the parameter for which the qualitative nature of the equilibrium solutions changes as $\lambda$ passes through $\lambda_0$. The actual change in the equilibrium solutions is called a **bifurcation**.

---

The real significance of bifurcations was first revealed in Euler's 1744 work on the buckling of an elastic straight beam or column under a compressive force. [The great Swiss mathematician Leonhard Euler (1707–1783) has been called "the Shakespeare of mathematics."] The normal upright position represents an equilibrium position. The parameter here is the force $F$ exerted on the top of the column. For certain values of $F$, say $F < F^*$, the column maintains its vertical position; but if the force is too great, say $F > F^*$, the vertical equilibrium position becomes unstable, and the column may buckle. The critical force $F^*$ is the bifurcation point. The equilibrium situation changes as the size of the force passes through the value $F^*$.

The next example reveals the bifurcation point for a simple first-order equation of a type we've discussed before (Example 1.2.1).

**FIGURE 2.24**
$\frac{dy}{dx} = f(y) = ay$ vs. $y$

## ■ Example 2.7.1 A Bifurcation Point for a Linear Equation

The equation $\frac{dy}{dt} = ay$ expresses the fact that at any time $t$, some quantity $y$ grows at a rate proportional to its size at time $t$. The parameter $a$ is the constant of proportionality that captures some growth characteristic of the quantity.

Setting $\frac{dy}{dt} = 0$, we find that the equilibrium solutions are described by $ay = 0$. If $a = 0$, then *every* value of $y$ is an equilibrium point. If $a \neq 0$, then $y \equiv 0$ is the only equilibrium point. For the equation $\frac{dy}{dt} = ay = f(y)$, we have $f'(y) \equiv a$; and we use the Derivative Test of Section 2.6 to conclude that if $a > 0$, then $y \equiv 0$ is a *source*, and if $a < 0$, then $y \equiv 0$ is a *sink*. Clearly, $a = 0$ is a bifurcation point, because the number and nature of the equilibrium solutions change as $a$ passes through 0. Figure 2.24 shows graphs of $f(y)$ against $y$ for the three possibilities for $a$ and the corresponding phase portraits.

We can show the dependence of the equilibrium points on $a$ by drawing a bifurcation diagram, plotting $y(t)$ against $a$ (Figure 2.25). The $y$-axis itself represents all the solutions $y = C$, where $C$ is any constant, for $a = 0$. It is usual in bifurcation diagrams to use solid curves to indicate stable equilibrium solutions (sinks) and dashed lines to denote unstable solutions (sources). Arrows indicate the directions of change of some solutions with time.

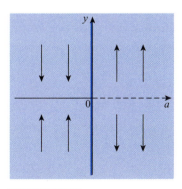

**FIGURE 2.25**
Bifurcation diagram for $\frac{dy}{dt} = ay$

## ■ Example 2.7.2 A Bifurcation Point for a Nonlinear Equation

Now let's look at the first-order nonlinear equation $\frac{dy}{dx} = \alpha y - y^3 = f(y)$. This is the *Landau equation*, which appeared in Problem C3 of Exercises 2.5; it arises in the study of one-dimensional patterns in fluid systems. Here $y = y(t)$ gives the amplitude of the patterns, and $\alpha$ is a small, dimensionless parameter that measures the distance from the bifurcation. [L. D. Landau (1908–1968) was a Russian physicist who won the Nobel Prize in 1962.]

We see that $\frac{dy}{dx} = 0$ implies that $\alpha y - y^3 = y(\alpha - y^2) = 0$, so $y = 0$, $y = \sqrt{\alpha}$, and $y = -\sqrt{\alpha}$ are the only equilibrium points. Looking at these points, we can see (because of the radical sign) that we have three cases to consider: (1) $\alpha = 0$, (2) $\alpha > 0$, and (3) $\alpha < 0$.

When $\alpha = 0$, there is the single equilibrium point $y = 0$. Then we have $f'(y) = \alpha - 3y^2 = -3y^2$, so $f'(0) = 0$ and we can't determine the nature of the equilibrium solution $y = 0$ from the Derivative Test. However, we can see that when $\alpha = 0$, the differential equation is $\frac{dy}{dx} = -y^3$, whose solution tends to zero as $x$ becomes infinite in the positive direction. (*Look at the slope field or a phase portrait.*) Thus, $y = 0$ is a *sink*.

If $\alpha$ is less than zero, then $y = 0$ is the only equilibrium point because $\sqrt{\alpha}$ and $-\sqrt{\alpha}$ are imaginary numbers. For this case, we see that $f'(0) = \alpha < 0$, so $y = 0$ is a *sink*.

However, if $\alpha$ is greater than zero, then the equation has three distinct equilibrium points: $y = 0$, $y = \sqrt{\alpha}$, and $y = -\sqrt{\alpha}$. We see that $f'(0) = \alpha > 0$, so $y = 0$ is a *source*; $f'(\sqrt{\alpha}) = -2\alpha < 0$, so $y = \sqrt{\alpha}$ is a *sink*; and $f'(-\sqrt{\alpha}) = -2\alpha < 0$, so $y = -\sqrt{\alpha}$ is also a *sink*.

Note how the value of $\alpha$ determines the number and the nature of the equilibrium solutions of our equation. Clearly, $\alpha = 0$ is the only bifurcation point for our original equation. Figure 2.26 shows two representations of our situation: graphs of $f(y)$ against $y$ for the three descriptions of $\alpha$ considered above and corresponding phase portraits.

We can also show this dependence of the equilibrium points on $\alpha$ by means of a bifurcation diagram, in which we plot $y$ against $\alpha$ (Figure 2.27).

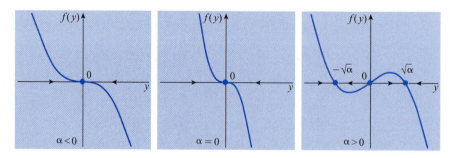

**FIGURE 2.26**

$f(y) = \alpha y - y^3$ vs. $y$

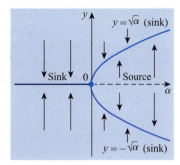

**FIGURE 2.27**
Bifurcation diagram for $\frac{dy}{dx} = \alpha y - y^3$

There are different kinds of bifurcations. Figure 2.27 shows a **pitchfork bifurcation**, named for obvious reasons. (See Problem C1 of Exercises 2.7 for a generalization of this example.) ∎

A laser—the word stands for **l**ight **a**mplification by **s**timulated **e**mission of **r**adiation—is a marvelous device that produces a beam of intense, concentrated pure light that can be used to cut diamonds, destroy cancerous cells, perform eye surgery, and enhance telecommunications when used in fiber optics. Basically, an external energy source is used to excite atoms and produce photons (light particles) that have the same frequency and phase. A. Schawlow and C. Townes received a patent for the invention of the laser in 1960, and the first laser was built by the American physicist T. H. Maiman in the same year. The mathematical model of a laser that follows is an important scientific example that illustrates another type of bifurcation. It is more complicated than the previous two examples because the bifurcation behavior depends on the values of *two* parameters.

### ■ Example 2.7.3 A Laser Model That Has a Transcritical Bifurcation

A simplified model of the basic physics behind a laser is given by the equation

$$\dot{n} = f(n) = Gn(N_0 - n) - kn = (GN_0 - k)n - Gn^2.$$

In this equation, $n = n(t)$ represents the number of photons at time $t$, $N_0$ is the (constant) number of "excited" atoms (in the absence of laser action), and $G$ and $k$ are positive parameters related to the gain and loss, respectively, of photons that have the same frequency and phase. We emphasize that we have *two* parameters in our equation, and we will see that our bifurcation analysis depends on the value of $N_0$ in relationship to them.

We can write the equation as $\dot{n} = n(GN_0 - k - Gn)$, so setting $\dot{n}$ equal to zero gives us $n \equiv 0$ or $GN_0 - k - Gn \equiv 0$. This tells us that the equilibrium solutions are $n \equiv 0$ and $n = (GN_0 - k)/G = N_0 - (k/G)$, where $N_0 \neq k/G$.

Looking at the first equilibrium solution, $n \equiv 0$, we will see that this equilibrium solution is a sink when $N_0 < k/G$—that is, when $GN_0 - k < 0$. From the original equation, we have $f(n) = (GN_0 - k)n - Gn^2$, so $f'(n) = (GN_0 - k) - 2Gn$. Then $f'(0) = GN_0 - k < 0$, so $n \equiv 0$ is indeed a sink by the Derivative Test. Physically, this means that there is no stimulated emission and no photons are produced that have the same frequency and phase. The laser device functions like a light bulb. Similarly, we can determine that $n \equiv 0$ is a source when $N_0 > k/G$.

Focusing on the second equilibrium point, $n = N_0 - (k/G)$, where $N_0 \neq k/G$, we see that

$$f'\left(N_0 - \frac{k}{G}\right) = (GN_0 - k) - 2G\left(N_0 - \frac{k}{G}\right) = -GN_0 + k.$$

If $N_0 < k/G$, then $-GN_0 + k > 0$ and therefore $n = N_0 - (k/G)$ is a source. If $N_0 > k/G$, then $-GN_0 + k < 0$ and therefore $n = N_0 - (k/G)$ is a sink. The physical interpretation of this last fact is that the external energy source has excited the atoms enough so that some atoms produce photons that have the same frequency and phase. The device is now producing coherent light.

Finally, if $N_0 = k/G$, then our original equation reduces to $\dot{n} = -Gn^2$, so we get only one equilibrium solution, $n \equiv 0$, which is a sink if we consider only positive values of $n$. Because of this change in the nature and number of equilibrium solutions, we can interpret $N_0 = k/G$ as our bifurcation point (called the *laser threshold*). The bifurcation diagram (Figure 2.28) summarizes this model.

The physical interpretation is that when the amount of energy supplied to the laser exceeds a certain threshold—that is, when $N_0 > k/G$—the "light bulb" has turned into a laser. Notice that at the bifurcation value $N_0 = k/G$, the two equilibrium solutions merge, and when they split apart, they have interchanged stability. Such a bifurcation is called a **transcritical**

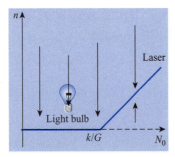

**FIGURE 2.28**
*Bifurcation diagram for the laser model* [6]

[6] Adapted from Figure 3.3.3 in S. H. Strogatz, *Nonlinear Dynamics and Chaos: With Applications to Physics, Biology, Chemistry, and Engineering* (Reading, MA: Addison-Wesley, 1994): 55.

bifurcation. For $N_0 > k/G$, the equilibrium solution $n \equiv 0$ becomes unstable by transferring its stability to another equilibrium solution, $n = N_0 - (k/G)$, the straight line with slope 1 in Figure 2.27. ∎

## Exercises 2.7

### A

For each of the following equations in Problems 1–6, (1) sketch all the qualitatively different graphs of $f(x)$ against $x$ as the parameter $c$ is varied; (2) determine the bifurcation point(s); and (3) sketch the bifurcation diagram of equilibrium solutions against $c$.

1. $\dfrac{dx}{dt} = x^2 - c$

2. $\dfrac{dx}{dt} = 1 + cx + x^2$

3. $\dfrac{dx}{dt} = x - cx(1 - x)$

4. $\dfrac{dx}{dt} = x^2 - 2x + c$

5. $\dfrac{dx}{dt} = x(x - c)$

6. $\dfrac{dx}{dt} = cx - x^2$

### B

1. Consider the logistic equation (see Example 2.5.1) with a constant harvesting (hunting, fishing, reaping, etc.) rate $h$ : $\dfrac{dP}{dt} = P(5 - P) - h$. Does there exist a maximum harvest rate $h^*$ beyond which the population will become extinct for every initial population $P_0 = P(0)$?

2. Construct the bifurcation diagram for the equation $x' = \alpha - e^{-x^2}$, where $\alpha > 0$.

3. Construct the bifurcation diagram for $\dfrac{dx}{dt} = x(c - x^2)$, where $c$ is a parameter.

4. Construct the bifurcation diagram for $\dfrac{dx}{dt} = x(x^2 - 1 - \alpha)$, $-\infty < \alpha < \infty$.

5. Construct the bifurcation diagram for the equation $x' = 3x - x^3 - \alpha$, where $\alpha$ is a parameter.

### C

1. The *Landau equation* $\dot{x} = (R - R_c)x - kx^3$, where $k$ and $R_c$ are positive constants and $R$ is a parameter that may take on various values, is important in the field of fluid mechanics.

   a. If $R < R_c$, show that there is only the equilibrium solution $x = 0$ and it is a sink.
   b. If $R > R_c$ , show that there are three equilibrium solutions, $x = 0$, $x = \sqrt{(R - R_c)/k}$, and $x = \sqrt{(R - R_c)/k}$, and that the first solution is a source while the other two are sinks.
   c. Sketch a graph in the $R$-$x$ plane showing all equilibrium solutions and label each one as a sink or a source. How would you describe the bifurcation point $R = R_c$?

2. The following equation occurs in the study of *gene activation*:

$$\frac{dx}{dt} = \alpha - x + \frac{4x^2}{1 + x^2}.$$

Here $x(t)$ is the concentration of gene product at time $t$.

**a.** Sketch the phase portrait for $\alpha = 1$.

**b.** There is a small value of $\alpha$, say $\alpha_0$, where a bifurcation occurs. Estimate $\alpha_0$ and sketch the phase portrait for some $\alpha$ in the open interval $(0, \alpha_0)$.

**c.** Draw the bifurcation diagram for this differential equation.

## *2.8 EXISTENCE AND UNIQUENESS OF SOLUTIONS

This is the time to acknowledge that we have been avoiding a very important question: When we're trying to solve a differential equation, how do we know whether there *is* a solution? We could be looking for something that doesn't exist—a waste of time, effort, and (these days) computer resources.

We've already noted in Section 1.2 that the equation $(y')^2 + 1 = 0$ has no real-valued solution. You can easily check that the IVP $y' = \frac{3}{2} y^{1/3}, y(0) = 0$ has three distinct solutions: $y \equiv 0, y = -x^{3/2}$, and $y = x^{3/2}$.

Calculators and computers can mislead. They may present us with a solution where there is none. If there are several possible solutions, our user-friendly device may make its own selection, whether or not it is the one that we want for our problem. A skeptical attitude and a knowledge of mathematical theory will protect us against inappropriate answers.

First, let's look at what can happen when we try to solve first-order initial-value problems. Then we'll discuss an important result guaranteeing when such IVPs have one and only one solution.

### ■ Example 2.8.1 An IVP with a Unique Solution on a Restricted Domain

We'll see that for each value of $x_0$, the initial-value problem $x' = 1 + x^2, x(0) = x_0$ has a unique solution but that this solution does not exist for all values of the independent variable $t$. The slope field for this equation (Figure 2.29) gives us some clues.

To see things clearly, we can focus on the initial condition $x(0) = 0$. There seems to be only one solution satisfying this condition, but the direction field suggests that the solution curve may have vertical asymptotes. Separating variables, we see that $\int \frac{dx}{1 + x^2} = \int dt$, which gives us $\arctan x = t + C$, or $x(t) = \tan(t + C)$. The initial condition $x(0) = 0$ implies that $C = 0$, so that the solution of the IVP is $x(t) = \tan t$. But this solution's domain is the open interval $(-\pi/2, \pi/2)$. Recall that the function approaches $\pm\infty$ as $t \to \pm\pi/2$. (We say the function "blows up in finite time.") Therefore, the unique solution of our IVP doesn't exist outside the (time) interval $(-\pi/2, \pi/2)$.   ■

---

* Denotes an optional section.

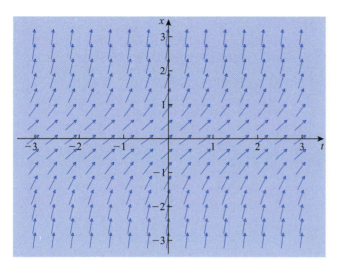

**FIGURE 2.29**
*Slope field for $x' = 1 + x^2$; $-3 \le t \le 3$, $-3 \le x \le 3$*

Now even if we have determined that a given equation *has* a solution, a second important concern is whether there is *only one* solution. This question is usually asked about solutions to initial-value problems.

### ■ Example 2.8.2 An IVP with Infinitely Many Solutions

The nonlinear separable differential equation $x' = x^{2/3}$ has *infinitely many* solutions satisfying $x(0) = 0$ on every interval $[0, \beta]$. To prove this claim, we actually construct the family of solutions of the IVP.

For each number $c$ such that $0 < c < \beta$, we can define the function

$$x_c(t) = \begin{cases} 0 & \text{for } 0 \le t \le c \\ \dfrac{1}{27}(t - c)^3 & c \le t \le \beta. \end{cases}$$

You should verify that each such function satisfies the differential equation with $x(0) = 0$. (You should even be able to show that such a function is differentiable at the break point $c$.) Because there are infinitely many values of the parameter $c$, our IVP has infinitely many solutions. Figure 2.30 shows a few of these solutions with $\beta = 7$. ■

## 2.8.1 An Existence and Uniqueness Theorem

For first-order differential equations, the answers to the existence and uniqueness questions we have just posed are fairly easy. We have an **Existence and Uniqueness Theorem**—simple conditions that guarantee one and only one solution of an initial-value problem.

## Existence and Uniqueness Theorem

Let $R$ be a rectangular region in the $x$-$y$ plane described by the two inequalities $a \leq x \leq b$ and $c \leq y \leq d$. Suppose that the point $(x_0, y_0)$ is inside $R$. Then if $f(x, y)$ and the partial derivative $\dfrac{\partial f}{\partial y}(x, y)$ are continuous functions on $R$, there is an interval $I$ centered at $x = x_0$ and a unique function $y(x)$ defined on $I$ such that $y$ is a solution of the initial-value problem $y' = f(x, y), y(x_0) = y_0$.

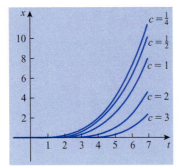

**Figure 2.30**
Solutions of the IVP $x' = x^{2/3}, x(0) = 0$
The functions $x_c(t)$ forc $= \frac{1}{2}, \frac{1}{4}, 1, 2,$ and
$3\beta = 7, -1 \leq t \leq 7$

The preceding statement may look a bit abstract, but it is the simplest and probably the most widely used result that guarantees the existence and uniqueness of a solution of a first-order initial-value problem. Using this theorem is simple. Take your IVP, write it in the form $y' = f(x, y), y(x_0) = y_0$, and then examine the functions $f(x, y)$ and $\frac{\partial f}{\partial y}$, the *partial derivative* of $f$ with respect to the dependent variable $y$. (If you don't know about partial derivatives, see Section A.7 for a quick introduction.)

Figure 2.31 gives an idea of what such a region $R$ and interval $I$ in the Existence and Uniqueness Theorem may look like.

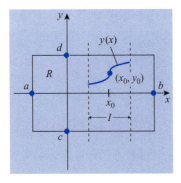

**FIGURE 2.31**
Region of existence and uniqueness: $R = \{(x, y) | a \leq x \leq b, c \leq y \leq d\}$

It's important to make the following comments about this fundamental theorem:

1. If the conditions of our result are satisfied, then solution curves for the IVP can never intersect. (*Do you see why?*)

2. If $f(x, y)$ and $\partial f/\partial y$ happen to be continuous for *all* values of $x$ and $y$, our result does *not* say that the unique solution must be valid for *all* values of $x$ and $y$.

3. The continuity of $f(x, y)$ and $\partial f/\partial y$ are *sufficient* for the existence of solutions, but they may not be *necessary* to guarantee existence. This means that you may have solutions even if the continuity condition is not satisfied.

4. Note that this is an *existence theorem*, which means that if the right conditions are satisfied, you can find a solution, but you are not told how to find it. In particular, you may not be able to describe the interval $I$ without actually solving the differential equation.

The significance of these remarks will be explored in some of the following examples and in some of the problems in Exercises 2.8. First, let's apply the Existence and Uniqueness Theorem to IVPs involving first-order linear ODEs.

### ■ Example 2.8.3 Any "Nice" Linear IVP Has a Unique Solution

Because linear equations model many important physical situations, it's important to know when such equations have unique solutions. We show that if $P(x)$ and $Q(x)$ are continuous ("nice") on an interval $(a, b)$ containing $x_0$, then any IVP of the form $\dfrac{dy}{dx} + P(x)y = Q(x)$, $y(x_0) = y_0$, has one and only one solution on $(a, b)$.

In terms of the Existence and Uniqueness Theorem, we have

$$f(x, y) = -P(x)y + Q(x) \quad \text{and} \quad \frac{\partial f}{\partial y}(x, y) = -P(x).$$

But both $P(x)$ and $Q(x)$ are assumed continuous on the rectangle $R = \{(x, y) | a \le x \le b, c \le y \le d\}$ for any values of $c$ and $d$, and $f(x, y)$ is a combination of continuous functions. (There are no values of $x$ and $y$ that give us division by zero or an even root of a negative number, for example.) The conditions of the theorem are satisfied, and so any IVP of the form described previously has a unique solution.

In Section 2.2 we showed how to find a solution of a linear differential equation explicitly. Now we see that, given an appropriate initial condition, we have learned how to find the *unique* solution. ■

Now let's go back to re-examine examples we discussed earlier.

### ■ Example 2.8.4 Example 2.8.1 Revisited

Assume that $x$ is a function of the independent variable $t$. If we look at the IVP $x' = 1 + x^2$, $x(0) = x_0$, in light of the Existence and Uniqueness Theorem, we see that $f(t, x) = 1 + x^2$, a function of $x$ alone that is clearly continuous at all points $(t, x)$, and $\dfrac{\partial f}{\partial x} = 2x$, also continuous for all $(t, x)$.

The conditions of the theorem are satisfied, and so the IVP has a unique solution. But even though both $f(t, x)$ and $\dfrac{\partial f}{\partial x}$ are continuous for *all* values of $t$ and $x$, we know that any unique solution is limited to an interval

$$\left( \frac{(2n-1)\pi}{2}, \frac{(2n+1)\pi}{2} \right), \quad n = 0, \pm 1, \pm 2, \pm 3, \ldots,$$

separating consecutive vertical asymptotes of the tangent function. (Go back to look at the one-parameter family of solutions for the equation, and see comment 2 that follows the statement of the Existence and Uniqueness Theorem.) ■

Next, we scrutinize Example 2.8.2 in light of the Existence and Uniqueness Theorem.

### ■ Example 2.8.5 Example 2.8.2 Revisited

Here, we have the form $x' = x^{2/3} = f(x)$, with $x(0) = 0$, so we must look at $f(x)$ and $\frac{\partial f}{\partial x}$. But $\frac{\partial f}{\partial x} = f'(x) = \frac{2}{3}x^{-1/3} = \frac{2}{3\sqrt[3]{x}}$, which is not continuous in any rectangle in the $t$-$x$ plane that includes $x = 0$ (that is, any part of the $t$-axis). Therefore, we shouldn't expect to have both existence and uniqueness on an interval of the form $[0, \beta]$—and in fact we don't have uniqueness, as we saw.

However, if we avoid the $t$-axis—that is, if we choose an initial condition $x(t_0) = x_0 \neq 0$—then the Existence and Uniqueness Theorem guarantees that there will be a unique solution for the IVP. Figure 2.32a shows the slope field for the autonomous equation $x' = x^{2/3}$ in the rectangle $-1 \le t \le 5$, $0 \le x \le 3$. This rectangle includes part of the $t$-axis, and it is easy to visualize many solutions starting at the origin, gliding along the $t$-axis for a little while, and then taking off. Figure 2.30 shows some of these solution curves.

**FIGURE 2.32a**
*Slope field for $x' = x^{2/3}$, $-1 \le t \le 5$, $0 \le x \le 3$*

**FIGURE 2.32b**
*Slope field for $x' = x^{2/3}$, $-1 \le t \le 5$, $1 \le x \le 3$*

Figure 2.32b, on the other hand, shows what happens if we choose a rectangle that avoids the $t$-axis. It should be clear that if we pick any point $(t_0, x_0)$ in this rectangle, there will be one and only one solution of the equation that passes through this point.  ∎

## Exercises 2.8

### A

For each of the following initial value Problems 1–8, determine a rectangle $R$ in the appropriate plane ($x$-$y$, $t$-$x$, etc.) for which the given differential equation would have a unique solution through a point in the rectangle. **Do not solve the equations**.

1. $\dfrac{dx}{dt} = \dfrac{1}{x}$, $x(0) = 3$

2. $\dfrac{dy}{dt} = \dfrac{5}{4}y^{1/5}$, $y(0) = 0$

3. $t\dfrac{dx}{dt} = x$, $x(0) = 0$

4. $y' = -\dfrac{t}{y}$, $y(0) = 0.2$

5. $y' = \dfrac{t}{1 + t + y}$, $y(-2) = 1$

6. $x' = \tan x$, $x(0) = \dfrac{\pi}{2}$

7. $(1 + t)\dfrac{dy}{dt} = 1 - y$

8. $y' = \dfrac{x + y}{x - y}$

9. What is the length of the largest interval $I$ on which the IVP $y' = 1 + y^2$, $y(0) = 0$ has a solution?
10. Show that $y \equiv -1$ is the only solution of the IVP $y' = t(1 + y)$, $y(0) = -1$.
11. What is the solution to the IVP $\dfrac{dx}{dt} = x^{2/3}$, $x(0) = x_0$ if $x_0 < 0$? Compare your answer to the answer(s) in Example 2.8.2. What has changed?

### B

1. Look at the IVP $Q' = |Q - 1|$, $Q(0) = 1$.

   a. Explain why the conditions of the Existence and Uniqueness Theorem do not hold for this equation.

   b. Guess at a solution of this initial value problem.

   c. Explain why the solution you found in part (b) is unique.

2. Consider the equation $\dot{y} = \sqrt{|y|} + k$, where $k$ is a positive constant.

   a. Solve the equation. (You will get an implicit solution.)

   b. For what initial values $(t_0, y_0)$ does the equation have a unique solution?

   c. For what values of $k \le 0$ does the equation have unique solutions?

3. The parabola $y = x^2$ and the line $y = 2x - 1$ are both solutions of the equation $y' = 2x - 2\sqrt{x^2 - y}$ and satisfy the initial condition $y(1) = 1$. Does this contradict the Existence and Uniqueness Theorem?

4. Consider the equation $\dfrac{dy}{dx} + x^2 y^3 = \cos x$.

   a. Does this equation have a unique solution passing through any point $(x_0, y_0)$?

   b. Try to solve the equation using the ODE solver in your CAS. Comment on the result.

**C**

1. Consider the initial value problem $\dfrac{dy}{dx} = P(x)y^2 + Q(x)y$, $y(2) = 5$, where $P(x)$ and $Q(x)$ are third-degree polynomials in $x$. Does this problem have a unique solution on some interval $|x - 2| \le h$ around $x = 2$? Explain why or why not.

2. Consider the nonlinear equation $\dfrac{dx}{dt} = (\alpha - x)(\beta - x)$, where $\alpha$ and $\beta$ are positive constants. (See Example 2.1.7.) Without solving the equation, show that the solution of any IVP involving this equation is unique.

3. Consider the equilibrium solution $P \equiv b$ of the logistic equation (Section 2.5) $\dfrac{dP}{dt} = kP(b - P)$, where $k$ and $b$ are positive constants. Is it possible for a solution near $P \equiv b$ to reach (i.e., equal) this solution for a finite value of $x$? [*Hint*: Use the uniqueness part of the Existence and Uniqueness Theorem.]

4. Why can't the family of curves shown in the following be the solution curves for the differential equation $y' = f(t, y)$, where $f$ is a polynomial in $t$ and $y$?

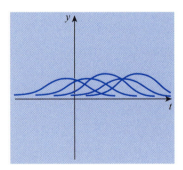

5. Example 2.8.3 indicates that the linear IVP $y' = y$, $y(0) = 1$ has a unique solution. Ignoring the fact that you can actually solve this equation, prove the following properties of $y(t)$:

   a. For all real values of $t$, $y(t)\,y(-t) = 1$.

   b. $y(t) > 0$ for all real numbers $t$.

   c. For all real numbers $t_1$ and $t_2$, $y(t_1 + t_2) = y(t_1) \cdot y(t_2)$.

6. Suppose that a differential equation is a model for a certain type of chemical reaction. Could the fact that the equation does *not* have a solution indicate that the reaction *cannot* take place? Would the fact that the equation *has* a solution guarantee that the reaction *does* take place?

## SUMMARY

Perhaps the easiest type of first-order ODE to solve is a **separable equation**, one that can be written in the form $\frac{dy}{dx} = f(x)g(y)$, where $f$ denotes a function of $x$ alone and $g$ denotes a function of $y$ alone. "Separating the variables" leads to the equation $\int \frac{dy}{g(y)} = \int f(x)dx$. It is possible that you cannot carry out one of the integrations in terms of elementary functions or you may wind up with an *implicit* solution. Furthermore, the process of separating variables may introduce **singular solutions**.

Another important type of first-order ODE is a **linear equation**, one that can be written in the form $a_1(x)y' + a_0(x)y = f(x)$, where $a_1(x)$, $a_0(x)$, and $f(x)$ are functions of the independent variable $x$ alone. The standard form of such an equation is $\frac{dy}{dx} + P(x)y = Q(x)$. The equation is called **homogeneous** if $Q(x) \equiv 0$ and **nonhomogeneous** otherwise. Any homogeneous linear equation is separable.

After writing a first-order linear equation in the standard form $\frac{dy}{dx} + P(x)y = Q(x)$, we introduce an **integrating** factor, $\mu(x) = e^{\int P(x)dx}$, multiply each side of the equation by $\mu(x)$, and see that the equation can be written as $\frac{d}{dx}(e^{\int P(x)dx} y) = e^{\int P(x)dx}Q(x)$. Integrating each side and then multiplying by $e^{-\int P(x)dx}$, we get an explicit formula:

$$y = e^{-\int P(x)dx} \cdot \int e^{\int P(x)dx}Q(x)dx + Ce^{-\int P(x)dx}.$$

A typical first-order differential equation can be written in the form $\frac{dy}{dx} = f(x, y)$. Graphically, this tells us that at any point $(x, y)$ on a solution curve of the equation, the slope of the tangent line is given by the value of the function $f$ at that point. We can outline the solution curves by using possible tangent line segments. Such a collection of tangent line segments is called a **direction field** or **slope field** of the equation. The set of points $(x, y)$ such that $f(x, y) = C$, a constant, defines an **isocline**, a curve along which the slopes of the tangent lines are all the same (namely, $C$). In particular, the **nullcline** (or **zero isocline**) is a curve consisting of points at which the slopes of solution curves are zero. A differential equation in which the independent variable does not appear explicitly is called an **autonomous** equation. If the independent variable *does* appear, the equation is called **nonautonomous**. For an autonomous equation, the slopes of the tangent line segments that make up the slope field depend only on the values of the dependent variable. Graphically, if we fix the value of the dependent variable, say $x$, by drawing a horizontal line $x = C$ for any constant $C$, we see that all the tangent line segments along this line have the same slope, no matter what the value of the independent variable, say $t$. Another way to look at this is to realize that we can generate infinitely many solutions by taking any one solution and translating (shifting) its graph left or right. Even when we can't solve an equation, an analysis of its slope field can be very instructive. However, such a graphical analysis may miss certain important features of the integral curves, such as vertical asymptotes.

An *autonomous* first-order equation can be analyzed qualitatively by using a **phase line** or **phase portrait.** For an autonomous equation, the points $x$ such that $\frac{dy}{dx} = f(x) = 0$ are called **critical points.** We also use the terms **equilibrium points, equilibrium solutions,** and **stationary points** to describe these key values. There are three kinds of equilibrium points for an autonomous first-order equation: **sinks, sources,** and **nodes.** An equilibrium solution $y$ is a **sink** (or **asymptotically stable solution**) if solutions with initial conditions "sufficiently close" to $y$ are asymptotic to $y$ as the independent variable tends to infinity. On the other hand, if solutions "sufficiently close" to an equilibrium solution $y$ are asymptotic to $y$ as the independent variable tends to negative infinity, then we call $y$ a **source** (or **unstable equilibrium solution**). An equilibrium solution that shows any other kind of behavior is called a **node** (or **semistable equilibrium solution**). A simple (but not always conclusive) test is as follows:

> If $x^*$ is an equilibrium point for the equation $\frac{dx}{dt} = f(x)$, it is true that (1) if $f'(x^*) > 0$, then $x^*$ is a source; (2) if $f'(x^*) < 0$, then $x^*$ is a sink; and (3) if $f'(x^*) = 0$, then we can't tell what sort of equilibrium point $x^*$ may be without further investigation.

Suppose that we have an autonomous differential equation with a parameter $\alpha$. A **bifurcation point** $\alpha_0$ is a value that causes a change in the nature of the equation's equilibrium solutions as $\alpha$ passes through the value $\alpha_0$.

When we are trying to solve a differential equation, especially an initial-value problem, it is important to understand whether the problem *has* a solution and whether any solution is *unique.* There are simple conditions that guarantee that there is one and only one solution of an initial-value problem:

> Let $R$ be a rectangular region in the $x$-$y$ plane described by the two inequalities $a \le x \le b$, $c \le y \le d$. Suppose that the point $(x_0, y_0)$ is inside $R$. Then, if $f(x, y)$ and $\frac{\partial f}{\partial y}$ are continuous functions on $R$, there is an interval $I$ centered at $x = x_0$ and a unique function $y(x)$ defined on $I$ such that $y$ is a solution of the initial-value problem $y' = f(x, y)$, $y(x_0) = y_0$.

## PROJECT 2-1

### The Price Is Right

Let $p(t)$, $s(t)$, and $d(t)$ denote the price, supply, and demand of a commodity at time $t$. **Allen's Speculative Model** in economics assumes that $s$ and $d$ are linear functions in $p(t)$ and $p'(t)$:

$$s(t) = a_1 p(t) + a_2 p'(t) + a_3 \tag{*}$$

$$d(t) = b_1 p(t) + b_2 p'(t) + b_3, \tag{**}$$

where the $a_i'$s and $b_i'$s are constants.

The **Economic Principle of Supply and Demand**, which guarantees a state of dynamic equilibrium, is

$$d(t) = s(t). \qquad\qquad (\text{***})$$

a. By combining (*), (* *), and (* * *), find a single linear differential equation involving $p(t)$.

b. Assuming that $a_1 \neq b_1$, $a_2 \neq b_2$, and $a_3 \neq b_3$, solve the equation you found in part (a) with the initial condition $p(0) = p_0$.

c. Interpret the solution in economic terms if $p_0 = \dfrac{a_3 - b_3}{b_1 - a_1}$.

d. Suppose that $\dfrac{b_1 - a_1}{b_2 - a_2} > 0$. What happens to the price as $t$ increases without bound?

e. Suppose that $\dfrac{b_1 - a_1}{b_2 - a_2} < 0$. Now what happens to the price as $t$ increases without bound?

f. Suppose that $s(t) = 30 + p(t) + 4p'(t)$ and $d(t) = 48 - 2p(t) + 3p'(t)$, where $s$ and $d$ are given in thousands of units. If $p(0) = 10$ monetary units, find the price at any later time $t$. What happens to the price as $t$ increases?

# PROJECT 2-2

## Cultured Perils

A continuous culture device, or **chemostat**, is a well-stirred vessel that contains microorganisms and into which fresh medium (nutrient) is pumped at a constant rate $F$. The contents of the growth vessel are pumped out at the same rate, so the volume $V$ remains constant. Microbiologists and ecologists use the chemostat as a laboratory simulation of an aquatic environment, and it has also been used to model the waste water treatment process. (See the accompanying schematic diagram.)

Around 1950, the biologist Jacob Monod developed a mathematical model[7] for the continuous culture of a single species of microorganism whose growth is dependent solely on a single nutrient supplied at a constant rate via the input to the growth vessel.

Chemostat experiments[8] with a certain strain of *Escherichia coli* (*E. coli*) led to the particular equation

$$\frac{dx}{dt} = \left(\frac{0.81(10 - x)}{3 + (10 - x)} - D\right)x \qquad (*)$$

where $x(t)$ denotes the concentration of the organism at time $t$; and $D = \frac{F}{V}$, the pump rate divided by the volume, is a parameter under control of the experimenter.

We want to study the effects of varying $D$. Intuition suggests that if the pump is allowed to run too fast, then the *E. coli* will eventually approach extinction in the chemostat because they are being pumped out at a faster rate than they can grow and reproduce. On the other hand, if the pump is run slowly enough, the *E. coli* should be able to grow at a rate sufficient to overcome washout and should be able to thrive in the growth vessel indefinitely.

Our problem is to determine which values of $D$ result in extinction and which result in survival. This can be done by studying Equation (*), treating $D$ as a bifurcation parameter (see Section *2.7).

**a.** Using technology, study solutions to Equation (*) for parameter values of 0.8, 0.7, 0.4, and 0.5. For each choice of $D$, use several different initial conditions $x_0$. What are your observations?

**b.** Find the equilibrium solutions $\hat{x}$ as a function of the parameter $D$ and determine whether they are sinks, sources, or nodes. (Assume that the only meaningful equilibrium solutions are those for which $0 \le \hat{x} \le 10$.) Construct the bifurcation diagram for Equation (*). (See Section *2.7.) At what value of $D$ does a bifurcation occur? Explain the significance of this bifurcation with regard to the fate of the *E. coli*.

**c.** Use technology to test the validity of your bifurcation analysis in part (b) by examining the solutions of Equation (*) again, choosing various values of $D$ very close to the bifurcation value you found in part (b). Are your observations as expected?

**d.** Assuming that the growth vessel is kept at a volume of 20 liters, at what speed should the chemostat pump be run in order to maintain a steady-state *E. coli* population of $8\mu g$/liter? A population of $4.5\mu g$/liter?

[7] J. Monod, "La technique de culture continue: Théorie et applications," *Annales de L'Institut Pasteur* **79** (1950): 390–410.

[8] S. R. Hansen and S. P. Hubbell, "Single Nutrient Microbial Competition: Agreement Between Experimental and Theoretically Forecast Outcomes," *Science* **20** (1980): 1491–1493.

# CHAPTER 3

# The Numerical Approximation of Solutions

## INTRODUCTION

Historically, numerical methods of working with differential equations were developed when some equations could not be solved analytically—that is, with their solutions expressed in terms of elementary functions. Over the past 300 years, mathematicians and scientists have learned to solve more and more types of differential equations. However, today there are still equations that are impossible to solve in closed form (for instance, Example 2.1.5). In fact, very few differential equations that arise in applications can be solved exactly; and, perhaps more important, even solution formulas often express the solutions *implicitly* via complicated combinations of the solution and the independent variable that are difficult to work with. Take a look back at the solution in Example 2.4.4, for instance. In this chapter we will describe some ways of getting an *approximate numerical solution* of a first-order IVP $y' = f(x, y), y(x_0) = y_0$. This means being able to calculate approximate values of the solution function $y$ by some process requiring a finite number of steps, so that at the end of this step-by-step process we are reasonably close to the "true" answer. Graphically, we are trying to approximate the solution curve with a simpler curve, usually a curve made up of straight line segments.

The very nature of what we will be trying to do contains the notion of *error*, the discrepancy between a true value and its approximate value. Error is what stands between reality and perfection. It is the static in our telephone line, the wobble in a kitchen chair, a slip of the tongue. Although there are various ways to measure error, we will focus on **absolute error**, which is defined by the quantity |true value − approximation|, the absolute value of the difference between the exact value and the approximate value. We'll have more to say about error in the following sections.

Let's see how all this applies to a first-order IVP $y' = f(x, y), y(x_0) = y_0$.

## 3.1 EULER'S METHOD

One of the easiest methods of obtaining an approximation to a solution curve is attributed to the mathematician Euler. He used this approach to solve differential equations around 1768.

A modern way of expressing his idea is to say that he used *local linearity*. Geometrically, this simply refers to the fact that if a function $F$ is differentiable at $x = x_0$ and we "zoom" in on the point $(x_0, F(x_0))$ lying on the curve $y = F(x)$, then we will think we're looking at a straight line segment. Numerically, we're saying that if we have a straight line tangent to a curve $y = F(x)$ at a point $(x_0, F(x_0)) = (x_0, y_0)$, then for a value of $x$ close to $x_0$, the corresponding value on the tangent line is approximately equal to the value on the curve. In other words, we can avoid the complexity of dealing with values on what may be a complicated curve by dealing with values on a straight line. (See Section A.1 for more details on this topic.)

Using the familiar "point-slope" formula for the equation of a straight line, we can derive the equation of the line $T$ tangent to the curve $y = F(x)$ at the point $(x_0, y_0)$:

$$T(x) = F'(x_0)(x - x_0) + F(x_0) = y'(x_0)(x - x_0) + y_0. \tag{3.1.1}$$

Now we can express the idea of local linearity by writing

$$\underbrace{y(x)}_{\text{value on the curve}} \approx \underbrace{y'(x_0)(x - x_0) + y_0}_{\text{value on the tangent line}}$$

where the symbol $\approx$ means "is approximately equal to." Figure 3.1 shows what we are saying.

Note that because the curve we're using as an illustration is concave down near the point $(x_0, y_0)$, the tangent line lies *above* the curve here, so the value $T(x)$ given by the tangent line is *greater than* the true value $y(x)$ for $x$ near $x_0$.

Now let's look at an IVP $y' = f(x, y)$, $y(x_0) = y_0$ so that we can write $y'_0 = y'(x_0) = f(x_0, y_0)$. **In what follows, we assume that there is a unique solution $\varphi$ in some interval containing $x_0$.** Suppose we want to know the height of the solution curve corresponding to a value $x_1$ that is close to $x_0$, but to the *right* of $x_0$. We can describe such a new value of the independent variable as $x_1 = x_0 + h$, where $h > 0$ is the size of a small "step." Now let's try to approximate

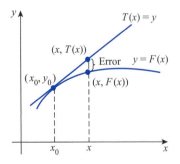

**FIGURE 3.1**
*Local linearity*

$\varphi(x_1)$, a value on the actual solution curve, by some value $y_1$ on the tangent line to $y = \varphi(x)$ at $x_0$:

$$\varphi(x_1) \approx y_1 = \varphi'(x_0)(x_1 - x_0) + y_0$$
$$= f(x_0, y_0)(x_0 + h - x_0) + y_0$$
$$= f(x_0, y_0) \cdot h + y_0.$$

Therefore, we can write

$$\varphi(x_1) \approx y_1 = f(x_0, y_0) \cdot h + y_0$$

and we have a good *local linear approximation* of $\varphi(x)$ at $x = x_1$ if we choose $h$ small enough. Figure 3.2 illustrates what's going on.

We can repeat the process using $(x_1, y_1)$ as our jumping-off point, realizing that the value $y_1$ is only an approximation. Using Equation (3.1.1) again, with $(x_0, y_0)$ replaced by $(x_1, y_1)$, we see that the line through $(x_1, y_1)$ with slope equal to $f(x_1, y_1)$ has $y$ values given by

$$f(x_1, y_1)(x - x_1) + y_1. \tag{3.1.2}$$

We should realize that the point $(x_1, y_1)$ is not expected to be on the actual solution curve, so in general $f(x_1, y_1) \neq f(x_1, \varphi(x_1))$, the slope of the actual solution at $x_1$.

For convenience, suppose that we want to approximate the solution curve's height corresponding to a value $x_2$ that is the same distance from $x_1$ as $x_1$ is from $x_0$. That is, we take a step to the right of size $h : x_2 = x_1 + h = (x_0 + h) + h = x_0 + 2h$. We can approximate $\varphi(x_2)$, the actual value of the solution function at $x = x_2$, by using Equation (3.1.2):

$$\varphi(x_2) \approx y_2 = f(x_1, y_1) \cdot h + y_1.$$

Similarly, for $x_3 = x_2 + h = (x_0 + 2h) + h = x_0 + 3h$, we approximate $\varphi(x_3)$ as follows:

$$\varphi(x_3) \approx y_3 = f(x_2, y_2) \cdot h + y_2.$$

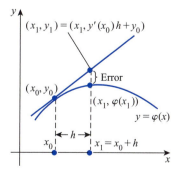

**FIGURE 3.2**
*A local linear approximation of a solution*

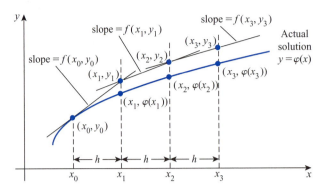

**FIGURE 3.3**
*A three-step linear approximation*

Figure 3.3 shows what we are doing.

Continuing in this way, we generate a sequence of approximate values $y_1, y_2, y_3, \ldots, y_n$ for the solution function $\varphi$ at various equally spaced points $x_1, x_2, x_3, \ldots, x_n$:

$$\underbrace{y_{k+1}}_{\text{new approx. value}} = \underbrace{y_k}_{\text{old approx. value}} + \underbrace{h}_{\text{step size}} \cdot f(x_k, y_k), \qquad (3.1.3)$$

where $x_k = x_0 + kh$, $k = 0, 1, \ldots, n$. If you go back through the derivation, you'll realize that Formula (3.1.3) is valid for $h < 0$ also. Note that if the points $x_k$ are equally spaced with step size $h$ and we want to get from $(x_0, y_0)$ to $(x^*, y^*)$ along the approximating polygonal curve, then we must have $n = \frac{x^* - x_0}{h}$ steps. For example, if we start at $x_0 = 2$ and want to approximate $\varphi(2.7)$ using steps of size $h = 0.1$, we can reach $x^* = 2.7$ by taking $n = \frac{2.7 - 2}{0.1} = 7$ steps. In practice, once we have chosen the step size $h$, the number of steps needed, $n$, will be obvious.

If we stand back from all these equations and look at Figures 3.2 and 3.3 again, we can see that what we are doing is using the slope field for our IVP as a set of stepping stones. We "walk" on a tangent line segment for a short distance, stop to look forward for the next step, jump to that step, and so on. We are approximating the flow of the solution curve by using flat rocks set into the "stream." If you play "connect the dots" with the points $(x_0, y_0), (x_1, y_1), (x_2, y_2), \ldots, (x_n, y_n)$, you see a polygonal line (called the **Euler polygon** or the **Cauchy-Euler polygon**) that approximates the actual solution curve. Figure 3.4 shows this for the initial-value problem $y' = x^2 + y, y(0) = 1$, where we try to approximate $y(1)$ with different step sizes $h = 1, 0.5$, and $0.25$.

We can look at this approximation process, Formula (3.1.3), in another geometrical way. Suppose we have the differential equation $y' = f(x, y)$. Then the Fundamental Theorem of Calculus tells us that

$$y_{k+1} - y_k \approx y(x_{k+1}) - y(x_k) = \int_{x_k}^{x_{k+1}} y'(x)\,dx = \int_{x_k}^{x_{k+1}} f(x, y)\,dx.$$

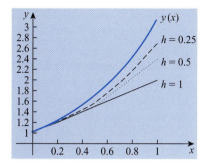

**FIGURE 3.4**

*The actual solution of the IVP $y' = x^2 + y, y(0) = 1$, and three Euler approximations ($h = 1, 0.5,$ and $0.25$) on the interval $[0, 1]$*

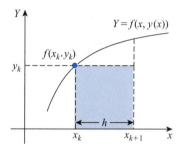

**FIGURE 3.5**

*Approximation of an integral by a rectangular area*

But Formula (3.1.3) requires us to replace $y_{k+1} - y_k$ by $h \cdot f(x_k, y_k)$. This means that we are approximating $\int_{x_k}^{x_{k+1}} f(x, y)dx$ by $h \cdot f(x_k, y_k)$. Figure 3.5 shows the geometry of the situation in the interval $[x_k, x_{k+1}]$.

We have approximated the area under the curve $Y = f(x, y(x))$ by the area of the shaded rectangle—a rectangle formed by using the height of the curve at the left-hand endpoint of the interval. Thus, Euler's method amounts to using a (left-hand) *Riemann sum* approximation to the area under a curve.

If $y$ is the solution of the equation $y' = f(x, y)$, we can view Euler's method in yet another way by considering the Taylor expansion of $y(x)$ about $x = x_k$ (see Section A.3):

$$y(x_{k+1}) = y(x_k) + y'(x_k)h + y''(\xi_k)\frac{h^2}{2} = \underbrace{y(x_k) + f(x_k, y(x_k))h}_{y_{k+1}} + y''(\xi_k)\frac{h^2}{2}$$

with $x_k < \xi_k < x_{k+1}$. Assuming that $y(x)$ has a bounded second derivative and realizing that $h^2 < h$ for small values of $h$, we see that Euler's method is essentially using a first-degree

Taylor polynomial to approximate the solution curve:

$$y(x_{k+1}) \approx y(x_k) + f(x_k, y(x_k))h.$$

The approximation processes we have been describing (and will describe later in this chapter) are subject to two basic kinds of error: **truncation error**, which occurs when we stop (or truncate) an approximation process after a certain number of steps, and **propagated error**, the accumulated error resulting from many calculations with rounded values. (See Section A.3 for further remarks about these types of errors.) We must be aware that there is usually a trade-off in dealing with error. If we try to reduce the *truncation* error and increase the accuracy of our approximation by carrying out more steps (for example, by taking more terms of a Taylor series or more steps in Euler's method), we are increasing our calculation load and consequently running the risk of increasing *propagated* error. Figure 3.6 shows the trade-off in general terms.

Clearly, at each stage of Euler's method, we choose to round off entries in a certain way. Even if we assume for the sake of simplicity that round-off error is negligible, our use of local linearity—using straight lines to approximate curves—introduces truncation error. Now suppose that we are given the value, $y(x_0)$, of a solution at an initial point and want to approximate the value, $y(b)$, at some later point $b = x_0 + nh$. First, there is **local truncation error** at each step, defined as $y(x_{k+1}) - y_{k+1}$ for each $k (k = 0, 1, 2, \ldots, n - 1)$. This is the error introduced in computing the value $y_{k+1}$ from the value $y_k$, *assuming that $y_k$ is exact*. Then we have the **cumulative truncation error**, defined as $y(b) - y_n = y(x_n) - y_n$ which is the (total) actual error in the value of $y(x_0 + nh)$, or $y(b)$, caused by all the previous approximations—that is, by the cumulative effect after $n$ steps of the local errors from previous steps. (This is not just the sum of all the local truncation errors. *Life isn't that simple.*)

In any case, a mathematically rigorous analysis of the errors produced shows that *the local truncation error at any step of Euler's method behaves like a constant multiple of $h^2$*, which is smaller than $h$ when $h$ is small:

$$|\text{local truncation error at step } k| = |y(x_{k+1}) - y_{k+1}| \le \frac{M}{2}h^2,$$

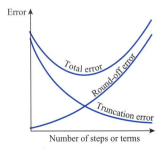

**FIGURE 3.6**
*Total error = round-off error + truncation error*

where $M = \max\limits_{x_k < x < x_{k+1}} |y''(x)|$. This follows easily from the Taylor series expansion

$$y(x_{k+1}) = \underbrace{y(x_k) + f(x_k, y(x_k))h}_{y_{k+1}} + y''(\xi_k)\frac{h^2}{2}$$

given previously.

It is also true that for Euler's method, the cumulative truncation error is no greater than a constant multiple of the step size $h$:

$$|\text{true value} - \text{approximation}| = |y(b) - y_n| \leq K \cdot h,$$

where $K$ is independent of $h$ but depends on $|y''(x)|$ and the interval $[x_0, b]$. Because the cumulative error is bounded by a constant multiple of the *first* power of the step size $h$, we say that Euler's method is a **first-order method**. (The number $K$ is a *maximum* bound. In practice, the actual error incurred in a problem will usually be less than this bound.) Intuitively, we can reason as follows: There are $n = \frac{b-x_0}{h}$ steps in the Euler method approximation, each having a local error less than or equal to some multiple of $h^2$. If $K^*$ is the largest of the multipliers, then the cumulative error is less than or equal to $\frac{b-x_0}{h} \cdot K^* h^2 = Kh$.

Therefore, if we ignore round-off error as essentially a statistical problem outside our range of interest right now, we can make the total error as small as we wish by making $h$ "sufficiently small"—that is, by making the *number* of steps "sufficiently large." This is not very satisfactory because a larger number of steps *does* require more calculating time by hand or by computer, and in real-life problems, the larger number of steps often leads to a "snowballing" of round-off error. Take another look at Figure 3.6.

If you want to understand and improve the accuracy of your approximations, here are two rules of thumb you can use: (1) Start your calculations with many more decimal places than you need. (2) Keep on redoing your calculations with a step size $h$ equal to one-half its previous value. If you reach a stage at which the new result agrees with the previous one to $d$ decimal places after appropriate rounding, then you can assume that you have $d$ decimal place accuracy. (Look at Example 3.1.4 for a slight variation of this rule.)

Euler's method is not very accurate and is not used widely in practice. But the method is simple and displays the essential characteristics of more sophisticated methods. In the next section we discuss an improved method, one that uses Euler's basic idea in a more efficient way.

This is enough theory for now. Let's see how this "method of tangents" works with a simple initial-value problem.

### ■ Example 3.1.1 Euler's Method with Error Analysis

Suppose we're given the IVP $\frac{dx}{dt} = t^2 + x$, $x(1) = 3$. We want to use Euler's method to approximate $x(1.5)$.

This is a first-order linear equation whose particular solution for the initial condition $x(1) = 3$ is $x(t) = -t^2 - 2t - 2 + 8e^{t-1}$. (*Verify this.*) Thus, the actual value of $x(1.5)$ is $-(1.5)^2 - 2(1.5) - 2 + 8e^{(1.5)-1} = 5.939770\ldots$. We'll use the actual value to see how good an approximation Euler's method gives us.

In our problem, $f(t, x) = t^2 + x$, so Euler's formula (3.1.3) becomes

$$x_{k+1} = x_k + h \cdot (t_k^2 + x_k), \tag{3.1.4}$$

where $t_k = t_0 + kh$, $k = 0, \ldots, n$, $t_0 = 1$, and $x_0 = 3$. (By now you should be comfortable with the switch from the traditional $x$-$y$ coordinates to $t$-$x$ coordinates.) Suppose we take $h = 0.1$—that is, our step size is one-tenth of a unit. Because our target $t = 1.5$ is 0.5 unit away from our initial point $t = 1$, we'll need $n = 5$ steps of size $h = 0.1$ to reach this with Euler's process (Figure 3.7).

Using Formula (3.1.4), let's generate our approximate values, stepping from $t = 1$ to $t = 1.5$:

$$x_1 = x_0 + (0.1)(t_0^2 + x_0) = 3 + (0.1)(1^2 + 3) = 3.40$$

$$x_2 = x_1 + (0.1)(t_1^2 + x_1) = 3.40 + (0.1)(1.1^2 + 3.4) = 3.861$$

$$x_3 = x_2 + (0.1)(t_2^2 + x_2) = 3.861 + (0.1)(1.2^2 + 3.861) = 4.3911$$

$$x_4 = x_3 + (0.1)(t_3^2 + x_3) = 4.3911 + (0.1)(1.3^2 + 4.3911) = 4.99921$$

$$x_5 = x_4 + (0.1)(t_4^2 + x_4) = 4.99921 + (0.1)(1.4^2 + 4.99921) = 5.695131.$$

Thus, Euler's method gives the approximation 5.695131 for the value $x(1.5)$. In this example, the *absolute error* is |true value − approximation| = $|5.939770 - 5.695131| = 0.244639$.

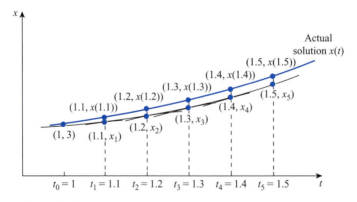

**FIGURE 3.7**
*A five-step approximation*

**Table 3.1** Euler's Method for $\frac{dx}{dt} = t^2 + x, x(1) = 3$, with $h = 0.05$

| k | $t_k$ | $x_k$ | True Value | Absolute Error |
|---|---|---|---|---|
| 0 | 1 | 3.000000 | 3.000000 | 0.00000 |
| 1 | 1.05 | 3.200000 | 3.207669 | 0.00767 |
| 2 | 1.1 | 3.415125 | 3.431367 | 0.01624 |
| 3 | 1.15 | 3.646381 | 3.672174 | 0.02579 |
| 4 | 1.2 | 3.894825 | 3.931222 | 0.03640 |
| 5 | 1.25 | 4.161567 | 4.209703 | 0.04814 |
| 6 | 1.3 | 4.447770 | 4.508870 | 0.06110 |
| 7 | 1.35 | 4.754658 | 4.830040 | 0.07538 |
| 8 | 1.4 | 5.083516 | 5.174598 | 0.09108 |
| 9 | 1.45 | 5.435692 | 5.543997 | 0.10831 |
| 10 | 1.5 | 5.812602 | 5.939770 | 0.12717 |

If we try again, using a step size only *half* the size of the step we used before—that is, using a step size $h = 0.05$—it will take *twice* as many steps to bridge the gap between the initial value $t = 1$ and the final value $t = 1.5$. Table 3.1 shows the result of a spreadsheet calculation of Euler's method for this new sequence of steps. If you have access to a spreadsheet program, you'll find it fairly easy to use it to set up Euler's method.

Note that the number of calculations (steps) has doubled, but the absolute error has been cut almost in half. Also note the cumulative growth of the error in the last column.

If we cut the step size in half again, working with $h = 0.025$ this time, we can see a pattern emerging (Table 3.2).

Note that the absolute error increases as a function of $k$, the number of steps. Furthermore, the differences between successive errors are increasing slightly. For example, if you subtract the error for $k = 9$ from the error corresponding to $k = 10$, you get 0.0030, whereas subtracting the $k = 10$ error from the $k = 11$ error yields 0.0032.

If this error pattern seems vaguely familiar, it may be because you have seen error analysis applied to left- and right-hand Riemann sum approximations in calculus.  ∎

Let's try another problem. Practice makes perfect—or at least we can approximate perfection.

**Table 3.2** Euler's Method for $\frac{dx}{dt} = t^2 + x, x(1) = 3$, with $h = 0.025$

| $k$ | $t_k$ | $x_k$ | True Value | Absolute Error |
|---|---|---|---|---|
| 0 | 1 | 3.000000 | 3.000000 | 0.0000 |
| 1 | 1.025 | 3.100000 | 3.101896 | 0.0019 |
| 2 | 1.05 | 3.203766 | 3.207669 | 0.0039 |
| 3 | 1.075 | 3.311422 | 3.317448 | 0.0060 |
| 4 | 1.1 | 3.423098 | 3.431367 | 0.0083 |
| 5 | 1.125 | 3.538926 | 3.549563 | 0.0106 |
| 6 | 1.15 | 3.65904 | 3.672174 | 0.0131 |
| 7 | 1.175 | 3.783578 | 3.799345 | 0.0158 |
| 8 | 1.2 | 3.912683 | 3.931222 | 0.0185 |
| 9 | 1.225 | 4.046500 | 4.067957 | 0.0215 |
| 10 | 1.25 | 4.185178 | 4.209703 | 0.0245 |
| 11 | 1.275 | 4.32887 | 4.356620 | 0.0277 |
| 12 | 1.3 | 4.477733 | 4.508870 | 0.0311 |
| 13 | 1.325 | 4.631926 | 4.666620 | 0.0347 |
| 14 | 1.35 | 4.791615 | 4.830040 | 0.0384 |
| 15 | 1.375 | 4.956968 | 4.999306 | 0.0423 |
| 16 | 1.4 | 5.128158 | 5.174598 | 0.0464 |
| 17 | 1.425 | 5.305362 | 5.356098 | 0.0507 |
| 18 | 1.45 | 5.488761 | 5.543997 | 0.0552 |
| 19 | 1.475 | 5.678543 | 5.738489 | 0.0599 |
| 20 | 1.5 | 5.874897 | 5.939770 | 0.0649 |

## ■ Example 3.1.2 Euler's Method with Error Analysis

Consider the IVP $\frac{dy}{dt} = \frac{1}{t}$, $y(1) = 0$, and suppose we want to approximate $y(2)$.

You should recognize the solution of this IVP as $y = \ln t$, so we're really trying to approximate $\ln 2 = 0.69314718056\ldots$. (*Hint*: If you check this "exact" answer on your calculator or CAS, realize that these devices are using very sophisticated approximation methods themselves!)

If we take $h = 0.05$, we'll need 20 steps to stretch from $t = 1$ to $t = 2$. In our example, Euler's method gives us the formula

$$y_{k+1} = y_k + \frac{0.05}{t_k}$$

for $t_k = 1 + 0.05\, k (k = 0, \ldots, 20)$. Table 3.3 gives the results.

**Table 3.3** Euler's Method for $\frac{dy}{dt} = \frac{1}{t}$, $y(1) = 0$, with $h = 0.05$

| $k$ | $t_k$ | $y_k$ | True Value | Absolute Error |
|---|---|---|---|---|
| 0 | 1 | 0.000000 | 0.000000 | 0.00000 |
| 1 | 1.05 | 0.050000 | 0.048790 | 0.00121 |
| 2 | 1.1 | 0.097619 | 0.095310 | 0.00231 |
| 3 | 1.15 | 0.143074 | 0.139762 | 0.00331 |
| 4 | 1.2 | 0.186552 | 0.182322 | 0.00423 |
| 5 | 1.25 | 0.228219 | 0.223144 | 0.00507 |
| 6 | 1.3 | 0.268219 | 0.262364 | 0.00585 |
| 7 | 1.35 | 0.306680 | 0.300105 | 0.00658 |
| 8 | 1.4 | 0.343717 | 0.336472 | 0.00724 |
| 9 | 1.45 | 0.379431 | 0.371564 | 0.00787 |
| 10 | 1.5 | 0.413914 | 0.405465 | 0.00845 |
| 11 | 1.55 | 0.447247 | 0.438255 | 0.00899 |
| 12 | 1.6 | 0.479506 | 0.470004 | 0.00950 |
| 13 | 1.65 | 0.510756 | 0.500775 | 0.00998 |
| 14 | 1.7 | 0.541059 | 0.530628 | 0.01043 |
| 15 | 1.75 | 0.570470 | 0.559616 | 0.01085 |
| 16 | 1.8 | 0.599042 | 0.587787 | 0.01126 |
| 17 | 1.85 | 0.626820 | 0.615186 | 0.01163 |
| 18 | 1.9 | 0.653847 | 0.641854 | 0.01199 |
| 19 | 1.95 | 0.680162 | 0.667829 | 0.01233 |
| 20 | 2 | 0.705803 | 0.693147 | 0.01266 |

The solution curve $y = \ln t$ is concave down, so the approximating tangent lines all lie *above* the solution curve, leading to an approximation of ln 2 that's too large. Just as in Example 3.1.1, the errors increase with the value of $k$, but this time, if you subtract successive errors (corresponding to successive values of $k$), you'll see that the differences are *decreasing*. (To gain some insight into this phenomenon, compare $f(x,y)$ in Examples 3.1.1 and 3.1.2.)

Changing to $h = 0.025$ and $n = 40$ yields the approximate value 0.699436, whereas setting $h = 0.01$ and $n = 100$ gives us an approximation of 0.695653. Of course, technology (a spreadsheet) was used to obtain the last two approximations. ■

Next, we'll see what happens when we are given an equation whose solution we don't know.

### ■ Example 3.1.3 Euler's Method—Unknown Exact Solution

Suppose we're given the IVP $y' = \sqrt{x+y}$, $y(5) = 4$, and we want to find $y(4)$.

The first thought that should occur to us is that the equation is neither separable nor linear. Are we in trouble here? *No*, not if we understand Euler's method.

In our problem, $f(x,y) = \sqrt{x+y}$, so Formula (3.1.3) takes the form

$$y_{k+1} = y_k + h\sqrt{x_k + y_k},$$

where $x_k = 5 + kh$, $k = 0, 1, \ldots, n$. As usual, $n$ denotes the number of steps we choose.

Let's start off by choosing five steps to get us from the initial point $x = 5$ to our destination $x = 4$. Each step has to have length 0.2, and because we are moving *backward* from the initial point, we must take $h = -0.2$ in the formula. We'll carry out this first attempt at approximation by hand and then use a spreadsheet when the calculations become more numerous (and more tedious).

The formula gives us

$$y_1 = y_0 + h\sqrt{x_0 + y_0} = 4 + (-0.2)\sqrt{5 + 4} = 3.4$$

$$y_2 = y_1 + h\sqrt{x_1 + y_1} = 3.4 + (-0.2)\sqrt{4.8 + 3.4} = 2.82728716$$

$$y_3 = y_2 + h\sqrt{x_2 + y_2} = 2.82728716 + (-0.2)\sqrt{4.6 + 2.82728716} = 2.28222616$$

$$y_4 = y_3 + h\sqrt{x_3 + y_3} = 2.28222616 + (-0.2)\sqrt{4.4 + 2.28222616} = 1.76522612$$

$$y_5 = y_4 + h\sqrt{x_4 + y_4} = 1.76522612 + (-0.2)\sqrt{4.2 + 1.76522612} = 1.27674987.$$

These calculations tell us that $y(4) \approx 1.27674987$. The *true answer* is **1.34042895566892....**[1] Therefore, when we round the "true" answer to eight places, the absolute error is $|1.34042896 - 1.27674987| = 0.06367909$. If we choose $h = -0.01$ and use 100 steps, a spreadsheet calculation gives us an approximation of 1.337296, with an absolute error of 0.0031. ∎

So far we've been cheating a bit, discussing the numerical solutions of equations for which we could find an analytic solution (even if implicit). Knowing the exact solution enabled us to analyze the error—the gap between the true solution value and the approximate value of a solution at a point. However, it's time that we consider a more typical example.

## ■ Example 3.1.4 Euler's Method—A Completely Unknown Solution

The initial-value problem $\frac{dy}{dt} = y^2 - t^2, y(0) = \frac{1}{2}$, cannot be solved by any of the methods we have discussed so far, although this special type of *Riccati equation*[2] does have a *series solution* in terms of *Bessel functions* (see Section D.3). Nevertheless, we can approximate the solution at $t = 1$ (for example) so that it is accurate to, say, three decimal places.

"Without the exact answer as a guide, how do we know that these three decimal places are accurate?" you may be asking. Let's skip the detailed formula and see what happens for different step sizes (Table 3.4). We have rounded the approximations in the last column to six decimal places.

**Table 3.4** The IVP $\frac{dy}{dt} = y^2 - t^2, y(0) = \frac{1}{2}$: Approximate Values of $y(1)$ for Various Step Sizes

| Step Size | Number of Steps | Approximate Value |
|---|---|---|
| 1/100 | 100 | 0.512113 |
| 1/1000 | 1000 | 0.506106 |
| 1/2000 | 2000 | 0.505769 |
| 1/4000 | 4000 | 0.505600 |
| 1/8000 | 8000 | 0.505515 |
| 1/16,000 | 16,000 | 0.505473 |
| 1/20,000 | 20,000 | 0.505464 |

[1] This answer is obtained by making a substitution to transform the given equation into a separable equation (see the explanation that precedes Problems A12–A14 of Exercises 2.1), solving the equation to get an implicitly defined solution to the IVP, and then solving for the value of $y$ when $x = 4$. Solving the implicit relation for $y$ requires a calculator or CAS with a "solve" function for general equations. Even this "true" answer is only an approximation (although presumably a very accurate one) because the algebraic equation can't be solved exactly.
[2] See Chapter 2, Section 5 of *Ordinary Differential Equations: A Brief Eclectic Tour* by David Sánchez (MAA, 2002) for an informative discussion of this important class of differential equations.

We have reached a stage at which the first three digits of the approximate values do not seem to be changing. The last approximate value agrees with the previous one to three decimal places after appropriate rounding, so we can assume that the approximation is 0.505, accurate to three significant digits. The idea—a rule of thumb based on mathematical analysis—is to keep on using smaller step sizes until there are changes only *past* the decimal place in which we are interested. Then we can be sure of those decimal places that do *not* change. ■

## *3.1.1 Stiff Differential Equations

We can encounter difficulty in applying Euler's method (and some other methods) to approximate solutions when these solutions have components whose time scales differ widely. For instance, the solution of the circuit problem in Example 2.2.5 had two components: (1) a *transient term* of the form $Ce^{-at}$, with $C > 0$ and $a > 0$, that decreased rapidly to zero as $t \to \infty$ and (2) a *steady-state term* of the form $A \sin(\omega t) + B \cos(\omega t)$ that oscillated with time. Instead of approximating the steady-state part of the solution, Euler's method may allow the error associated with the transient part to dominate, producing meaningless results.

Equations exhibiting this characteristic behavior include many that arise in electrical circuit theory and in the study of chemical reactions. The term *stiff* is used because these numerical difficulties occur in analyzing the motion of spring-mass systems with large spring constants—that is, systems with "stiff" springs. (The stiffness of a spring depends on the materials of which it is made and on the specific manufacturing processes used.) In Chapters 4 and 5, we'll discuss spring-mass problems in greater detail.

For now, let's look at an example that highlights the difficulty.

### ■ Example 3.1.5 A Stiff Differential Equation

Suppose we look at the IVP $\frac{dI}{dt} + 50I = \sin(\pi t)$, $I(0) = 0$, which is just the equation in Example 2.2.5 with $L = 1$, $R = 50$, $v_0 = 1$, and $\omega = \pi$. According to that example, the solution is

$$I(t) = \frac{1}{(2500 + \pi^2)} \left\{ 50 \sin(\pi t) - \pi \cos(\pi t) + \pi e^{-50t} \right\}.$$

(*Check this for yourself.*) Note both the transient component and the steady-state part.

Now suppose that we want to approximate the solution at $t = 2$. To understand the accuracy of the approximation, we can first use the solution formula to find the exact answer,

$$I(2) = \frac{-\pi}{2500 + \pi^2} \left( 1 - \frac{1}{e^{100}} \right) = -0.001251695566\ldots.$$

For further comparison with approximations, here are some actual values of $I$ at intermediate points between 0 and 2:

$$I(0.5) = 0.01992135365\ldots \qquad\qquad I(1.0) = 0.001251695566\ldots$$

$$I(1.5) = -0.1992135365\ldots.$$

---

* Denotes an optional section.

**Table 3.5a** $h = 0.1$

| $k$ | $t_k$ | $i_k$ |
|---|---|---|
| 0 | 0 | 0 |
| 5 | 0.5 | −1.26575 |
| 10 | 1.0 | 1316.73504 |
| 15 | 1.5 | $-0.13483 \times 10^7$ |
| 20 | 2.0 | $0.13807 \times 10^{10}$ |

**Table 3.5b** $h = 0.05$

| $k$ | $t_k$ | $i_k$ |
|---|---|---|
| 0 | 0 | 0 |
| 10 | 0.5 | 0.09262 |
| 20 | 1.0 | 4.18748 |
| 30 | 1.5 | 241.37834 |
| 40 | 2.0 | 13,920.24471 |

**Table 3.5c** $h = 0.01$

| $k$ | $t_k$ | $i_k$ |
|---|---|---|
| 0 | 0 | 0 |
| 50 | 0.5 | 0.01994 |
| 100 | 1.0 | 0.00125396 |
| 150 | 1.5 | −0.01994 |
| 200 | 2.0 | −0.00125396 |

Euler's method yields the formula $i_{k+1} = i_k + h(\sin(\pi t_k) - 50i_k)$. Table 3.5a displays the results of using Euler's method with $h = 0.1$, Table 3.5b shows the results when $h = 0.05$, and Table 3.5c shows what happens when $h = 0.01$. We omit the error column in each table because the discrepancies or agreements between actual values of $I$ and approximate values are fairly obvious in each table.

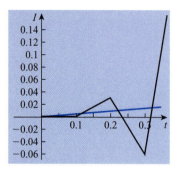

**FIGURE 3.8a**

$h = 0.1$

**FIGURE 3.8b**

$h = 0.05$

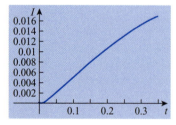

**FIGURE 3.8c**

$h = 0.01$

You can see that the errors in approximating $I(2)$ are horribly large for $h = 0.1$ and $h = 0.05$, whereas there is very little error when we have reduced $h$ to 0.01. Comparing the graph of the actual solution curve for $I(t)$ with the graphs of the approximation curves given by Euler's method for these three values of $h$ is a real eye-opener (see Figures 3.8a, 3.8b, and 3.8c). In each graph, the blue line is the actual solution curve, and the black line is the approximation. Note that the scales are different from graph to graph.

Figure 3.8c shows that you can hardly distinguish between the actual solution curve and its Euler method approximation when $h = 0.01$. The choice of the interval [0, 0.35] for $t$ was made after some experimentation. Using the technology available to you, you should look at the graphs of the approximations on larger intervals. For larger values of $t$, beginning around $t = 1$, you will find that the approximation curves are rather alarming distortions of the steady-state solution. ∎

See Section A.3 for a further discussion of approximation error. In subsequent sections of this chapter, we will investigate improved algorithms.

## Exercises 3.1

In the following problems, being asked to do a problem "by hand" or "manually" means that each step should be written out and that, although calculators may be used to do arithmetic, no calculator routine or CAS program for Euler's method should be used. This is the opposite of being allowed to "use technology."

### A

For each of Problems 1–3, use Euler's method by hand with the given step sizes to approximate the solution to the given initial value problem over the specified interval. Include a table of values, and give a sketch of the approximate solution by plotting the values you have calculated.

1. $\dfrac{dy}{dt} = t^2 - y^2, y(0) = 1; 0 \leq t \leq 1, h = 0.25$

2. $\dfrac{dy}{dt} = e^{(2/y)}, y(0) = 2; 0 \leq t \leq 2, h = 0.5$

3. $\dfrac{dy}{dt} = e^{(2/y)}, y(1) = 2; 1 \leq t \leq 3, h = 0.5$

4. Compare your answers to Problems 2 and 3 and explain what you see.

5. If $y$ is the solution of the IVP $\dfrac{dy}{dt} = \cos t, y(0) = 0$, use Euler's method manually with $h = \pi/10$ to approximate $y(\pi/2)$. What is the absolute error?

6. Approximate $y(1.4)$ by hand if $y$ is the solution to the IVP $\dfrac{dy}{dx} = x^3, y(1) = 1$. Use $h = 0.1$.

7. Given $\dfrac{dy}{dx} = \dfrac{x}{y}, y(0) = 1$, use $h = 0.1$ to approximate $y(1)$ manually.

8. Given the IVP $y' = y \sin 3t, y(0) = 1$, use technology to approximate $y(4)$ using 20 steps.

9. Given $y' = 1/(1 + x^2), y(0) = 0$, use $h = 0.1$ to approximate $y(1)$ manually. How can you use your result to compute $\pi$?

10. Consider the IVP $y' = x^2 + y, y(0) = 1$. By hand, approximate $y(0.1), y(0.2)$, and $y(0.3)$ using both $h = 0.1$ and $h = 0.05$ for each approximation.

11. Consider the IVP $y' = y^2, y(0) = 1$.

   a. Using $h = 0.2$, approximate the solution $y$ over the interval $[0, 1.2]$ by hand.
   b. Show that the exact solution is given by $y = \dfrac{1}{1-t}$.
   c. Compare the values found in part (a) with values given by the formula in part (b). Explain any strange numerical behavior. [*Hint*: A slope field or solution graph may help.]

**B**

1. In Problem B1 of Exercises 2.3, you were given the following model for the population of Botswana: $\frac{dP}{dt} = 0.0355\,P - 0.00160625\,t$, with $P(0) = 1.285$ (million). The value $t = 0$ corresponds to 1990.

    a. Use technology and Euler's method with $h = 0.01$ to approximate $P(1)$, the population in 1991.
    b. Using the approximation for $P(1)$ found in part (a) as your starting point and $h = -0.01$, approximate $P(0)$.

2. In the area of pharmacokinetics, the *Michaelis-Menten equation* $\frac{dx}{dt} = \frac{-Kx}{A+x}$ describes the rate at which a body processes a drug. Here $x(t)$ is the concentration of the drug in the body at time $t$, and $K$ and $A$ are positive constants. [The equation was developed by the biochemical/medical researchers Leonor Michaelis (1875–1949) and Maud Menten (1879–1960).]

    a. For a particular controlled substance, let $A = 6$, $K = 1$, and $x(0) = 0.0025$. Use technology and Euler's method with $h = 0.1$ to evaluate $x$ for $t = 1, 2, 3, 10$, and 20. Estimate how long it takes for the concentration to be half of its initial value.
    b. For alcohol, let $A = 0.005$, $K = 1$, and $x(0) = 0.025$. Use technology and Euler's method with $h = 0.01$ to evaluate $x$ for $t = 0.01, 0.02, 0.03, 0.04$, and 0.05. Estimate how long it takes for the concentration to be half its initial value.

3. In modeling aircraft speed and altitude loss in a pull-up from a dive, basic laws of physics yield the differential equation

$$\frac{dV}{d\theta} = \frac{-gV \sin\theta}{kV^2 - g \cos\theta},$$

where $\theta$ denotes the dive angle (in radians), $V = V(\theta)$ is the speed of the plane, $g = 9.8\,\text{m/s}^2$ is the acceleration constant, and $k$ is a constant related to the wing surface area. For a particular plane, $k = 0.00145$, $\theta_0 = -0.786$, and $V(\theta_0) = V_0 = 150\,\text{m/s}$. Use $h = 0.006$ (which divides $\theta_0$ evenly) and $n = 131$ to estimate $V(0)$, the plane's speed at the completion of its pull-up—that is, when it levels out to $\theta = 0$. (Of course, use technology!)

4. Consider the IVP $y' = 1 - t + 4y$, $y(0) = 1$. Using technology and $h = 0.1$, approximate the solution on the interval $0 \le t \le 1$. What error is made at $t = 1/2$ and $t = 1$?

5. Use Euler's method manually with both $h = 0.5$ and $h = 0.25$ to approximate $x(2)$, where $x(t)$ is the solution of the IVP $\frac{dx}{dt} = \frac{3t^2}{2x}$, $x(0) = 1$. Solve the equation exactly and compare the absolute errors you get with the different values of $h$.

6. Consider the IVP $y' = y(1 - y^2)$, $y(0) = 0.1$. Note that the equation has three equilibrium solutions.

    a. Use a phase portrait analysis or a direction field to predict what *should* happen to the solution.
    b. Use technology and Euler's method with $h = 0.1$ to step out to $x = 3$. What happens to the numerical solution?

7. Suppose $x' = x^3$.

    a. Find an expression for $x''$ in terms of $x$, assuming that $x$ is a function of $t$.
    b. Suppose $x(0) = 1$. Is the solution curve concave up or concave down? Use the result in part (a) to justify your answer.

    **c.** Does Euler's method overestimate or underestimate the true value of the solution at $t = 0.1$? Explain. (Don't actually carry out Euler's method.)

**8.** Consider the IVP $y' = y^\alpha, \alpha < 1, y(0) = 0$.

    **a.** Find the *exact* solution of the IVP.

    **b.** Show that Euler's method fails to determine an approximate solution to the IVP.

    **c.** Show what happens if the initial condition is changed to $y(0) = 0.01$.

**9. a.** Sketch the direction field for $y' = \sqrt{1 - y^2}$.

    **b.** Verify that a solution for this equation satisfying the initial condition $y(0) = 0$ is given by

$$y = \begin{cases} \sin t & 0 \le t < \frac{1}{2}\pi \\ 1 & \frac{1}{2}\pi \le t \end{cases}.$$

    **c.** Describe the behavior of Euler's method when $h = 0.4$. Could you have predicted this behavior without any calculations?

## C

**1.** Describe a class of differential equations for which Euler's method gives a *completely accurate* numerical solution—that is, for which $y_k$ exactly equals the true solution $\varphi(x_k)$ for every $k$. [*Hint*: Try to think of differential equations for which all solution curves coincide with the tangent line segments.]

**2.** Consider the stiff differential equation $\frac{dy}{dt} = -100y + 1$, with $y(0) = 1$.

    **a.** Solve this IVP and calculate the exact value of $y(1)$.

    **b.** Use technology and Euler's method to approximate $y(1)$ with $h = 0.1, 0.05$, and $0.01$.

    **c.** Use technology to plot the exact solution and an approximate solution of the equation over the interval $[0, 0.03]$ on the same set of axes. Do this for each of the three values of $h$ mentioned in this problem.

**3.** The equation $y' = -50(y - \cos x)$ is stiff.

    **a.** Use software to solve the equation with the initial condition $y(0) = 0$. Then calculate the exact value $y(0.2)$.

    **b.** Use Euler's method and technology to approximate $y(0.2)$ with step size $h = 1.974/50$. What is the absolute error?

    **c.** Use Euler's method and technology to approximate $y(0.2)$ with step size $h = 1.875/50$. What is the absolute error?

    **d.** Use Euler's method and technology to approximate $y(0.2)$ with step size $h = 2.1/50$. What is the absolute error now?

    **e.** Using technology, plot the three approximation curves found in parts (b), (c), and (d) on the same axes. Use the interval $[0, 1]$. Would you call the Euler method solution of the equation "sensitive to step size"?

4. The second-order IVP $y'' = F(x, y, y')$, $y(a) = c_1$, $y'(a) = c_2$ may be written as two simultaneous first-order equations: $y' = u$, $u' = F(x, y, u)$, where $y(a) = c_1$, $u(a) = c_2$.

   a. Devise a procedure for approximating $y$ and $y'$ when $x = a + h$.
   b. Use the method found in part (a) to approximate the solution of the IVP
      $y'' = x + y$, $y(0) = y'(0) = 0$ at $x = 1$.
   c. Given that the exact solution of the IVP in part (b) is $y = \frac{1}{2}e^x - \frac{1}{2}e^{-x} - x$, compare the approximate value of $x(1)$ found in part (b) to the exact value.

## 3.2 THE IMPROVED EULER METHOD

In Euler's original method, the slope $f(x, y)$ over any interval $x_k \leq x \leq x_{k+1}$ of length $h$ is replaced by $f(x_k, y_k)$, so that $x$ always takes the value of the left endpoint of the interval. (As noted just before Example 3.1.1, if $y' = f(x)$, a function of $x$ alone, then Euler's method is equivalent to using a left-hand Riemann sum to approximate a definite integral.)

Now instead of always using the slope at the *left* endpoint of the interval $[x_k, x_{k+1}]$, we can think of using an *average* derivative value over the interval. The **improved Euler method** involves two stages that will be combined into one approximation formula. The first stage involves moving tentatively across the interval $[x_k, x_{k+1}]$ using Euler's original method, thereby producing a guess, or trial value, $\hat{y}_{k+1} = y_k + h \cdot f(x_k, y_k)$. Note that the values $f(x_k, y_k)$ and $f(x_{k+1}, \hat{y}_{k+1})$ approximate the slopes of the solution curve at $(x_k, y(x_k))$ and $(x_{k+1}, y(x_{k+1}))$, respectively. Now the second stage looks at the *average* of the derivative $f(x_k, y_k)$ and the guess $f(x_{k+1}, \hat{y}_{k+1}) = f(x_{k+1}, y_k + hf(x_k, y_k))$ and uses this average to take the *real* step across the interval.

**Guess (tentative step):** $\hat{y}_{k+1} = y_k + h \cdot f(x_k, y_k)$

$$\textbf{Real step: } y_{k+1} = y_k + h \left\{ \frac{f(x_k, y_k) + f(x_{k+1}, \hat{y}_{k+1})}{2} \right\}$$

$$= y_k + h \left\{ \frac{f(x_k, y_k) + f(x_{k+1}, y_k + h \cdot f(x_k, y_k))}{2} \right\} \qquad (3.2.1)$$

$$= y_k + \frac{h}{2} \left\{ f(x_k, y_k) + f(x_{k+1}, y_k + h \cdot f(x_k, y_k)) \right\}.$$

Formula (3.2.1) describes the **improved Euler method** [or **Heun's method**, named for Karl Heun (1859–1929), a German applied mathematician who devised this scheme around 1900]. It is an example of a **predictor-corrector method:** We use $\hat{y}_{k+1}$ (via Euler's method) to *predict* a value of $y(x_{k+1})$ and then use $y_{k+1}$ to *correct* this value by averaging.

Look carefully at Equation (3.2.1). If $f(x, y)$ is really just $f(x)$, a function of $x$ alone, then solving the IVP $y' = f(x, y)$, $y(x_0) = x_0$, amounts to solving the equation $y' = f(x)$, which is a matter of simple integration. In Section 1.3 [Equation (1.3.1)] we saw that we can write the solution as

$$y(x) = \int_{x_0}^{x} f(t)dt + y_0 = \sum_{k=0}^{n-1} \int_{x_k}^{x_{k+1}} f(t)dt + y_0,$$

where $x_n = x$. In this case, Formula (3.2.1) reduces to

$$y_{k+1} = y_k + \frac{h}{2} \{f(x_k) + f(x_{k+1})\}$$

and the Fundamental Theorem of Calculus tells us that on the interval $[x_k, x_{k+1}]$,

$$\int_{x_k}^{x_{k+1}} \overbrace{f(t)}^{y'} dt = y(x_{k+1}) - y(x_k) \approx y_{k+1} - y_k = \frac{h}{2} \{f(x_k) + f(x_{k+1})\}.$$

In other words, in this situation we are using the *Trapezoid Rule* from calculus to approximate each integral on $[x_k, x_{k+1}]$.

Next, we see some illustrations of why this method is called "improved."

### ■ Example 3.2.1 The Improved Euler Method

Let's use the improved Euler formula, Equation (3.2.1), to calculate an approximate value of the solution of the IVP $y' = y$, $y(0) = 1$, at $x = 1$. Of course, you realize that this is just a roundabout way of asking for an approximation of that important mathematical constant $e$. (*Right?*)

Let's start with $h = 0.1$, so we'll need 10 steps to reach $x = 1$ from the initial point $x = 0$. Thus, in Formula (3.2.1) we have $x_0 = 0$, $y_0 = 1$, $h = 0.1$, $x_k = 0 + kh = kh$ ($k = 0, \ldots, 10$), and $f(x, y) = y$. When we put all this information together, we see that the formula takes a simplified form:

$$y_{k+1} = y_k + \frac{h}{2} \{y_k + (y_k + hy_k)\}$$

$$= y_k + \frac{h}{2} \{(2 + h)y_k\} = y_k + (0.05)(2.1)y_k = 1.105 \, y_k.$$

Therefore, the calculations are

$$y_1 = 1.105y_0 = 1.105(1) = 1.105$$
$$y_2 = 1.105y_1 = (1.105)^2 = 1.221025$$
$$y_3 = 1.105y_2 = (1.105)^3 = 1.349232625$$
$$\vdots \qquad\qquad \vdots$$
$$y_{10} = 1.105y_9 = (1.105)^{10} = 2.71408084661.$$

Comparing this approximation to the actual value 2.71828182846 (rounded to 11 decimal places), we find that the absolute error is 0.00420098185. (In Problem A4 of Exercises 3.2, you'll be asked to try this with the original Euler method.)

Using 20 steps, a CAS gives the approximate value 2.71719105435, so the absolute error is now 0.00109077410. Note that when we doubled the number of steps from 10 to 20, the result was that the absolute error was roughly one-fourth what it was before.    ■

Now let's revisit Example 3.1.1 to see how the improved method compares with the original process of approximation.

### ■ Example 3.2.2 The Improved Euler Method—Example 3.1.1 Revisited

We want to approximate $x(1.5)$, given the IVP $\frac{dx}{dt} = t^2 + x$, $x(1) = 3$. The actual value is 5.939770.... We'll start with $h = 0.1$, so we'll need five steps to stretch between $t = 1$ and $t = 1.5$.

For this problem, the improved Euler formula is

$$x_{k+1} = x_k + \frac{h}{2}\left\{(t_k^2 + x_k) + t_{k+1}^2 + x_k + h\left(t_k^2 + x_k\right)\right\}$$

$$= x_k + \frac{h}{2}\left\{t_{k+1}^2 + (1+h)t_k^2 + (2+h)x_k\right\}$$

$$= x_k + (0.05)\left\{t_{k+1}^2 + 1.1t_k^2 + 2.1x_k\right\},$$

where $t_0 = 1$, $t_1 = 1.1$, $t_2 = 1.2$, $t_3 = 1.3$, $t_4 = 1.4$, and $t_5 = 1.5$. Therefore,

$$x_1 = 3 + (0.05)\{(1.1)^2 + 1.1(1)^2 + 2.1(3)\} = 3.4305$$

$$x_2 = 3.4305 + (0.05)\{(1.2)^2 + 1.1(1.1)^2 + 2.1(3.4305)\} = 3.9292525$$

$$x_3 = 3.9292525 + (0.05)\{(1.3)^2 + 1.1(1.2)^2 + 2.1(3.9292525)\}$$

$$= 4.5055240125$$

$$x_4 = 4.5055240125 + (0.05)\{(1.4)^2 + 1.1(1.3)^2 + 2.1(4.5055240125)\}$$

$$= 5.16955403381$$

$$x_5 = 5.16955403381 + (0.05)\{(1.5)^2 + 1.1(1.4)^2 + 2.1(5.16955403381)\}$$

$$= 5.93265720736.$$

To five decimal places, we have $x(1.5) \approx 5.93266$. The absolute error is 0.00711. When we employed Euler's method in Example 3.1.1, the error was 0.244639.

If we use 10 steps in the improved Euler method, then we get $x(1.5) \approx 5.943455$, with absolute error 0.00369, compared to the Euler method's error of 0.12717.    ■

Now let's go back and redo another earlier example with the new method.

**Table 3.6** Improved Euler Method with $h = -0.2$

| $k$ | $x_k$ | $y_k$ | True Value | Absolute Error |
|---|---|---|---|---|
| 0 | 5.0 | 4.000000 | 4.000000 | 0.000000 |
| 1 | 4.8 | 3.413644 | 3.413384 | 0.000260 |
| 2 | 4.6 | 2.854277 | 2.853750 | 0.000527 |
| 3 | 4.4 | 2.322249 | 2.321444 | 0.000805 |
| 4 | 4.2 | 1.817952 | 1.816857 | 0.001100 |
| 5 | 4.0 | 1.341827 | 1.34043 | 0.00140 |

### ■ Example 3.2.3 Improved Euler Method—Example 3.1.3 Revisited

In Example 3.1.3 we discussed the IVP $y' = \sqrt{x + y}$, $y(5) = 4$, with the goal of approximating $y(4)$. If we apply the improved method to the problem, with five backward steps, each of length 0.2—that is, with $h = -0.2$—we get the values shown in Table 3.6.

Thus, $y(4) \approx 1.341827$ by the improved method, compared to the "true" answer 1.34042895566892 and the original Euler method approximate value 1.27674987.    ■

An analysis of error shows that the *local* truncation error at any stage of the improved Euler method behaves like a constant multiple of $h^3$ and that *the cumulative truncation error is no greater than a constant multiple of the square of the step size h*: |true value − approximation| ≤ $K \cdot h^2$, where $K$ is a constant that depends on the function $f(x, y)$, on its partial derivatives, and on the interval involved but does not depend on $h$. We say that the improved Euler method is a **second-order method**.

In the next section, we'll look at a fourth-order method and a powerful combination of fourth- and fifth-order techniques.

### Exercises 3.2

**A**

Use the following table to enter the data from Problems 1 and 2.

| | TRUE VALUE | Euler's Method | Absolute Error | Improved Euler Method | Absolute Error |
|---|---|---|---|---|---|
| $h = 0.1$ | | | | | |
| $h = 0.05$ | | | | | |
| $h = 0.025$ | | | | | |

1. Use the improved Euler method to redo Example 3.1.1 with $h = 0.1, 0.05$, and $0.025$.
2. Use the improved Euler method to redo Example 3.1.2 with $h = 0.1, 0.05$, and $0.025$. (You'll also have to use Euler's method for $h = 0.1$.)
3. **a.** Find the exact solution to the IVP $\frac{dx}{dt} = t + x, x(0) = 1$.
   **b.** Apply the improved Euler method with step size $h = 0.1$ to approximate the value $x(1)$.
   **c.** Calculate the absolute error at each step of part (b).
4. Using technology, redo Example 3.2.1 with both Euler's method and the improved Euler method, using a step size of $h = 0.01$—that is, using 100 steps. For each method, calculate the absolute errors incurred in approximating $y(0.01), y(0.02), \ldots, y(0.99), y(1.0)$. (A spreadsheet program can be particularly useful here.)

**B**

1. Redo Problem B1 in Exercises 3.1 using the improved Euler method.
2. Redo Problem B3 in Exercises 3.1 using the improved Euler method.
3. Redo Problem B8 in Exercises 3.1 using the improved Euler method.

**C**

1. Redo Problem C2 in Exercises 3.1 using the improved Euler method.

## 3.3 MORE SOPHISTICATED NUMERICAL METHODS: RUNGE-KUTTA AND OTHERS

Modern computers (and even hand-held calculators) have many algorithms for solving differential equations numerically. Some of these are highly specialized and are meant to handle very particular types of ODEs (such as stiff equations—see Section 3.1) and systems of ODEs. Euler's method and its improved version are useful for illustrating the idea behind numerical approximation, but they are not very efficient in terms of approximating a solution of an IVP very accurately and with a minimum number of steps.

A very good method, implemented in many computer algebra systems and in calculator firmware, is the **fourth-order Runge-Kutta method (RK4)**, which was developed in an 1895 paper by Carl Runge (1856–1927), a German applied mathematician, and was generalized to *systems* of ODEs in 1901 by M. Wilhelm Kutta (1867–1944), a German mathematician and aerodynamicist. As the description indicates, in this method the total accumulated error is proportional to $h^4$, so reducing the step size by a factor of $\frac{1}{10}$ produces four more digits of accuracy—for example, reducing the step size from $h = 0.1$ to $h = 0.01$ generally decreases the total error by a factor of $0.0001$. (The *local* truncation error behaves like $h^5$.) There are also second- and third-order Runge-Kutta methods. (Euler's method can be called a first-order Runge-Kutta method.)

Now suppose we have an IVP $y' = f(x, y), y(x_0) = y_0$. The RK4 formula is a bit strange-looking, but not if we realize that it is approximating the value $y(x_{k+1})$ by a *weighted average*, $y_{k+1}$, of values of $f(x, y)$ calculated at different points in the interval $[x_k, x_{k+1}]$. For each interval $[x_k, x_{k+1}]$, we calculate the following slopes in the order given:

$$m_1 = f(x_k, y_k)$$

$$m_2 = f\left(x_k + \frac{h}{2}, y_k + \frac{h}{2}m_1\right)$$

$$m_3 = f\left(x_k + \frac{h}{2}, y_k + \frac{h}{2}m_2\right)$$

$$m_4 = f(x_k + h, y_k + hm_3) = f(x_{k+1}, y_k + hm_3). \tag{3.3.1}$$

Then the classical fourth-order Runge-Kutta formula is

$$y_{k+1} = y_k + \frac{h}{6}(m_1 + 2m_2 + 2m_3 + m_4), \tag{3.3.2}$$

where the sum $(m_1 + 2m_2 + 2m_3 + m_4)/6$ is a weighted average of slopes.

The value $m_1$ is the slope at $x_k$ calculated by Euler's method. Then $m_2$ is an estimate of the slope at the midpoint of the interval $[x_k, x_{k+1}]$, where Euler's method has been used to estimate the $y$ value there. Now $m_3$ is a value for the slope at the midpoint of $[x_k, x_{k+1}]$ using the improved Euler's method. Finally, $m_4$ is the slope at $x_{k+1}$ calculated by Euler's method, using the improved slope $m_3$ at the midpoint to step to $x_{k+1}$.

Perhaps this formula won't be so alarming if we look at the simplified situation when $f(x, y)$ is independent of $y$ in the equation $y' = f(x, y)$. If $f(x, y) = g(x)$, then the Formula (3.3.1) for $m_1, m_2, m_3,$ and $m_4$ reduce to

$$m_1 = g(x_k)$$

$$m_2 = g\left(x_k + \frac{h}{2}\right)$$

$$m_3 = g\left(x_k + \frac{h}{2}\right)$$

$$m_4 = g(x_k + h) = g(x_{k+1})$$

so Formula (3.3.2) becomes

$$y_{k+1} = y_k + \frac{h}{6}\left\{g(x_k) + 2g\left(x_k + \frac{h}{2}\right) + 2g\left(x_k + \frac{h}{2}\right) + g(x_{k+1})\right\}$$

$$= y_k + \frac{h}{6}\left\{g(x_k) + 4g\left(x_k + \frac{h}{2}\right) + g(x_{k+1})\right\}$$

and you may recognize the expression $\frac{h}{6}\left\{g(x_k) + 4g\left(x_k + \frac{h}{2}\right) + g(x_{k+1})\right\}$ as a form of *Simpson's Rule* for approximating $\int_{x_k}^{x_{k+1}} g(x)dx$. (Note that $x_k + \frac{h}{2}$ in the expression is the *midpoint* of the interval $[x_k, x_{k+1}]$ because $h = x_{k+1} - x_k$.)

To get a feel for the calculations, let's choose an example that we've seen before.

### ■ Example 3.3.1 RK4—Example 3.1.1 Yet Again

Let's approximate $x(1.5)$ by the Runge-Kutta method if we are given the IVP $\frac{dx}{dt} = t^2 + x$, $x(1) = 3$. We'll use $h = 0.1$, so we need five steps.

Just to get the idea, let's focus on the interval $[t_0, t_1] = [1, 1.1]$. We calculate

$$m_1 = f(t_0, x_0) = f(1, 3) = (1^2 + 3) = 4$$

$$m_2 = f\left(t_0 + \frac{h}{2}, x_0 + \frac{h}{2}m_1\right) = f(1 + 0.05, 3 + 0.05(4))$$

$$= 1.05^2 + 3.2 = 4.3025$$

$$m_3 = f\left(t_0 + \frac{h}{2}, x_0 + \frac{h}{2}m_2\right) = f(1 + 0.05, 3 + 0.4317625)$$

$$= 1.05^2 + 3.215125 = 4.317625$$

$$m_4 = f(t_0 + h, x_0 + hm_3) = f(t_1, x_0 + hm_3) = f(1.1, 3 + 0.4317625)$$

$$= 1.1^2 + 3.4317625 = 4.6417625$$

so

$$x(1.1) \approx x_1 = 3 + \frac{0.1}{6}(4 + 2(4.3025) + 2(4.317625) + 4.6417625)$$

$$= 3 + \frac{0.1}{6}(25.8820125) = 3.431366875.$$

The actual value of $x(1.1)$ (from the solution formula $x(t) = -t^2 - 2t - 2 + 8e^{t-1}$) is $3.4313673446\ldots$ . Here, the absolute error is $0.0000004696$. (*This is an amazingly close approximation!*)

Table 3.7 shows, for the same value $h = 0.1$, the exact values and the approximate values given for this problem by Euler's method, the improved Euler method, and the Runge-Kutta method. We can see how accurate the Runge-Kutta method is at each step. ■

As accurate as the classical Runge-Kutta method is, improvements are possible. For example, a very popular method, the **Runge-Kutta-Fehlberg algorithm**, combines fourth-order and fifth-order methods in a clever way announced by E. Fehlberg in 1969. The **rkf45** method, as its computer implementation is known, uses *variable* step sizes, choosing the step size at each stage to try to achieve a predetermined degree of accuracy. Such clever numerical techniques

**Table 3.7** Comparison of Methods with $h = 0.1$

| $t_k$ | True Value of $x(t_k)$ | Euler's Method | Improved Euler Method | Runge-Kutta Method |
|---|---|---|---|---|
| 1 | 3.00000 | 3.00000 | 3.00000 | 3.00000 |
| 1.1 | 3.43137 | 3.40000 | 3.43050 | 3.43137 |
| 1.2 | 3.93122 | 3.86100 | 3.92925 | 3.93122 |
| 1.3 | 4.50887 | 4.39110 | 4.50552 | 4.50887 |
| 1.4 | 5.17460 | 4.99921 | 5.16955 | 5.17460 |
| 1.5 | 5.93977 | 5.69513 | 5.93266 | 5.93977 |

are called **adaptive methods**. They may be useful, for example, in handling stiff differential equations (Section 3.1).

A more dramatic problem that can be handled by adaptive methods is that of calculating a flight path from the earth to the moon and back,[3] which involves solving a system of differential equations numerically. In deep space, the driving force (the gravitational potential gradients of earth, Venus, and the sun) varies slowly, so a relatively large step size can be used in solving the system. However, near the earth or the moon a relatively small step size is needed to achieve the same accuracy. (If this small step size is used for the entire flight, the calculation will be unnecessarily long; but if the spacecraft gets too close to the earth and the step size is too large, the calculation results in giving the craft too great a kinetic energy and the craft zips out of the earth-moon system at the speed of light.)

## Exercises 3.3

In the problems that follow, it is assumed that you have versions of the Runge-Kutta Fourth-Order method (RK4) and Runge-Kutta-Fehlberg (rkf45) method available to you. Use the following table to enter the data from Problems 1 and 2. You may go back to earlier examples to find needed values.

**A**

| | TRUE VALUE | Euler's Method | Improved Euler Method | RK4 Method |
|---|---|---|---|---|
| $h = 0.1$ | | | | |
| $h = 0.05$ | | | | |
| $h = 0.025$ | | | | |

---

[3] The astronauts of Apollo 13, whose engine failed on the flight to the moon, had to follow such a flight path. The spacecraft had to be swung around by the gravity of the moon and return to the vicinity of the earth without further thrust by its rocket motors.

1. Use the RK4 method to redo Example 3.2.1 with $h = 0.1, 0.05$, and $0.025$. (Using the improved Euler method, the cases $h = 0.1$ and $h = 0.05$ have been done for you in the example.)
2. Use the RK4 method to redo Example 3.2.2 with $h = 0.1, 0.05$, and $0.025$.
3. Use the rkf45 method to approximate the solution of $y' = y, y(0) = 1$, at $t = 1$, with $h = 0.1$. (That is, approximate the value of the constant $e$. See Example 3.3.1.)
4. **a.** Find the exact solution of the IVP $\frac{dx}{dt} = t + x, x(0) = 1$.
   **b.** Apply the rkf45 method with step size $h = 0.1$ to approximate $x(1)$, calculating the absolute error at each step.
5. **a.** Find the closed-form solution of the equation $\frac{dx}{dt} = -tx^2$.
   **b.** Using the rkf45 method with $h = 0.1$, approximate the value $x(1)$ if $x$ is the solution of the IVP $\frac{dx}{dt} = -tx^2, x(0) = 2$.
6. Approximate $y(0.8)$ using the rkf45 method with $h = 0.01$ if $y$ is the solution of the IVP $\frac{dy}{dx} = \sin(xy)$, $y(0) = 0$.

## B

1. A daredevil named Patrice goes skydiving, jumping from a plane at an initial altitude of 10,000 feet. At time $t$ her velocity $v(t)$ satisfies the initial value problem $\frac{dv}{dt} = f(v), v(0) = 0$, where

$$f(v) = 32 - (0.000025) \cdot (100v + 10v^2 + v^3).$$

If she does not open her parachute, she will reach a *terminal velocity* when the forces of gravity and air resistance balance.

   **a.** Use the rkf45 method to approximate her velocity at times $t = 5, 10, 15, 16, 17, 18, 19$, and $20$, and so guess at her terminal velocity (accurate to three decimal places).
   **b.** Use technology to graph Patrice's velocity over the interval $[0, 30]$.

2. In 1927, British scientists Kermack and McKendrick laid the foundations for the theory of epidemiology by presenting data on the number of deaths resulting from a rat-spread plague in Bombay (now Mumbai) during the period December 1905–July 1906. They gave the following equation for the total number of deaths as of week $t$:

$$\frac{dR}{dt} = 890 \operatorname{sech}^2(0.20t - 3.4),$$

where **sech** $t$ denotes the *hyperbolic secant function*.

   **a.** Assuming that $R(0) = 0$, use the rkf45 method to fill in the following table.

| $t$ (weeks) | Actual Deaths | Predicted Deaths |
|---|---|---|
| 1 | 4 | |
| 5 | 68 | |
| 10 | 425 | |
| 20 | 6339 | |
| 30 | 9010 | |

**b.** According to the Kermack-McKendrick model, what is the asymptotic value of $R$—that is, $\lim_{t \to \infty} R(t)$?

**3.** In using Runge-Kutta methods (including Euler's method and Heun's method), it is important to realize that the error depends on the *form* of the equation as well as on the solution itself. To see an example of this, note that $y(x) = (x + 1)^2$ is the solution of each of the two problems

$$y' = 2(x + 1), \quad y(0) = 1$$
$$y' = 2y/(x + 1), \quad y(0) = 1.$$

**a.** Show that Heun's method is exact for the first equation.
**b.** Show that the method is *not* exact when applied to the second equation, even though it has the same solution as the first equation.

**4.** Consider the IVP $\frac{dx}{dt} = x^2, x(0) = 2$.

**a.** Use Euler's method with $h = 0.1$ to approximate $x(1)$. Does your answer seem strange?
**b.** Use the rkf45 method with $h = 0.1$ to approximate $x(1)$. Compare your answer to the answer in part (a)—if your calculator or CAS gives you a meaningful answer in both cases.
**c.** To help explain your difficulties in parts (a) and (b), find the closed-form solution of the IVP.
**d.** Use your answer to part (c) to explain why your answers to parts (a) and (b) are both wrong.
**e.** How do you think you may be able to avoid the difficulty uncovered in part (d)? Maybe by changing step size? Try to solve the problem again using the rkf45 method.

**C**

**1.** Consider the generalized logistic equation

$$\frac{dP}{dt} = kP^{\alpha}\left(1 - \frac{P^{\beta}}{M}\right).$$

**a.** Let $k = 1, M = 5$, and $P(0) = 1$. Find numerical approximations to the solution in the range $0 \le t \le 10$ for the parameter pairs $(\alpha, \beta) = (0.5, 1), (0.5, 2), (1.5, 1), (1.5, 2), (2, 2)$.
**b.** Estimate a parameter pair $(r, q)$ that yields approximately the values $P(0) = 1, P(2) = 2.4$, $P(4) = 2.9$.

## SUMMARY

Even if we can solve a first-order differential equation, we may not be able to find a closed-form solution. This difficulty has led to the development of numerical methods to *approximate* a solution to any degree of accuracy. Leaving aside *input error*, there are two main sources of error in numerical calculations done by hand, calculator, or computer: *round-off error* and *truncation error*. **Round-off error** is the kind of inaccuracy we get by taking a certain number of decimal places instead of taking the entire number. In particular, remember that our calculator or computer is limited in the number of decimal places it can handle. **Truncation error** occurs when we stop (or truncate) an approximation process after a certain number of steps. Finally, we must be aware that there is usually a trade-off in dealing with error. If we try to reduce the truncation error and increase the accuracy of our approximation by carrying out more steps (for example, by taking more terms of a Taylor series), we are increasing calculation load and consequently running the risk of increasing *propagated* (cumulative) *error*. (See Section A.3.)

**Euler's method** uses the idea that values near a point on a curve can be approximated by values on the tangent line drawn to that point. If we want to approximate the solution of the IVP $y' = f(x, y), y(x_0) = y_0$ on an interval $[a, b]$, we first partition $[a, b]$ by using $n + 1$ equally spaced points:

$$a = x_0 < x_1 < x_2 < \cdots < x_{n-1} < x_n = b,$$

where $x_{i+1} - x_i = \frac{b-a}{n} = h$ for $i = 0, 1, \ldots, n - 1$. Then, if $y_i$ is an approximate value for $y(x_i)$, we can define the sequence of approximate solution values as follows: $y_{k+1} = y_k + hf(x_k, y_k)$. Generally speaking, you can increase the accuracy of the approximation (reduce the error) by making the step size $h$ smaller—that is, by making the *number* of steps $n$ larger. For Euler's method, a **first-order method**, the cumulative truncation error is bounded by a constant multiple of the step size: |true value − approximation| $\leq K \cdot h$, where $K$ is independent of $h$ but depends on $|y''(x)|$ and the interval $[x_0, b]$. In practice, the actual error incurred in a problem will usually be less than this bound.

An improvement of Euler's method called **Heun's method** guesses a value of $y(x_k)$ and then uses $y_k$ to correct this guess by an averaging process. The algorithm can be expressed as follows:

**Guess (tentative step):** $\hat{y}_{k+1} = y_k + h \cdot f(x_k, y_k)$

$$\text{\textbf{Real Step:} } y_{k+1} = y_k + h \left\{ \frac{f(x_k, y_k) + f(x_{k+1}, \hat{y}_{k+1})}{2} \right\}$$

$$= y_k + h \left\{ \frac{f(x_k, y_k) + f(x_{k+1}, y_k + h \cdot f(x_k, y_k))}{2} \right\}$$

$$= y_k + \frac{h}{2} \left\{ f(x_k, y_k) + f(x_{k+1}, y_k + h \cdot f(x_k, y_k)) \right\}.$$

For the improved Euler method, the cumulative truncation error is no greater than a constant multiple of the square of the step size $h$: |true value − approximation| $\leq K \cdot h^2$, where $K$ is a constant that depends on the function $f(x, y)$, on its partial derivatives, and on the interval involved but not on $h$. We say that the improved Euler method is a **second-order method**.

There are many more sophisticated algorithms for solving differential equations numerically. Two very effective methods implemented in many computer algebra systems and even some calculators are the **fourth-order Runge-Kutta method** and the **Runge-Kutta-Fehlberg algorithm**. The *rkf45* method, as the computer implementation of this last algorithm is known, uses *variable* step sizes, choosing the step size at each stage to try to achieve a predetermined degree of accuracy. Such clever numerical techniques are called **adaptive methods**.

# PROJECT 3-1

## Euler Backwards Is More Than reluE

A *stiff* differential equation, such as the one discussed in Example 3.1.5, does not respond well to Euler's method unless the step size is small, in which case the number of steps (and the accumulated round-off error) may be large. The solution of a problem that is stiff is impractical with numerical methods not designed specifically for such problems.

Suppose we have used the points $t_0, t_1, \ldots, t_n = b$ to divide the interval from $t_0$ to $b$ into $n$ equal subintervals of length $h$, as we would for Euler's method. Then the differential equation $\frac{dy}{dt} = f(t, y)$ at the point $t_k$ can be written in the form $\frac{dy}{dt}(t_k) = f(t_k, y(t_k))$. Instead of approximating the derivative in the last equation by the *forward* difference quotient $\frac{y(t_{k+1}) - y(t_k)}{h}$, as we did for Euler's method, we use the *backward* difference quotient $\frac{y(t_k) - y(t_{k-1})}{h}$, so we get the formula

$$\frac{dy}{dt}(t_k) = f(t_k, y(t_k)) \approx \frac{y(t_k) - y(t_{k-1})}{h}$$

or $y(t_k) = y(t_{k-1}) + hf(t_k, y(t_k))$. Replacing $k$ by $k + 1$, we get the **backward Euler formula**:

$$y_{k+1} = y_k + hf(t_{k+1}, y_{k+1}).$$

This is also called the *implicit* Euler method because the quantity $y_{k+1}$ appears on *both* sides of the equation and has to be solved for.

a.  By hand, use $h = 0.1$ in the backward Euler method on the stiff problem $y' = -2y$, $y(0) = 3$, to approximate $y(1)$. Compare your values to those given by the usual Euler method.

b.  By hand, use $h = 0.1$ in the backward Euler method to approximate $y(0.5)$ if $y' = 25 \cos(y)$, $y(0) = 1$. Use your calculator or CAS equation solver to find $y_{k+1}$ at each step, keeping all digits shown for use in the next step.

c.  Find out whether you have the backward Euler method available on a computer. If not, look for a numerical method described as being designed for stiff differential equations (maybe under the name LSODE) or perhaps a "multi-step" algorithm. Use such a method to check your answers to part (b).

# CHAPTER 4

# Second- and Higher-Order Equations

## INTRODUCTION

In Chapters 2 and 3, we analyzed first-order equations graphically, numerically, and analytically and introduced qualitative concepts that will be useful in later chapters.

In this chapter, we will make the jump from first-order equations to higher-order equations, especially second- and third-order equations. We'll start by investigating types of second-order equations that occur frequently in science and engineering applications. These equations have a fully developed theory that generalizes to higher-order equations of the same type.

Most of the chapter, however, will be devoted to a *systems* approach to higher-order equations. In particular, we will see how *any higher-order differential equation can be written as a system of first-order differential equations* and then learn how to handle such systems qualitatively and numerically. In fact, if we use a graphing calculator in our study of differential equations, this device will require us to input a higher-order equation as a system of first-order equations. As we'll see in Section 4.10, the numerical methods studied in Sections 3.1, 3.2, and 3.3 can be applied in a natural way to the systems representation of any higher-order differential equation.

To illustrate the systems approach, we'll analyze some very interesting and important examples such as spring-mass problems, predator-prey relations, and arms races.

## 4.1 HOMOGENEOUS SECOND-ORDER LINEAR EQUATIONS WITH CONSTANT COEFFICIENTS

A very important application of differential equations is the analysis of an *RLC circuit* containing a resistance $R$, an inductance $L$, and a capacitance $C$. (We have already seen some first-order examples in Chapter 2.) In electrical circuit theory, if $I = I(t)$ represents the current, *Kirchhoff's Voltage Law* leads to the equation $L\frac{d^2 I}{dt^2} + R\frac{dI}{dt} + \frac{1}{C}I = 0$ when the voltage applied to the circuit is constant (for example, when a battery is used). We describe any equation of the form $ay'' + by' + cy = 0$, where $a$, $b$, and $c$ are constants, $a \neq 0$, as a **homogeneous**

**second-order linear equation with constant coefficients**. In this section, we are going to develop a technique for solving any equation of this type.

Extending the way we considered a first-order linear equation in Section 2.2, we see that a linear second-order equation with constant coefficients can be viewed in terms of an *operator* $L$ transforming functions that have two derivatives: $L(y) = ay'' + by' + cy$. To solve a homogeneous equation, we must find a function $y$ such that $L(y) = 0$. There is a natural extension of the *Superposition Principle* (see Section 2.2) for homogeneous equations:

---

**SUPERPOSITION PRINCIPLE FOR HOMOGENEOUS EQUATIONS:**

Suppose $y_1$ and $y_2$ are solutions of the homogeneous second-order linear differential equation

$$ay'' + by' + cy = 0$$

on an interval $I$. Then the linear combination $c_1 y_1(x) + c_2 y_2(x)$, where $c_1$ and $c_2$ are arbitrary constants, is also a solution on this interval.

---

(You'll be asked to prove this in Problem B1 of Exercises 4.1.)

If we consider a homogeneous *first*-order linear equation with constant coefficients, $ay' + by = 0$, where $a \neq 0$, we know that the general solution is $y = Ce^{-\frac{b}{a}t}$. In 1739, aware of this solution, Euler[1] thought of solving an $n$th-order homogeneous linear equation with constant coefficients by looking for solutions of the form $y = e^{\lambda t}$, where $\lambda$ is a constant to be determined. Let's see how this works for the equation

$$ay'' + by' + cy = 0, \tag{4.1.1}$$

where $a$, $b$, and $c$ are constants. But we should be aware that, although (for example) the combination of exponentials $y(t) = 3e^t - 2e^{-t}$ is a solution of the equation $y'' - y = 0$, the similar equation $y'' + y = 0$ has solutions that are combinations of $\sin t$ and $\cos t$. As we'll see, if we start by focusing on exponential solutions, the trigonometric possibilities will appear also. The exponential and trigonometric solutions are related in an important way, through the use of complex numbers.

If we assume that $y = e^{\lambda t}$ is a solution of Equation (4.1.1), then $y' = \lambda e^{\lambda t}$ and $y'' = \lambda^2 e^{\lambda t}$. Substituting these derivatives into (4.1.1), we get $a\left(\lambda^2 e^{\lambda t}\right) + b\left(\lambda e^{\lambda t}\right) + c\left(e^{\lambda t}\right) = 0$, which simplifies to $\left(a\lambda^2 + b\lambda + c\right)e^{\lambda t} = 0$. Because the exponential factor is never zero, we must have $\left(a\lambda^2 + b\lambda + c\right) = 0$.

## 4.1.1 The Characteristic Equation and Eigenvalues

We have just concluded that if $y = e^{\lambda t}$ is a solution of Equation (4.1.1), then $\lambda$ must satisfy the equation $a\lambda^2 + b\lambda + c = 0$, which is called the **characteristic equation** (or **auxiliary**

---

[1] In a letter to John (Johannes) Bernoulli, who first solved the important type of differential equation devised by his brother Jakob. See the start of the "B" problems in Exercises 2.2.

**equation**) of the differential Equation (4.1.1). The roots of this characteristic equation will reveal to us the nature of the solution(s) of (4.1.1). Note that we can go straight from the ODE to the characteristic equation as follows:

$$\underbrace{ay''}_{\substack{a \cdot \text{2nd derivative}}} \quad + \quad \underbrace{by'}_{\substack{b \cdot \text{1st derivative}}} \quad + \quad \underbrace{cy}_{\substack{c \cdot \text{0th derivative}}} \quad = \quad 0$$

$$\updownarrow \qquad\qquad \updownarrow \qquad\qquad \updownarrow$$

$$\underbrace{a \cdot \text{2nd-degree term}}_{\displaystyle a\lambda^2} \quad + \quad \underbrace{b \cdot \text{1st-degree term}}_{\displaystyle b\lambda} \quad + \quad \underbrace{c \cdot \text{0th-degree term}}_{\displaystyle c} \quad = \quad 0$$

Because the characteristic equation of our second-order ODE is a quadratic equation, we know that there are two roots, called **characteristic values** or **eigenvalues**,[2] say $\lambda$ and $\lambda_2$. There are only three possibilities for these eigenvalues: (1) The eigenvalues are both real numbers with $\lambda_1 \neq \lambda_2$; (2) the eigenvalues are real numbers with $\lambda_1 = \lambda_2$; or (3) the eigenvalues are complex numbers: $\lambda_1 = p + qi$ and $\lambda_2 = p - qi$, where $p$ and $q$ are real numbers (called the *real part* and the *imaginary part*, respectively) and $i = \sqrt{-1}$. In possibility (3), we say that $\lambda_1$ and $\lambda_2$ are *complex conjugates* of each other. (Now would be a good time to review the quadratic formula and its implications. See Appendix C, especially Section C.3, for more information about complex numbers.)

### 4.1.2 Real but Unequal Eigenvalues

In possibility (1), where $\lambda_1$ and $\lambda_2$ are unequal real numbers, then both $y_1(t) = e^{\lambda_1 t}$ and $y_2(t) = e^{\lambda_2 t}$ are solutions of (4.1.1). By the extension of the Superposition Principle given earlier in this section, any *linear combination* of the form $y(t) = c_1 e^{\lambda_1 t} + c_2 e^{\lambda_2 t}$ is also a solution, where $c_1$ and $c_2$ are arbitrary constants. It can be shown (see Section 4.2 for the details) that this is the *general* solution of (4.1.1)—that is, if the eigenvalues of (4.1.1) are real and distinct, then *any* solution of (4.1.1) must have the form $y(t) = c_1 e^{\lambda_1 t} + c_2 e^{\lambda_2 t}$ for some constants $c_1$ and $c_2$. The next example shows how to solve equations of the form (4.1.1) using eigenvalues.

### ■ Example 4.1.1 The Characteristic Equation—Unequal Eigenvalues

Let's solve the homogeneous linear second-order equation with constant coefficients $6y'' + 13y' - 5y = 0$. We find that the characteristic equation of this ODE is $6\lambda^2 + 13\lambda - 5 = 0$:

$$\underbrace{6y''}_{\substack{6 \cdot \text{2nd derivative}}} \quad + \quad \underbrace{13y'}_{\substack{13 \cdot \text{1st derivative}}} \quad + \quad \underbrace{-5y}_{\substack{-5 \cdot \text{0th derivative}}} \quad = \quad 0$$

$$\updownarrow \qquad\qquad \updownarrow \qquad\qquad \updownarrow$$

$$\underbrace{6 \cdot \text{2nd-degree term}}_{\displaystyle 6\lambda^2} \quad + \quad \underbrace{13 \cdot \text{1st-degree term}}_{\displaystyle 13\lambda} \quad + \quad \underbrace{-5 \cdot \text{0th-degree term}}_{\displaystyle (-5)} \quad = \quad 0$$

---

[2] In German, the word *eigen* means "own, proper, inherent, special, characteristic," etc.

Using the quadratic formula, we find

$$\lambda = \frac{-13 \pm \sqrt{13^2 - 4(6)(-5)}}{2(6)} = \frac{-13 \pm \sqrt{289}}{12} = \frac{-13 \pm 17}{12} = \frac{1}{3} \text{ or } -\frac{5}{2},$$

so that we have two distinct real eigenvalues, $\lambda_1 = \frac{1}{3}$ and $\lambda_2 = -\frac{5}{2}$, and we can write the general solution of our equation as $y(t) = c_1 e^{\frac{t}{3}} + c_2 e^{-\frac{5t}{2}}$. ∎

### 4.1.3 Real but Equal Eigenvalues

Next, we consider possibility (2), that the eigenvalues are real numbers with $\lambda_1 = \lambda_2$. In this situation, we get only one solution, $y = e^{\lambda t}$, where $\lambda$ is the value of the repeated eigenvalue. To obtain the general solution in this case, we have to find another solution that is not merely a constant multiple of $e^{\lambda t}$ (or else the "two" solutions can be merged into a single solution requiring only one arbitrary constant). Again, Euler comes to the rescue (this time in 1743), suggesting that an independent[3] second solution might be found by considering functions of the form $y_2(t) = u(t)e^{\lambda t}$, where $u(t)$ is an unknown function that must be determined.

Rather than deriving the consequences of Euler's assumption in the general case (see Problem C3 in Exercises 4.1), we'll illustrate his ingenious technique by an example.

### ∎ Example 4.1.2 The Characteristic Equation—Equal Eigenvalues

The equation $y'' - 4y' + 4y = 0$ has the characteristic equation $\lambda^2 - 4\lambda + 4 = (\lambda - 2)^2 = 0$, so $\lambda = 2$ is a repeated eigenvalue. We know that $y_1 = e^{2t}$ is one solution of the differential equation. Taking Euler's advice, we consider $y_2(t) = u(t)e^{2t}$.

Now, by the Product Rule and the Chain Rule, $y_2' = 2ue^{2t} + u'e^{2t}$ and $y_2'' = 4ue^{2t} + 4u'e^{2t} + u''e^{2t}$. Substituting $y_2$ and its derivatives into our original differential equation, we obtain

$$y_2'' - 4y_2' + 4y_2 = \left(4ue^{2t} + 4u'e^{2t} + u''e^{2t}\right) - 4\left(2ue^{2t} + u'e^{2t}\right) + 4\left(ue^{2t}\right)$$

$$= u''e^{2t} = 0.$$

Therefore, we must have $u''(t) = 0$, and two successive integrations give us $u'(t) = A$ and $u(t) = At + B$, where $A$ and $B$ are arbitrary constants. Our conclusion is that $y_2(t) = (At + B)e^{2t}$ is a solution of the original ODE that is *not* a constant multiple of $y_1 = e^{2t}$. The Superposition Principle tells us that the general solution is given by

$$y(t) = c_1 y_1(t) + c_2 y_2(t) = c_1 e^{2t} + c_2(At + B)e^{2t} = (C_1 t + C_2)\, e^{2t},$$

where $C_1 = c_2 A$ and $C_2 = c_1 + c_2 B$ are arbitrary constants. ∎

---

[3] Two functions $f_1$ and $f_2$ are called **(linearly) independent** on an interval $I$ if one is not a constant multiple of the other. Equivalently, if $c_1$ and $c_2$ are constants, then the only way for $c_1 f_1 + c_2 f_2$ to be the zero function on $I$ is if $c_1 = c_2 = 0$. It can be shown that Euler's technique produces a new solution independent of the first.

### 4.1.4 Complex Conjugate Eigenvalues

When the eigenvalues are complex numbers—$\lambda_1 = p + qi$ and $\lambda_2 = p - qi$, where $p$ and $q$ are real numbers—the two corresponding solutions of the differential equation $ay'' + by' + cy = 0$ are $y_1(t) = e^{(p+qi)t}$ and $y_2(t) = e^{(p-qi)t}$. At this point, a crucial fact to know is **Euler's formula**,[4] which defines the exponential function for complex values of the argument (exponent):

$$e^{p+qi} = e^p \left( \cos(q) + i\, \sin(q) \right).$$

(If we let $p = 0$ and $q = \pi$, we get a particularly elegant formula connecting five of the most significant constants in all of mathematics: $e^{\pi i} + 1 = 0$. Also see Section C.4.)

Using Euler's formula, we can write the solutions as

$$y_1(t) = e^{(p+qi)t} = e^{pt} e^{(qt)i} = e^{pt} \left( \cos(qt) + i\, \sin(qt) \right)$$

and

$$y_2(t) = e^{(p-qi)t} = e^{pt} e^{-(qt)i} = e^{pt} \left( \cos(-qt) + i\, \sin(-qt) \right)$$
$$= e^{pt} \left( \cos(qt) - i\, \sin(qt) \right),$$

where we have simplified $y_2(t)$ by recognizing that the cosine is an even function and the sine is an odd function. If we combine these complex-valued solutions carefully (see Problem C1 in Exercises 4.1), we find that

$$y(t) = e^{pt} \left( C_1 \cos(qt) + C_2 \sin(qt) \right),$$

a real-valued function, is a solution of $ay'' + by' + cy = 0$ for all constants $C_1$ and $C_2$. **In fact, $y(t) = e^{pt}(C_1 \cos(qt) + C_2 \sin(qt))$ is the general solution of the homogeneous equation when the characteristic equation has complex conjugate roots $p \pm qi$.**

Now let's practice with complex eigenvalues.

### ■ Example 4.1.3 Complex Conjugate Eigenvalues

The equation $\ddot{x} + 8\dot{x} + 25x = 0$ models the motion of a steel ball suspended from a spring, where $x(t)$ is the ball's distance (in feet) from its rest (equilibrium) position at time $t$ seconds. Distance below the rest position is considered positive, and distance above is considered negative. We want to describe the motion of the ball by finding a formula for $x(t)$.

The characteristic equation is $\lambda^2 + 8\lambda + 25 = 0$. The quadratic formula gives us

$$\lambda = \frac{-8 \pm \sqrt{8^2 - 4(1)(25)}}{2} = \frac{-8 \pm \sqrt{-36}}{2} = \frac{-8 \pm 6i}{2} = -4 \pm 3i,$$

---

[4] Euler discovered this formula in 1740, while investigating solutions of the equation $y'' + y = 0$. For a marvelous account of this formula and its consequences, see *Dr. Euler's Fabulous Formula: Cures Many Mathematical Ills* by Paul J. Nahin (Princeton: Princeton University Press, 2006).

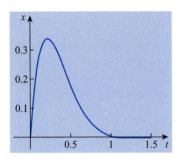

**FIGURE 4.1**
Graph of $\frac{4}{3}e^{-4t}\sin(3t)$, the solution of the IVP
$\ddot{x} + 8\dot{x} + 25x = 0$; $x(0) = 0, \dot{x}(0) = 4$; $0 \le t \le 1.5$

so the eigenvalues are $\lambda_1 = -4 + 3i$ and $\lambda_2 = -4 - 3i$. Using the solution formula derived previously, with $p = -4$ and $q = 3$, we see that $x(t) = e^{-4t}(C_1 \cos(3t) + C_2 \sin(3t))$ for arbitrary constants $C_1$ and $C_2$.

Suppose that we specify initial conditions, say $x(0) = 0$ and $\dot{x}(0) = 4$. These conditions say that the ball is at its equilibrium position at the beginning of our investigation and that the ball is started in motion from its equilibrium position with an initial velocity of 4 ft/sec in the downward direction. Applying these conditions, we have

$$x(0) = e^{-4(0)}(C_1 \cos(0) + C_2 \sin(0)) = C_1 = 0$$

and

$$\dot{x}(0) = e^{-4(0)}(-3C_1 \sin(0) + 3C_2 \cos(0)) - 4e^{-4(0)}(C_1 \cos(0) + C_2 \sin(0))$$

$$= 3C_2 - 4C_1 = 4.$$

Therefore, $C_1 = 0, C_2 = \frac{4}{3}$, and the solution of our IVP is $x(t) = \frac{4}{3}e^{-4t}\sin(3t)$. The graph of this solution (Figure 4.1) shows that the motion is dying out as time passes—that is, $x \to 0$ as $t \to \infty$.

As we'll see later in this chapter, the differential equation has a term in it that represents air resistance, and this results in what is called *damped* motion. ∎

## 4.1.5 The Amplitude-Phase Angle Form of a Solution

In working with a differential equation part of whose solution is a linear combination of $\cos \omega t$ and $\sin \omega t$, where $\omega$ is a parameter, we can use a basic identity to write the trigonometric part of the solution as a single trigonometric function. This will enable us to "read" the solution more clearly and visualize its graph more easily.

Specifically, we show that

$$c_1 \cos \omega t + c_2 \sin \omega t = M \cos(\omega t - \phi), \tag{4.1.2}$$

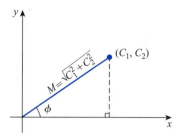

**FIGURE 4.2**
*Amplitude-phase angle*

where $M$ is a constant depending on $c_1$ and $c_2$ and $\phi$ is an angle depending on $c_1$ and $c_2$, where not both $c_1$ and $c_2$ are zero. To see this, we write

$$c_1 \cos \omega t + c_2 \sin \omega t = \sqrt{c_1^2 + c_2^2} \left[ \frac{c_1}{\sqrt{c_1^2 + c_2^2}} \cos \omega t + \frac{c_2}{\sqrt{c_1^2 + c_2^2}} \sin \omega t \right]. \qquad (4.1.3)$$

We note three things about the right-hand side of (4.1.3): (1) $\left| \frac{c_1}{\sqrt{c_1^2 + c_2^2}} \right| \leq 1$ and $\left| \frac{c_2}{\sqrt{c_1^2 + c_2^2}} \right| \leq 1$;

(2) $\left( \frac{c_1}{\sqrt{c_1^2 + c_2^2}} \right)^2 + \left( \frac{c_2}{\sqrt{c_1^2 + c_2^2}} \right)^2 = 1$; and (3) the right-hand side resembles the trigonometric identity $\cos(A - B) = \cos A \cos B + \sin A \sin B$, where $B = \omega t$. To see the similarity indicated in (3) more clearly, we choose an angle $\phi$ (measured in radians) such that $\cos \phi = \frac{c_1}{\sqrt{c_1^2 + c_2^2}}$ and

$\sin \phi = \frac{c_2}{\sqrt{c_1^2 + c_2^2}}$. We may be tempted to define $\phi = \tan^{-1}\left( \frac{c_2}{c_1} \right)$, but as we'll see in Example 4.1.4(b) the value of the quotient $c_2/c_1$ does not determine the quadrant in which $\phi$ lies. The range of the inverse tangent is assumed to be the open interval $(-\pi/2, \pi/2)$, which may not be consistent with the true location of $\phi$. The necessary adjustments are indicated in the summary later in this section.

The angle $\phi$, called the **phase angle**, always exists because of properties (1) and (2). The quantity $M = \sqrt{c_1^2 + c_2^2}$ is the **amplitude**. The expression $M \cos(\omega t - \phi)$ is called the **amplitude-phase angle** form of the solution $c_1 \cos \omega t + c_2 \sin \omega t$. We see that $M \cos(\omega t - \phi) = M \cos\left( \omega \left[ t - \frac{\phi}{\omega} \right] \right)$ has period $2\pi/\omega$ and that the graph of $M \cos(\omega t - \phi)$ is the graph of $M \cos(\omega t)$ shifted $\phi/\omega$ units to the right.

Figure 4.2 illustrates the situation for a first-quadrant point $(c_1, c_2)$.

### ■ Example 4.1.4 The Amplitude-Phase Angle Forms of Solutions

(a) The initial-value problem $\ddot{x} + 256x = 0, x(0) = 1/4, x'(0) = 1$ has the solution $x(t) = \frac{1}{4} \cos 16t + \frac{1}{16} \sin 16t$. To write this solution in the amplitude-phase angle form $M \cos(16t - \phi)$, we first calculate $M = \sqrt{c_1^2 + c_2^2} = \sqrt{\left( \frac{1}{4} \right)^2 + \left( \frac{1}{16} \right)^2} = \frac{\sqrt{17}}{16}$. Next, we note that

$(c_1, c_2) = (1/4, 1/16)$ is in the first quadrant, and we determine the phase angle $\phi$ by the equations $\cos\phi = \frac{1/4}{\sqrt{17}/16} = \frac{4\sqrt{17}}{17}$ and $\sin\phi = \frac{1/16}{\sqrt{17}/16} = \frac{\sqrt{17}}{17}$, which indicate that $\phi = 0.2450$ radian. Therefore, we can write the solution as $\frac{\sqrt{17}}{16}\cos(16t - 0.2450)$.

(b) The solution of the IVP $y'' + 9y = 0, y(0) = -0.3, y'(0) = 1.2$ is $y(t) = -0.3\cos 3t + 0.4\sin 3t$. The amplitude of this solution is $M = \sqrt{c_1^2 + c_2^2} = \sqrt{(-0.3)^2 + (0.4)^2} = 0.5$. The point $(c_1, c_2) = (-0.3, 0.4)$ lies in the second quadrant, and the unique solution of the equations $\cos\phi = \frac{-0.3}{0.5} = -0.6$ and $\sin\phi = \frac{0.4}{0.5} = 0.8$ is approximately 2.2143 radians $(= \pi - \sin^{-1}(0.8))$. Thus, $y(t) = 0.5\cos(3t - 0.2143)$. Note that in this example $\tan^{-1}(c_2/c_1) = \tan^{-1}(0.4/-0.3) = -0.9273$, an incorrect answer that places $\phi$ in the fourth quadrant. However, $\phi$ can be calculated correctly as $\tan^{-1}(0.4/-0.3) + \pi$. ■

We can summarize the process of expressing a solution in amplitude-phase angle form as follows.

**AMPLITUDE-PHASE ANGLE FORM**

$$c_1\cos\omega t + c_2\sin\omega t = M\cos(\omega t - \phi)$$

$$M = \sqrt{c_1^2 + c_2^2}$$

$$\tan\phi = \frac{c_2}{c_1}$$

$$\phi = \begin{cases} \tan^{-1}(c_2/c_1) & \text{if } c_1 \geq 0 \\ \tan^{-1}(c_2/c_1) + \pi & \text{if } c_1 < 0 \end{cases}$$

### 4.1.6 Summary

We can summarize the situation for *homogeneous* linear second-order equations with constant coefficients as follows:

Suppose that we have the equation $ax'' + bx' + cx = 0$, where $a$, $b$, and $c$ are constants, $a \neq 0$, and $\lambda_1$ and $\lambda_2$ are the zeros of the characteristic equation $a\lambda^2 + b\lambda + c = 0$. Then

1. If there are two distinct real eigenvalues—$\lambda_1, \lambda_2$, with $\lambda_1 \neq \lambda_2$—corresponding to our equation, the general solution is

$$x(t) = c_1 e^{\lambda_1 t} + c_2 e^{\lambda_2 t}.$$

2. If there is a repeated real eigenvalue $\lambda$, the general solution has the form

$$x(t) = c_1 e^{\lambda t} + c_2 t\, e^{\lambda t} = (c_1 + c_2 t)\, e^{\lambda t}.$$

3. If the eigenvalues form a complex conjugate pair $p \pm qi$, then Euler's formula can be used to show that the (real-valued) general solution has the form

$$x(t) = e^{pt}\big(c_1\cos(qt) + c_2\sin(qt)\big).$$

## Exercises 4.1

### A

Find the general solution of each of the equations in Problems 1–10.

1. $y'' - 4y' + 4y = 0$
2. $\ddot{x} + 4\dot{x} - 5x = 0$
3. $x'' - 2x' + 2x = 0$
4. $x'' + 5x' + 6x = 0$
5. $\ddot{x} + 2\dot{x} = 0$
6. $\ddot{x} - x = 0$
7. $y'' + 4y = 0$
8. $6\ddot{x} - 11\dot{x} + 4x = 0$
9. $\ddot{r} - 4\dot{r} + 20r = 0$
10. $y'' + 4ky' - 12k^2 y = 0$ ($k$ is a parameter)
11. Solve the IVP $\ddot{x} - 3\dot{x} + 2x = 0$, $x(0) = 1, \dot{x}(0) = 0$.
12. Solve the IVP $y'' - 2y' + y = 0$, $y(0) = 0, y'(0) = 0$.
13. Solve the IVP $y'' - 4y' + 20y = 0$, $y(\pi/2) = 0, y'(\pi/2) = 1$.
14. Write each of the following functions in amplitude-phase angle form.

   a. $x(t) = 3\cos 5t - 7\sin 5t$
   b. $y(t) = \sqrt{3}\cos 14t + \sin 14t$
   c. $x(t) = -6\cos 5t + 6\sin 5t$
   d. $y(t) = \sqrt{3}\cos 6t - \sin 6t$

### B

1. a. Show that if $a$, $b$, and $c$ are constants and $y$ is any function having at least two derivatives, then the *differential operator* $L$ defined by the relation $L(y) = ay'' + by' + cy$ is linear: $L\left(c_1 y_1 + c_2 y_2\right) = c_1 L\left(y_1\right) + c_2 L\left(y_2\right)$ for any twice-differentiable functions $y_1$ and $y_2$ and any constants $c_1$ and $c_2$.
   b. Show that if $y_1$ and $y_2$ are two solutions of $L(y) = 0$, then the function $c_1 y_1 + c_2 y_2$ is also a solution of $L(y) = 0$.

As we noted at the beginning of Section 4.1, if $I = I(t)$ represents the current in an electrical circuit, then *Kirchhoff's Voltage Law* gives us the equation $L\frac{d^2 I}{dt^2} + R\frac{dI}{dt} + \frac{1}{C}I = 0$ when the voltage applied to the circuit is constant. In this equation, $L$ is the inductance, $R$ is the resistance, and $C$ is the capacitance. Use this equation in Problems 2–3.

2. An *RLC* circuit with $R = 10$ ohms, $L = 0.5$ henry, and $C = 0.01$ farad has a constant voltage of 12 volts. Assume no initial current and that $\frac{dI}{dt} = 60$ when the voltage is first applied. Find an expression for the current in the circuit at time $t > 0$.
3. An *RLC* circuit with $R = 6$ ohms, $L = 0.1$ henry, and $C = 0.02$ farad has a constant voltage of 6 volts. Assume no initial current and that $\frac{dI}{dt} = 60$ when the voltage is first applied.

   a. Find an expression for the current in the circuit at time $t > 0$.
   b. Use technology to graph the answer found in part (a) for $0 \le t \le 0.5$.

c. From the graph in part (b), estimate the maximum value of $I$ and find the exact value by calculus techniques applied to the expression found in part (a).

d. At what time is the maximum value found in part (c) achieved? (You can use a calculator or a CAS for this.)

According to *Newton's Second Law of Motion* (see Section 4.8 for a further discussion), if an object with mass $m$ is suspended from a spring attached to the ceiling, then the motion of the object is governed by the equation $m\ddot{x} + a\dot{x} + kx = 0$. In this equation, $x(t)$ is the object's distance from its rest (equilibrium) position at time $t$ seconds. Distance below the rest position is considered positive, while distance above is considered negative. Also, $a$ is a constant representing the air resistance and/or friction present in the system and $k$ is the spring constant, describing the "give" in the spring. (Recall that mass = weight/$g$, where $g$ is the gravitational constant—32 ft/sec$^2$ or 9.8 m/sec$^2$.) Use this equation to do Problems 4–7.

4. An object of mass 4 slugs (= 128 lbs/32 ft/sec$^2$) is suspended from a spring having spring constant 64 lbs/ft. The object is started in motion, with no initial velocity, by pulling it **6 inches** (*watch the units!*) below the equilibrium position and then releasing it. If there is no air resistance, find a formula for the position of the object at any time $t > 0$. (Note that the problem statement contains two initial conditions.)

5. A 20-g mass hangs from the end of a spring having a spring constant of **2880 dynes/cm** and is allowed to come to rest. It is then set in motion by stretching the spring **3 cm** from its equilibrium position and releasing the mass with an initial velocity of **10 cm/sec** in the downward (positive) direction. Find the position of the mass at time $t > 0$ if there is no air resistance.

6. A $\frac{1}{2}$-kg mass is attached to a spring having a spring constant of 6 lbs/ft. The mass is set in motion by displacing it 6 inches below its equilibrium position with no initial velocity. Find the subsequent motion of the mass if $a$, the constant representing air resistance, is 4 lbs.

7. A $\frac{1}{2}$-kg mass is attached to a spring having a spring constant of 8 Newtons/m. The mass is set in motion by displacing it 10 cm above its equilibrium position with an initial velocity of 2 m/sec in the upward direction.

a. Find the subsequent motion of the mass if the constant representing air resistance is 2 Newtons.

b. Graph the function $x(t)$ found in part (a) for $0 \le t \le 3$, $2 \le t \le 3$, and $3 \le t \le 4$. Describe the motion of the mass in your own words.

c. Estimate the greatest distance of the mass above its equilibrium position.

8. The equation $\theta'' = -4\theta - 5\theta'$ represents the angle $\theta(t)$ made by a swinging door, where $\theta$ is measured from the equilibrium position of the door, which is the closed position. The initial conditions are $\theta(0) = \frac{\pi}{3}$ and $\theta'(0) = 0$.

a. Determine the angle $\theta(t)$ as a function of time ($t > 0$).

b. What does your solution tell you is going to happen as $t$ becomes large?

c. Use technology to graph the solution $\theta(t)$ on the interval $[0, 5]$.

## C

1. We know that $y_1(t) = e^{pt}(\cos(qt) + i\sin(qt))$ and $y_2(t) = e^{pt}(\cos(qt) - i\sin(qt))$ are complex-valued solutions of the homogeneous Equation (4.1.1) when the eigenvalues are complex conjugate

numbers $p \pm qi$. In what follows, you may assume that complex constants are valid in the Superposition Principle.

**a.** Calculate $Y_1 = \frac{y_1 + y_2}{2}$ and show that $Y_1$ is a real-valued solution of (4.1.1).

**b.** Calculate $Y_2 = \frac{y_1 - y_2}{2i}$ and show that $Y_2$ is a real-valued solution of (4.1.1).

**c.** Calculate $Y = c_1 Y_1 + c_2 Y_2$ and conclude that $Y$ is a real-valued solution of (4.1.1) for arbitrary real constants $c_1$ and $c_2$.

**2.** Given the equation $L\frac{d^2 I}{dt^2} + R\frac{dI}{dt} + \frac{1}{C}I = 0$,

**a.** For what values of $R$ will the current subside to zero without oscillating? [*Hint*: Oscillation is equivalent to having trigonometric functions in the solution.]

**b.** For what values of $R$ will the current oscillate before subsiding to zero?

**3.** Suppose that we have a constant-coefficient equation $ay'' + by' + cy = 0$ whose characteristic equation has a repeated root $r$. Then we know that $y_1(t) = e^{rt}$ is a solution of the equation. If we form the new function $y_2(t) = u(t)e^{rt}$, where $u(t)$ is unknown, we want to determine $u(t)$ so that $y_2$ is a solution of the differential equation, but is not a constant multiple of $y_1$.

**a.** Show that any constant-coefficient equation $ay'' + by' + cy = 0$ whose characteristic equation has a double root $r$ must have the form $y'' - 2ry' + r^2 y = 0$.

**b.** Find $y_2'$ and $y_2''$ and then substitute $y_2$ and these derivatives into the equation $y'' - 2ry' + r^2 y = 0$. Simplify the result.

**c.** Solve the equation you get in part (b) for $u(t)$.

**4.** Consider the equation $ay'' + by' + cy = 0$. Another approach (c. 1748) to the situation in which the characteristic equation has a double real root $\lambda^*$ is due to the French mathematician d'Alembert (1717–1783). He proposed splitting this root into two "neighboring" roots $\lambda^*$ and $\lambda^* + \varepsilon$, where $\varepsilon$ is small.

**a.** Show that $e^{\lambda^* t}$, $e^{(\lambda^* + \varepsilon)t}$, and also the combination $y_\varepsilon(t) = \frac{e^{(\lambda^* + \varepsilon)t} - e^{\lambda^* t}}{\varepsilon}$ are solutions of the "perturbed" equation $ay'' + (b - \varepsilon a)y' + (c + \varepsilon a\lambda^*)y = 0$. [Use the result of Problem C3a.]

**b.** Show that as $\varepsilon \to 0$, $y_\varepsilon(t)$ becomes the function $te^{\lambda^* t}$, which is a solution of the original equation.

# 4.2 NONHOMOGENEOUS SECOND-ORDER LINEAR EQUATIONS WITH CONSTANT COEFFICIENTS

## 4.2.1 The Structure of Solutions

If we take the same *RLC* circuit that we considered at the beginning of the preceding section and hook up a generator supplying alternating current to it, Kirchhoff's Voltage Law will now take the form $L\frac{d^2 I}{dt^2} + R\frac{dI}{dt} + \frac{1}{C}I = \frac{dE}{dt}$, where $E$ is the applied nonconstant voltage. Such an equation is called a **nonhomogeneous second-order linear equation with constant coefficients**. (The nonzero right-hand side of such an equation is often called the **forcing function** or the **input**. The solution of the equation is the **output**. See Section 2.2.)

To get a handle on solving a nonhomogeneous linear equation, let's think a bit about the difference between a nonhomogeneous equation $ay'' + by' + cy = f(t)$ and its **associated**

**homogeneous equation** $ay'' + by' + cy = 0$. If $y$ is the general solution of the *homogeneous* system, then $y$ doesn't quite "reach" all the way to $f(t)$ under the transformation $ay'' + by' + cy$. It stops short at 0. Perhaps we could enhance the solution $y$ in some way so that operating on this new function *does* give us all of $f$. We have to be able to capture the "leftover" term $f(t)$.

For nonhomogeneous second-order equations, the proper form of the Superposition Principle is the following: If $y_1$ is a solution of $ay'' + by' + cy = f_1(t)$ and $y_2$ is a solution of $ay'' + by' + cy = f_2(t)$, then $y = c_1 y_1 + c_2 y_2$ is a solution of $ay'' + by' + cy = c_1 f_1(t) + c_2 f_2(t)$ for any constants $c_1$ and $c_2$. (See Section 2.2 for the first-order Superposition Principle.)

Here's a fundamental fact about linear equations:

> The general solution, $y_{GNH}$, of a linear nonhomogeneous equation $ay'' + by' + cy = f(t)$ is obtained by finding a *particular* solution, $y_{PNH}$, of the nonhomogeneous equation and adding it to the *general* solution, $y_{GH}$, of the associated homogeneous equation.

We can prove this easily using operator notation, where $L(y) = ay'' + by' + cy$:

1. First note that $L(y_{GH}) = 0$ and $L(y_{PNH}) = f(t)$ by definition.

2. Then if $y = y_{GH} + y_{PNH}$, we have $L(y) = L(y_{GH} + y_{PNH}) = L(y_{GH}) + L(y_{PNH}) = 0 + f(t) = f(t)$, so $y$ is a solution of the nonhomogeneous equation.

3. Now we must show that *every* solution of the nonhomogeneous equation has the form $y = y_{GH} + y_{PNH}$. To do this, we assume that $y^*$ is any solution of $L(y) = f(t)$ and let $z = y^* - y_{PNH}$. Then

$$L(z) = L\left(y^* - y_{PNH}\right) = L\left(y^*\right) - L\left(y_{PNH}\right) = f(t) - f(t) = 0,$$

which shows that $z$ is a solution to the homogeneous equation $L(y) = 0$. Because $z = y^* - y_{PNH}$, it follows that $y^* = z + y_{PNH}$, where $z$ is a solution of $L(y) = 0$. (See Problem C5 of Exercises 1.3 and Problem C3 of Exercises 2.2 for related results.)

Let's go through some simple examples to get a feel for the solutions of nonhomogeneous equations.

## ■ Example 4.2.1 Solving a Nonhomogeneous Equation

If we are given the nonhomogeneous equation $y'' + 4y' + 5y = 10e^{-2x} \cos x$, the general solution will be made up of the general solution of the associated homogeneous equation and a particular solution of the nonhomogeneous equation: $y_{GNH} = y_{GH} + y_{PNH}$. The characteristic equation $\lambda^2 + 4\lambda + 5 = 0$ has roots $-2 \pm i$, so we know that $y_{GH} = e^{-2x}(c_1 \cos x + c_2 \sin x)$. We can verify that a particular solution of the nonhomogeneous equation is $5xe^{-2x} \sin x$. Therefore, the general solution of the nonhomogeneous equation is $y = e^{-2x}(c_1 \cos x + c_2 \sin x) + 5xe^{-2x} \sin x$. ■

In the preceding example, a particular solution appeared magically. The next example hints at how we may find $y_{PNH}$ by examining the forcing function on the right-hand side of the equation. Sections 4.3 and 4.4 will provide systematic procedures for determining a particular solution of a nonhomogeneous equation.

## ■ Example 4.2.2 Solving a Nonhomogeneous Equation

Suppose we want to find the general solution of $y'' + 3y' + 2y = 12 e^t$. Because the characteristic equation of the associated homogeneous equation is $\lambda^2 + 3\lambda + 2 = 0$, with roots $-1$ and $-2$, we know that the general solution of the homogeneous equation is $y_{GH} = c_1 e^{-t} + c_2 e^{-2t}$.

Now we look carefully at the form of the nonhomogeneous equation. In looking for a particular solution $y_{PNH}$, we can ignore any terms of the form $e^{-t}$ or $e^{-2t}$ because these are part of the homogeneous solution and won't contribute anything new. But somehow, after differentiations and additions, we have to wind up with the term $12e^t$. We guess that $y = ce^t$ for some undetermined constant $c$. Substituting this expression into the left-hand side of the nonhomogeneous equation, we get $(ce^t) + 3(ce^t) + 2(ce^t) = 6ce^t$. If we choose $c = 2$, then $y_{PNH} = 2e^t$ is a particular solution of the nonhomogeneous equation.

Putting these two components together, we can write the general solution of the nonhomogeneous equation as $y_{GNH} = y_{GH} + y_{PNH} = c_1 e^{-t} + c_2 e^{-2t} + 2e^t$.   ■

The intelligent guessing used in the preceding example can be formalized into the **method of undetermined coefficients**, which will be discussed in the next section. But, as we'll see, this method is effective only when the forcing function $f(t)$ in the equation $ay'' + by' + cy = f(t)$ is of a special type.

## Exercises 4.2

### A

For each of the nonhomogeneous differential equations in Problems 1–5, verify that the given function $y_p$ is a particular solution.

1. $y'' + 3y' + 4y = 3x + 2; y_p = \frac{3}{4}x - \frac{1}{16}$
2. $y'' - 4y = 2e^{3x}; y_p = \frac{3}{5}e^{3x}$
3. $3y'' + y' - 2y = 2\cos x; y_p = -\frac{5}{13}\cos x + \frac{1}{13}\sin x$
4. $y'' + 5y' + 6y = x^2 + 2x; y_p = \frac{1}{6}x^2 + \frac{1}{18}x - \frac{11}{108}$
5. $y'' + y = \sin x; y_p = -\frac{1}{2}x\cos x$
6. If $x_1(t) = \frac{1}{2}e^t$ is a solution of $\ddot{x} + \dot{x} = e^t$ and $x_2(t) = -te^{-t}$ is a solution of $\ddot{x} + \dot{x} = e^{-t}$, find a particular solution of $\ddot{x} + \dot{x} = e^t + e^{-t}$ and verify that your solution is correct.
7. Given that $y_p = x^2$ is a solution of $y'' + y' - 2y = 2\left(1 + x - x^2\right)$, use the Superposition Principle to find a particular solution of $y'' + y' - 2y = 6\left(1 + x - x^2\right)$ and verify the correctness of your solution.
8. If $y_1 = 1 + x$ is a solution of $y'' - y' + y = x$ and $y_2 = e^{2x}$ is a solution of $y'' - y' + y = 3e^{2x}$, find a particular solution of $y'' - y' + y = -2x + 4e^{2x}$. Verify the correctness of your solution.

### B

1. Find the general solution of the equation given in Problem A1.
2. Find the general solution of the equation given in Problem A2.
3. Find the general solution of the equation given in Problem A3.
4. Find the general solution of the equation given in Problem A4.
5. Find the general solution of the equation given in Problem A5.

6. Find the general solution of the equation $y'' + y' - 2y = 6(1 + x - x^2)$ given in Problem A7.

7. Find the general solution of the equation $\ddot{x} + \dot{x} = e^t + e^{-t}$ given in Problem A6.

8. Find the form of a particular solution of $y'' - y = x$ by intelligent guessing and use this information to solve the IVP $y'' - y = x, y(0) = y'(0) = 0$.

**C**

1. Suppose $x(t)$ satisfies the initial-value problem

$$\ddot{x} + \pi^2 x = f(t) = \begin{cases} \pi^2, & 0 \le t \le 1 \\ 0, & t > 1 \end{cases}$$

with $x(0) = 1$ and $\dot{x}(0) = 0$. Determine the continuously differentiable solution for $t \ge 0$. (This means that the solution has a continuous derivative function. Note that it will have a discontinuous second derivative at $t = 1$.)

## 4.3 THE METHOD OF UNDETERMINED COEFFICIENTS

As we saw in Section 4.2, to find the general solution of a linear nonhomogeneous equation with constant coefficients $ay'' + by' + cy = f(t)$, we must find a particular solution $y_{PNH}$ of this equation and add it to the general solution $y_{GH}$ of the associated homogeneous equation $ay'' + by' + cy = 0$.

The **method of undetermined coefficients** is the systematic version of the "intelligent guessing" discussed in the preceding section. It was developed by Euler in his 1753 study of the motion of the moon. This technique uses the forcing function $f(t)$ on the right-hand side of the differential equation to suggest a form for $y_{PNH}$. This trial solution (guess) will contain undetermined constants that can be evaluated by substituting the suggested function $y_{PNH}$ into the nonhomogeneous equation. (Look back at Example 4.2.1 and Example 4.2.2.)

For example, if the forcing function $f(t)$ is a polynomial of degree $n$, it is reasonable to suspect that $y_{PNH}$ is also an $n$th degree polynomial $a_n t^n + a_{n-1} t^{n-1} + \cdots + a_1 t + a_0$ whose coefficients $a_n, a_{n-1}, \ldots, a_1, a_0$ we must determine. As we'll see, sometimes it's more intelligent to assume the degree of the trial solution is $n + 1$ or $n + 2$.

### ■ Example 4.3.1 Finding $y_{PNH}$ by the Method of Undetermined Coefficients

Consider the equation $y'' - y' - 2y = 3x^2 - 2x + 1$. The characteristic equation of the associated homogeneous equation is $\lambda^2 - \lambda - 2 = 0$, with roots $\lambda_1 = -1$ and $\lambda_2 = 2$, so $y_{GH}$ is $c_1 e^{-x} + c_2 e^{2x}$. Because the forcing function is $3x^2 - 2x + 1$, we guess that

$$y_{PNH} = a_2 x^2 + a_1 x + a_0.$$

Then we calculate $y'_{PNH} = 2a_2 x + a_1$ and $y''_{PNH} = 2a_2$. Substituting these derivatives into the differential equation, we get

$$2a_2 - (2a_2 x + a_1) - 2(a_2 x^2 + a_1 x + a_0) = 3x^2 - 2x + 1,$$

or, after collecting like terms,

$$-2a_2x^2 + (-2a_2 - 2a_1)\,x + (2a_2 - a_1 - 2a_0) = 3x^2 - 2x + 1.$$

Equating the coefficients of like powers of $x$, we find

$$-2a_2 = 3, \; -2a_2 - 2a_1 = -3, \; 2a_2 - a_1 - 2a_0 = 1.$$

The first equation yields $a_2 = -3/2$. Substituting this value for $a_2$ into the second equation gives $a_1 = 5/2$. Finally, we solve the third equation for $a_0$ in terms of $a_1$ and $a_2$ to find $a_0 = (1 + a_1 - 2a_2)/(-2) = -13/4$.

Now that we have determined the coefficients of our polynomial $y_{\text{PNH}}$, we see that $y_{\text{PNH}} = -\frac{3}{2}x^2 + \frac{5}{2}x - \frac{13}{4}$. Thus, the general solution of the nonhomogeneous equation is

$$y_{\text{GH}} + y_{\text{PNH}} = c_1e^{-x} + c_2e^{2x} - \frac{3}{2}x^2 + \frac{5}{2}x - \frac{13}{4}.$$

∎

The guesswork involved in the preceding example was a bit misleading. We assumed that the degree of the undetermined polynomial was less than or equal to the degree of the polynomial on the right-hand side of the equation. (It is possible that some of the coefficients were zero.) This assumption was valid in this example because the left-hand side of the differential equation $y'' - y' - 2y = 3x^2 - 2x + 1$ contained a term of the form $cy$. If we suppose that $y$ is an $n$th degree polynomial, then the calculation of $y''$ reduces the degree of $y$ by 2, but the term $-2y$ restores a polynomial of degree $n$ to the result. However, if this constant multiple of $y$ were missing (that is, if $c = 0$), we should guess that $y$ is a polynomial of degree $n + 1$. If the $y'$ term were also missing, then the degree of our trial solution should be $n + 2$.

If the equation is $ay'' + by' + cy = a_nt^n + a_{n-1}t^{n-1} + \cdots + a_1t + a_0$, assuming that $a \neq 0$, Table 4.1 summarizes our choice of a polynomial trial solution.

**Table 4.1**  Degree of Polynomial Trial Solutions

|  | Degree of Trial Solution |
| --- | --- |
| $b \neq 0, c \neq 0$ | $n$ |
| $b = 0, c \neq 0$ | $n$ |
| $b \neq 0, c = 0$ | $n + 1$ |
| $b = 0, c = 0$ | $n + 2$ |

### ■ Example 4.3.2 A Particular Solution of $ay'' + by' + cy = f(t)$ When $c = 0$

If we want to find a particular solution of the equation $y'' - 7y' = t^2 + 2t + 3$, we realize that choosing a trial solution of the form $a_2t^2 + a_1t + a_0$ won't work: Substituting such a second-degree polynomial into the left-hand side of the equation yields a *linear* result because of the differentiations. Therefore, we guess (see Table 4.1) that $y_{\text{PNH}} = a_3t^3 + a_2t^2 + a_1t + a_0$. Then

$y'_{\text{PNH}} = 3a_3 t^2 + 2a_2 t + a_1$ and $y''_{\text{PNH}} = 6a_3 t + 2a_2$. Substituting $y_{\text{PNH}}$ into the equation and collecting terms, we find

$$-21a_3 t^2 + (6a_3 - 14a_2)t + (2a_2 - 7a_1) = t^2 + 2t + 3.$$

Equating coefficients of equal powers of $t$, we have

$$-21a_3 = 1, \, 6a_3 - 14a_2 = 2, \quad 2a_2 - 7a_1 = 3.$$

Solving this system of equations, we conclude that $a_3 = -1/21, a_2 = -8/49$, and $a_1 = -163/343$, so $y_{\text{PNH}} = -\frac{1}{21}t^3 - \frac{8}{49}t^2 - \frac{163}{343}t$. We can take $a_0$ to be zero because it is a "free" variable and cannot be determined. (Also, noting that $y_{\text{GH}} = c_1 e^{7t} + c_2$, we see that any nonzero value of $a_0$ would be absorbed by $c_2$ in assembling $y_{\text{GNH}}$.) ∎

In general, the key idea behind the method of undetermined coefficients is that all the derivatives of the forcing function $f(t)$ should have the same form as $f(t)$ itself. If this is true, the method will work. If this is not true, we should not use the method.

If we think about functions whose derivatives have the same forms as themselves, we realize that we are limited to polynomials, exponential functions, linear combinations of sines and cosines, or combinations of sums and products of these functions. For example, the method of undetermined coefficients applies to equations whose forcing terms are

$$-3,$$
$$-2t^5 - 6t^3 + 4,$$
$$2\cos 4t - \frac{7}{3}\sin 4t,$$
$$25e^t \sin t,$$
$$t^2 e^{3t} + (1 - t^3)\cos 5t.$$

Table 4.2 provides suggestions for forms of particular solutions.

By the Superposition Principle (see Section 4.2), forcing functions that are linear combinations of the forms on the left side of Table 4.2 require the same linear combinations of the

| **Table 4.2** Trial Particular Solutions for Nonhomogeneous Equations | |
|---|---|
| $f(t)$ | **Form of Trial Solution** |
| $c \neq 0$, a constant | $K$, a constant |
| $P_n(t) = a_n t^n + a_{n-1}t^{n-1} + \cdots + a_1 t + a_0$ | $Q_m(t) = c_m t^m + c_{m-1}t^{m-1} + \cdots + c_1 t + c_0$ (See Table 4.1) |
| $ce^{at}$ | $K e^{at}$ |
| $a\cos rt + b\sin rt$ | $\alpha \cos rt + \beta \sin rt$ |
| $e^{Rt}(a\cos rt + b\sin rt)$ | $e^{Rt}(\alpha \cos rt + \beta \sin rt)$ |
| $P_n(t)e^{at}$ | $Q_m(t)e^{at}$ |

corresponding trial solutions: **If forcing function $f_1(t)$ suggests a trial solution of form $F_1(t)$ and forcing function $f_2(t)$ suggests a trial solution of form $F_2(t)$, then a forcing function of the form $c_1 f_1(t) + c_2 f_2(t)$ requires a trial solution of the form $c_1 F_1(t) + c_2 F_2(t)$.** This result generalizes to any finite linear combination $c_1 f_1(t) + c_2 f_2(t) + \cdots + c_k f_k(t)$ of forcing functions.

### ■ Example 4.3.3 Undetermined Coefficients with an Exponential Forcing Function

We can easily verify that $x_{GH} = c_1 \cos 2t + c_2 \sin 2t$ for the equation $\ddot{x} + 4x = 3e^{2t}$. To find a particular solution of the nonhomogeneous equation, we choose a function of the form $x = Ke^{2t}$, where $K$ is an undetermined constant. We calculate $\dot{x} = 2Ke^{2t}, \ddot{x} = 4Ke^{2t}$; and substitution into the nonhomogeneous equation gives us $4Ke^{2t} + 4Ke^{2t} = 4e^{2t}$, or $8Ke^{2t} = 4e^{2t}$, so $K = 1/2$.

Thus, a particular solution is $x_p(t) = \frac{1}{2}e^{2t}$, and the general solution of the nonhomogeneous equation is $x(t) = c_1 \cos 2t + c_2 \sin 2t + \frac{1}{2}e^{2t}$. ■

### ■ Example 4.3.4 Undetermined Coefficients with a Trigonometric Forcing Function

A 2560 lb car supported by a *MacPherson strut* (a particular type of shock absorbing system) is traveling over a bumpy road at a constant velocity $v$. The equation modeling the motion is

$$80\ddot{x} + 10{,}000x = 2500 \cos\left(\frac{\pi v t}{6}\right),$$

where $x$ represents the vertical position of the car's axle relative to its equilibrium position, and the basic units of measurement are feet and feet per second where appropriate. (Note that the coefficient of $\ddot{x}$ is $2560/g = 2560/32 = 80$, the *mass* of the car.) We want to determine how the velocity affects the way the car vibrates.

The general solution of the associated homogeneous equation is $x_{GH} = c_1 \sin\left(5\sqrt{5}t\right) + c_2 \cos\left(5\sqrt{5}t\right)$. Table 4.2 suggests we choose a trial solution of the form $x_P = A \sin\left(\frac{\pi v t}{6}\right) + B \cos\left(\frac{\pi v t}{6}\right)$. If we examine the left-hand side of the differential equation, we can simplify our work. We notice that if the trial solution $x$ contains a sine term, then $\ddot{x}$ yields a sine term and $10{,}000x$ contributes another sine. Because there is no multiple of $\sin\left(\frac{\pi v t}{6}\right)$ in the forcing function, we conclude that $A = 0$, giving the trial solution the simpler form $B \cos\left(\frac{\pi v t}{6}\right)$.

Now $\dot{x}_P = -\frac{\pi v}{6}B \sin\left(\frac{\pi v t}{6}\right)$ and $\ddot{x}_P = -\left(\frac{\pi v}{6}\right)^2 B \cos\left(\frac{\pi v t}{6}\right)$. Substituting these derivatives into the nonhomogeneous equations and collecting terms, we find

$$\left[10{,}000 - 80\left(\frac{\pi v}{6}\right)^2\right] B \cos\left(\frac{\pi v t}{6}\right) = 2500 \cos\left(\frac{\pi v t}{6}\right),$$

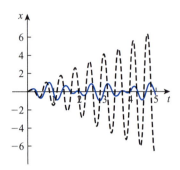

**FIGURE 4.3**
$x(t) = c_1 \sin\left(5\sqrt{5}t\right) + c_2 \cos\left(5\sqrt{5}t\right) + \dfrac{2500}{10{,}000 - 80\left(\frac{\pi v}{6}\right)^2} \cos\left(\frac{\pi v t}{6}\right); x(0) = 0, \dot{x}(0) = 0; 0 \le t \le 5$
$v = 15$ *(solid curve)*; $v = 21$ *(dashed curve)*

so $B = 2500/\left[10{,}000 - 80\left(\frac{\pi v}{6}\right)^2\right]$. Therefore, $x_{\text{PNH}} = \dfrac{2500}{10{,}000 - 80\left(\frac{\pi v}{6}\right)^2} \cos\left(\frac{\pi v t}{6}\right)$ and the general solution of our equation of motion is

$$x(t) = c_1 \sin\left(5\sqrt{5}t\right) + c_2 \cos\left(5\sqrt{5}t\right) + \frac{2500}{10{,}000 - 80\left(\frac{\pi v}{6}\right)^2} \cos\left(\frac{\pi v t}{6}\right).$$

Clearly, the first two trigonometric terms help describe the bumpy ride, but they have fixed amplitudes $|c_1|$ and $|c_2|$, so the ride can't get alarmingly bumpy. However, the amplitude of the last term is given by $\dfrac{2500}{10{,}000 - 80\left(\frac{\pi v}{6}\right)^2}$, which grows larger and larger as the denominator expression $10{,}000 - 80\left(\frac{\pi v}{6}\right)^2$ gets closer and closer to zero. Thus, we get *unbounded* vibrations when $10{,}000 - 80\left(\frac{\pi v}{6}\right)^2 = 0$—that is, when $v = \sqrt{\frac{4500}{\pi^2}} \approx 21.35$ ft/sec. The dimensional equation $\frac{\text{miles}}{\text{hr}} = \frac{\text{miles}}{\text{ft}} \cdot \frac{\text{ft}}{\text{sec}} \cdot \frac{\text{sec}}{\text{hr}}$ allows us to express our answer as $\frac{1}{5280} \cdot \frac{21.35}{1} \cdot \frac{3600}{1} \approx 14.56$ miles per hour.

Assuming the initial conditions $x(0) = 0$ and $\dot{x}(0) = 0$, Figure 4.3 shows two graphs of $x(t)$ against $t$, the solid graph having $v = 15$ ft/sec and the dashed line graph using $v = 21$ ft/sec. You can see that the car's vibrations become wilder over time for a speed close to 21.35 feet per second. ∎

The preceding example illustrates the phenomenon of **resonance**, the presence of oscillations of unbounded amplitude. We will encounter resonance again in Example 4.8.4 and discuss it further in the section following that example.

As indicated earlier, the Superposition Principle allows us to handle more complicated combinations of basic forcing functions.

## ■ Example 4.3.5 Undetermined Coefficients with a Linear Combination of Forcing Functions

Suppose we have the equation $y'' + y' - 2y = x^2 + 2\sin x - \cos x + e^{3x}$. In this case, the forcing function is a linear combination of familiar terms from the left-hand side of Table 4.2, so we choose a trial solution $y_\text{p}$ that is a combination of the second, third, and fourth entries of Table 4.2 (second column):

$$y_\text{P} = (Ax^2 + Bx + C) + (D\sin x + E\cos x) + Fe^{3x}.$$

Consequently, $y_\text{p}' = 2Ax + B + D\cos x - E\sin x + 3Fe^{3x}$ and $y_\text{p}'' = 2A + B - D\sin x - E\cos x + 9Fe^{3x}$. When we substitute these derivatives in the nonhomogeneous equation and collect terms, we get the equation

$$-2Ax^2 + (2A - 2B)x + (2A + B - 2C) + (-3D - E)\sin x + (-3E + D)\cos x + 10Fe^{3x}$$

$$= x^2 + 2\sin x - \cos x + e^{3x}.$$

Matching the coefficients of like terms on each side, we get the system

**1.** $-2A = 1$           [The coefficients of $x^2$ must be equal.]

**2.** $2(A - B) = 0$       [The coefficients of $x$ must be equal.]

**3.** $2A + B - 2C = 0$   [The constant terms must be equal.]

**4.** $-3D - E = 2$       [The coefficients of $\sin x$ must be equal.]

**5.** $-3E + D = -1$      [The coefficients of $\cos x$ must be equal.]

**6.** $10F = 1$            [The coefficients of $e^{3x}$ must be equal.]

Working from the top down, we find $A = -1/2, B = A = -1/2, C = -3/4, D = -7/10, E = 1/10$, and $F = 1/10$. Therefore,

$$y_\text{PNH} = -\frac{1}{2}x^2 - \frac{1}{2}x - \frac{3}{4} - \frac{7}{10}\sin x + \frac{1}{10}\cos x + \frac{1}{10}e^{3x}.$$

The general solution of the associated homogeneous equation is $c_1 e^x + c_2 e^{3x}$, so the general solution of the nonhomogeneous equation is

$$y(x) = c_1 e^x + c_2 e^{3x} - \frac{1}{2}x^2 - \frac{1}{2}x - \frac{3}{4} - \frac{7}{10}\sin x + \frac{1}{10}\cos x + \frac{1}{10}e^{3x}.$$

∎

There is an exception to the neatness of Table 4.2, however. If any term of the initial trial solution is also a term (or a multiple of a term) of $y_\text{GH}(t)$, then the trial solution must be modified by multiplying it by $t^m$, where $m$ is the smallest positive integer such that the product of $t^m$ and the trial solution has no terms in common with $y_\text{GH}(t)$. The next example illustrates this change of strategy.

### ■ Example 4.3.6 Undetermined Coefficients—An Exception to the General Rule

Let's solve the equation $y'' + 4y = 3\cos 2x + 2x^2$. The general solution of the homogeneous equation $y'' + 4y = 0$ is $c_1 \sin 2x + c_2 \cos 2x$. Ordinarily, we would choose a trial solution of the form $A\sin 2x + B\cos 2x + Cx^2 + Dx + E$. However, the expression $A\sin 2x + B\cos 2x$

duplicates the terms of $y_{GH}$. To deal with this, we choose a modified trial solution of the form

$$y_P = x(A \sin 2x + B \cos 2x) + Cx^2 + Dx + E,$$

noting that the second-degree polynomial does not have to be changed. Then $y_P' = x(2A \cos 2x - 2B \sin 2x) + (A \sin 2x + B \cos 2x) + 2Cx + D$ and

$$y_P'' = x(-4A \sin 2x - 4B \cos 2x) + (2A \cos 2x - 2B \sin 2x)$$
$$+ (2A \cos 2x - 2B \sin 2x) + 2C.$$

Substituting these derivatives in the nonhomogeneous equation and collecting terms, we get

$$-4B \sin 2x + 4A \cos 2x + 4Cx^2 + 4Dx + (4E + 2C) = 3 \cos 2x + 2x^2.$$

Equating coefficients of equal terms give us the equations $-4B = 0, 4A = 3, 4C = 2, 4D = 0$, and $4E + 2C = 0$, with solutions $A = 3/4, B = 0, C = 1/2, D = 0$, and $E = -1/4$. Therefore, after some simplification, $y_{PNH} = \frac{3}{4}x \sin 2x + \frac{1}{2}x^2 - \frac{1}{4}$ and $y_{GNH} = c_1 \sin 2x + c_2 \cos 2x + \frac{3}{4}x \sin 2x + \frac{1}{2}x^2 - \frac{1}{4}$. ■

## Exercises 4.3

### A

For each of the equations in Problems 1–10, find $y_{GH}$ and the expression in terms of undetermined coefficients that you would use to find $y_{PNH}$. **Do not actually determine the coefficients**.

1. $y'' + 3y' = 3$
2. $y'' - 7y' = (x - 1)^2$
3. $y'' + 7y' = e^{-7x}$
4. $y'' - 8y' + 16y = (1 - x)e^{4x}$
5. $y'' + 25y = \cos 5x$
6. $y'' + y = xe^{-x}$
7. $y'' + 6y' + 13y = e^{-3x} \cos 2x$
8. $y'' - 4y' + 3y = 3e^x + 2e^{-x} + x^3 e^{-x}$
9. $y'' + k^2 y = k$, where $k$ is a parameter
10. $4y'' + 8y' = x \sin x$

Find the general solution of each of the equations in Problems 11–20 by using the method of undetermined coefficients.

11. $y'' - 2y' - 3y = e^{4t}$
12. $\ddot{x} - 3\dot{x} + 2x = \sin t$
13. $x'' - 2x' + 2x = e^t + t \cos t$
14. $x'' + x' = 4t^2 e^t$
15. $\ddot{x} + \dot{x} = 4 \sin t$

16. $\ddot{x} - x = 2e^t - t^2$
17. $y'' + 10y' + 25y = 4e^{-5x}$
18. $6\ddot{x} - 11\dot{x} + 4x = t$
19. $\ddot{x} + 3\dot{x} + 2x = t\sin t$
20. $y'' + 5y' + 6y = 10(1-x)e^{-2x}$

## B

1. Solve the IVP $y'' - 3y' - 4y = 3e^{4x}$; $y(0) = 0, y'(0) = 0$.
2. Solve the IVP $y'' + \omega^2 y = t(\sin \omega t + \cos \omega t)$; $y(0) = 0, y'(0) = 0$.
3. Solve the IVP $y'' + y' + y = t^2 e^{-t}\cos t$, $y(0) = 1, y'(0) = 0$ using a CAS. [*Warning*: Serious mental injury may result from attempting to do this manually.]

As mentioned at the beginning of Section 4.2, if $I = I(t)$ represents the current in an electrical circuit, then *Kirchhoff's Voltage Law* gives us the nonhomogeneous equation $L\frac{d^2I}{dt^2} + R\frac{dI}{dt} + \frac{1}{C}I = \frac{dE}{dt}$, where $E$ is the applied nonconstant voltage. In this equation, $L$ is the inductance, $R$ is the resistance, and $C$ is the capacitance. Use this equation in Problems 4–6.

4. An *RLC* circuit has a resistance of 5 ohms, an inductance of 0.05 henry, a capacitor of 0.0004 farad, and an applied alternating voltage of $200\cos(100\,t)$ volts.

   a. Without using technology, find an expression for the current flowing through this circuit if the initial current is zero and $\frac{dI}{dt}(0)$ is 4000.
   b. Check your answer to part (a) by using technology.

5. Look for a particular solution of $y'' + 0.2y' + y = \sin(\omega x)$ and investigate its amplitude as a function of $\omega$. Use technology to graph the particular solution for values of $\omega$ that seem significant to you and describe the behavior of this solution.

6. In her dorm room, a student attaches a weight to a spring hanging from the ceiling. She starts the mass in motion from the equilibrium position with an initial velocity in the upward direction. But during this experiment, there is rhythmic stomping (dancing or pest control?) from the student upstairs that causes the ceiling and the entire spring-mass system to vibrate. Taking into account air resistance and this "external force," she determines that the equation of motion is $\ddot{x} + 9\dot{x} + 14x = \frac{1}{2}\sin t$, with $x(0) = 0$ and $\dot{x}(0) = -1$.

   a. Solve this equation for $x(t)$, the position of the weight relative to its rest position.
   b. Use technology to graph $x(t)$ for $0 \le t \le 10$.

7. Consider the equation

$$y'' + y = F(t), \quad \text{where } F(t) = \begin{cases} t & 0 \le t \le \pi, \\ 0 & t > \pi \end{cases}$$

$y(0) = y'(0) = 0$, and $y$ and $y'$ are continuous at $\pi$.

   a. Plot the forcing function against $t$.
   b. Solve the initial-value problem for $0 \le t \le \pi$.

    **c.** Solve the initial-value problem for $t > \pi$, determining the constants from the continuity conditions at $t = \pi$.

    **d.** Combine your answers to parts (b) and (c) into a single solution $Y$ to the original IVP and plot $Y$ against $t$.

**C**

**1.** Find the general solution of $y'' - 3y' + 2y = 4 \sin^3 3x$. [*Hint*: Use trigonometric identities to reduce the forcing function to a linear combination of functions in Table 4.2.]

**2.** Find the general solution of $y'' + 4y = \sin^4 x$. (See the hint for the preceding problem.)

**3.** Find the general solution of $y'' + 4y = \cos x \cos 2x \cos 3x$. (See the hint for Problem C1.)

**4.** Find the general solution of $y'' + \lambda^2 y = \sum_{k=1}^{N} a_k \sin k\pi t$, where $\lambda > 0$ and $\lambda \neq k\pi$ for $k = 1, 2, \ldots, N$.

**5.** Consider the equation $ay'' + by' + cy = g(t)$, where $a$, $b$, and $c$ are positive constants.

    **a.** If $Y_1(t)$ and $Y_2(t)$ are solutions of the given equation, show that $Y_1(t) - Y_2(t) \to 0$ as $t \to \infty$. [*Hint*: Show that $\sqrt{b^2 - 4ac} \leq b$.]

    **b.** Is the result of part (a) true if $b = 0$?

    **c.** If $g(t) = K$, a constant, show that every solution of the given equation approaches $K/c$ as $t \to \infty$. What happens if $c = 0$? What if $b = 0$ also?

A special type of second-order differential equation with variable coefficients is the **Cauchy-Euler equation** (or **Euler's differential equation**): $x^2 y'' + axy' + by = 0$, where $a$ and $b$ are real constants. The substitution $x = e^z$ reduces the equation to a second-order linear ODE with constant coefficients. Use this method to transform each of the following equations and then solve the resulting constant-coefficient equation.

**6.** $x^2 y'' + xy' + 4y = 1$

**7.** $x^2 y'' + 5xy' - 6y = \frac{4}{x^2} - 12$

## 4.4 VARIATION OF PARAMETERS

There are various techniques for finding a particular solution of the nonhomogeneous equation. The **method of variation of parameters** (or **variation of constants**) was developed by the French-Italian mathematician Joseph Louis Lagrange (1736–1813) in 1775.

Let's look at the nonhomogeneous equation $ay'' + by' + cy = f(t)$ and assume that $y_1(t)$ and $y_2(t)$ are two known solutions of the homogeneous equation $ay'' + by' + cy = 0$ that are independent of each other. Then we know that $c_1 y_1(t) + c_2 y_2(t)$ is also a solution of the homogeneous equation for any constants $c_1$ and $c_2$. Lagrange's idea was to look for a particular solution of the *nonhomogeneous* equation of the form $c_1(t)y_1(t) + c_2(t)y_2(t)$, where $c_1(t)$ and $c_2(t)$ are unknown *functions* that must be determined. By substituting this trial solution into the original nonhomogeneous differential equation, we will obtain one equation that must be satisfied by $c_1(t)$ and $c_2(t)$; but, because we have *two* unknown functions, this single equation is not enough. We need two equations to determine these functions completely, and Lagrange's method imposes an additional condition, one chosen to simplify the calculations involved.

Rather than go through this method in complete generality, we'll illustrate the technique in specific examples.

## ■ Example 4.4.1 Using Variation of Parameters

Suppose we want to solve $y'' + 3y' + 2y = 3e^{-2x} + x$. The characteristic equation of the associated homogeneous equation is $\lambda^2 + 3\lambda + 2 = 0$, with roots $-2$ and $-1$. Then $y_1(x) = e^{-2x}$ and $y_2(x) = e^{-x}$ are two independent solutions of the homogeneous equation, and the general solution of the homogeneous equation is $y_{GH} = c_1 e^{-2x} + c_2 e^{-x}$, where $c_1$ and $c_2$ are arbitrary constants. Now assume that $y = C_1 y_1 + C_2 y_2 = C_1 e^{-2x} + C_2 e^{-x}$ is a particular solution of the nonhomogeneous equation, where $C_1 = C_1(x)$ and $C_2 = C_2(x)$ are unknown *functions*. Differentiating, we obtain

$$y' = -2C_1 e^{-2x} + C_1' e^{-2x} - C_2 e^{-x} + C_2' e^{-x}$$
$$= \left(-2C_1 e^{-2x} - C_2 e^{-x}\right) + \left(C_1' e^{-2x} + C_2' e^{-x}\right).$$

To avoid messy higher derivatives (and being aware that we're looking for a *particular* solution), Lagrange's method requires that we impose the condition

$$C_1' e^{-2x} + C_2' e^{-x} = 0. \tag{*}$$

Accepting this condition, we have $y' = -2C_1 e^{-2x} - C_2 e^{-x}$, from which we calculate $y'' = 4C_1 e^{-2x} - 2C_1' e^{-2x} + C_2 e^{-x} - C_2' e^{-x}$. Substituting these expressions for $y$, $y'$, and $y''$ into the equation $y'' + 3y' + 2y = 3e^{-2x} + x$, we find that

$$\left(4C_1 e^{-2x} - 2C_1' e^{-2x} + C_2 e^{-x} - C_2' e^{-x}\right) + 3\left(-2C_1 e^{-2x} - C_2 e^{-x}\right) + 2\left(C_1 e^{-2x} + C_2 e^{-x}\right)$$
$$= 3e^{-2x} + x,$$

or

$$-2C_1' e^{-2x} - C_2' e^{-x} = 3e^{-2x} + x. \tag{**}$$

Equations (*) and (**) form a system of equations that we must solve for $C_1'$ and $C_2'$:

$$C_1' e^{-2x} + C_2' e^{-x} = 0 \tag{*}$$
$$-2C_1' e^{-2x} - C_2' e^{-x} = 3e^{-2x} + x. \tag{**}$$

Adding (*) and (**) gives us $-C_1' e^{-2x} = 3e^{-2x} + x$, so that $C_1' = -3 - xe^{2x}$.

Integration (by parts, manually, or by CAS) yields $C_1(x) = -3x - \frac{x}{2}e^{2x} + \frac{1}{4}e^{2x}$. In using variation of parameters, we make all constants of integration 0 because we want only a *particular* solution. Next, we use (*) to find that $C_2' = e^x \left(-C_1' e^{-2x}\right) = e^x \left(3e^{-2x} + x\right) = 3e^{-x} + xe^x$. Integration gives us $C_2(x) = -3e^{-x} + xe^x - e^x$.

Finally,

$$y_{PNH} = C_1 y_1 + C_2 y_2 = \left(-3x - \frac{x}{2}e^{2x} + \frac{1}{4}e^{2x}\right)\left(e^{-2x}\right) + \left(-3e^{-x} + xe^x - e^x\right)\left(e^{-x}\right)$$

$$= -3xe^{-2x} - \frac{x}{2} + \frac{1}{4} - 3e^{-2x} + x - 1 = -3xe^{-2x} - 3e^{-2x} + \frac{x}{2} - \frac{3}{4},$$

so that $y_{GNH} = c_1 e^{-2x} + c_2 e^{-x} + \frac{x}{2} - \frac{3}{4} - 3xe^{-2x}$ is the general solution of the original nonhomogeneous equation. (Note that the term $-3e^{-2x}$ in $y_{PNH}$ has been absorbed by the term $c_1 e^{-2x}$ in $y_{GH}$.) ∎

### ■ Example 4.4.2 Using Variation of Parameters

The equation $\ddot{x} + x = \tan t$ cannot be solved by using the method of undetermined coefficients because the forcing function $f(t) = \tan t$ cannot be expressed as a linear combination of one of the basic forms given in Table 4.2. However, Lagrange's method works.

The general solution of the associated homogeneous equation $\ddot{x} + x = 0$ is $y_{GH}(t) = c_1 \sin t + c_2 \cos t$, where $c_1$ and $c_2$ are arbitrary constants. Therefore, we try to find functions $C_1 = C_1(t)$ and $C_2 = C_2(t)$ such that $x_P(t) = C_1(t) \sin t + C_2(t) \cos t$ is a particular solution of the nonhomogeneous equation. To do this, we must first calculate

$$\dot{x} = C_1 \cos t + \dot{C}_1 \sin t - C_2 \sin t + \dot{C}_2 \cos t$$
$$= (C_1 \cos t - C_2 \sin t) + (\dot{C}_1 \sin t + \dot{C}_2 \cos t).$$

Before calculating $\ddot{x}$ and increasing the complexity of the expressions we have to use, we impose the condition

$$\dot{C}_1 \sin t + \dot{C}_2 \cos t = 0. \tag{*}$$

In particular, this condition ensures that $\ddot{x}$ will contain no second derivatives of $C_1$ or $C_2$.

Now differentiating the simplified expression for $\dot{x}$ gives us $\ddot{x} = -C_1 \sin t + \dot{C}_1 \cos t - C_2 \cos t - \dot{C}_2 \sin t$. Substituting for $x$ and $\ddot{x}$ in our original equation yields

$$\ddot{x} + x = \left(-C_1 \sin t + \dot{C}_1 \cos t - C_2 \cos t - \dot{C}_2 \sin t\right) + (C_1 \sin t + C_2 \cos t) = \tan t,$$

or

$$\dot{C}_1 \cos t - \dot{C}_2 \sin t = \tan t. \tag{**}$$

Now (*) and (**) give us a system of simultaneous equations for $\dot{C}_1$ and $\dot{C}_2$:

$$\left\{ \begin{array}{l} \dot{C}_1 \sin t + \dot{C}_2 \cos t = 0 \\ \dot{C}_1 \cos t - \dot{C}_2 \sin t = \tan t \end{array} \right\}.$$

Multiplying the first equation by $\sin t$ and the second by $\cos t$ and then adding, we find $\dot{C}_1 \left(\sin^2 t + \cos^2 t\right) = \tan t \cos t$, or $\dot{C}_1 = \sin t$. Thus, because we need only a particular solution, we can take the constant of integration to be zero and get $C_1 = -\cos t$. Using Equation (*), we derive $\dot{C}_2 = -\dot{C}_1 \sin t / \cos t = -\sin^2 t / \cos t = (\cos^2 t - 1)/\cos t = \cos t - 1/\cos t$. Then $C_2 = \int (\cos t - 1/\cos t)\, dt = \sin t - \int \sec t\, dt = \sin t - \ln|\sec t + \tan t|$.

Therefore, a particular solution of our original nonhomogeneous equation is given by $x_{PNH} = C_1(t) \sin t + C_2(t) \cos t = -\cos t \sin t + (\sin t - \ln|\sec t + \tan t|) \cos t = -\cos t \sin t + (\sin t - \ln|\sec t + \tan t|) \cos t = -\ln|\sec t + \tan t| \cos t$, and $y_{GNH}(t) = c_1 \sin t + c_2 \cos t -$

$\ln |\sec t + \tan t| \cos t$. (Note that because the last term is not a *constant* multiple of $\cos t$, it doesn't get absorbed by the term $c_2 \cos t$.) ∎

## ■ Example 4.4.3 Using Variation of Parameters

We cannot use the method of undetermined coefficients to solve the equation $y'' - 3y' + 2y = \sin(e^{-x})$ because the forcing function $f(x) = \sin(e^{-x})$ does not fit any of the patterns in Table 4.2. So we try variation of parameters.

We find that $y_{GH} = c_1 e^x + c_2 e^{2x}$, so we assume that a particular solution of the nonhomogeneous equation has the form $y_P = C_1 e^x + C_2 e^{2x}$, where $C_1$ and $C_2$ are undetermined functions of $x$. Then $y'_P = (C_1 e^x + 2C_2 e^{2x}) + (C'_1 e^x + C'_2 e^{2x})$ and, assuming

$$C'_1 e^x + C'_2 e^{2x} = 0, \tag{*}$$

$y''_P = C_1 e^x + C'_1 e^x + 4C_2 e^{2x} + 2C'_2 e^{2x}$. Substituting in the nonhomogeneous equation, we get

$$\left[ (C_1 + C'_1) e^x + (4C_2 + 2C'_2) e^{2x} \right] - 3 \left[ C_1 e^x + 2C_2 e^{2x} \right]$$
$$+ 2 \left[ C_1 e^x + C_2 e^{2x} \right] = \sin(e^{-x})$$

or, simplifying,

$$e^x C'_1 + 2e^{2x} C'_2 = \sin(e^{-x}). \tag{**}$$

Subtracting Equation (*) from (**) gives us $e^{2x} C'_2 = \sin(e^{-x})$, or $C'_2 =; e^{-2x} \sin(e^{-x})$. Then, from Equation (*), $C'_1 = e^{-x}(-e^{2x} C'_2) = -e^x(e^{-2x} \sin(e^{-x})) = -e^{-x} \sin(e^{-x})$. Therefore, $C_1 = \int \sin(e^{-x})(-e^{-x}) dx$ and $C_2 = -\int e^{-x} \sin(e^{-x})(-e^{-x}) dx$. Making the substitutions $u = e^{-x}, du = -e^{-x} dx$ in each integral, we find $C_1 = -\cos(e^{-x})$ and $C_2 = -\sin(e^{-x}) + e^{-x} \cos(e^{-x})$. Therefore, $y_{PNH} = C_1 e^x + C_2 e^{2x} = \left[ -\cos(e^{-x}) \right] e^x + \left[ -\sin(e^{-x}) + e^{-x} \cos(e^{-x}) \right] e^{2x} = -e^{2x} \sin(e^{-x})$, and the general solution is

$$y_{GNH} = y_{GH} + y_{PNH} = c_1 e^x + c_2 e^{2x} - e^{2x} \sin(e^{-x}).$$

∎

The preceding three examples involved quite a bit of algebra and calculus, but the method of variation of parameters is guaranteed to work. Even if the integrations of $C'_1(x)$ and $C'_2(x)$ can't be done in closed form, we can still use numerical methods such as Simpson's Rule to approximate the solution. The next example illustrates this kind of integration difficulty.

## ■ Example 4.4.4 Variation of Parameters—No Closed Form Solution

Let's solve the equation $y'' + y' - 2y = \ln x$. The characteristic equation of the associated homogeneous equation is $\lambda^2 + \lambda - 2 = 0$, with roots $-2$ and $1$, so we know that $y_{GH} = c_1 e^{-2x} + c_2 e^x$, where $c_1$ and $c_2$ are constants.

Next, we consider $y = C_1 e^{-2x} + C_2 e^x$, where $C_1$ and $C_2$ are unknown functions of $x$. Differentiating, we get

$$y' = \left(-2C_1 e^{-2x} + C_2 e^x\right) + \left(C_1' e^{-2x} + C_2' e^x\right) = -2C_1 e^{-2x} + C_2 e^x$$

because we must assume that

$$C_1' e^{-2x} + C_2' e^x = 0. \tag{#}$$

Then $y'' = 4C_1 e^{-2x} - 2C_1' e^{-2x} + C_2 e^x + C_2' e^x$.

After substituting these expressions for $y$, $y'$, and $y''$ into the equation $y'' + y' - 2y = \ln x$, we get

$$-2C_1' e^{-2x} + C_2' e^x = \ln x. \tag{##}$$

Now we must solve the following system for $C_1'$ and $C_2'$:

$$C_1' e^{-2x} + C_2' e^x = 0 \tag{#}$$

$$-2C_1' e^{-2x} + C_2' e^x = \ln x. \tag{##}$$

Subtracting (##) from (#) gives us $3C_1' e^{-2x} = -\ln x$, so that $C_1' = -\frac{1}{3} e^{2x} \ln x$ and $C_1(x) = -\frac{1}{3} \int e^{2x} \ln x \, dx = -\frac{1}{6} e^{2x} \ln x + \frac{1}{6} \int \frac{e^{2x}}{x} dx$. This integration was done manually (integration by parts: $u = \ln x$, $dv = e^{2x} dx$, etc.). A CAS might give an answer in terms of the "exponential integral," which you may not recognize. In any case, the integral $\int \frac{e^{2x}}{x} dx$ cannot be expressed in closed form.

Equation (#) tells us that $C_2' = e^{-x}\left(-C_1' e^{-2x}\right) = -e^{-3x}\left(-\frac{1}{3} e^{2x} \ln x\right) = \frac{1}{3} e^{-x} \ln x$, and an integration by parts leads to the conclusion that $C_2(x) = -\frac{1}{3} e^{-x} \ln x + \frac{1}{3} \int \frac{e^{-x}}{x} dx$.

The next to the last step is to calculate

$$y_{\text{PNH}} = c_1 y_1 + c_2 y_2 = \left(-\frac{1}{6} e^{2x} \ln x + \frac{1}{6} \int \frac{e^{2x}}{x} dx\right)\left(e^{-2x}\right)$$

$$+ \left(-\frac{1}{3} e^{-x} \ln x + \frac{1}{3} \int \frac{e^{-x}}{x} dx\right)\left(e^x\right)$$

$$= -\frac{\ln x}{2} + \frac{e^{-2x}}{6} \int \frac{e^{2x}}{x} dx + \frac{e^x}{3} \int \frac{e^{-x}}{x} dx.$$

Finally, the general solution is given by the formula

$$y_{\text{GNH}} = y_{\text{GH}} + y_{\text{PNH}} = c_1 e^{-2x} + c_2 e^x - \frac{\ln x}{2} + \frac{e^{-2x}}{6} \int \frac{e^{2x}}{x} dx + \frac{e^x}{3} \int \frac{e^{-x}}{x} dx. \quad \blacksquare$$

Another important feature of variation of parameters is that the method remains valid for a linear equation whose coefficients are continuous functions of the independent variable. Specifically,

Suppose $p(t)$ and $q(t)$ are continuous functions and $f(t)$ is continuous or piecewise continuous. If $y_{GH} = c_1 y_1(t) + c_2 y_2(t)$ is the general solution of $y'' + p(t)y' + q(t)y = 0$, then we can express the general solution of $y'' + p(t)y' + q(t)y = f(t)$ as $y = y_{GH} + C_1(t)y_1(t) + C_2(t)y_2(t)$, where $C_1(t)$ and $C_2(t)$ can be found by the method of variation of parameters.

Of course, in the context of an equation with nonconstant coefficients, finding linearly independent functions such that $y_{GH} = c_1 y_1(t) + c_2 y_2(t)$ can be difficult. Rather than focus on specialized techniques, we will be content to illustrate what to do once we have $y_{GH}$.

### ■ Example 4.4.5 Variation of Parameters—Nonconstant Coefficients

It is easy to see that $y_1(x) = x$ and $y_2(x) = 1/x$ are linearly independent solutions of the differential equation $x^2 y'' + xy' - y = 0$. If we suppose the functions are *not* independent, then $x = c/x$ for some constant $c$ and every $x \neq 0$. Now let $x = 1$ and $x = 2$, forcing the contradiction that $c$ must be equal to both 1 and 4. Thus, the functions are independent and $y_{GH} = c_1 x + c_2/x$.

Now we use the method of variation of parameters to find the general solution of $x^2 y'' + xy' - y = x, x \neq 0$.

We start with a trial solution $y_P = C_1 x + C_2/x$, where $C_1 = C_1(x)$ and $C_2 = C_2(x)$. Then $y_P' = C_1 + C_1' x - C_2/x^2 + C_2'/x = (C_1 - C_2/x^2) + (C_1' x + C_2'/x)$, where we impose the condition

$$C_1' x + \frac{C_2'}{x} = 0 \qquad (*)$$

before calculating the next derivative.

Now $y_P'' = C_1' + 2C_2/x^3 - C_2'/x^2$. Substituting in the nonhomogeneous equation, we get

$$x^2 \left[ C_1' + 2\,C_2/x^3 - C_2'/x^2 \right] + x \left[ C_1 - C_2/x^2 \right] - \left[ C_1 x + C_2/x \right] = x,$$

or

$$x^2 C_1' - C_2' = x. \qquad (**)$$

Equations $(*)$ and $(**)$ must be solved for $C_1'$ and $C_2'$: Multiplying $(*)$ by $x$ gives us $x^2 C_1' + C_2' = 0$, and adding this result to $(**)$ yields $2x^2 C_1' = x$, from which it follows that $C_1(x) = \frac{1}{2} \ln |x|$. Then Equation $(**)$ yields $C_2' = x^2 C_1' - x$, or $C_2' = x^2(1/(2x)) - x = -x/2$, so $C_2 = -x^2/4$.

Therefore, $y_{PNH} = C_1 x + C_2/x = (x/2) \ln |x| - x/4$ and $y_{GNH} = c_1 x + c_2/x + (x/2) \ln |x|$, where the term $-x/4$ of $y_{PNH}$ has been absorbed into the term $c_1 x$ from $y_{GH}$. ■

In summary, the method of variation of parameters works for all second-order linear differential equations provided that the coefficients are continuous functions of the independent variable. In those (limited) situations in which the method of undetermined coefficients can be used (Section 4.3), that method is usually easier than variation of parameters.

## Exercises 4.4

### A

Find the general solution of each of the equations in Problems 1–10 by using the method of variation of parameters.

1. $x'' - 2x' + x = \frac{e^t}{t}$
2. $y'' + 4y = 2\tan x$
3. $\ddot{r} + r = \frac{1}{\sin t}$
4. $y'' + 2y' + y = \frac{e^{-x}}{x}$
5. $y'' + 4y' + 4y = 3xe^{-2x}$
6. $y'' + y = \sec x$
7. $y'' + 2y' + y = e^{-x}\ln x$
8. $y'' - y = \sin^2 x$
9. $y'' - 3y' + 2y = \cos\left(e^{-x}\right)$
10. $y'' + 3y' + 2y = \frac{1}{1+e^x}$

### B

Find the general solution of each of the equations in Problems 1–5. Linearly independent solutions for the associated homogeneous equation are shown next to each nonhomogeneous equation.

1. $x^2 y'' - xy' + y = x; y_1 = x, y_2 = x\ln x$
2. $2x^2 y'' + 3xy' - y = \frac{1}{x}; y_1 = \sqrt{x}, y_2 = \frac{1}{x}$
3. $y'' - \frac{2}{x}y' + \frac{2}{x^2}y = x\ln x; y_1 = x, y_2 = x^2$
4. $y'' + \frac{2}{x}y' + y = \frac{1}{x}, x \neq 0; y_1 = \frac{\sin x}{x}, y_2 = \frac{\cos x}{x}$
5. $y'' - 2(\tan x)y' = 1; y_1 = C(\text{a constant}), y_2 = \tan x$
6. Consider the equation $x^2 y'' - 4xy' + 6y = 0$.

   a. Show that the general solution of this equation is $y = c_1 x^3 + c_2 x^2$.
   b. Use the result of part (a) to find the general solution of

   $$x^2 y'' - 4xy' + 6y = x^4.$$

7. Find the general solution of $x^2 y'' - 4xy' + 6y = x^4$ [part (b) of the previous problem] by reducing this *Cauchy-Euler equation* to one with constant coefficients via the substitution $x = e^z$. (See Section C of Exercises 4.3.)

### C

1. Show that the solution of the IVP $y'' + a^2 y = F(x), y(0) = y'(0) = 0$, is

$$y = \frac{1}{a}\int_0^x F(u)\sin a(x-u)du.$$

**2.** Suppose $y'' + a(x)y' + b(x)y = 0$ has continuous coefficient functions $a(x)$ and $b(x)$ on an interval $I$. Assume $y_1(x)$ is a solution of the equation and is nonzero on a subinterval $J$ of $I$.

**a.** Let $y_2(x) = y_1(x)u(x)$, where $u(x)$ is a nonconstant function. Assuming that $y_2(x)$ is a solution of the differential equation, show that

$$u(x) = c_1 \int \frac{e^{-\int a(x)dx}}{y_1^2(x)} dx + c_2,$$

where $c_1$ and $c_2$ are arbitrary constants.

**b.** Show that $y_2(x) = y_1(x) \int \frac{e^{-\int a(x)dx}}{y_1^2(x)} dx$ defines a solution on $J$ that is independent of $y_1(x)$.

**3.** If the solution of $y'' + p(x)y' + q(x)y = 0$ is $\alpha y_1(x) + \beta y_2(x)$, show that the general solution of $y'' + p(x)y' + q(x)y = r(x)$ is

$$y = c_1 y_1(x) + c_2 y_2(x) + y_2(x) \int \frac{r(x)y_1(x)}{W(y_1, y_2)} dx - y_1(x) \int \frac{r(x)y_2(x)}{W(y_1, y_2)} dx,$$

where $W(y_1, y_2) = y_1(x)y_2'(x) - y_2(x)y_1'(x)$ is called the *Wronskian* of $y_1(x)$ and $y_2(x)$ and is not the zero function. Discuss the case in which $W(y_1, y_2)$ is identically equal to zero.

**4.** An alternative method of solving a *Cauchy-Euler differential equation* $x^2y'' + axy' + by = 0$ (see Section C of Exercises 4.3), where $a$ and $b$ are real constants, requires the substitution $y = x^r$, where $x > 0$.

**a.** If $y = x^r$, calculate $y'$ and $y''$ and show that for $y$ to be a solution of the Cauchy-Euler equation, $r$ must satisfy the *indicial equation*

$$r^2 + (a - 1)r + b = 0.$$

**b.** Show that if the indicial equation has real roots $r_1 \neq r_2$, then $y_1 = x^{r_1}$ and $y_2 = x^{r_2}$ are linearly independent solutions of the differential equation.

**c.** Show that if the indicial equation has complex conjugate roots $r_1 = \alpha + \beta i$ and $r_2 = \alpha - \beta i$, then $y_1 = x^\alpha \cos(\beta \ln x)$ and $y_2 = x^\alpha \sin(\beta \ln x)$ are solutions. [*Note:* $x^{\alpha \pm \beta i} = x^\alpha e^{\pm i\beta \ln x}$ for $x > 0$.]

## 4.5 HIGHER-ORDER LINEAR EQUATIONS WITH CONSTANT COEFFICIENTS

Linearity is such a marvelous property that we can generalize our work in the preceding few sections in a very natural way. The details may get a bit complicated, but the theory is crisp and clear.

If $y$ is a function that is $n$-times differentiable and $a_0, a_1, a_2, \ldots, a_n$ are constants, $a_n \neq 0$, then we can define the *nth-order linear operator* $L$ as follows:

$$L(y) = a_n y^{(n)} + a_{n-1} y^{(n-1)} + \cdots + a_2 y'' + a_1 y' + a_0 y.$$

Any $n$th-order linear differential equation with constant coefficients can be expressed concisely as $L(y) = f(t)$. If $f(t) \equiv 0$, then the equation is called a **homogeneous** *nth-order*

**linear equation with constant coefficients**. If $f(t)$ is not the zero function, then we have a **nonhomogeneous $n$th-order linear equation with constant coefficients**.

An important property of such $n$th-order equations is the (extended) Superposition Principle:

### Superposition Principle

If $y_j$ is a solution of $L(y) = f_j$ for $j = 1, 2, \ldots, n$, and $c_1, c_2, \ldots, c_n$ are arbitrary constants, then $c_1 y_1 + c_2 y_2 + \cdots + c_n y_n$ is a solution of $L(y) = c_1 f_1 + c_2 f_2 + \cdots + c_n f_n$—that is,

$$L(c_1 y_1 + c_2 y_2 + \cdots + c_n y_n) = c_1 L(y_1) + c_2 L(y_2) + \cdots + c_n L(y_n)$$

$$= c_1 f_1 + c_2 f_2 + \cdots + c_n f_n.$$

First, let's look at an $n$th-order *homogeneous* linear equation with constant coefficients

$$a_n y^{(n)} + a_{n-1} y^{(n-1)} + \cdots + a_2 y'' + a_1 y' + a_0 y = 0.$$

For such an equation, there's a neat algorithm for finding the general solution, a generalization of the procedure we've already seen: First, find the roots of the characteristic equation $a_n \lambda^n + a_{n-1} \lambda^{n-1} + \cdots + a_1 \lambda + a_0 = 0$. *You should see how to form this equation.*

Focus on the fact that the characteristic equation of an $n$th-order linear equation is an $n$th-degree polynomial equation. Realize that once a polynomial has degree greater than or equal to 5, there is no longer a general formula that gives the zeros. (Even the formulas that exist for the zeros of third- and fourth-degree polynomials are very unwieldy.) In general, the only practical way to tackle such equations is to use *approximation* methods. A CAS or a graphing calculator should have various algorithms implemented to solve (or approximate the solutions of) polynomial equations.

Next, group these roots as follows:

1. Distinct real roots

2. Distinct complex conjugate pairs $p \pm qi$

3. Multiple real roots

4. Multiple complex roots

Then the general solution is a sum of $n$ terms of the forms

1. $c_i e^{\lambda_i t}$ for each distinct real root $\lambda_i$

2. $e^{pt}(c_1 \cos qt + c_2 \sin qt)$ for each distinct complex pair $p \pm qi$

3. $\left(c_1 + c_2 t + \cdots + c_k t^{k-1}\right) e^{\lambda_i t}$ for each multiple real root $\lambda_i$, where $k$ is the multiplicity of that root

4. $e^{pt}(c_1 \cos qt + c_2 \sin qt) + t\, e^{pt}(c_3 \cos\ qt + c_4 \sin qt) + \cdots + t^{k-1} e^{pt}(c_{2k-1} \cos qt + c_{2k} \sin qt)$ for each multiple complex pair of roots $p \pm qi$, where $k$ is the multiplicity of the pair $p \pm qi$

Now let's see how to use this procedure to solve some higher-order homogeneous linear equations with constant coefficients.

### ■ Example 4.5.1 Solving a Fourth-Order Homogeneous Linear Equation

Let's find the general solution of the fourth-order equation

$$x^{(4)} - 3x'' + 2x' = 0.$$

The characteristic equation is $\lambda^4 - 3\lambda^2 + 2\lambda = \lambda(\lambda^3 - 3\lambda + 2) = 0$, whose roots are 0, 1, 1, and $-2$. (*Verify this.*) Thus, we have two distinct real roots and another real root of multiplicity 2.

According to the process described here, the general solution is

$$x = c_1 e^{0 \cdot t} + c_2 e^{-2t} + (c_3 + c_4 t)\, e^{1 \cdot t} = c_1 + c_2 e^{-2t} + (c_3 + c_4 t)\, e^{t}.$$

(*You should check that this is a solution, manually or by using a CAS.*)  ■

### ■ Example 4.5.2 Solving an Eighth-Order Homogeneous Linear Equation

The equation $64y^{(8)} + 48y^{(6)} + 12y^{(4)} + y'' = 0$ should be interesting to tackle. The characteristic equation is $64\lambda^8 + 48\lambda^6 + 12\lambda^4 + \lambda^2 = 0$. A CAS gives the roots 0, 0, $i/2$, $-i/2$, $i/2$, $-i/2$, $i/2$, and $-i/2$. Grouping these, we see that 0 is a real root of multiplicity 2, whereas the complex conjugate pair $\pm i/2$ ($= 0 \pm i/2$) has multiplicity 3. Therefore, the form of the general solution of this eighth-order equation is

$$
\begin{aligned}
y(t) =&\ (c_1 + c_2 t)\, e^{0 \cdot t} + e^{0 \cdot t} \left( c_3 \cos\left(\frac{t}{2}\right) + c_4 \sin\left(\frac{t}{2}\right) \right) + t e^{0 \cdot t} \\
&\left( c_5 \cos\left(\frac{t}{2}\right) + c_6 \sin\left(\frac{t}{2}\right) \right) + t^2 e^{0 \cdot t} \left( c_7 \cos\left(\frac{t}{2}\right) + c_8 \sin\left(\frac{t}{2}\right) \right) \\
=&\ c_1 + c_2 t + \left( c_3 + c_5 t + c_7 t^2 \right) \cos\left(\frac{t}{2}\right) + \left( c_4 + c_6 t + c_8 t^2 \right) \sin\left(\frac{t}{2}\right).
\end{aligned}
$$
■

For the nonhomogeneous case, once again the theory is simple:

> The general solution, $y_{\text{GNH}}$, of an $n$th-order linear nonhomogeneous equation $a_n y^{(n)} + a_{n-1} y^{(n-1)} + \cdots + a_2 y'' + a_1 y' + a_0 y = f(t)$ is obtained by finding a particular solution, $y_{\text{PNH}}$, of the nonhomogeneous equation and adding it to the general solution, $y_{\text{GH}}$, of the associated homogeneous equation: $y_{\text{GNH}} = y_{\text{GH}} + y_{\text{PNH}}$.

As before, the challenge is to find a particular solution of the nonhomogeneous equation. But once again we can use the method of variation of parameters or the method of undetermined coefficients ("educated guessing").

If we look back at Section 4.4 to see the number of calculations required to implement variation of parameters, we realize that the work can be formidable for equations of order 3 and above. But there is no need to do problems of higher order manually because any CAS will use the appropriate method efficiently to give us a general solution or solve an IVP. (One of the best methods for handling single linear equations and systems of linear equations is the *Laplace transform*, which we'll study in Chapter 6.) For now we'll just give an example of solving a higher-order linear equation, with some of the gory details left out.

### ■ Example 4.5.3 Solving a Nonhomogeneous Third-Order Equation

Suppose we want to find the general solution of $y''' - y'' - 6y' = 3t^2 + 2$. The first thing to do is to find the general solution of the associated homogeneous equation $y''' - y'' - 6y' = 0$. The characteristic equation is $\lambda^3 - \lambda^2 - 6\lambda = \lambda(\lambda^2 - \lambda - 6) = \lambda(\lambda - 3)(\lambda + 2) = 0$, with roots 0, 3, and $-2$, so the general solution of the homogeneous equation is $c_1 e^{0 \cdot t} + c_2 e^{3t} + c_3 e^{-2t}$, or $c_1 + c_2 e^{3t} + c_3 e^{-2t}$.

Next, we look for a particular solution of the original nonhomogeneous equation. Examining the right-hand side of the equation, we can guess that a particular solution will be a polynomial in $t$. If the degree of this guessed-at polynomial is $n$, then the three individual derivative terms making up the differential equation will leave behind polynomials of degrees $n-3$, $n-2$, and $n-1$. In order for the combination $y''' - y'' - 6y'$ to produce the second-degree polynomial $3t^2 + 2$, we must have $n - 1 = 2$—that is, the polynomial we're looking for must be a third-degree polynomial, say $y(t) = At^3 + Bt^2 + Ct + D$, where $A$, $B$, $C$, and $D$ are *undetermined coefficients*. (Think about the reasoning that led to this form for $y$.)

Substituting this guess into the nonhomogeneous equation, we find that

$$-18At^2 - (12B + 6A)t + (6A - 2B - 6C) = 3t^2 + 2.$$

Equating coefficients of terms of equal degree on both sides, we get the algebraic equations

$$-18A = 3 \quad \text{[Second-degree terms must match.]}$$

$$-(12B + 6A) = 0 \quad \text{[First-degree terms must match.]}$$

$$6A - 2B - 6C = 2 \quad \text{[Constant terms must match.]}$$

Starting from the top, we can solve the equations successively to obtain $A = -\frac{1}{6}$, $B = \frac{1}{12}$, and $C = -\frac{19}{36}$.

Therefore, $y_{\text{PNH}} = -\frac{1}{6}t^3 + \frac{1}{12}t^2 - \frac{19}{36}t$ and the general solution of the nonhomogeneous equation is given by

$$y = y_{\text{GNH}} = y_{\text{GH}} + y_{\text{PNH}} = c_1 + c_2 e^{3t} + c_3 e^{-2t} - \frac{1}{6}t^3 + \frac{1}{12}t^2 - \frac{19}{36}t.$$

■

As we might suspect from the discussion at the end of Section 4.4, variation of parameters can be used for all higher-order linear differential equations, provided that the coefficients are continuous functions of the independent variable.

## Exercises 4.5

### A

Find the general solution of each of the higher-order equations in Problems 1–10, using a graphing calculator or CAS only to solve each characteristic equation.

1. $y''' - 2y'' - 3y' = 0$
2. $y''' - 3y'' + 3y' - y = 0$
3. $y''' + 2y'' + y' = 0$
4. $y''' + 4y'' + 13y' = 0$
5. $y''' - 12y'' + 22y' - 20y = 0$
6. $y^{(4)} + 2y'' + y = 0$
7. $y^{(4)} - 13y'' + 36y = 0$
8. $y^{(4)} + 13y'' + 36y = 0$
9. $y^{(4)} - 3y'' + 2y' = 0$
10. $y^{(5)} + 2y''' + y' = 0$

### B

1. Find the general solution of the following equation, using technology only to solve a characteristic equation:

$$y^{(7)} - 14y^{(6)} + 80y^{(5)} - 242y^{(4)} + 419y^{(3)} - 416y'' + 220y' - 48y = 0.$$

2. Apply your CAS solver to find the general solution of the equation in the preceding problem.
3. The author of a classic differential equations text[5] once wrote

   *In preparing problems and examinations … teachers (including the author) must use some restraint. It is not reasonable to expect students in this course to have computing skill and equipment necessary for efficient solving of equations such as*

$$4.317\frac{d^4y}{dx^4} + 2.179\frac{d^3y}{dx^3} + 1.416\frac{d^2y}{dx^2} + 1.295\frac{dy}{dx} + 3.169y = 0.$$

   Demonstrate that technology has advanced in the past five decades by feeding this equation into your CAS and obtaining the general solution. (You may have to use some "simplify" commands to get a neat answer.)
4. Solve the IVP $3y''' + 5y'' + y' - y = 0$; $y(0) = 0, y'(0) = 1, y''(0) = -1$.
5. A uniform horizontal beam sags by an amount $y = y(x)$ at a distance $x$ from one end. For a fairly rigid beam with uniform loading, $y(x)$ typically satisfies an equation of the form $d^4y/dx^4 = R$, where $R$ is a constant depending on the load being carried and on the characteristics of the beam itself. If the ends of the beam are supported at $x = 0$ and $x = L$, then $y(0) = y(L) = 0$. The extended beam also behaves as though its profile had an inflection point at each support so that

$$y''(0) = y''(L) = 0.$$

---

[5] Ralph P. Agnew, *Differential Equations*, 2nd ed. (New York: McGraw-Hill, 1960): 176.

**a.** Use the multiple eigenvalue of the associated homogeneous equation to find the general solution of the homogeneous equation.

**b.** Show that the sag (vertical deflection) at point $x$ is

$$\frac{1}{24}R\left(x^4 - 2Lx^3 + L^3x\right), \quad 0 \le x \le L.$$

**6.** Solve the IVP $y^{(5)} = y'$; $y(0) = 0, y'(0) = 1, y''(0) = 0, y'''(0) = 1, y^{(4)}(0) = 2$.

**7.** For all positive integers $n \ge 2$, find the general solution of the equation $x^{(n)} = x^{(n-2)}$.

**8.** Find the general solution of the equation $y''' - 2y'' - y' + 2y = e^x$.

**9.** Find the general solution of the equation $\dddot{y} + 5\ddot{y} - 6\dot{y} = 9e^{3t}$.

**10.** Find the general solution of the equation $y''' + 6y'' + 11y' + 6y = 6x - 7$.

**11.** Find the general solution of $y''' - 5y'' - 2y' + 24y = x^2 e^{3x}$.

**12.** Solve the initial value problem $y''' + 5y'' - 6y' = 3e^x, y(0) = 1, y'(0) = 3/7, y''(0) = 6/7$.

## C

**1.** Use the method of undetermined coefficients to solve the equation

$$y^{(4)} + y'' = 3x^2 + 4\sin x - 2\cos x.$$

**2.** Consider the IVP $y^{(4)} + 8y'' + 16y = 0$, with $y^{(k)}(0)$ given for $k = 0, 1, 2, 3$. For what initial values $y''(0)$ and $y'''(0)$ will solutions of this equation be periodic?

**3.** Consider a third-order differential equation $y''' + py'' + qy' + ry = g$, where $p, q, r$, and $g$ are continuous functions. Suppose $y_{GH} = c_1y_1 + c_2y_2 + c_3y_3$ is known.

**a.** Write down a form for $y_{PNH}$ in terms of the known solutions of the homogeneous problem and unknown coefficient functions $C_1, C_2$, and $C_3$.

**b.** Derive a system of equations that determines $C_1, C_2$, and $C_3$. This system should involve $C_1', C_2', C_3', y_i, y_i'$, and $y_i'', i = 1, 2, 3$.

**c.** Solve $y''' - 2y'' - y' + 2y = e^x$ by using the results of parts (a) and (b).

**4.** Find a formula involving integrals for a particular solution of the equation

$$x^3 y''' - 3x^2 y'' + 6xy' - 6y = g(x), x > 0.$$

[*Hint*: Verify that $x, x^2$, and $x^3$ are solutions of the homogeneous equation.]

## 4.6 HIGHER-ORDER EQUATIONS AND THEIR EQUIVALENT SYSTEMS

To see where we're headed, think back to the first time you had to solve the following kind of word problem:

> Lenston has 21 coins, all nickels and dimes, in his pockets. They amount to $1.75. How many dimes does he have?

The first time you saw this problem, you were probably shown a solution like this one:

> Let $x$ be the number of dimes. Then the total *amount* corresponding to dimes is $10x$ cents. The *number* of nickels must be $21 - x$, so the *amount* corresponding to nickels is $5(21 - x)$ cents. Because the total amount of money in Lenston's pockets is \$1.75—or 175 cents—we have the equation $10x + 5(21 - x) = 175$, equivalent to $5x + 105 = 175$, which has the solution $x = 14$. Thus, Lenston has 14 dimes (and $21 - 14 = 7$ nickels).

A bit later in your algebra course, you could have seen the same problem again, but this time you were probably shown how to turn this problem into a *system* problem:

> Let $x$ be the number of dimes and let $y$ be the number of nickels. Then the words of the problem tell us two things, one fact about the *number* of coins and one fact about the *amount* of money: (1) $x + y = 21$ and (2) $10x + 5y = 175$. In other words, viewed this way, the problem gives us the *system* of equations

$$x + y = 21$$
$$10x + 5y = 175.$$

> This system can be solved by elimination (multiply the first equation by $-5$ and then add the result to the second equation) or by substitution (solve the first equation for $x$, for example, and then substitute for $x$ in the second equation).

The most important consequence of looking at our problem as a system problem is that the system has a very nice geometrical interpretation as a set of two straight lines (Figure 4.4). The solution of the system (and of our original problem) is given by the coordinates of the point where the lines intersect: $x = 14$, $y = 7$.

Similarly, for differential equations, a systems approach has certain advantages, especially the graphical interpretation of a problem and its solution. Also, certain problems may naturally occur in system form. For example, we may want to compute the trajectory of a baseball.

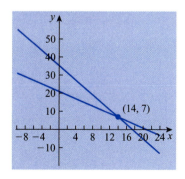

**FIGURE 4.4**
*Graphs of* $x + y = 21$ *and* $10x + 5y = 175$

In this case, it is natural to consider the components, $u$ and $v$, of the ball's velocity in both its horizontal $(x)$ and vertical $(y)$ directions, respectively. A system[6] arising from this problem is

$$mu\frac{du}{dx} = -F_L \sin\theta - F_D \cos\theta$$

$$mv\frac{dv}{dy} = F_L \cos\theta - F_D \sin\theta - mg.$$

Similarly, in an ecological study, we may want to analyze the interaction of two or more biological species, each of which needs its own equation to represent its growth rate and its relationship to the other species.

### 4.6.1 Conversion Technique I: Converting a Higher-Order Equation into a System

Now that our previous discussion has prepared us to see even simple problems as systems, we can tackle some higher-order differential equations. The key here is the following result:

> Any single $n$th-order differential equation can be converted into an equivalent system of first-order equations. More precisely, any $n$th-order differential equation of the form
>
> $$x^{(n)} = F\left(t, x, x', x'', \ldots, x^{(n-1)}\right)$$
>
> can be converted into an equivalent system of $n$ first-order equations by letting
>
> $$x_1 = x, x_2 = x', x_3 = x'', \ldots, x_n = x^{(n-1)}.$$

Here, *equivalent* means that a function $x = u(t)$ is a solution of $x^{(n)} = F(t, x, x', \ldots, x^{(n-1)})$ if and only if the ordered $n$-tuple of functions $\left(u(t), u'(t), \ldots, u^{(n-1)}(t)\right)$ is a solution of the system $x_1' = x_2, x_2' = x_3, \ldots, x_n' = F(t, x_1, x_2, \ldots, x_n)$. In particular, our substitution scheme indicates that any solution of the single $n$th-order equation is the first component of the $n$-tuple that's the solution of the system and vice versa.

After looking at some examples of how this conversion technique works, we'll introduce the geometric/graphical significance of this method.

### ■ Example 4.6.1 Converting a Second-Order Linear Equation

As we saw in Section 4.1 and will see again in Section 4.7 and its exercises, the second-order linear equation $2\frac{d^2x}{dt^2} + 3\frac{dx}{dt} + x = 0$ could represent the motion of a weight attached to a spring, the flow of electricity through a circuit, or other important phenomena.

---

[6] Robert B. Banks, *Towing Icebergs, Falling Dominoes, and Other Adventures in Applied Mathematics* (Princeton, NJ: Princeton University Press, 1998).

Using the substitutions described previously, we introduce new variables $x_1$ and $x_2$: Let $x_1 = x$ and $x_2 = \frac{dx}{dt}$. Now isolate the highest derivative (the second) in the original equation, and then substitute the new variables in the right-hand side:

$$(1) \quad \frac{d^2x}{dt^2} + \frac{3}{2}\frac{dx}{dt} + \frac{1}{2}x = 0$$

$$(2) \quad \frac{d^2x}{dt^2} = -\frac{3}{2}\frac{dx}{dt} - \frac{1}{2}x$$

$$(3) \quad \frac{d^2x}{dt^2} = -\frac{3}{2}x_2 - \frac{1}{2}x_1$$

In terms of the new variables, we see that $\frac{dx_1}{dt} = \frac{dx}{dt} = x_2$ and $\frac{dx_2}{dt} = \frac{d}{dt}\left(\frac{dx}{dt}\right) = \frac{d^2x}{dt^2} =$ [from step (3) above] $-\frac{3}{2}x_2 - \frac{1}{2}x_1$. From this, we see that our original second-order equation leads to the following system of linear first-order equations in two unknown functions $x_1$ and $x_2$:

$$(A) \quad \frac{dx_1}{dt} = x_2$$

$$(B) \quad \frac{dx_2}{dt} = -\frac{3}{2}x_2 - \frac{1}{2}x_1$$

This system is *equivalent* to the original single differential equation in the sense that any solution $x(t)$ of the original equation yields solutions $x_1(t) = x(t)$ and $x_2(t) = \frac{d}{dt}x(t)$ of the system, and any solution $(x_1(t), x_2(t))$ of the system gives us a solution $x(t) = x_1(t)$ of the original equation.

Let's follow up on the first part of that statement. From our work in Section 4.1, we know that $x(t) = e^{-t/2} + 2e^{-t}$ is a solution of the original second-order equation. Then the pair $x_1(t) = x(t) = e^{-t/2} + 2e^{-t}$ and $x_2(t) = \frac{d}{dt}x(t) = -\frac{1}{2}e^{-t/2} - 2e^{-t}$ constitutes a solution of the system. (*Verify this!*) ■

Let's look at a few more examples of this technique of converting a higher-order equation into a system of first-order equations.

### ■ Example 4.6.2 Converting a Second-Order Nonlinear Equation

Suppose we have the second-order nonlinear equation $y'' = y^3 + (y')^3$. Let $x_1 = y$ and $x_2 = y'$. Then $x_1' = y' = x_2$, $y'' = x_2'$, $y^3 = x_1^3$, and $(y')^3 = x_2^3$, so we can rewrite $y'' = y^3 + (y')^3$ as $x_2' = x_1^3 + x_2^3$.

Finally, putting these pieces together, we can write the original equation as the following equivalent nonlinear system in $x_1$ and $x_2$:

$$x_1' = x_2$$
$$x_2' = x_1^3 + x_2^3.$$

■

## ■ Example 4.6.3 Converting a Third-Order Equation

The nonautonomous third-order linear equation

$$\frac{d^3x}{dt^3} - \frac{d^2x}{dt^2} + 2t\frac{dx}{dt} - 3x + 6 = 0$$

can be changed into a system of first-order equations as follows: Let $x_1 = x, x_2 = \frac{dx}{dt}$, and $x_3 = \frac{d^2x}{dt^2}$. Then $\frac{dx_1}{dt} = \frac{dx}{dt} = x_2$, $\frac{dx_2}{dt} = \frac{d}{dt}\left(\frac{dx}{dt}\right) = \frac{d^2x}{dt^2} = x_3$, and $\frac{dx_3}{dt} = \frac{d}{dt}\left(\frac{d^2x}{dt^2}\right) = \frac{d^3x}{dt^3}$.

Solving the original equation for $\frac{d^3x}{dt^3}$ and then substituting the new variables $x_1, x_2$, and $x_3$, we have

$$\frac{d^3x}{dt^3} = \frac{d^2x}{dt^2} - 2t\frac{dx}{dt} + 3x - 6 = x_3 - 2tx_2 + 3x_1 - 6.$$

Putting all the information together, we see that the original third-order equation is equivalent to the system of three first-order equations

$$\frac{dx_1}{dt} = x_2$$

$$\frac{dx_2}{dt} = x_3$$

$$\frac{dx_3}{dt} = x_3 - 2tx_2 + 3x_1 - 6.$$

To be mathematically precise, we can describe this system as a *three-dimensional nonautonomous linear system with independent variable t and dependent variables $x_1, x_2$, and $x_3$.* ■

As we'll see later in this chapter, an autonomous system has a nice graphical interpretation that gives us a neat qualitative analysis. We lose some of this power when we are dealing with a nonautonomous system. But even when we are confronted with a nonautonomous equation, a simple variation of the conversion technique we've been illustrating will allow us to transform the equation into an autonomous system. To convert a single *nonautonomous* nth-order equation into an equivalent *autonomous* system (one whose equations do not explicitly contain the independent variable t), we need $n + 1$ first-order equations: $x_1 = x, x_2 = x', x_3 = x'', \dots, x_n = x^{(n-1)}, x_{n+1} = t$. We see this in the next example.

## ■ Example 4.6.4 Converting a Nonautonomous Equation into an Autonomous System

The nonautonomous second-order linear equation $2\frac{d^2x}{dt^2} + 3\frac{dx}{dt} + x = 50\sin t$ could be handled in the same way as the equation in Example 4.6.3, but instead we'll demonstrate the extension of the conversion technique.

Start by letting $x_1 = x$ and $x_2 = \frac{dx}{dt}$ as before, but also introduce $x_3 = t$. Then

$$\frac{dx_1}{dt} = \frac{dx}{dt} = x_2,$$

$$\frac{dx_2}{dt} = \frac{d^2x}{dt^2} = \frac{1}{2}\left(-3\frac{dx}{dt} - x + 50\sin t\right) = \frac{1}{2}(-3x_2 - x_1 + 50\sin x_3),$$

and

$$\frac{dx_3}{dt} = 1,$$

so the equivalent system is

$$\frac{dx_1}{dt} = x_2$$

$$\frac{dx_2}{dt} = \frac{1}{2}(-x_1 - 3x_2 + 50\sin x_3)$$

$$\frac{dx_3}{dt} = 1.$$

Our second-order equation has been replaced by an equivalent *autonomous three-dimensional* system. If we had not used the third variable $x_3$ and had written our equation as a system of *two* equations, the second equation would have been nonautonomous. We would have had

$$\frac{dx_1}{dt} = x_2 \quad \text{and} \quad \frac{dx_2}{dt} = \frac{1}{2}(-x_1 - 3x_2 + 50\sin t)$$

as the system, with the explicit presence of $t$ in the second equation making this equation (and therefore the system) nonautonomous. ∎

Of course, we should be able to convert an initial-value problem into a system IVP as well. If you think about this, we would expect that the original initial conditions would have to expand to cover each first-order equation in the system. The next example shows how this works.

### ■ Example 4.6.5 Converting a Second-Order Initial-Value Problem

The nonautonomous second-order linear IVP $y'' - xy' - x^2y = 0$; $y(0) = 1, y'(0) = 2$ can be transformed into a system IVP as follows. Let $u_1 = y$ and $u_2 = y'$. (We're using a different letter for the new variables to avoid confusion with the original independent variable $x$.) We see that $u_1' = y' = u_2$ and $u_2' = y'' = xy' + x^2y = xu_2 + x^2u_1$. Then, because $u_1 = y, y(0) = 1$ implies that $u_1(0) = 1$; and $y'(0) = 2$ implies that $u_2(0) = 2$ because $u_2 = y'$. Therefore, the original IVP becomes the system IVP

$$u_1' = u_2$$

$$u_2' = xu_2 + x^2u_1; \quad u_1(0) = 1, \quad u_2(0) = 2.$$

Note that because each equation in the system is first-order, we need only one initial condition for each new variable. *What would the equivalent autonomous system look like?* ■

### 4.6.2 Conversion Technique II: Converting a System into a Higher-Order Equation

We showed in Examples 4.6.1–4.6.5 how higher-order equations can always be transformed into equivalent first-order systems. However, the reverse process may be difficult or impossible for some systems.[7] The next example illustrates how a linear system can be represented by a single higher-order equation.

### ■ Example 4.6.6 Converting a System into a Single Equation

Can you convert the system

$$(1) \quad y' = z$$

$$(2) \quad z' = w$$

$$(3) \quad w' = x - 3y - 6z - 3w,$$

where $y$, $z$, and $w$ are functions of $x$, into an equivalent single higher-order equation?

Sure, you can. Just look back at what we did in our earlier examples, but start with the *last* equation and work backward: Differentiating Equation (2) gives us $z'' = w'$. But Equation (1) says that $z'' = (y')'' = y'''$, so that $w' = y'''$. Now we use this last fact to rewrite Equation (3) as

$$y''' = x - 3y - 6z - 3w$$

$$= x - 3y - 6y' - 3z' \quad \text{[from (1) and (2)]}$$

$$= x - 3y - 6y' - 3y'' \quad \text{[from (1)]}$$

or $y''' + 3y'' + 6y' + 3y = x$, a third-order linear nonautonomous differential equation. ■

### 4.6.3 Looking Ahead

Now that we've seen how to transform any differential equation of order greater than 1 into a system of first-order equations, how can we use this information to gain insight into the behavior of solutions of higher-order equations?

---

[7] See, for example, Section 6.4 of *Differential Equations: A Dynamical Systems Approach: Higher-Dimensional Systems* by J. H. Hubbard and B. H. West (New York: Springer-Verlag, 1995).

The next section will exploit the geometric (graphical) aspects of an autonomous system of equations and give us qualitative tools for analysis. We'll see that the qualitative approach will give us useful information not easily obtained otherwise. We'll also discuss important applied examples, both linear and nonlinear. Later in Chapter 4, we'll deal with numerical approximations to solutions of systems of equations. Chapter 5 will explore linear autonomous systems thoroughly, and Chapter 7 will introduce valuable methods for analyzing nonlinear systems of equations.

## Exercises 4.6

### A

Write each of the higher-order ODEs or systems of ODEs in Problems 1–11 as a system of first-order equations. If initial conditions are given, rewrite them in terms of the first-order system.

1. $\frac{d^2x}{dt^2} - x = 1$
2. $(x'')^2 - (\sin t)x' = x \cos t$
3. $x^2 y'' - 3xy' + 4y = 5 \ln x$
4. $\ddot{x} + (\dot{x})^2 + x(x-1) = 0$
5. $x''' - tx'' + x' - 5x + t^2 = 0$
6. $y^{(4)} + y = 0$
7. $w^{(4)} - 2w''' + 5w'' + 3w' - 8w = 6 \sin(4t)$
8. $\ddot{y} + y = t;\quad y(0) = 1, y'(0) = 0$
9. $x'' + 3x' + 2x = 1;\quad x(0) = 1, x'(0) = 0$
10. $\frac{d^2x}{dt^2} = -x,\ \frac{d^2y}{dt^2} = y$ [Hint: Write each second-order equation as two first-order equations.]
11. $x\frac{d^2y}{dt^2} - y = 4t,\ 2\frac{d^2x}{dt^2} + \left(\frac{dy}{dt}\right)^2 = x$
    (Convert each second-order equation into two first-order equations.)

Write each of the systems of equations in Problems 12–16 as a single second-order equation, rewriting any initial conditions as necessary.

12. $\frac{dy}{dt} = x,\ \frac{dx}{dt} = -y;\quad y(0) = 0, x(0) = 1$
13. $\frac{du}{dx} = 2v - 1,\ \frac{dv}{dx} = 1 + 2u$
14. $x' = x + y,\ y' = x - y$
15. $\frac{dx}{dt} = 7y - 4x - 13,\ \frac{dy}{dt} = 2x - 5y + 11;\quad x(0) = 2, y(0) = 3$
16. $x' = y + \sin x,\ y' = \cos(x + y)$

### B

1. The equation $\frac{d^2x}{dt^2} + 4\frac{dx}{dt} + 4x = 0$ describes the position, $x(t)$, of a particular mass attached to a spring and set in motion by pulling it down 2 ft below its equilibrium position $(x = 0)$ and giving it an initial velocity of 2 ft/sec in the upward direction. Some air resistance is assumed. Express this equation as a system of first-order equations and describe what each equation of the system represents.

2. In electrical circuit theory, the current $I$ is the derivative of the charge $Q$. By making this natural substitution $Q' = I$ in the equation

$$LQ'' + RQ' + \frac{1}{C}Q = E(t),$$

transform the equation into an equivalent system of two first-order equations.

3. An object placed in water, pushed down a certain distance below the water and then released, has its bobbing motion described by the equation

$$\frac{d^2y}{dt^2} + \left(\frac{g}{s_0}\right)y = 0,$$

where $y$ is the vertical displacement from its equilibrium position, $g$ is the acceleration due to gravity, and $s_0$ is the initial depth. Express this equation as a system of first-order equations.

4. The second-order nonlinear equation $\frac{d^2x}{dt^2} + \frac{g}{L}\sin x = 0$ describes the swinging of a pendulum, where $x$ is the angle the pendulum makes with the vertical, $g$ is the acceleration due to gravity, and $L$ is the pendulum's length. Convert this equation into a nonlinear system of first-order equations.

5. The equation $y''' + y' - \cos y = 0$ describes a geometrical model of crystal growth. Express this third-order equation as a system of three first-order equations.

6. The equation $y^{(4)} + \lambda(yy''' - y'y'') - y' = 0$, where $\lambda$ is a positive parameter, arises in a nonlinear "boundary layer" problem in physical oceanography. Write this equation as a system of four first-order equations.

7. Rewrite the system IVP given in Example 4.6.5 as an equivalent *autonomous* system.

8. Consider the equation $y'' + y = 0$.

   a. Convert this equation into a system with variables $u$ and $v$.
   b. Use the result of part (a) and the Chain Rule to conclude that $u^2 + v^2$ is a constant.

**C**

1. Write the following system of equations as a single fourth-order equation, with appropriate initial conditions:

$$\frac{d^2x}{dt^2} + 2\frac{dy}{dt} + 8x = 32t$$

$$\frac{d^2y}{dt^2} + 3\frac{dx}{dt} - 2y = 60e^{-t}; \quad x(0) = 6, x'(0) = 8, y(0) = -24, \text{ and } y'(0) = 0.$$

2. Suppose you are given the linear system of first-order equations

$$t\frac{dx}{dt} = -3x + 4y$$

$$t\frac{dy}{dt} = -2x + 3y.$$

Introduce a new independent variable $w$ by the substitution $w = \ln t$ (or $t = e^w$) and show that this substitution allows you to write the system as a new system with constant coefficients.

**3.** Consider the system

$$x' = f(x, y)$$
$$y' = g(x, y),$$

where $x$ and $y$ are functions of $t$. Assume you can use the first equation to express $y$ explicitly as a function of $x$ and $x'$, say $y = F(x, x')$ for some function $F(u, v)$ of two variables.

**a.** Find an expression for $y' = dy/dt$ by differentiating the equation $y = F(x, x')$ via the Chain Rule for functions of two variables (see Section A.7).

**b.** Substitute the expression for $y'$ found in part (a) into the second equation of the original system and set the right-hand side equal to $g(x, y) = g(x, F(x, x'))$.

**c.** Observe how the results of parts (a) and (b) yield a second-order equation solely in terms of $x$ and $x'$.

## 4.7 THE QUALITATIVE ANALYSIS OF AUTONOMOUS SYSTEMS

In this section, we investigate the graphical representation of a system of first-order equations. Because many systems—especially nonlinear systems—cannot be solved in closed form, the ability to analyze systems graphically is very important. The first thing we have to realize is that the very useful graphical tool of *slope fields* can't be applied directly to higher-order equations; this technique depends on a knowledge of the first derivative alone. However, there's a clever way of using our knowledge of first-order qualitative methods in the analysis of higher-order differential equations.

For convenience, we'll spend most of our time analyzing autonomous two-dimensional systems, although we will also tackle some nonautonomous systems and some three-dimensional systems toward the end of this section.

### 4.7.1 Phase Portraits for Systems of Equations

Suppose we have an autonomous system of the form

$$\frac{dx}{dt} = f(x, y)$$

$$(4.7.1)$$

$$\frac{dy}{dt} = g(x, y).$$

For example, let's take the system

$$\frac{dx}{dt} = y$$

$$\frac{dy}{dt} = -17x - 2y$$

and work with it throughout our initial discussions.

First, we can eliminate the variable $t$ by dividing the second equation by the first equation:

$$\frac{dy}{dx} = \frac{dy}{dt} \cdot \frac{dt}{dx} = \frac{\frac{dy}{dt}}{\frac{dx}{dt}} = \frac{g(x,y)}{f(x,y)}. \tag{4.7.2}$$

(See Section A.2 for a reminder of the Chain Rule used in this process.) For our example, $g(x, y) = -17x - 2y$ and $f(x, y) = y$ in (4.7.2), and we get

$$\frac{dy}{dx} = \frac{-17x - 2y}{y}.$$

Now we have a single first-order differential Equation $\frac{dy}{dx} = \frac{g(x,y)}{f(x,y)}$ in the variables $y$ and $x$. If we could solve Equation (4.7.2) for $y$ in terms of $x$, or even implicitly, we would have a solution curve in the $x$-$y$ plane. The plane of the variables $x$ and $y$ (with $x$- and $y$-axes) is called the **phase plane** of the original system of differential equations. As we saw in Section 1.3, each individual solution curve in the phase plane, $x = x(t), y = y(t)$, is called a **trajectory** (or **orbit**) of the system of equations. Although the independent variable $t$ is not present explicitly, the passage of time is represented by the *direction* that a point $(x(t), y(t))$ takes on a particular trajectory. The way the curve is followed as the values of $t$ increase (offstage) is called the **positive direction** on the trajectory. The collection of plots of the trajectories is called the system's **phase portrait** or **phase-plane diagram**. (You may want to review the qualitative analysis for first-order equations in Section 2.5.)

Even if we can't solve the system, we can look at the slope field of the single Equation (4.7.2), the outline of the phase portrait of the system. If we give some initial points $\left(x_0^i, y_0^i\right) = \left(x^i(t_0), y^i(t_0)\right), i = 1, 2, \ldots, n$, through which we want the trajectories to pass, we can plot a few specific trajectories and get a less complicated view of the phase plane. Let's do this for the system we've been discussing.

## ■ Example 4.7.1 Phase Portrait—One Trajectory

Our system is

$$\frac{dx}{dt} = y$$

$$\frac{dy}{dt} = -17x - 2y,$$

which gives us the first-order equation $\frac{dy}{dx} = \frac{-17x - 2y}{y}$ when we eliminate the variable $t$. Using a calculator or CAS to draw a piece of the slope field for this first-order equation can result in an erroneous plot (see Problem B5 in Exercises 4.7). Your technological device may have a problem at points $(x, y)$ with $y = 0$. If you draw the slope field by hand, be sure to place vertical tangent line segments along the $x$-axis (where $y = 0$), corresponding to an undefined (or infinite) slope when $y = 0$. It is better to use technology that takes the *pair* of equations given previously as input.

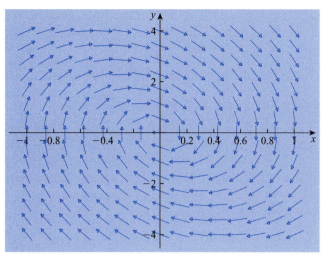

**FIGURE 4.5a**
*Slope field for* $\frac{dy}{dx} = \frac{-17x-2y}{y}$
$0 \leq t \leq 5; -1 \leq x \leq 1, -4 \leq y \leq 4$

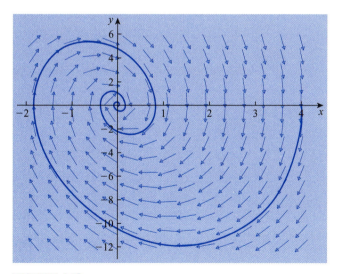

**FIGURE 4.5b**
*Trajectory for* $\left\{ \frac{dx}{dt} = y, \frac{dy}{dt} = -17x - 2y; \, x(0) = 4, y(0) = 0 \right\}$
$0 \leq t \leq 5; -2 \leq x \leq 4, -12 \leq y \leq 7$

Figure 4.5a shows the slope field, and Figure 4.5b shows a single trajectory satisfying the initial condition $x(0) = 4$, $y(0) = 0$—that is, a trajectory passing through the point $(4, 0)$ in the $x$-$y$ (phase) plane—superimposed on the slope field.

Because the trajectory starts at $(4, 0)$, you can see that the positive direction on the trajectory is clockwise, and the curve seems to spiral into the origin. (Try using technology to draw the trajectory for $0 \le t \le b$, letting $b$ get larger and larger.) To get an accurate phase portrait, you may want to use the slope field to suggest good initial points to use. Each dynamical system has its own appropriate range for $t$. ∎

Now let's look at a more elaborate phase portrait, one showing several trajectories.

### ∎ Example 4.7.2 Phase Portrait—Several Trajectories

The system consists of the two equations (1) $\frac{dx}{dt} = x + y$ and (2) $\frac{dy}{dt} = -x + y$. Whatever quantities these equations describe, certain facts should be obvious from the nature of the equations. First of all, from Equation (1), the growth of quantity $x$ depends on itself and on the other quantity $y$ in a positive way. On the other hand, Equation (2) indicates that quantity $y$ depends on itself positively, but its growth is hampered by the presence of quantity $x$—a larger value of $x$ leads to a slowdown in the growth of $y$.

Let's look at the phase portrait corresponding to this problem. For our system, Equation (4.7.2) looks like

$$\frac{dy}{dx} = \frac{\frac{dy}{dt}}{\frac{dx}{dt}} = \frac{-x + y}{x + y}.$$

This first-order equation is neither separable nor linear, but it is *homogeneous* and can be solved implicitly. (See the explanation for Problems A15–A18 of Exercises 2.1.) Figure 4.6a shows several trajectories, obtained by specifying nine initial points $(x(0), y(0))$, superimposed on

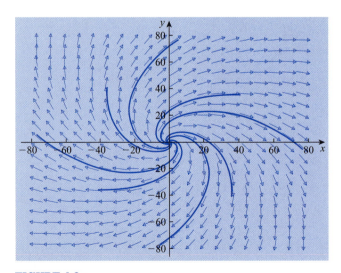

**FIGURE 4.6a**
*Trajectories for* $\left\{ \frac{dx}{dt} = x + y, \frac{dy}{dt} = -x + y \right\}$
$(x(0), y(0)) = (-1, -1), (-1, 0), (-1, 1), (0, -1), (0, 0), (0, 1), (1, -1), (1, 0),$ *and* $(1, 1)$
$0 \le t \le 4$

**FIGURE 4.6b**

Trajectories for $\left\{\frac{dx}{dt} = x + y, \frac{dy}{dt} = -x + y\right\}$

$(x(0), y(0)) = (0, 60), (25, 100), (50, 140),$ and $(0, 160)$

$0 \leq t \leq 1.2$

the slope field for $\frac{dy}{dx} = \frac{-x+y}{x+y}$. Because points on a trajectory are calculated by numerical methods, your CAS may allow you (or require you) to specify a step size and the actual numerical approximation method to be used. Numerical methods for systems of differential equations will be discussed in Section 4.10.

Each point on a particular curve in Figure 4.6a represents a *state* of the system: For each value of $t$, the point $(x(t), y(t))$ on the curve provides a snapshot of this dynamical system. If the variables $x$ and $y$ are supposed to represent animal or human populations, for example, then the proper place to view the trajectories is the first quadrant. Figure 4.6b describes the first quadrant of the phase plane for our problem, with four trajectories determined by four initial points.

These trajectories tell us that for the initial points chosen, the quantity $y$ increases to a maximum value and then decreases to zero, while the quantity $x$ also increases until it reaches its maximum level after quantity $y$ has disappeared. ∎

## 4.7.2 Other Graphical Representations

Using technology again in our preceding example, we can graph $x(t)$ and $y(t)$ in the $t$-$x$ and $t$-$y$ planes, respectively. Figures 4.7a and 4.7b show solution curves with $x(0) = 50$ and $y(0) = 140$, respectively.

These graphs show clearly that the quantity $y$ reaches a maximum of about 164 when $t \approx 0.4$ and that the $x$ quantity hits a peak of about 800 when $t \approx 2$. Note that the horizontal and vertical scales are different for Figures 4.7a and 4.7b.

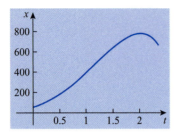

**FIGURE 4.7a**

$x(t); x(0) = 50$

$0 \le t \le 2.355$

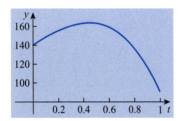

**FIGURE 4.7b**

$y(t); y(0) = 140$

$0 \le t \le 1$

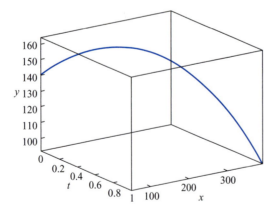

**FIGURE 4.8**

*Solution of* $\left\{ \frac{dx}{dt} = x + y, \frac{dy}{dt} = -x + y; x(0) = 50, y(0) = 140 \right\}$

$0 \le t \le 1$

Viewed another way, the system in the preceding example had *three* variables—the independent variable $t$ and the dependent variables $x$ and $y$. To be precise about all this, we state that *a solution of our system is a pair of functions* $x = x(t), y = y(t)$, *and the graphical representation of such a solution is a curve in three-dimensional t-x-y space—a set of points of the form* $(t, x(t), y(t))$. Figure 4.8 shows what the solution with initial point $(0, 50, 140)$ looks like for our problem.

Your CAS will probably allow you to manipulate the axes and get different views of this space curve. Figure 4.6a represents the *projection* of several such space curves onto the *x-y* plane, a much less confusing way of viewing the behavior of the system. These projections can be thought of as the shadows that would be cast by the space curves if a very bright light were shining on them from the front (the *x-y* face) of Figure 4.8.

Note that because the system we started with in the preceding example is autonomous, the solution curves are *independent of the starting time*. This means that if you pick a starting point $(x^*, y^*)$ at time $t^*$, then the path of a population starting at this point is the same as the path of a population starting at the same point at any other time $t^{**}$. Geometrically, this says that there is only one path (trajectory) through each point of the *x-y* plane. This is a consequence of an Existence and Uniqueness Theorem for systems that we'll see in Section 4.9. (Look back at Section 2.8 for the theorem that applies to first-order ODEs.)

From the slope field and phase portrait in Figure 4.6a, it seems clear that all trajectories (solution curves of the single differential equation) are escaping from the origin as $t$ increases. The variable $t$ is behind the scenes in a phase portrait, but you should experiment with different ranges of $t$ in your CAS or graphing calculator to verify the preceding statement. The **critical point** or **equilibrium point** $(0, 0)$—where both $dx/dt = 0$ and $dy/dt = 0$—is called a **source** in this case. We have used this terminology before (for the one-dimensional case, in Section 2.6), we will use it in this chapter, and we'll see it again as part of the discussion of systems in Chapter 5.

The algebra of finding equilibrium points (or equilibrium *solutions*) is trickier now because we must solve a *system* of equations. For example, if we want to find the equilibrium solutions of the nonlinear system of differential equations $\{\dot{x} = x - y,\ \dot{y} = 1 - xy\}$, we must solve the algebraic system

$$(1) \quad x - y = 0$$

$$(2) \quad 1 - xy = 0.$$

We can solve Equation (1) for $y$, finding that $y = x$, and then substitute for $y$ in the second equation. We get $1 - x^2 = 0$, which implies that $x = \pm 1$. Because $y = x$, the only equilibrium points are $(-1, -1)$ and $(1, 1)$.

Before we move on, let's look at the system $\{\dot{x} = 4 - 4x^2 - y^2,\ \dot{y} = 3xy\}$. Any equilibrium solution has to satisfy the equations

$$(A) \quad 4 - 4x^2 - y^2 = 0$$

$$(B) \quad 3xy = 0.$$

Equation (B) tells us that we have two possibilities: (i) $x = 0$ or (ii) $y = 0$. [We can eliminate $x = y = 0$ because (A) wouldn't be satisfied with this choice.] If $x = 0$, substituting in (A) gives us $4 - y^2 = 0$, so $y = \pm 2$. Then we have two equilibrium solutions, $(0, 2)$ and $(0, -2)$. Alternatively, if $y = 0$, substituting in (A) yields $4 - 4x^2 = 0$, so $x = \pm 1$. Now we have the remaining two equilibrium solutions, $(1, 0)$ and $(-1, 0)$.

The next example presents a simple system model of an *arms race*. Models of this general form were proposed by the English scientist Lewis F. Richardson (1881–1953) in the 1930s. As a Quaker, he was greatly interested in the causes and avoidance of war. We'll see how a qualitative analysis helps us to understand the situation being modeled.

### ■ Example 4.7.3 An Arms Race Model

Let's look at an autonomous linear system:

$$\frac{dx}{dt} = 7y - 4x - 13$$
$$\frac{dy}{dt} = 2x - 5y + 11.$$

The functions $x(t)$ and $y(t)$ could represent the readiness for war of two nations, X and Y, respectively. This readiness might be measured, for example, in terms of the level of expenditures for weapons for each country at time $t$. To get the first equation, this model assumes that the rate of increase of $x$ is a linear function of both $x$ and $y$. In particular, if $y$ increases, then so does the rate at which $x$ increases. *This makes sense, doesn't it?* But the cost of building up and maintaining a supply of weapons also puts the brakes on *too* much expansion. The term $-4x$ in the first equation suggests a sense of restraint proportional to the arms level of nation X. Finally, the constant term $-13$ can represent some basic, constant relationship of nation X to nation Y—probably some underlying feelings of good will that diminish the threat and therefore decrease dependence on weaponry. The second equation can be interpreted in a similar way, but here the positive constant 11 probably signifies a grievance by Y against X that results in an accumulation of arms. Now what does this model tell us about the situation? We don't know how to solve such a system, but we can still learn a lot about the arms race between the two nations.

As in Example 4.7.1, we can start constructing the phase portrait of the system by eliminating the variable $t$:

$$\frac{dy}{dx} = \frac{\dfrac{dy}{dt}}{\dfrac{dx}{dt}} = \frac{2x - 5y + 11}{7y - 4x - 13}.$$

Now we can look at the slope field and some trajectories corresponding to this single equation (Figure 4.9). Several initial points were chosen. (*Try a smaller set of initial points yourself.*) For this to be a realistic model of an arms race, the values of $x$ and $y$ should be positive; hence, our focus on the first quadrant.

First of all, note that one solution of the system is the pair of functions $x(t) \equiv 2, y(t) \equiv 3$. In this phase portrait, if we look hard enough, we may notice that the points $(x(t), y(t))$ on every trajectory seem to be moving toward the point $(2, 3)$ as $t$ increases. (To verify this last statement, you should plot the phase portrait for $0 \leq t \leq b$ and let $b$ increase.) The point $(2, 3)$ is an equilibrium point—as we've seen previously, a point $(x, y)$ at which both $dx/dt$ and $dy/dt$ equal 0. The behavior of trajectories near this point entitles it to be called a **sink**. In

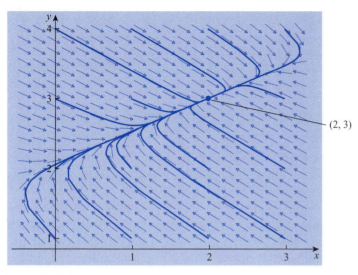

**FIGURE 4.9**
Trajectories for $\left\{\frac{dx}{dt} = 7y - 4x - 13, \frac{dy}{dt} = 2x - 5y + 11\right\}$
Initial points $(i, j), i = 0, 1, 2, 3; j = 1, 2, 3, 4$
$0 \leq t \leq 5$

real-life terms, this means that the arms race represented by this system would *stabilize* as time passes, approaching a state in which the level of military expenditures for nation X would be 2 and the level for nation Y would be 3, where the units could be millions or billions.  ■

### 4.7.3 A Predator-Prey Model: The Lotka-Volterra Equations

An important type of real-life problem that can be modeled by a system of differential equations is a **predator-prey problem**, in which we assume that there are two species of animals, X and Y, in a small geographical region such as an island. One species (the **predator**) thinks of the other species (the **prey**) as food and is very dependent on this food supply for survival.

Let $x(t)$ and $y(t)$ represent the populations of the two species at time $t$. We can make the following reasonable assumptions:

1. If there are no predators, the prey species will grow at a rate proportional to its own population (assuming an unlimited food supply).

2. If there are no prey, the predator species will decline at a rate proportional to the predator population.

3. The presence of both predators and prey is beneficial to the growth of the predator species and is harmful to the growth of the prey species.

The third assumption says that interactions (or close encounters of the hungry kind) between the predator and prey lead to a decrease in the prey population and to a resulting increase in the predator population. As we will see, these contacts are indicated mathematically by a *multiplication* of the variables that represent predator and prey. These assumptions lead to a system of *nonlinear* first-order differential equations such as the following:

$$\frac{dx}{dt} = 0.2x - 0.002xy, \quad \frac{dy}{dt} = -0.1y + 0.001xy. \tag{4.7.3}$$

For this system, how can we see that $x(t)$ is the size of the *prey* population at any time $t$ and $y(t)$ is the number of *predators* at time $t$?

First of all, note that if there are *no* predators—that is, if $y$ is always 0—the system reduces to $dx/dt = 0.2x$, $dy/dt = 0$. This says that the prey population would increase at a rate that is proportional to the actual prey population at any time. Also, the predator population is constant—at zero. This is realistic and consistent with assumption 1. Furthermore, if there are no *prey*—that is, if $x \equiv 0$—the system becomes $dx/dt = 0$, $dy/dt = -0.1y$, which means that the number of predators would decrease at a rate proportional to the predator population, where 0.1 is the constant of proportionality, the predator's *intrinsic death rate*. Again, this is realistic because in the absence of a crucial food supply, the bottom line would be starvation and a net decline in the predator population.

The intriguing terms in (4.7.3) are the terms involving the product $xy$. We've already suggested that these terms represent *the number of possible interactions* between the two species. To illustrate this point, suppose there were four foxes and three rabbits on an island. If we label the foxes $F_1, F_2, F_3$, and $F_4$ and the rabbits $R_1, R_2$, and $R_3$, then we have the following possible one-on-one encounters between foxes and rabbits: $(F_1, R_1)$, $(F_1, R_2)$, $(F_1, R_3)$, $(F_2, R_1)$, $(F_2, R_2)$, $(F_2, R_3)$, $(F_3, R_1)$, $(F_3, R_2)$, $(F_3, R_3)$, $(F_4, R_1)$, $(F_4, R_2)$, and $(F_4, R_3)$. Note that there are $4 \times 3 = 12$, or $x$ times $y$, possible interactions. Of course, we can have two foxes meeting up with one rabbit or one fox coming upon three rabbits, and so on, but the idea is that *the number of interactions is proportional to the product of the two populations*. The coefficient of $xy$ in the first equation, $-0.002$, is a measure of the predator's effectiveness in terms of prey capture, whereas the coefficient 0.001 in the second equation is an indicator of the predator's efficiency in terms of prey consumption.

This nonlinear system is a particular example of a system called the **Lotka-Volterra equations**:

$$\frac{dx}{dt} = a_1 x - a_2 xy$$
$$\frac{dy}{dt} = -b_1 y + b_2 xy,$$

where $a_2$ and $b_2$ are positive constants. [Alfred Lotka (1880–1949) was a chemist and demographer and Vito Volterra (1860–1940) was a mathematical physicist. In the 1920s they derived these equations independently—Lotka from a chemical reaction problem and Volterra from a problem concerned with fish catches in the Adriatic Sea.] In general, there is no explicit solution of the Lotka-Volterra equations in terms of elementary functions. We'll discuss numerical solutions of systems in Section 4.10.

However, as the next example shows, we can understand the relationship between the two species by using a *qualitative* analysis.

### ■ Example 4.7.4 Qualitative Analysis of a Predator-Prey Model

Figure 4.10 shows the trajectory corresponding to our system

$$\frac{dx}{dt} = 0.2x - 0.002xy$$

$$\frac{dy}{dt} = -0.1y + 0.001xy,$$

with $x(0) = 100, y(0) = 25$, and $0 \leq t \leq 52$. What does this picture tell us? First, realize that the horizontal axis ($x$) represents the prey; the vertical axis ($y$), the predators. Our starting point, corresponding to $t = 0$, is $(100, 25)$, and the direction of the trajectory is counterclockwise. To see the direction, use technology to look at partial trajectories such as those given by $0 \leq t \leq 10, 0 \leq t \leq 15$, or $0 \leq t \leq 25$.

Figure 4.10 illustrates a cyclic behavior that seems a bit too neat to be found in the wild. However, regular population cycles do seem to occur in nature.[8] In our graph, both prey and

**FIGURE 4.10**

*Trajectory for* $\left\{ \frac{dx}{dt} = 0.2x - 0.002xy, \frac{dy}{dt} = -0.1y + 0.001xy; x(0) = 100, y(0) = 25 \right\}$

$0 \leq t \leq 52$

---

[8] Examination of the records of the Hudson's Bay Company, which trapped fur-bearing animals in Canada for almost 200 years, suggests a periodic pattern in the number of lynx pelts harvested from about 1845 to the 1930s. The lynx, a cat-like predator, has the snowshoe hare as its main prey. For an analysis of the data, see J. D. Murray, *Mathematical Biology I: An Introduction* (*Third Edition*) (New York: Springer-Verlag, 2002): 83–84.

predator populations increase as the number of prey increases, but when the prey population exceeds about 350, the predators seem to overwhelm their prey to the extent that there are more and more predators but a declining prey population. The predators continue to increase until their number is about 260, at which time the effect of a dwindling food supply catches up to the predators and their population begins to decline. The predators may starve or start killing each other as competition for diminishing resources grows fierce. Finally, the predator population is low enough for the prey population to recover, and the cycle begins again.

Figure 4.10 highlights the point $(100, 100)$ because $x = 100$, $y = 100$ is an equilibrium solution of the system, called a **center** in this case. (*Verify the preceding statement.*) If this system were to have initial point $(100, 100)$, neither population would move from this state. The origin is also an equilibrium point. ∎

### 4.7.4 Other Graphical Representations

With the aid of technology, we can look at plots of $x(t)$ against $t$ and $y(t)$ against $t$ separately (Figures 4.11a and 4.11b). Compare these graphs, noting the way in which one population lags behind the other over time. The trajectory (Figure 4.10) gives the big picture, the state

**FIGURE 4.11a**

$x(t)$, *prey population*

$x(0) = 100; 0 \le t \le 200$

**FIGURE 4.11b**

$y(t)$, *predator population*

$y(0) = 25; 0 \le t \le 220$

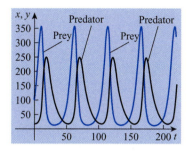

**FIGURE 4.12**

*Predator and prey population vs. t*

$(x(t), y(t))$ of the ecological system as time marches on, whereas Figures 4.11a and 4.11b show the individual population fluctuations. Figure 4.12 exhibits the cyclic nature of the predator fluctuation and that of the prey fluctuation on the same set of axes. Each graph in this example was done by a CAS using a numerical approximation to the actual system solution.

### 4.7.5 Three-Dimensional Systems

We have been focusing on second-order equations and their equivalent systems, but the techniques we have discussed apply to any differential equation of order greater than 1. The main difficulty with equations of order 3 and higher is that we lose some aspects of the graphical interpretation of the solution. Let's look at the next example, which presents us with a three-dimensional system.

### ■ Example 4.7.5 A System of Three First-Order Equations

We want to examine the behavior of the three-dimensional system

$$\dot{x} = -0.1x - y$$
$$\dot{y} = x - 0.1y$$
$$\dot{z} = -0.2z.$$

### A Three-Dimensional Trajectory

The complete picture of this linear system is given by the set of points $(t, x(t), y(t), z(t))$, a *four-dimensional* situation. Assuming that $x$, $y$, and $z$ are functions of the parameter $t$ and that we have the initial condition $x(0) = 5$, $y(0) = 5$, and $z(0) = 10$, we get the three-dimensional trajectory shown in Figure 4.13. This is a *projection* of the four-dimensional picture onto three-dimensional space, just as the two-dimensional phase portraits we've seen previously are projections of three-dimensional curves onto two-dimensional planes.

By plotting this with your CAS and rotating the axes (if possible), you should be able to see that the solution spirals into the origin in the $x$-$y$ plane, while it moves toward the origin in the variable $z$ as well.

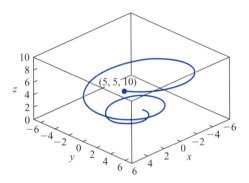

**FIGURE 4.13**

*An x-y-z plane trajectory for the system*
$$\{\dot{x} = -0.1x - y, \dot{y} = x - 0.1y, \dot{z} = -0.2z; \ x(0) = 5, y(0) = 5, z(0) = 10\}$$
$0 \le t \le 15$

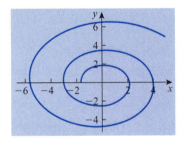

**FIGURE 4.14**

*An x-y plane trajectory for the system*
$$\{\dot{x} = -0.1x - y, \dot{y} = x - 0.1y, \dot{z} = -0.2z; \ x(0) = 5, y(0) = 5, z(0) = 10\}$$
$0 \le t \le 15$

## A Two-Dimensional Trajectory

Now we can, for example, project the three-dimensional spiral (in *x-y-z* space) onto the *x-y* plane (Figure 4.14).

If you increase the range of *t*, you will get a tighter spiral and see that the origin is a *sink* for this system. ∎

In this section, we have seen how any differential equation of order greater than 1 can be turned into an equivalent system of first-order equations. We've looked at different ways to view such systems graphically. In later sections, we will discuss other aspects of first-order equations that can be extended to systems. In particular, we'll investigate questions of the existence and uniqueness of solutions and the numerical approximation of solutions of systems.

## Exercises 4.7

### A

Assume that each function in Problems 1–7 is a function of time, $t$. For each of these initial value problems, (a) convert to a system, (b) use technology to find the graph of the solution in the phase plane, and (c) show a graph of the two components of the solution relative to the $t$-axis.

1. $x'' + x' = 0;$    $x(0) = 1, x'(0) = 2$
2. $\ddot{r} - r = 0;$    $r(0) = 0, \dot{r}(0) = 1$
3. $\ddot{y} + y = 0;$    $y(0) = 2, \dot{y}(0) = 0$
4. $y'' = -4;$    $y(0) = y'(0) = 0$
5. $\ddot{x} - \dot{x} = 0;$    $x(0) = 1, \dot{x}(0) = 1$
6. $x'' - 2x' + x = 0;$    $x(0) = -1, x'(0) = -1$
7. $x'' = x - x^3;$    $x(0) = 0, x'(0) = 1$

### B

1. Read the explanation before Problem A15 of Exercises 2.1 and solve the equation

$$\frac{dy}{dx} = \frac{-x + y}{x + y}$$

   that arises in Example 4.7.2.

2. Consider the specific Lotka-Volterra Equations (4.7.3) in Example 4.7.4.

   a. Find the *first-order* differential equation that defines the *trajectories* of this system in the phase plane.

   b. Solve this separable equation to find the implicit algebraic equation of the trajectories.

3. The equation $\ddot{Q} + 9\dot{Q} + 14\,Q = \frac{1}{2}\,\sin t$ models an electric circuit with resistance of 180 ohms, capacitance of 1/280 farad, inductance of 20 henries, and an applied voltage given by $E(t) = 10 \sin t$. $Q = Q(t)$ denotes the capacitance, the charge on the capacitor at time $t$, and $\dot{Q}(t)$ denotes the current in the circuit. Assume $Q(0) = 0$ and $\dot{Q}(0) = 0.1$.

   a. Express this IVP as a system of two first-order equations, with the appropriate initial conditions.

   b. Use technology to graph the solution of the system in the phase plane, with $0 \leq t \leq 8$.

   c. Use technology to graph the solution of the original second-order equation relative to the $t$-axis, considering first the interval $0 \leq t \leq 2$ and then $0 \leq t \leq 8$.

   d. Describe the behavior of the capacitance as $t \to \infty$.

4. Find all equilibrium solutions of each of the following systems.

   a. $\dot{x} = x - 3y, \dot{y} = 3x + y$
   b. $x' = 2x + 4y, y' = 3x + 6y$
   c. $\dot{r} = -2rs + 1, \dot{s} = 2rs - 3s$
   d. $x' = \cos y, y' = \sin x - 1$
   e. $\dot{x} = x - y^2, \dot{y} = x^2 - y$
   f. $r' = 1 - s, s' = r^3 + s$

5. Find all the equilibrium points of the system

$$x' = x^2 y^3$$
$$y' = -x^3 y^2$$

and sketch the phase-plane diagram for this system.

**C**

1. Convert the equation $x'' + x' - x + x^3 = 0$ to a system and find all equilibrium points.
2. The equation $\frac{d^2\theta}{dt^2} + k^2 \sin\theta = 0$ describes the motion of an *undamped pendulum*, where $\theta$ is the angle the pendulum makes with the vertical. Convert this equation to a system and describe all its equilibrium points.
3. Consider the differential equation $x'' + \lambda - e^x = 0$, where $\lambda$ is a parameter.

   a. Sketch the phase-plane diagram for $\lambda > 0$.
   b. Sketch the phase-plane diagram for $\lambda < 0$.
   c. Describe the significance of the value $\lambda = 0$.

## 4.8 SPRING-MASS PROBLEMS

### 4.8.1 Simple Harmonic Motion

We have seen spring-mass problems before (for example, in the exercises for Section 4.1 and Section 4.2). The treatment in this section represents a systematic exposition of this important model and demonstrates the usefulness of a qualitative approach.

To start with, suppose we have a spring attached to the ceiling and a weight (mass) hanging from the bottom of the spring, as in Figure 4.15a.

If we set the mass in motion by giving it an upward or downward push, we can use Newtonian mechanics and the qualitative analysis of systems of ODEs to investigate the forces acting on the mass during its motion. We want to describe the state of this system, giving the mass's position and velocity at any time $t$. First, we'll assume that there's no air resistance, friction, or other impeding force. The resulting situation is called **simple harmonic motion** or **free undamped motion**.

**FIGURE 4.15a**
*Spring-mass system, mass in the equilibrium position*

**FIGURE 4.15b**

*Spring-mass system, mass displaced from the equilibrium position*

Fundamental to understanding the mass's movement is **Newton's Second Law of Motion**, which can be stated as $F = m \cdot a$, where $F$ is a force (or sum of forces) acting on a body (such as the weight hanging from the spring), $m$ is the body's mass, and $a$ is the acceleration of the body. If $x$ denotes the displacement (distance) of the mass from its equilibrium (rest) position, where a move downward is considered a *positive* displacement (Figure 4.15b), we can write this expression for the force as $m \cdot \frac{d^2x}{dt^2}$.

Now note that if you pull *down* on the weight (stretching the spring in the process), you can feel a certain tension—a tendency for the spring to pull the weight back *up*. Similarly, if you push *up* on the weight, thereby compressing the spring, you feel a force that tends to push the weight *down*. This behavior is described by **Hooke's Law**: The force $F$ (called the restoring force) exerted by a spring, tending to restore the weight to the equilibrium position, is proportional to the distance $x$ of the weight from the equilibrium position. Stated simply, *force is proportional to stretch*. Mathematically, we write $F = -kx$, where $k$ is a positive constant called the *spring constant*. Note that if $x$ is *positive*, then the restoring force is *negative*, whereas if $x$ is *negative*, then $F$ is *positive*.

Because we are ignoring any other kind of force acting on the weight, we can equate the two expressions for the force to get

$$m \cdot \frac{d^2x}{dt^2} = -kx,$$

which we can write in the form

$$\frac{d^2x}{dt^2} + \beta x = 0, \text{ where } \beta = \frac{k}{m}. \tag{4.8.1}$$

We saw this kind of homogeneous second-order linear equation in Section 4.1; and from our work in Section 4.6 we know how to convert this equation into an equivalent system of first-order equations. Earlier in this section, we learned how to understand what a phase portrait is telling us. Now let's analyze this problem qualitatively.

### ■ Example 4.8.1 A Spring-Mass System—Simple Harmonic Motion

Given the equation $\frac{d^2x}{dt^2} + \beta x = 0$, where $\beta = \frac{k}{m}$, we let $x_1 = x$ and $x_2 = \dot{x}$. We see that $\dot{x}_1 = \dot{x} = x_2$ and $\dot{x}_2 = \ddot{x} = -\beta x$ (by solving the second-order equation for the second derivative) $= -\beta x_1$, so we have the two-dimensional system

$$\dot{x}_1 = x_2$$
$$\dot{x}_2 = -\beta x_1. \tag{4.8.2}$$

First of all, note that $x_2$ represents the *velocity* of the mass: $x_2 = \dot{x}$, the rate of change of the position, or displacement of the mass. Using the language developed in Example 4.7.1, we say that if we could solve the System (4.8.2) for $x_1$ and $x_2$, then the ordered pair $(x_1(t), x_2(t))$, consisting of the mass's current position and velocity, would give the *state* of the system at time $t$.

Now we can look at some trajectories in the phase plane of (4.8.2)—that is, some solution curves in the $x_1$-$x_2$ plane. Using initial points $(x_1(0), x_2(0)) = (1, 0)$, $(0, 1)$, and $(2, 0)$, Figure 4.16 shows what these curves look like when $\beta = \frac{2}{5}$ and we take the interval $0 \le t \le 10$. You should use technology to plot your own trajectories, with different initial points and smaller ranges for $t$. ■

### 4.8.2  Analysis

Is this the behavior you would expect from a bouncing mass? First of all, note that the origin is a special point, an equilibrium solution, because both equations of our system vanish at

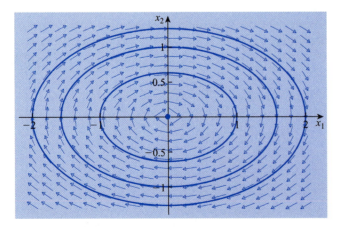

**FIGURE 4.16**
*Trajectories for* $\left\{ \dot{x}_1 = x_2, \dot{x}_2 = -\frac{2}{5}x_1 \right\}$*, initial points* $(1, 0)$, $(0, 1)$, $(2, 0)$; $0 \le t \le 10$

$(x_1, x_2) = (0, 0)$. Physically, this means that a mass-spring system that starts at its equilibrium position ($x(0) = 0$) and has no initial push or pull ($\dot{x}(0) = x_2(0) = 0$) will remain at rest forever, which makes sense.

Now look closely at a typical **closed orbit**, as one of these elliptical trajectories is called. Assume that $x_1 = 0$ and $x_2$ is positive—that is, the mass is at its equilibrium position and is given an initial tug downward. When the mass is at rest ($x_1 = 0$) and it is pushed or pulled in a downward direction ($dx_1/dt = x_2 > 0$), the flow moves in a clockwise direction (note the direction of the slope field arrows), with $x_2$ decreasing and $x_1$ increasing until the trajectory is at the $x_1$-axis. Physically, this means that the mass moves downward until the spring reaches its maximum extension ($x_1$ is at its most positive value), depending on how much force was applied initially to pull the mass downward, at which time the mass has lost all its initial velocity (that is, $x_2 = 0$). Then the energy stored in the spring serves to pull the mass back up toward its equilibrium position, so that $x_1$ is decreasing at the same time that the velocity $x_2$ is increasing—*but in a negative direction* (*upward*). Graphically, this is taking place in the fourth quadrant of the phase plane. When the flow has reached the state $(0, x_2)$, where $x_2$ is negative, the mass has reached its original position and has attained its maximum velocity upward.

As the trajectory takes us into the third quadrant, the mass is overshooting its original position but is slowing down: $x_1 < 0$ and $x_2 < 0$. When the trajectory has reached the point $(x_1, 0)$, where $x_1$ is negative, the spring is most compressed and the mass is (for an instant) not moving.

As the trajectory moves through the second quadrant, the mass is headed back toward its initial position with increasing velocity in a downward (positive) direction: $x_1 < 0$ and $x_2 > 0$. Finally, the mass reaches its initial position with its initial velocity in the positive (downward) direction—$x_1 = 0, x_2 > 0$—and the cycle begins all over again.

This analysis seems to say that the mass will never stop, bobbing up and down forever. This apparently nonsensical conclusion is perfectly reasonable when you realize that a real mass-spring system is always subject to some air resistance and some sort of friction that slows the system down and eventually forces the mass to stop moving. Our analysis assumes no such impeding force, so the conclusion is rational, even though the assumption is unrealistic.

### 4.8.3 Another View—Solution Curves

As we did in Examples 4.7.1 and 4.7.3, we can use technology to plot each solution of our system against $t$. Figures 4.17a and 4.17b show the solution with $\beta = \frac{2}{5}, x_1(0) = 1$, and $x_2(0) = 1$, corresponding to a spring-mass system that starts 1 unit below its equilibrium position and has been given an initial velocity of 1 in a downward direction. We should not be surprised at the appearance of these solution curves. The closed orbits in Figure 4.16 reflect the periodic nature of the motion of the mass. Such motions are called **oscillations**.

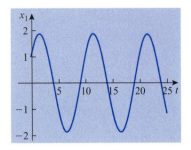

**FIGURE 4.17a**

$x_1(t)$, *displacement*

$x_1(0) = 1, 0 \le t \le 25$

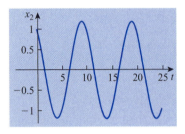

**FIGURE 4.17b**

$x_2(t)$, *velocity*

$x_2(0) = 1, 0 \le t \le 25$

Using methods that we saw in Section 4.1, we can determine that when $\beta = \frac{2}{5}$, the general solution of System (4.8.2) is

$$x_1(t) = \frac{C_1}{2}\sqrt{10}\sin\left(\frac{1}{5}\sqrt{10}t\right) + C_2\cos\left(\frac{1}{5}\sqrt{10}t\right)$$

$$x_2(t) = C_1\cos\left(\frac{1}{5}\sqrt{10}t\right) - \frac{C_2}{5}\sqrt{10}\sin\left(\frac{1}{5}\sqrt{10}t\right),$$

and we can see that the explicit source of the oscillations is the trigonometric terms. The particular system solution shown in Figures 4.17a and 4.17b corresponds to the initial conditions $x_1(0) = 1$, $x_2(0) = 1$, so $C_1 = C_2 = 1$. (*Verify this.*)

Remembering the discussion of the *equivalence* of a second-order equation and a system in Example 4.6.1, we realize that

$$x_1(t) = \frac{C_1}{2}\sqrt{10}\sin\left(\frac{1}{5}\sqrt{10}t\right) + C_2\cos\left(\frac{1}{5}\sqrt{10}t\right)$$

is the general solution of the original single differential equation $\frac{d^2x}{dt^2} + \beta x = 0$, where $\beta = \frac{2}{5}$. (Review Section 4.1. It happens that $x_2(t) = dx_1/dt$ is also a solution, but this is true only because the equation is *homogeneous*.)

### 4.8.4 Free Damped Motion

Now let's look at a more realistic version of a spring-mass system. This time we'll assume the existence of a combination of air resistance and some friction in the spring-mass system, called a **damping force**, to slow the mass down. To dramatize the situation, you may think of the mass as being immersed in a bucket of water, oil, or maple syrup, so that any initial force imparted to the mass is opposed by a force in the opposite direction as the mass meets resistance. The motion that results is called **free damped motion**. For instance, the damping produced by automobile shock absorbers provides a more comfortable ride.

The damping force works *against* the motion of the mass, so when the mass is moving *down* (the positive direction), the damping force acts in an *upward* direction, and when the mass is moving *up* (the negative direction), the damping force acts in a *downward* direction. In algebraic terms, this damping force's sign must be opposite to the sign of the direction of the velocity. For small velocities, experiments have shown that the damping force is proportional to the velocity of the mass. We can express the last two sentences mathematically as $F = -\alpha \frac{dx}{dt}$, where $\alpha$ is a positive constant of proportionality called the **damping constant**. Realizing that both the spring's restoring force and this damping force are opposed to the mass's motion, we can use *Newton's Second Law of Motion* to derive the equation

$$m \cdot \frac{d^2x}{dt^2} = -\alpha \frac{dx}{dt} - kx,$$

which we can write in the form

$$\frac{d^2x}{dt^2} + b\frac{dx}{dt} + cs = 0, \quad \text{where} \quad b = \frac{\alpha}{m} \text{ and } c = \frac{k}{m}. \tag{4.8.3}$$

Now we can convert this second-order differential equation into a system and analyze our problem qualitatively.

### ■ Example 4.8.2 A Spring-Mass System—Free Damped Motion

The second-order linear equation $\frac{d^2x}{dt^2} + b\frac{dx}{dt} + cx = 0$ is equivalent to the two-dimensional system

$$\frac{dx_1}{dt} = x_2$$

$$\frac{dx_2}{dt} = -bx_2 - cx_1. \tag{4.8.4}$$

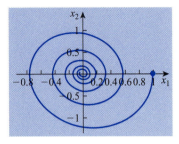

**FIGURE 4.18**
*Trajectory for the system*

*Damped free motion* $\left\{ \dfrac{dx_1}{dt} = x_2, \dfrac{dx_2}{dt} = -\dfrac{1}{4}x_2 - 2x_1; x_1(0) = 1, x_2(0) = 0 \right\}$

$0 \le t \le 25$

## Phase Portrait Analysis

To understand the motion of the mass, we'll look first at the trajectory we get when we take $b = \frac{1}{4}, c = 2, x_1(0) = 1$, and $x_2(0) = 0$ (Figure 4.18). In particular, you should see that the mass starts off 1 unit *below* its equilibrium position with *no* initial velocity in any direction.

The direction of the trajectory in Figure 4.18 indicates very dramatically that the state of the system is spiraling into the origin—that is, $x_1(t) \to 0$ and $x_2(t) \to 0$ as $t \to \infty$. Every time the spiral trajectory in Figure 4.18 crosses the $x_2$-axis (so that $x_1 = 0$), the mass is at its equilibrium position—on its way up when the velocity $x_2$ is negative and on its way down when $x_2$ is positive. (Remember our agreement on which direction is positive and which direction is negative.) This type of spiral clearly indicates why we can say that the origin is a **sink** for the system.

## Another View

We can also look at the graphs of $x_1(t)$ against $t$ (Figure 4.19a) and $x_2(t)$ against $t$ (Figure 4.19b) for the same system. The oscillations shown in Figures 4.17a and 4.17b reflect the behavior of the system in a different way.

The mass reaches its equilibrium position when the $x_1(t)$ curve crosses the $t$-axis. If $x_1(t^*) = 0$, then look at Figure 4.19b to see what the value of $x_2(t^*)$ is. If $x_2(t^*) > 0$, for example, the mass is on its way *down*. Also note how Figures 4.18, 4.19a, and 4.19b show that the successive rises and falls get progressively smaller.

The figures all reflect the initial conditions and seem to say that the mass eventually comes to rest at its equilibrium position. If you were to hit a brass gong with a special ceremonial hammer, the vibrations would be loud at the beginning but would gradually fade to nothing. This is roughly what we are seeing here.

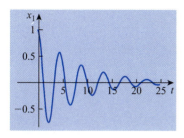

**FIGURE 4.19a**
$x_1(t)$, *displacement*
$x_1(0) = 1$
$0 \leq t \leq 25$

**FIGURE 4.19b**
$x_2(t)$, *velocity*
$x_2(0) = 0$
$0 \leq t \leq 25$

## A Look at the Actual Solution

The curves in Figures 4.19a and 4.19b are *not* periodic, despite their resemblance to familiar trigonometric curves that are. In Section 4.1 we saw how to determine that the solution of our IVP $\frac{d^2x}{dt^2} + \frac{1}{4}\frac{dx}{dt} + 2x = 0$, with $x(0) = 1$ and $\frac{dx}{dt}(0) = 0$, is given by

$$x(t) = e^{\left(-\frac{1}{8}t\right)}\left(\cos\left(\frac{1}{8}\sqrt{127}t\right) + \frac{\sqrt{127}}{127}\sin\left(\frac{1}{8}\sqrt{127}t\right)\right),$$

which is not a pure trigonometric function because of the exponential factor. You should verify that this is a solution. In terms of our System (4.8.4), we have $x_1(t) = x(t)$ and $x_2(t) = \frac{dx}{dt}$. (Do the differentiation to see what $x_2(t)$ looks like.)

The exponential factor $e^{\left(-\frac{1}{8}t\right)}$, called the **time-varying amplitude**, forces the decay of the oscillations indicated by the trigonometric terms. Figure 4.20 shows the graph of the solution $x(t)$, together with the graphs of $e^{\left(-\frac{1}{8}t\right)}$ and $-e^{\left(-\frac{1}{8}t\right)}$.

**FIGURE 4.20**
*The graphs of $x(t)$, $e^{\left(-\frac{1}{8}t\right)}$, and $-e^{\left(-\frac{1}{8}t\right)}$*

### 4.8.5 Different Kinds of Damping

You should be aware that there are different kinds of damped motion. The behavior of a damped system described by the equation $m\frac{d^2x}{dt^2} + \alpha\frac{dx}{dt} + kx = 0$ depends on the relationship among the three constants $m, \alpha$, and $k$—the mass, the damping coefficient, and the spring constant, respectively. The example we've just analyzed is a case of **underdamped motion**, occurring when the damping coefficient is relatively small compared to the other constants: $\alpha^2 < 4mk$, technically. The other two possibilities, **overdamped motion** ($\alpha^2 > 4mk$) and **critically damped motion** ($\alpha^2 = 4mk$), are explored in Problems A1 and A2 in Exercises 4.8. We'll give a detailed explanation of the significance of the relationship among $m, \alpha$, and $k$ in Chapter 5.

### 4.8.6 Forced Motion

Sometimes a physical system is subject to external forces, which must appear in its mathematical representation. For example, the motion of an automobile (whose body-suspension combination can be considered a spring-mass system) is influenced by irregularities in the road surface. Similarly, a tall building may be subjected to strong winds that will cause it to sway in an uncharacteristic way.

We're going to look at an initial-value problem related to Example 2.2.5 and to Problems B9–B11 in Exercises 2.2. This discussion will involve an important type of second-order linear equation with a forcing term.

### ■ Example 4.8.3 Forced Damped Motion

Suppose we have an electrical circuit with an inductance of 0.5 henry, a resistance of 6 ohms, a capacitance of 0.02 farad, and a generator providing alternating voltage given by 24 sin (10t) for $t \geq 0$. The alternating voltage is the external force applied to the circuit, and the resistance is a damping coefficient. Then, letting Q denote the instantaneous charge on the capacitor, *Kirchhoff's Law* gives us the equation

$$0.5\frac{d^2Q}{dt^2} + 6\frac{dQ}{dt} + 50Q = 24\sin 10t,$$

or

$$\frac{d^2Q}{dt^2} + 12\frac{dQ}{dt} + 100Q = 48\sin 10t.$$

Let's assume that $Q(0) = 0$ and $\frac{dQ}{dt}(0) = 0$.

This second-order nonhomogeneous equation is equivalent to the nonautonomous system

$$\frac{dx_1}{dt} = x_2$$

$$\frac{dx_2}{dt} = 48 \sin 10t - 12x_2 - 100x_1,$$

with initial conditions $x_1(0) = 0$ and $x_2(0) = 0$. (*You should work this out for yourself.*)

## Phase Portrait

The phase portrait (Figure 4.21a) corresponding to this system, for $0 \le t \le 0.94$, is interesting. At first, we suspect that we may get a spiral opening outward. But with an expanded range for $t$—say, from 0 to 5—the phase portrait resembles a closed orbit around the origin (Figure 4.21b). We can understand the initial "blip" by using the explicit solution found by the techniques discussed in Section 4.2:

$$Q(t) = \frac{1}{10}e^{-6t}(4\cos 8t + 3\sin 8t) - \frac{2}{5}\cos 10t.$$

As in Example 2.2.5, we see that there is a *transient term*, $\frac{1}{10}e^{-6t}(4\cos 8t + 3\sin 8t)$, that becomes negligible as $t$ grows large (*Why?*), and a *steady-state term*, $\frac{2}{5}\cos 10t$, that controls the behavior of $Q(t)(=x_1)$ eventually. This steady-state term is periodic with the same period $\left(\frac{2\pi}{10} = \frac{\pi}{5}\right)$ as the forcing term and has the amplitude $\frac{2}{5}$. The *current* in the circuit is given by $I = \frac{dQ}{dt} = x_2$. ∎

Let's look at one other example of a spring-mass system. First, we suppose that there is no air resistance or friction. Next, we assume that the spring to which the mass is attached is supported by a board. Now we set the mass into motion by moving the supporting board up and down in a periodic manner. This situation is described as **driven undamped motion** or

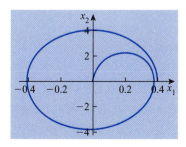

**FIGURE 4.21a**
*Trajectory for the system*
$\left\{ \frac{dx_1}{dt} = x_2, \frac{dx_2}{dt} = 48\sin 10t - 12x_2 - 100x_1; x_1(0) = 0 = x_2 \right\}$
$0 \le t \le 0.94$

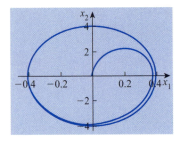

**FIGURE 4.21b**
*Trajectory for the system*
$\left\{ \frac{dx_1}{dt} = x_2, \frac{dx_2}{dt} = 48 \sin 10\, t - 12x_2 - 100x_1; x_1(0) = 0 = x_2(0) \right\}$
$0 \le t \le 5$

**forced undamped motion**. As in the preceding example, a force external to the spring-mass system itself is being applied to the system, and we want to understand the behavior of the system.

When we apply Newton's Second Law of Motion, an analysis similar to that provided in Example 4.8.1 gives us the equation

$$m \cdot \frac{d^2x}{dt^2} = -kx + f(t),$$

which we can write as

$$\frac{d^2x}{dt^2} + \beta x = F(t) \quad \text{where} \quad \beta = \frac{k}{m} \text{ and } F(t) = \frac{f(t)}{m}. \tag{4.8.5}$$

The **forcing function** $f(t)$ (or $F(t)$) describes the external force that jiggles the supporting board up and down rhythmically. Remember that we are assuming that this force is *periodic*, so $f(t)$ is sometimes positive and sometimes negative—that is, sometimes the board is moved downward, and sometimes it is moved upward. (Did you ever see the toy consisting of a paddle with a rubber ball attached to it by an elastic cord?)

The next example gives us the qualitative analysis of this problem.

### ■ Example 4.8.4 Forced Undamped Motion

The system equivalent to our problem is

$$\frac{dx_1}{dt} = x_2$$

$$\frac{dx_2}{dt} = F(t) - \beta x_1. \tag{4.8.6}$$

Let's take $\beta = 4$ and assume that the forcing function is $F(t) = \cos(2t)$. Furthermore, let's assume that the mass starts from its equilibrium position, $x_1(0) = x(0) = 0$, and that it has no initial motion before the external force is applied—that is, $x_2(0) = \frac{dx}{dt}(0) = 0$. Figure 4.22 shows the phase portrait corresponding to this IVP for $0 \leq t \leq 20$.

## Analysis

Note that because the initial point is the origin, it is obvious that the spiral trajectory is moving *outward*—that is, in a clockwise direction. (You should contrast this with Figure 4.18 in Example 4.8.2.) Figure 4.22 indicates that both the displacement of the mass and its velocity are growing without bound. The graphs of $x_1(t)$ and $x_2(t)$ against $t$ (Figures 4.23a and b) confirm this.

## The Actual Solution

The solution of the system we have chosen as an example is $x_1(t) = x(t) = \frac{t}{4}\sin(2t)$. (Check that this is a solution of the IVP.) The sine term contributes an oscillation between $-1$ and $1$,

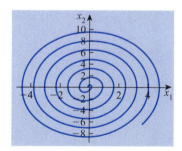

**FIGURE 4.22**
*Trajectory for the system*
$\left\{ \frac{dx_1}{dt} = x_2, \frac{dx_2}{dt} = \cos 2t - 4x_1; x_1(0) = 0 = x_2(0) \right\}$
$0 \leq t \leq 20$

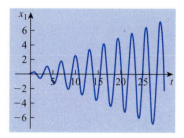

**FIGURE 4.23a**
$x_1(t)$, *displacement*
$x_1(t)$ *in Example 4.8.4*, $0 \leq t \leq 30$

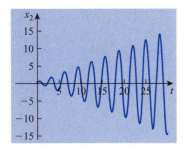

**FIGURE 4.23b**
$x_2(t)$, *velocity*
$x_2(t)$ *in Example 4.8.4*, $0 \leq t \leq 30$

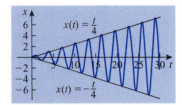

**FIGURE 4.24**
$x_1(t) = x(t) = \frac{t}{4}\sin(2t)$, $0 \leq t \leq 30$

but the factor $\frac{t}{4}$ affects the *amplitude* of the oscillations: $|x(t)| = \left|\frac{t}{4}\right| |\sin(2t)|$, so that $-\frac{t}{4} \leq x(t) \leq \frac{t}{4}$ for $t \geq 0$, and $x(t)$ gets larger and larger in both the positive and negative directions as $t$ gets larger.

Figure 4.24 shows how the linear factor $\frac{t}{4}$ magnifies the oscillation caused by the trigonometric factor. ∎

## 4.8.7 Resonance

A situation in which we have unbounded oscillation, as shown in the preceding example, is called **resonance**. This is particularly important because all mechanical systems have **natural** or **characteristic frequencies**—that is, each atom making up the system is vibrating at a particular frequency, and the composite system has its own characteristic frequency. Recall that if a function $g$ is periodic with period $T$ (so that $T$ is the smallest number for which $g(t + T) = g(t)$ for all $t$), then its **frequency** $f$ is the number of cycles per unit of time: $f = \frac{1}{T}$. Resonance occurs when the frequency of an external force coincides with the natural frequency of the system, thereby amplifying it. You may have experienced having the windows in your home rattle when a heavy vehicle drives by. Going faster than a certain speed in a car may cause a disturbing rattling. In the preceding example, the natural frequency of the system is $\frac{1}{\pi}$ cycles per unit of time, which is equal to the frequency of the forcing function $F(t) = \cos(2t)$.

An unfortunately frequent physical consequence of such amplified vibration is the destruction of the system. In a spring-mass system, the spring could break. A serious situation can occur when numbers of people march in step over a bridge and the frequency of the vibrations set up by the marching feet causes resonance and the collapse of the bridge. (This is why military columns and parade marchers "break step" when crossing a bridge.) As another example, in 1959 and 1960, several models of the same plane crashed, seeming to explode in midair. The Civil Aeronautics Board (CAB) determined that the disintegration of the planes was due to mechanical resonance: A component within the planes, when not fastened securely, generated oscillations that acted as an excessive external force on the wings, breaking them within 30 seconds.[9] Similarly, resonance occurs when the ocean's waves hit a human-made barrier or when wind swirls around a bridge support or tower.

A less disastrous example of resonance is the shattering of a glass by a powerful singer hitting a very high note. The external force here is the sound wave that amplifies the natural frequency of the glass.

It should be pointed out, however, that resonance can also be our friend. The great scientist Galileo (1564–1642) made the following observation about resonance used in the ringing of heavy, free-swinging bells in a tower[10]:

> Even as a boy, I observed that one man alone by giving these impulses at the right instant was able to ring a bell so large that when four, or even six, men seized the rope and tried to stop it they were lifted from the ground, all of them together being unable to counterbalance the momentum which a single man, by properly timed pulls, had given it.

A parent pushing a child's swing, timing the pushes to coincide with the swing's motion, is using resonance to increase the amplitude of each swing.[11] A motorist rocking his or her car to get it out of a muddy rut or a snow bank is applying an external force to amplify the car's natural frequency. Tuning a radio depends on resonance.

## Exercises 4.8

### A

1. The IVP $\ddot{x} + 20\dot{x} + 64x = 0$, with $x(0) = 1/3$ and $\dot{x}(0) = 0$, models the motion of a spring-mass system with a damping force. The initial conditions indicate that the mass has been pulled below its equilibrium position and released.

   a. Express this IVP as a system of two first-order equations, with the appropriate initial conditions.
   b. Use technology to graph the solution of the system in the phase plane.

---

[9] For examples of resonance, see Alice B. Dickinson, *Differential Equations: Theory and Use in Time and Motion* (Reading, MA: Addison-Wesley, 1972): 100 ff.

[10] Galileo Galilei, *Dialogues Concerning Two New Sciences*, translated by H. Crew and A. DeSalvio (New York: Macmillan, 1914), "First Day," 98.

[11] See, for example, "How to Pump a Swing" by S. Wirkus, R. Rand, and A. Ruina, *College Math. J.* **29** (1998): 266–275.

    **c.** Use technology to graph the solution of the original second-order equation relative to the $t$-axis.

    **d.** Comparing the results of parts (b) and (c) to the appropriate graphs in Examples 4.8.1 and 4.8.2, why do you think that the motion shown in this problem should be called *overdamped*?

**2.** Consider the spring-mass system modeled by the IVP $\ddot{x} + c\dot{x} + 0.25x = 0$, with $x(0) = \frac{1}{2}$ and $\dot{x}(0) = \frac{7}{4}$. Here, $c$ is a positive parameter.

    **a.** Express the IVP in terms of a system of first-order equations, including initial conditions.

    **b.** For each of the values $c = 0.5, 1$, and $1.5$, use technology to graph the solution of the system in the phase plane, $0 \le t \le 20$.

    **c.** For each of the values $c = 0.5, 1$, and $1.5$, use technology to graph the solution of the original equation with respect to $t$ on the interval

$$0 \le t \le 20.$$

    **d.** Based on your answers to parts (b) and (c), describe how the nature of the solution changes as the value of $c$ passes through the value $1$? (When $c = 1$, the system is *critically damped*.)

**3.** Consider the following model of a spring-mass system: $\ddot{x} + 64x = 16\cos 8t$, with $x(0) = 0$ and $\dot{x}(0) = 0$.

    **a.** Express the IVP in terms of a system of first-order equations, including initial conditions.

    **b.** Use technology to graph the solution of the system in the phase plane.

    **c.** Use technology to graph the solution of the original second-order equation relative to the $t$-axis.

    **d.** What is the relationship of the graph in (c) to the two half-lines $x = t$ and $x = -t$ for $t \ge 0$?

## B

**1.** A spring having a spring constant of 250 is used in a simple set of scales to measure the weights of objects placed on the pan. The pan (of mass 0.5 kg) rests on top of the spring (see the following illustration).

A block of mass $M$ kg is placed on the pan, causing the spring to oscillate. There is a damping force of $10v$ Newtons, where $v$ m/sec is the speed of the pan.

    **a.** Determine the differential equation of motion for the subsequent damped oscillations.

    **b.** Show that the general solution of the equation found in part (a) has the form $x = A_0 e^{-\frac{10t}{2M+1}} \cos(nt + \varepsilon)$, where $A_0$ is the initial amplitude. Hence, find an expression for $n$ in terms of $M$.

    **c.** Find, in terms of $M$, the time it takes for the system to settle down to oscillations of only 25% of the initial amplitude.

    **d.** What effect would removing the damping force have on the system?

**2.** A particle is in simple harmonic motion along the $y$-axis. At $t = 0$, $y = 3$ and $v = dy/dt = 0$. Exactly 1/2 second later, these values repeat themselves. Find $y(t)$ and $v(t)$.

**C**

**1.** Convert each of the following systems to a single second-order equation. Then interpret each equation to determine which (if any) *cannot* represent a spring-mass system. Explain your reasoning.

    **a.** $Q' = -6Q + 3R$
       $R' = -Q - 2R$

    **b.** $\dot{x} = 3x - y$
       $\dot{y} = x + 3y$

## *4.9  EXISTENCE AND UNIQUENESS

Now that we've learned how to convert higher-order equations to equivalent systems of first-order equations and we've seen some qualitative analyses of these systems, it's time to ask that important question we first considered in Section 2.8 in the context of first-order equations: How do we know that a given higher-order equation or equivalent system *has* a solution—and do we know that any such solution is *unique*?

We don't want to waste human and computer resources searching for a solution that may not exist or that may merely be one of many solutions. For now we'll focus on second-order equations and their corresponding systems. In Chapter 5 we'll look at generalizations to higher-order equations and larger systems.

The first example shows that when there is one solution of a system, there may be many.

### ■ Example 4.9.1 A System IVP with Many Solutions

Let's look at the initial-value problem

$$t^2 x'' - 2tx' + 2x = 0, \text{ with } x(0) = 0 \text{ and } x'(0) = 0.$$

This is equivalent to the system IVP

$$x_1' = x_2$$
$$x_2' = \frac{2}{t} x_2 - \frac{2}{t^2} x_1, \quad \text{with } x_1(0) = 0 = x_2(0).$$

Then $x(t) \equiv 0$ and any function of the form $x(t) = Kt^2$ (where $K$ is any constant) are solutions of the original IVP. (*Verify this.*) With respect to the equivalent system of equations, $x_1(t) \equiv 0$, $x_2(t) \equiv 0$, is a solution, and any pair of functions $x_1(t) = Kt^2$, $x_2(t) = 2Kt$ is a solution. What we are saying here is that *our IVP has infinitely many solutions.* ■

---

\* Denotes an optional section.

In contrast to the IVP in the preceding example, we can have a system of differential equations with *no* solution.

### ■ Example 4.9.2 A System IVP with No Solution

Let's look at the IVP

$$x_1' = \frac{1}{x_1^2}, x_2' = 2x_1 - x_2, \text{ with } x_1(0) = 0 \text{ and } x_2(0) = 1.$$

When we examine this situation carefully, we see that if $x_1(t)$ is part of a solution pair for this IVP, then $x_1'$ doesn't exist for $t = 0$ because $x_1'(0) = \frac{1}{[x_1(0)]^2}$ and $x_1(0) = 0$. This says that there is no solution to this IVP. ■

What we want in most real-life situations is one and only one solution to an initial-value problem. The next example shows such a case.

### ■ Example 4.9.3 A System IVP with a Unique Solution

The IVP $\left\{ \frac{dx}{dt} = y, \frac{dy}{dt} = x; \ x(0) = 1, \ y(0) = 0 \right\}$ has the *unique* solution $x(t) = \frac{1}{2} \left( e^t + e^{-t} \right)$, $y(t) = \frac{1}{2} \left( e^t - e^{-t} \right)$. You may recognize $x$ and $y$ as the *hyperbolic cosine* (cosh) and *hyperbolic sine* (sinh), respectively.

This system is equivalent to the single equation $\ddot{x} - x = 0$, or $\ddot{x} = x$, with $x(0) = 1$ and $\dot{x}(0) = 0$, and it isn't too difficult to guess what kind of function is equal to its own second derivative. Problem A1 in Exercises 4.9 will ask you to explore this further. ■

### 4.9.1 An Existence and Uniqueness Theorem

At this point we have seen that the possibilities for second-order IVPs are similar to those we saw in Section 2.8 for first-order IVPs. We can have *no* solution, *infinitely many solutions*, or *exactly one solution*. Once again we would like to determine when there is one and only one solution of an initial-value problem.

The simplest Existence and Uniqueness Theorem for second-order differential equations or two-dimensional systems of first-order equations is one that is a natural extension of the result we saw in Section 2.8. We'll state two forms of this.

#### *Existence and Uniqueness Theorem*

Suppose we have a second-order IVP $\frac{d^2y}{dt^2} = f(t, y, \dot{y})$, with $y(t_0) = y_0$ and $\dot{y}(t_0) = \dot{y}_0$. If $f$, $\frac{\partial f}{\partial y}$, and $\frac{\partial f}{\partial \dot{y}}$ are continuous in a closed box $B$ in three-dimensional space ($t$-$y$-$\dot{y}$ space) and the point $(t_0, y_0, \dot{y}_0)$ lies inside $B$, then the IVP has a unique solution $y(t)$ on some $t$-interval $I$ containing $t_0$.

Equivalently,

### Existence and Uniqueness Theorem

Suppose we have a two-dimensional system of first-order equations

$$\frac{dx_1}{dt} = f(t, x_1, x_2)$$

$$\frac{dx_2}{dt} = g(t, x_1, x_2),$$

where $x_1(t_0) = x_1^0$ and $x_2(t_0) = x_2^0$. If $f, g, \frac{\partial f}{\partial x_1}, \frac{\partial g}{\partial x_1}, \frac{\partial f}{\partial x_2}$, and $\frac{\partial g}{\partial x_2}$ are all continuous in a box $B$ in $t$-$x_1$ – $x_2$ space containing the point $(t_0, x_1^0, x_2^0)$, then there is an interval $I$ containing $t_0$ in which there exists a unique solution $x_1 = y_1(t), x_2 = y_2(t)$ of the IVP.

## 4.9.2 Many Solutions

We can write the equation in Example 4.9.1 in the form $\ddot{x} = f(t, x, \dot{x}) = \frac{2t\dot{x}-2x}{t^2}$, so we see that $f$ does not exist in any box in which $t = 0$. Therefore, we should not expect exactly one solution, and, in fact, although there *is* a solution to the IVP with initial conditions $x(0) = 0$ and $\dot{x}(0) = 0$, any such solution is not unique.

## 4.9.3 No Solution

In Example 4.9.2, we can use the system form of our Existence and Uniqueness Theorem to see that the function $f(t, x_1, x_2) = 1/x_1^2$ does not exist at the point $(t, x_1^0, x_2^0) = (0, 0, 1)$, so once again we are not guaranteed exactly one solution—and, in fact, there is *no* solution of the IVP.

## 4.9.4 Exactly One Solution

Finally, if we examine the IVP in Example 4.9.3 from either the single-equation or the systems point of view, we should see that in this situation we are guaranteed the existence of one and only one solution of the initial-value problem. (*Check this.*)

The nice thing about these questions is that in most common applied problems, the functions and their derivatives are well behaved (continuous, etc.), so that we *do* have both existence and uniqueness.

## Exercises 4.9

### A

1. In Example 4.9.3 you saw an IVP for a system of equations that was equivalent to the single equation IVP $x'' - x = 0$, or $x'' = x$, with $x(0) = 1$ and $x'(0) = 0$. Using the technique of Section 4.1, show that $x(t) = \frac{1}{2}(e^t + e^{-t})$ is the solution of the IVP $x'' - x = 0$ with $x(0) = 1$ and $x'(0) = 0$.
2. Verify that each of the following initial value problems has a solution that is guaranteed unique *everywhere* in three-dimensional space.
   a. $x_1' = x_2, x_2' = 3x_1 - 5x_2;$   $x_1(0) = 1, x_2(0) = 0$
   b. $x_1' = x_1^2, x_2' = \sin x_1 - x_2^2;$   $x_1(0) = 0, x_2(0) = 0$
   c. $x_1' = x_2^3, x_2' = tx_1 - x_2;$   $x_1(0) = 0, x_2(0) = 1$

3. For each of the following equations, determine intervals in which solutions are guaranteed to exist.

   **a.** $y^{(iv)} + 4y''' + 3y = t$
   **b.** $ty''' + (\sin t)y'' + 3y = \cos t$
   **c.** $t(t-1)y^{(iv)} + e^t y'' + 4t^2 y = 0$
   **d.** $y''' + ty'' + t^2 y' + t^3 y = \ln t$

**B**

1. Show that the initial value problem

$$\left\{ yx' = y - 4t, (x - 3)y' = -4x + \sin t; \; x(0) = 3, y(0) = 0 \right\}$$

has no solution. Does this contradict the existence part of the result we have given in this section? Explain.

2. **a.** Show that $\left\{ x_1(t) = e^{-t}\sin(3t), y_1(t) = e^{-t}\cos(3t) \right\}$ and

$$\left\{ x_2(t) = e^{-(t-1)}\sin(3(t-1)), y_2(t) = e^{-(t-1)}\cos(3(t-1)) \right\}$$

   are solutions of the system

$$\frac{dx}{dt} = -x + 3y$$

$$\frac{dy}{dt} = -3x - y.$$

   **b.** Use technology to draw the graphs of each of the solutions in part (a) in the $x$-$y$ phase plane.
   **c.** Explain why the solutions in part (a) don't contradict the uniqueness part of the result in this section.

**C**

1. Consider the equation

$$5x^2 y^{(5)} - (6\sin x)y''' + 2xy'' + \pi x^3 y' + (3x - 5)y = 0.$$

   Suppose that $Y(x)$ is a solution of this equation such that $Y(1) = 0, Y'(1) = 0, Y''(1) = 0, Y'''(1) = 0, Y^{(4)}(1) = 0,$ and $Y^{(5)}(1) = 0.$ Why must $Y(x)$ be equal to 0 for *all* values of $x$?

2. Use technology to plot some trajectories of the nonautonomous system

$$\frac{dx}{dt} = (1 - t)x - ty$$

$$\frac{dy}{dt} = tx + (1 - t)y.$$

   Your graph should show some intersecting curves. Does the graph contradict the existence and uniqueness theorem? *Explain.*

**3.** Consider the system

$$\dot{x} = y$$

$$\dot{y} = -x + (1 - x^2 - y^2)y.$$

   **a.** Let $D$ denote the region in the phase plane defined by $x^2 + y^2 < 4$. Verify that the given system satisfies the hypotheses of the Existence and Uniqueness Theorem throughout $D$.

   **b.** By substitution, show that $x(t) = \sin t$, $y(t) = \cos t$ is a solution of the system.

   **c.** Now consider a different solution, in this case starting from the initial conditions $x(0) = 1/2$, $y(0) = 0$. Without doing any calculations, explain why this solution *must* satisfy $x(t)^2 + y(t)^2 < 1$ for all real values of $t$.

## 4.10 NUMERICAL SOLUTIONS

The difficulty of finding closed-form solutions of single differential equations is compounded when it comes to systems of equations. We've already seen some useful ways in which systems are analyzed qualitatively. However, you should realize that a graphing calculator or computer produces phase portraits by using numerical methods. As for any computer graph, individual points are calculated and then connected by a series of small line segments that give the impression of a continuous curve.

Now it is time to see that any of the numerical techniques introduced for first-order equations in Sections 3.1, 3.2, and 3.3 can be extended to *systems* of first-order equations in a natural way. In this section, we'll work with two-dimensional systems, leaving the obvious generalizations to Chapters 5 and 7. Even though it is important to be able to solve simple numerical problems by hand, most systems of differential equations are solved using numerical methods implemented on computers.

### 4.10.1 Euler's Method Applied to Systems

Let's start by recalling *Euler's method* for solving the first-order initial-value problem $y' = f(x, y)$, $y'(x_0) = y_0$. This algorithm was originally given as Formula (3.1.3):

$$y_{k+1} = y_k + h \cdot f(x_k, y_k).$$

Here, $h$ is the step size and $y_k$ denotes the approximate value of the solution at the point $x_k = x_0 + kh$.

Now suppose we have a system of two first-order differential equations

$$\frac{dx}{dt} = f(t, x, y)$$

$$\frac{dy}{dt} = g(t, x, y),$$

with $x(t_0) = x_0$ and $y(t_0) = y_0$. If we let $t_k = t_0 + kh$, $x_k \approx x(t_k)$, and $y_k \approx y(t_k)$, we can apply Euler's algorithm to each equation separately to get the result

$$x_{k+1} = x_k + h \cdot f(t_k, x_k, y_k)$$

$$y_{k+1} = y_k + h \cdot f(t_k, x_k, y_k). \tag{4.10.1}$$

Let's see how this method works on a system we've already seen.

### ■ Example 4.10.1 Euler's Method for a System—by Hand

As a simple illustration of Euler's method applied to a system, let's approximate the solution of the IVP of Example 4.9.3 at $t = 0.5$. The system, which we know has a unique solution, is

$$\frac{dx}{dt} = y, \frac{dy}{dt} = x; \ x(0) = 1, y(0) = 0.$$

### Equations

When we use a step size $h = 0.1$, the algorithm given by Equation (4.10.1) looks like

$$x_{k+1} = x_k + (0.1)y_k$$

$$y_{k+1} = y_k + (0.1)x_k,$$

where $x_0 = x(0) = 1$ and $y_0 = y(0) = 0$.

### Calculation

We approximate the solution at $t = 0.5$ by taking five steps:

$$x_1 = x_0 + (0.1)y_0 = 1 + (0.1)(0) = 1$$

$$y_1 = y_0 + (0.1)x_0 = 0 + (0.1)(1) = 0.1$$

$$x_2 = x_1 + (0.1)y_1 = 1 + (0.1)(0.1) = 1.01$$

$$y_2 = y_1 + (0.1)x_1 = 0.1 + (0.1)(1) = 0.2$$

$$x_3 = x_2 + (0.1)y_2 = 1.01 + (0.1)(0.2) = 1.03$$

$$y_3 = y_2 + (0.1)x_2 = 0.2 + (0.1)(1.01) = 0.301$$

$$x_4 = x_3 + (0.1)y_3 = 1.03 + (0.1)(0.301) = 1.0601$$

$$y_4 = y_3 + (0.1)x_3 = 0.301 + (0.1)(1.03) = 0.404$$

$$x_5 = x_4 + (0.1)y_4 = 1.0601 + (0.1)(0.404) = 1.1005$$

$$y_5 = y_4 + (0.1)x_4 = 0.404 + (0.1)(1.0601) = 0.51001$$

## Result

These calculations indicate that $x(0.5) \approx 1.1005$ and $y(0.5) \approx 0.5100$. But to four decimal places, the *exact* solution is $x(0.5) = \cosh(0.5) = \left(\frac{1}{2}\right)(\exp(0.5) + \exp(-0.5)) = 1.1276$ and $y(0.5) = \sinh(0.5) = \left(\frac{1}{2}\right)(\exp(0.5) - \exp(-0.5)) = 0.5211$. Thus, the absolute error is 0.0271 for $x$ and 0.0111 for $y$.

If we cut our step size in half, letting $h = 0.05$ and using technology, we need 10 steps and find that our approximations are $x(0.5) \approx 1.1138$ and $y(0.5) \approx 0.5151$, to four decimal places. Now the error—0.0130 for $x$ and 0.006 for $y$—is roughly half of what these errors were when $h = 0.1$. Having computer resources at our command, it's hard to resist another run, this time with $h = 0.01$. Taking 50 steps, we have $x(0.5) \approx 1.1248$ and $y(0.5) \approx 0.5198$, with errors 0.0028 and 0.0013 for $x$ and $y$, respectively. You should experiment with a few other values of $h$ on your own CAS or graphing calculator. ∎

Problem A1 in Exercises 4.10 asks you to write the system form of the *improved Euler method* (*Heun's method*).

### 4.10.2  The Fourth-Order Runge-Kutta Method for Systems

As an additional example, let's look at the system form of the Runge-Kutta algorithm introduced in Section 3.3. (As we mentioned in that discussion, it was Kutta who generalized the basic method to *systems* of ODEs in 1901.)

We start with the same general first-order system we considered before:

$$\frac{dx}{dt} = f(t, x, y)$$

$$\frac{dy}{dt} = g(t, x, y),$$

with $x(t_0) = x_0$ and $y(t_0) = y_0$. Again, let $t_k = t_0 + kh$, $x_k \approx x(t_k)$, and $y_k \approx y(t_k)$. Then the system version of the classic Runge-Kutta Formula (3.3.2) is

$$x_{k+1} = x_k + \frac{1}{6}(m_1 + 2m_2 + 2m_3 + m_4)$$

$$y_{k+1} = y_k + \frac{1}{6}(M_1 + 2M_2 + 2M_3 + M_4),$$

where

$$m_1 = hf(t_k, x_k, y_k)$$

$$m_2 = hf\left(t_k + \frac{h}{2}, x_k + \frac{m_1}{2}, y_k + \frac{M_1}{2}\right)$$

$$m_3 = hf\left(t_k + h, x_k + \frac{m_2}{2}, y_k + \frac{M_2}{2}\right)$$

$$m_4 = hf(t_k + h, x_k + m_3, y_k + M_3) = hf(t_{k+1}, x_k + m_3, y_k + M_3).$$

and

$$M_1 = hg(t_k, x_k, y_k)$$

$$M_2 = hg\left(t_k + \frac{h}{2}, x_k + \frac{m_1}{2}, y_k + \frac{M_1}{2}\right)$$

$$M_3 = hg\left(t_k + \frac{h}{2}, x_k + \frac{m_2}{2}, y_k + \frac{M_2}{2}\right)$$

$$M_4 = hg(t_k + h, x_k + m_3, y_k + M_3) = hg(t_{k+1}, x_k + m_3, y_k + M_3).$$

Now let's put this algorithm to use—with the aid of technology, of course.

### ■ Example 4.10.2 Using Runge-Kutta (RK4) and a CAS

Let's look again at the initial-value problem analyzed in Example 4.7.1. The system IVP is $\frac{dx}{dt} = y$, $\frac{dy}{dt} = x$; $x(0) = 1$, $y(0) = 0$, and we want to approximate $x(0.5)$ and $y(0.5)$. Rather than wearing ourselves out trying to implement the fourth-order Runge-Kutta method by hand, we can enter the equations and initial conditions into our CAS, specify the method (in whatever way you must describe the RK4 method), and choose a step size $h = 0.1$.

What we get is an approximation for $x(0.5)$ of 1.1276 and an approximation for $y(0.5)$ of 0.5211, both rounded to four decimal places. To four decimal places, the absolute error for each approximation is 0! ■

Our final example shows how the Runge-Kutta-Fehlberg fourth- and fifth-order algorithm works on an interesting system application.

### ■ Example 4.10.3 Using Runge-Kutta-Fehlberg (rkf45) and a CAS

The Japanese-born British mathematician E. C. Zeeman (1925– ) developed a simple nonlinear model of the human heartbeat:

$$\varepsilon \frac{dx}{dt} = -(x^3 - Ax + c)$$

$$\frac{dc}{dt} = x,$$

where $x(t)$ is the displacement from equilibrium of the heart's muscle fiber, $c = c(t)$ is the concentration of a chemical control at time $t$, and $\varepsilon$ and $A$ are positive constants. Because the levels of $c$ determine the contraction and expansion (relaxation) of the muscle fibers, we can think of $c$ as a *stimulus* and of $x$ as a *response*.

We want to investigate the nature of the model's solution, and for convenience we'll assume that $\varepsilon \approx 1.0$ and $A \approx 3$. Also, let $x(0) \approx 1.7$ and $c(0) \approx 0.3$. (The initial conditions were determined after experimenting with various values on a CAS.) The calculations producing the graphs and the values discussed next were carried out using the rkf45 method for the system.

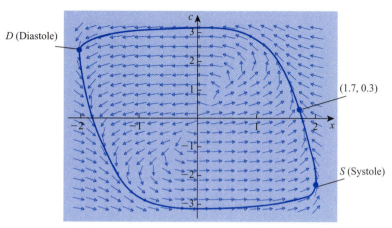

**FIGURE 4.25**
Slope field and trajectory for the system IVP
$$\left\{ \frac{dx}{dt} = -(x^3 - 3x + c), \frac{dc}{dt} = x; \; x(0) = 1.7, c(0) = 0.3 \right\}$$
$0 \le t \le 30$

| **Table 4.3** Solution Values of $\left\{ \frac{dx}{dt} = -(x^3 - 3x + c), \frac{dc}{dt} = x; x(0) = 1.7, \; c(0) = 0.3 \right\}$ | | |
|---|---|---|
| $t$ | $x(t)$ | $c(t)$ |
| 0 | 1.7000 | 0.3000 |
| 2 | 0.7499 | 2.9990 |
| 4 | −1.8417 | 0.4728 |
| 6 | −1.1132 | −2.5911 |
| 8 | 1.9436 | −1.2862 |
| 10 | 1.3384 | 2.0618 |

Because one important feature of a heartbeat is that it is periodic (lub-*dub*, lub-*dub*, …), the solution should reveal this in the $x$-$c$ phase plane—and in fact it does (Figure 4.25). Both the *systole*, corresponding to a fully relaxed heart muscle, and the *diastole*, indicating a state of full contraction, are labeled on Figure 4.25.

We see that the heart muscle starts at $(1.7, 0.3)$ and, under the influence of increasing $c$, contracts until it is fully contracted at $D$. Then the muscle begins to relax until it attains systole at $S$, returns to the initial point, and (we hope) begins the cycle again. Superimposing the trajectory on the slope field makes it easy to see the direction of the trajectory, but the numerical values in Table 4.3 also tell the story.

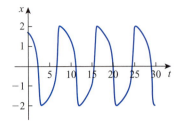

**FIGURE 4.26a**
*Graph of x(t) vs. t for the system IVP*
$$\left\{\frac{dx}{dt} = -(x^3 - 3x + c), \frac{dc}{dt} = x; x(0) = 1.7, c(0) = 0.3\right\}$$
$0 \le t \le 30, -2.5 \le x \le 2.5$

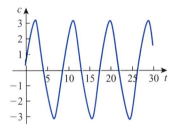

**FIGURE 4.26b**
*Graph of c(t) vs. t for the system IVP*
$$\left\{\frac{dx}{dt} = -\left(x^3 - 3x + c\right), \frac{dc}{dt} = x; x(0) = 1.7, c(0) = 0.3\right\}$$
$0 \le t \le 30, -4 \le c \le 4$

If we examine the signs of $x$ and $c$ as $t$ increases, we see that the points $(x, c)$ are moving *counterclockwise* through the quadrants of the $x$-$c$ plane. Looking carefully at the data in the table, we can see that the trajectory returns to its initial point $(1.7, 0.3)$ sometime between 8 and 10. In fact, a more detailed analysis reveals that the solution of our IVP has period approximately equal to 8.88. (See Problem C2 in Exercises 4.10.)

Solving the system with rkf45 and then plotting $x$ against $t$ (Figure 4.26a), we see the periodic nature of the heart muscle's expansions and contractions. Figure 4.26b shows how the electrochemical activity represented by the variable $c$ also varies periodically.

We will investigate interesting nonlinear systems again in Chapter 7.  ■

Just as for a single first-order equation, we can use spreadsheet commands to carry out the calculations needed to approximate the solutions of systems. Systems versions of the standard numerical techniques may be a bit more difficult to program, may require more intermediate storage, and may take a little more time, but they work well. Graphing calculators also handle systems of differential equations. In fact, as we remarked in the Introduction for this chapter, they usually deal with a single higher-order equation by requiring the user to write it in terms of a system and then solving the system numerically.

Whatever technology you use, try to understand what methods have been implemented by reading your documentation or checking out your software's "Help" features.

## Exercises 4.10

All problems are to be done using technology, unless otherwise indicated.

### A

1. **a.** Extend the improved Euler method given by Formula (3.2.1) to a system of two first-order equations.
   **b.** *By hand*, re-do Example 4.10.1, using $h = 0.1$ to find approximations to $x(0.5)$ and $y(0.5)$.
   **c.** Calculate the absolute error in part (b).
   **d.** Use technology and the improved Euler method with $h = 0.1$ to check your answers to part (b).
2. Consider the system $x' = x - 4y$, $y' = -x + y$, with $x(0) = 1$ and $y(0) = 0$. The exact solution is $x(t) = \left(e^{-t} + e^{3t}\right)$, $y(t) = \left(e^{-t} - e^{3t}\right)$.

   **a.** Verify that the exact solution of the IVP is the solution given above.
   **b.** Approximate the value of the solution at the point $t = 0.2$ using Euler's method with $h = 0.1$. Compare your result with the values of the exact solution, calculating the absolute error.
   **c.** Approximate the value of the solution at the point $t = 0.2$ using a fourth-order Runge-Kutta method with $h = 0.2$. Calculate the absolute error.
3. Consider the initial value problem $y'' + y' - 2y = 2x$, with $y(0) = 1$ and $y'(0) = 1$.

   **a.** Convert this problem into a system of two first-order equations. (*Choose your new variables carefully.*)
   **b.** Determine approximate values of the solution at $x = 0.5$ and $x = 1.0$ by using Euler's method with $h = 0.1$.
   **c.** Determine approximate values of the solution at $x = 0.5$ and $x = 1.0$ by using the fourth-order Runge-Kutta method with $h = 0.1$.
4. In Example 4.8.2 you were told that the solution to the IVP

$$\frac{d^2x}{dt^2} + \frac{1}{4}\frac{dx}{dt} + 2x = 0, \quad \text{with} \quad x(0) = 1,\ \dot{x}(0) = 0,$$

is

$$x(t) = \frac{1}{127}e^{\left(-\frac{1}{8}t\right)}\left(127\cos\left(\frac{1}{8}\sqrt{127}t\right) + \sqrt{127}\sin\left(\frac{1}{8}\sqrt{127}t\right)\right).$$

   **a.** Convert this IVP into a system of first-order equations.
   **b.** Determine the approximate value of the solution at $t = 0.6$ by using the Runge-Kutta-Fehlberg method (*rkf45*), if available. Otherwise, use the highest-order Runge-Kutta method available to you, with $h = 0.01$. Compare your values with the exact solution above.

**B**

1. A particle moves in three-dimensional space according to the equations

$$\frac{dx}{dt} = yz, \quad \frac{dy}{dt} = zx, \quad \frac{dz}{dt} = xy.$$

   **a.** Assuming that $x(0) = 0$, $y(0) = 5$, and $z(0) = 0$, use the Runge-Kutta-Fehlberg method, if available, to approximate the solution at $t = 0.1, 0.2, 0.3, 0.4, 0.5, 0.6, 0.7, 0.8, 0.9, 1.0, 5.0$, and 37. (Otherwise use the highest-order Runge-Kutta method available to you, with $h = 0.01$.) Describe what these values seem to be telling you about the motion of the particle.

   **b.** Now assume that $x(0) = y(0) = 1$ and $z(0) = 0$. Approximate the solution at $t = 0.1, 0.2, 0.3$, 1.5, 1.6, 1.7, 1.8, and 1.9 using the same procedure you used in part (a). What seems to be happening?

2. The system

$$\frac{dx}{dt} = 7y - 4x - 13$$

$$\frac{dy}{dt} = 2x - 5y + 11$$

   appeared in Example 4.7.3, where it was described as a possible arms race model.

   **a.** Suppose that $x(0) = 1$ and $y(0) = 1$. Use technology and the Runge-Kutta-Fehlberg fourth-fifth-order method (or a reasonable substitute) to estimate $x$ and $y$ for $t = 1, 2, 3, 4, 5, 10, 15$, and 20.

   **b.** On the basis of the values found in part (a), guess at $\lim_{t \to \infty} x(t)$ and $\lim_{t \to \infty} y(t)$.

3. The Lotka-Volterra system (Section 4.7)

$$\dot{x} = 3x - 2xy$$

$$\dot{y} = 0.5xy - y$$

   has solutions $(x(t), y(t))$ that are periodic because a given trajectory always returns to its initial point in some finite time $t^*$ : $x(t + t^*) = x(t)$ and $y(t + t^*) = y(t)$. By using technology and the rkf45 method, estimate the smallest value of $t^*$ to two decimal places if $x(0) = 3$ and $y(0) = 2$. (Try different values of $t \neq 0$ until you get $x(t) \approx 3$ and $y(t) \approx 2.0$.)

4. The equations

$$\dot{x} = y$$

$$\dot{y} = -0.25y - 2x; \quad x(0) = 1, y(0) = 0$$

   represent a certain spring-mass system with damping. As usual, assume that the positive direction for $x(t)$ and $y(t)$ is downward and time is measured in seconds.

   **a.** Using technology, approximate $x(t)$ and $y(t)$ for $t = 1, 2, 3, 4$ and interpret the position and velocity in each case.

   **b.** Estimate (to the nearest hundredth of a second) the time when the mass *first* reaches its equilibrium position, $x = 0$.

**5.** *Emden's equation* $\frac{d^2y}{dx^2} + \frac{2}{x}\frac{dy}{dx} + y^n = 0$, where $n$ is a parameter, has been used to model the thermal behavior of a spherical cloud of gas.[12] When the cloud of gas is a star, the first zero of the solution, multiplied by $10^{10}$ cm, represents the radius of the star. In modeling the bright component of Capella, the astrophysicist A. S. Eddington took $n = 3$ in Emden's equation and used the initial conditions $y(0) = 1, y'(0) = 0$.

   **a.** Express Eddington's version of Emden's equation as an equivalent system of two first-order equations in the variables $u$ and $v$.

   **b.** Approximate the radius of Capella by determining (approximating) the first value of $x$ for which $y(x) = 0$.

## C

**1.** A famous model for the spread of a disease is the *S-I-R* model. At a given time $t$, $S$ represents the population of *susceptibles*, those who have never had the disease and can get it; $I$ stands for the *infected*, those who have the disease now and can give it to others; and $R$ denotes the *recovered*, people who have already had the disease and are immune. Suppose these populations are related by the system

$$\frac{dS}{dt} = (-0.00001)SI$$

$$\frac{dI}{dt} = (0.00001)SI - \frac{I}{14}$$

$$\frac{dR}{dt} = \frac{I}{14},$$

with $S(0) = 45,400, I(0) = 2100, R(0) = 2500$.

   **a.** Add the three differential equations and interpret the result in terms of a population.

   **b.** Use your CAS to plot $S$, $I$, and $R$ as functions of $t$ on separate graphs. [*Warning*: Some mathematical software (such as *Maple*) may reserve the letter $I$ for the imaginary unit $\sqrt{-1}$. If this is your situation, use *IN* to denote the infected population.]

   **c.** Use your CAS to plot phase portraits in the *S-I*, *S-R*, and *I-R* planes.

   **d.** Use a powerful numerical method (with $h = 0.1$ if appropriate) to approximate the values of $S$, $I$, and $R$ at $t = 1, 2, 3, 10, 15, 16$, and 17 days. What do you see?

   **e.** Approximate the value of $t$ at which $I = 0$.

**2.** Use the rkf45 method to show why the period of the trajectory in Figure 4.25 is approximately 8.88. (Use the method suggested in Problem B3.)

**3.** Investigate the Zeeman heartbeat model in Example 4.10.3 with $\varepsilon = 0.025, A = 0.1575$, and $(x_0, c_0) = (0.45, -0.02025)$.

   **a.** Use the rkf45 method to approximate $x(t)$ and $c(t)$ for $t = 0.01, 0.02, \ldots, 0.10$ seconds. What do your calculations tell you about the direction of the solution curve in the $x$-$c$ plane?

   **b.** Draw the trajectory corresponding to the initial conditions given above.

---

[12] See H. T. Davis, *Introduction to Nonlinear Differential Equations and Integral Equations* (New York: Dover, 1962): 371ff.

   **c.** Approximate the period of the trajectory found in part (b).

   **d.** Estimate the coordinates of the diastole and the systole.

**4.** A lunar lander is falling freely toward the surface of the moon. If $x(t)$ represents the distance of the lander from the *center* of the moon (in meters, with $t$ in seconds), then $x(t)$ satisfies the initial-value problem

$$\frac{d^2x}{dt^2} = 4 - \frac{4.9044 \times 10^{12}}{x^2},$$

with $x(0) = 1{,}781{,}870$ and $x'(0) = -450$. (The value $x(0)$ represents the fact that the retro rockets are fired at $t = 0$—when the lander is at a height of 41,870 meters from the moon's surface, or 1,781,870 meters from the moon's center.)

   **a.** Determine the value of $t$ when $x(t) = 1{,}740{,}000$—that is, when the craft has landed on the lunar surface.

   **b.** What is the lunar lander's velocity at touchdown?

## SUMMARY

For second-order homogeneous linear equations with constant coefficients—equations of the form

$$ax'' + bx' + cx = 0,$$

where $a$, $b$, and $c$ are constants, $a \neq 0$—we can describe the solutions explicitly in terms of the roots of the associated **characteristic equation** $a\lambda^2 + b\lambda + c = 0$ as follows:

   **1.** If there are two distinct real roots—$\lambda_1, \lambda_2$ with $\lambda_1 \neq \lambda_2$—then the general solution is

$$x(t) = c_1 e^{\lambda_1 t} + c_2 e^{\lambda_2 t}.$$

   **2.** If there is a repeated real root $\lambda$, then the general solution has the form

$$x(t) = c_1 e^{\lambda t} + c_2 t e^{\lambda t} = (c_1 + c_2 t)\, e^{\lambda t}.$$

   **3.** If the roots form a complex conjugate pair $p \pm qi$, then the (real) general solution has the form $x(t) = e^{pt}\big(c_1 \cos(qt) + c_2 \sin(qt)\big)$. Here, we need Euler's formula to deal with complex exponentials.

The general solution, $y_{GNH}$, of a linear nonhomogeneous system is obtained by finding a particular solution, $y_{PNH}$, of the nonhomogeneous system and adding it to the general solution, $y_{GH}$, of the homogeneous system: $y_{GNH} = y_{GH} + y_{PNH}$. A particular solution can be found using the method of **undetermined coefficients** or **variation of parameters**.

For a linear equation of any order, we have the **Superposition Principle**: If $y_j$ is a solution of $L(y) = f_j$ for $j = 1, 2, \ldots, n$, and $c_1, c_2, \ldots, c_n$ are arbitrary constants, then $c_1 y_1 + c_2 y_2 + \cdots + c_n y_n$ is a solution of $L(y) = c_1 f_1 + c_2 f_2 + \cdots + c_n f_n$. That is,

$$L\left(c_1 y_1 + c_2 y_2 + \cdots + c_n y_n\right) = c_1 L\left(y_1\right) + c_2 L\left(y_2\right) + \cdots + c_n L\left(y_n\right) = c_1 f_1 + c_2 f_2 + \cdots + c_n f_n.$$

As a consequence of the Superposition Principle, the formula $y_{\text{GNH}} = y_{\text{GH}} + y_{\text{PNH}}$ is valid for a linear equation of any order $n$. We have an algorithm to find the general solution $y_{\text{GH}}$ of the associated $n$th-order homogeneous equation $a_n y^{(n)} + a_{n-1} y^{(n-1)} + \cdots + a_2 y'' + a_1 y' + a_0 y = 0$, where $y$ is a function of the independent variable $t$ and $a_n, a_{n-1}, \ldots, a_1, a_0$ are constants.

First, find the roots of the characteristic equation

$$a_n \lambda^n + a_{n-1} \lambda^{n-1} + \cdots + a_1 \lambda + a_0 = 0.$$

Use a CAS to solve the equation if $n$ is greater than or equal to 3. Next, group these roots as follows: (a) distinct real roots; (b) distinct complex conjugate pairs $p \pm qi$; (c) multiple real roots; (d) multiple complex roots. Then the general solution is a sum of $n$ terms of the forms

1. $c_i e^{\lambda_i t}$ for each distinct real root $\lambda_i$

2. $e^{pt}\left(c_1 \cos qt + c_2 \sin qt\right)$ for each distinct complex pair $p \pm qi$

3. $\left(c_1 + c_2 t + \cdots + c_k t^{k-1}\right) e^{\lambda_i t}$ for each multiple real root $\lambda$, where $k$ is the multiplicity of that root

4. $e^{pt}\left(c_1 \cos qt + c_2 \sin qt\right) + t e^{pt}\left(c_3 \cos qt + c_4 \sin qt\right) + \cdots + t^{k-1} e^{pt}\left(c_{2k-1} \cos qt + c_{2k} \sin qt\right)$ for each multiple complex pair of roots $p \pm qi$, where $k$ is the multiplicity of the pair $p \pm qi$

To find a particular solution of the $n$th-order nonhomogeneous equation, we can use the method of **undetermined coefficients** or **variation of parameters** as we did in the second-order case (although more work is involved).

The most important fact in this chapter is that **any single $n$th-order differential equation can be converted into an equivalent system of first-order equations. More precisely, any $n$th-order differential equation**

$$x^{(n)} = F\left(t, x, x', x'', \ldots, x^{(n-1)}\right)$$

**can be converted into an equivalent system of first-order equations by letting $x_1 = x$, $x_2 = x', x_3 = x'', \ldots, x_n = x^{(n-1)}$.** However, to convert a single *nonautonomous* $n$th-order equation into an equivalent *autonomous* system (one whose equations do not explicitly contain the independent variable $t$), we need $n+1$ first-order equations: $x_1 = x, x_2 = x', x_3 = x'', \ldots, x_n = x^{(n-1)}, x_{n+1} = t$. The system is linear or nonlinear, autonomous or nonautonomous, according to the nature of the individual equations in the system. Linear systems are easier to calculate with and understand than nonlinear systems. Similarly, autonomous systems are nicer than nonautonomous systems.

Converting a single higher-order equation to a system often provides graphical insights that cannot be obtained from the one equation. This conversion also allows us to use the first-order methods in Chapters 2 and 3 to understand higher-order equations.

A two-dimensional system has the form

$$x' = F(t, x, y)$$
$$y' = G(t, x, y).$$

A particular **solution** of such a system consists of a *pair* of functions $x(t)$, $y(t)$ that, when substituted into the equations of the system, give true statements. The proper graphical representation of a solution is a curve in three-dimensional $t$-$x$-$y$ space, the set of points of the form $(t, x(t), y(t))$; but often it is useful to think of the points $(x(t), y(t))$ as tracing out a path (also called an **orbit** or a **trajectory**) in the $x$-$y$ plane (called the **phase plane**) as the parameter $t$ varies "offstage." The *positive* direction of the path is the direction that corresponds to increasing values of $t$. The collection of all trajectories is the **phase portrait** of the system. Technology also enables us to study the graphs of $x$ vs. $t$ and $y$ vs. $t$.

For *autonomous* systems $x' = f(x, y)$, $y' = g(x, y)$, we can eliminate any explicit reliance on the parameter $t$ by using the Chain Rule to form the first-order differential equation

$$\frac{dy}{dx} = \frac{dy}{dt} \cdot \frac{dt}{dx} = \frac{\frac{dy}{dt}}{\frac{dx}{dt}} = \frac{g(x, y)}{f(x, y)}.$$

This gives the slope of the tangent line at points of the phase plane. The slope field of this first-order equation outlines the phase portrait of the system.

Given any two-dimensional autonomous system $x' = f(x, y)$, $y' = g(x, y)$, an **equilibrium point** is a point $(x, y)$ such that $f(x, y) = 0 = g(x, y)$. This means, for example, that a particle at this point in the phase plane is not moving. The language of **sinks** and **sources** used in Section 2.6 can be extended to apply to equilibrium solutions of systems. The behavior of trajectories near equilibrium points of linear systems will be discussed systematically in Chapter 5. Trajectories for nonlinear systems are treated in Chapter 7.

As examples of these ideas, we discussed **predator-prey systems**, in particular the **Lotka-Volterra equations**. Several examples of **spring-mass problems** were also analyzed, including one exhibiting the phenomenon of **resonance**.

Before getting too immersed in trying to solve higher-order equations or their equivalent systems, we have to determine when solutions *exist*—and whether existing solutions are *unique*. A useful result applies to a second-order IVP, $\frac{d^2y}{dt^2} = f\left(t, y, \frac{dy}{dt}\right)$ with $y(t_0) = y_0$ and $\frac{dy}{dt}(t_0) = \dot{y}_0$.

*If $f$, $\frac{\partial f}{\partial y_1}$, and $\frac{\partial f}{\partial \dot{y}}$ are continuous in a closed box $B$ in three-dimensional space ($t$-$y$-$\dot{y}$ space) and* the point $(t_0, y_0, \dot{y}_0)$ lies inside $B$, then the IVP has a unique solution $y(t)$ on some $t$-interval $I$ containing $t_0$.

Equivalently, suppose we have a two-dimensional system of first-order equations

$$\frac{dx_1}{dt} = f(t, x_1, x_2)$$

$$\frac{dx_2}{dt} = g(t, x_1, x_2),$$

where $x_1(t_0) = x_1^0$ and $x_2(t_0) = x_2^0$. Then if $f, g, \frac{\partial f}{\partial x_1}, \frac{\partial g}{\partial x_1}, \frac{\partial f}{\partial x_2}$, and $\frac{\partial g}{\partial x_2}$ are all continuous in a box $B$ in $t - x_1 - x_2$ space containing the point $(t_0, x_1^0, x_2^0)$, there is an interval $I$ containing $t_0$ in which there exists a unique solution $x_1 = y_1(t), x_2 = y_2(t)$ of the IVP.

Once we are confident that an IVP involving a higher-order equation or its system equivalent *has* a unique solution, we can apply natural two-dimensional generalizations of the numerical solution methods introduced in Sections 3.1, 3.2, and 3.3: Euler's method; the improved Euler method; and higher-order techniques such as the fourth-order Runge-Kutta and Runge-Kutta-Fehlberg methods. Technology is indispensable in the numerical solution of both single equations and systems.

## PROJECT 4-1

### Get the Lead Out

In analyzing the flow of lead pollution in a human body among the three compartments bone, blood, and tissue, the following system was developed[13]:

$$\dot{x}_1 = -\frac{65}{1800}x_1 + \frac{1088}{87,500}x_2 + \frac{7}{200,000}x_3 + \frac{6162}{125}$$

$$\dot{x}_2 = \frac{20}{1800}x_1 - \frac{20}{700}x_2$$

$$\dot{x}_3 = \frac{7}{1800}x_1 - \frac{7}{200,000}x_3.$$

Here, $x_1(t)$ is the amount of lead in the blood at time $t$ (in years), $x_2(t)$ is the amount of lead in tissue, and $x_3(t)$ is the amount of lead in bone. Assume that $x_1(0) = x_2(0) = x_3(0) = 0$.

**a.** Use technology to graph the three-dimensional trajectory in $x_1$-$x_2$-$x_3$ space with $0 \leq t \leq 250$. (Move the axes around to get a good view.)

**b.** Use technology to graph the solution in the $t$-$x_1$ plane, $0 \leq t \leq 150$. What seems to be the equilibrium level of lead in the blood?

---

[13] E. Batschelet, L. Brand, and A. Steiner, "On the Kinetics of Lead in the Human Body," *Journal of Mathematical Biology* **8** (1979): 15–23.

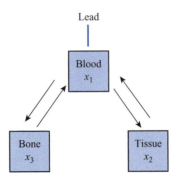

c. Use technology to graph the solution in the $t$-$x_2$ plane, $0 \le t \le 250$. What seems to be the equilibrium level of lead in tissue?

d. Use technology to graph the solution in the $t$-$x_3$ plane, $0 \le t \le 70{,}000$. In your CAS, specify a step size of 50 if you can. (*Warning: It may take a long time for your CAS to produce the graph.*) What seems to be the equilibrium level of lead in bone?

e. What do the graphs in parts (b), (c), and (d) say about the comparative times it takes blood, tissue, and bone to reach their equilibrium levels of lead?

# CHAPTER 5

# Systems of Linear Differential Equations

## INTRODUCTION

In Chapter 4, we saw how any higher-order differential equation can be written as an equivalent system of first-order differential equations. The examples we discussed introduced some algebraic manipulations and some geometric aspects of second- and third-order systems such as the *phase plane*, but there was no attempt to give a systematic approach.

In this chapter, we will explore (for the most part) autonomous systems of first-order *linear* differential equations, for which the theory is neat and complete. An important component of this theory is the *Superposition Principle*, which we discussed in Chapters 2 and 4 and which is the distinguishing characteristic of linear systems, as we will see in the sections to come. This fundamental principle will help us to determine the general solution of linear systems in essentially the same way in which we solved single second-order linear equations in Sections 4.1 and 4.2.

To understand the important ideas underlying the theory and application of linear systems, we'll introduce some of the language and concepts from the area of mathematics called *linear algebra* without probing too deeply into the intricacies of this valuable and useful subject. For the most part, we'll stick to two-dimensional systems (two equations in two unknown functions) for the sake of geometric intuition, but we will also look at some higher-order systems. Chapter 6 will enhance our ability to handle linear systems, and in Chapter 7 we'll see how *nonlinear* systems can be analyzed in terms of certain related linear systems.

## 5.1 SYSTEMS AND MATRICES

### 5.1.1 Matrices and Vectors

Suppose we look at the linear system

$$\begin{aligned}\dot{x} &= ax + by \\ \dot{y} &= cx + dy,\end{aligned}$$

(5.1.1)

where $x$ and $y$ are functions of $t$, and $a$, $b$, $c$, and $d$ are constants.

There is a useful notation for linear systems that was invented by the English mathematician Arthur Cayley and named by his fellow countryman James Sylvester around 1850. This notation allows us to pick out the coefficients $a$, $b$, $c$, and $d$ in System (5.1.1) and write them in a square array $A = \begin{bmatrix} a & b \\ c & d \end{bmatrix}$ called a **matrix**—in this case, the **matrix of coefficients** of the linear system. (The plural of *matrix* is *matrices*.) In general, a matrix is just a rectangular array of mathematical objects (numbers or functions in this book) and can describe linear systems of all sizes. The size of a matrix is given in terms of the number of its **rows** and **columns**. For example, $A = \begin{bmatrix} a & b \\ c & d \end{bmatrix}$ is described as a 2 × 2 matrix because it has two rows, $(a\,b)$ and $(c\,d)$, and two columns, $\begin{bmatrix} a \\ c \end{bmatrix}$ and $\begin{bmatrix} b \\ d \end{bmatrix}$. The matrix $B = \begin{bmatrix} -4 & 0 & 1 & 5 \\ 2 & 6 & 7 & -\pi \\ 0 & \sqrt{5} & 3 & 5/9 \end{bmatrix}$ is a 3 × 4 matrix because it has three rows and four columns.

In describing the linear System (5.1.1), we can also introduce a **column matrix** or **vector** $X = \begin{bmatrix} x \\ y \end{bmatrix}$. (This is a 2 × 1 matrix.) If $x(t)$ and $y(t)$ are solutions of the System (5.1.1), we call $X = \begin{bmatrix} x \\ y \end{bmatrix}$ a **solution vector** of the system. We can view $X$ as a point in the $x$-$y$ plane, or phase plane, whose coordinates are written vertically instead of in the usual horizontal ordered-pair configuration. If a vector is made up of constants, then the *direction* of the vector is taken as the direction of an arrow from the origin to the point $(x, y)$ in the $x$-$y$ plane. (See Section B.1 for more information.)

If $A = \begin{bmatrix} a & b \\ c & d \end{bmatrix}$ and $B = \begin{bmatrix} e & f \\ g & h \end{bmatrix}$, then we say the matrices are **equal** and write $A = B$ if $a = e$, $b = f$, $c = g$, and $d = h$. We say that "corresponding elements must be equal." Similarly, if $V = \begin{bmatrix} x_1 \\ y_1 \end{bmatrix}$ and $W = \begin{bmatrix} x_2 \\ y_2 \end{bmatrix}$, we say that $V = W$ if and only if $x_1 = x_2$ and $y_1 = y_2$.

If a vector (or, more generally, a matrix) is made up of objects (**elements** or **entries**) that are *functions*, we can define the *derivative* of such a vector as the vector whose elements are the derivatives of the original elements, provided that all these individual derivatives exist. For example, if $X = \begin{bmatrix} -t^2 \\ \sin t \end{bmatrix}$, then

$$\frac{d}{dt}X = \begin{bmatrix} \frac{d}{dt}(-t^2) \\ \frac{d}{dt}(\sin t) \end{bmatrix} = \begin{bmatrix} -2t \\ \cos t \end{bmatrix}.$$

## 5.1.2 The Matrix Representation of a Linear System

We can write the system

$$\dot{x} = ax + by$$
$$\dot{y} = cx + dy$$

compactly and symbolically as

$$\begin{bmatrix} \dot{x} \\ \dot{y} \end{bmatrix} = \begin{bmatrix} a & b \\ c & d \end{bmatrix} \begin{bmatrix} x \\ y \end{bmatrix},$$

or $\dot{X} = AX$, where $X = \begin{bmatrix} x \\ y \end{bmatrix}$ and $A = \begin{bmatrix} a & b \\ c & d \end{bmatrix}$. The juxtaposition ("product") $\begin{bmatrix} a & b \\ c & d \end{bmatrix} \begin{bmatrix} x \\ y \end{bmatrix}$ repre-

sents the vector $\begin{bmatrix} ax + by \\ cx + dy \end{bmatrix}$. For example, $\begin{bmatrix} 3 & -2 \\ 1 & 4 \end{bmatrix} \begin{bmatrix} x \\ y \end{bmatrix} = \begin{bmatrix} 3x - 2y \\ x + 4y \end{bmatrix}$. There is a way to define

and interpret this product of a matrix and a vector in the context of linear algebra (see Section B.3 for details), but **we will take this product as a symbolic representation of the system, highlighting the matrix of coefficients and the solution vector**. Soon we will see how a linear system's solutions—its behavior in the phase plane—are determined by the matrix of coefficients. For now, let's look at some examples of the use of matrix notation.

### ■ Example 5.1.1 Matrix Form of a Two-Dimensional Linear System

We can write the linear system of ODEs

$$\dot{x} = -3x + 5y$$
$$\dot{y} = x - 4y$$

in matrix terms as

$$\begin{bmatrix} \dot{x} \\ \dot{y} \end{bmatrix} = \begin{bmatrix} -3 & 5 \\ 1 & -4 \end{bmatrix} \begin{bmatrix} x \\ y \end{bmatrix}.$$

■

The next example demonstrates how we have to be careful in extracting the right matrix of coefficients from a linear system problem.

### ■ Example 5.1.2 Matrix Form of a Two-Dimensional Linear System

The linear system $\frac{dx}{dt} = y$, $\frac{dy}{dt} = -x$ should be written as

$$\frac{dx}{dt} = 0 \cdot x + 1 \cdot y$$
$$\frac{dy}{dt} = -1 \cdot x + 0 \cdot y$$

first. Then it is clear that the matrix representation of the system is

$$\frac{d}{dt}\begin{bmatrix} x \\ y \end{bmatrix} = \begin{bmatrix} \frac{dx}{dt} \\ \frac{dy}{dt} \end{bmatrix} = \begin{bmatrix} 0 & 1 \\ -1 & 0 \end{bmatrix} \begin{bmatrix} x \\ y \end{bmatrix}.$$

■

### 5.1.3 Some Matrix Algebra

Before we continue, we should discuss some properties of matrix algebra that we'll be using in the rest of this chapter. For example, if we're given the system

$$x' = -4x + 6y = 2(-2x + 3y)$$

$$y' = 2x - 8y = 2(x - 4y)$$

it is natural to write $\begin{bmatrix} x' \\ y' \end{bmatrix} = \begin{bmatrix} -4 & 6 \\ 2 & -8 \end{bmatrix} = 2\begin{bmatrix} -2 & 3 \\ 1 & -4 \end{bmatrix}$, or $X' = 2AX$, where $A = \begin{bmatrix} -2 & 3 \\ 1 & -4 \end{bmatrix}$.

More generally, if $A = \begin{bmatrix} a & b \\ c & d \end{bmatrix}$ and $k$ is a constant (called a **scalar** to distinguish it from a vector or a matrix), then $kA = k\begin{bmatrix} a & b \\ c & d \end{bmatrix} = \begin{bmatrix} ka & kb \\ kc & kd \end{bmatrix}$. In particular, for vectors, we have $k\begin{bmatrix} u \\ v \end{bmatrix} = \begin{bmatrix} ku \\ kv \end{bmatrix}$.

Put simply, *multiplying a matrix by a number requires multiplying each element of that matrix by the number.* For example, if $A = \begin{bmatrix} 2 & -3 \\ 5 & 0 \end{bmatrix}$ and $k = -2$, then

$$kA = -2A = \begin{bmatrix} -2(2) & -2(-3) \\ -2(5) & -2(0) \end{bmatrix} = \begin{bmatrix} -4 & 6 \\ -10 & 0 \end{bmatrix}.$$

Two matrices, $A$ and $B$, of the same size (that is, having the same number of rows and the same number of columns) can be added in an element-by-element way. For example, if $A = \begin{bmatrix} -2 & 3 \\ 4 & -1 \end{bmatrix}$ and $B = \begin{bmatrix} 1 & 2 \\ 3 & 4 \end{bmatrix}$, then

$$A + B = \begin{bmatrix} -2 + 1 & 3 + 2 \\ 4 + 3 & (-1) + 4 \end{bmatrix} = \begin{bmatrix} -1 & 5 \\ 7 & 3 \end{bmatrix} \text{ and}$$

$$A - B = A + (-1)B = \begin{bmatrix} -2 & 3 \\ 4 & -1 \end{bmatrix} + \begin{bmatrix} -1 & -2 \\ -3 & -4 \end{bmatrix} = \begin{bmatrix} -3 & 1 \\ 1 & -5 \end{bmatrix}.$$

Similarly, if $V = \begin{bmatrix} x_1 \\ y_1 \end{bmatrix}$ and $W = \begin{bmatrix} x_2 \\ y_2 \end{bmatrix}$, then $V + W = \begin{bmatrix} x_1 + x_2 \\ y_1 + y_2 \end{bmatrix}$. The vector defined as $\mathbf{0} = \begin{bmatrix} 0 \\ 0 \end{bmatrix}$, which is called the **zero vector**, behaves in the world of vectors the way the *number* 0 acts in arithmetic: $V + \mathbf{0} = V = \mathbf{0} + V$ for any vector $V$. Similarly, we can define the **zero matrix**, $Z = \begin{bmatrix} 0 & 0 \\ 0 & 0 \end{bmatrix}$, having the same property for matrix addition. Note that $X = \begin{bmatrix} x \\ y \end{bmatrix}$ is an *equilibrium point* for the system $\begin{bmatrix} \dot{x} \\ \dot{y} \end{bmatrix} = \begin{bmatrix} a & b \\ c & d \end{bmatrix}\begin{bmatrix} x \\ y \end{bmatrix}$ if and only if $\begin{bmatrix} a & b \\ c & d \end{bmatrix}\begin{bmatrix} x \\ y \end{bmatrix} = \mathbf{0}$—that is, if and only if $X$ is a solution of the matrix equation $AX = \mathbf{0}$.

A particularly useful idea for our future work is a *linear combination* of vectors. Given two vectors $V = \begin{bmatrix} x_1 \\ y_1 \end{bmatrix}$ and $W = \begin{bmatrix} x_2 \\ y_2 \end{bmatrix}$, any vector of the form $k_1 V + k_2 W$, where $k_1$ and $k_2$ are scalars, is called a **linear combination** of $V$ and $W$. In terms of our given vectors, a linear combination of $V$ and $W$ is any vector of the form $k_1 V + k_2 W = \begin{bmatrix} k_1 x_1 \\ k_1 y_1 \end{bmatrix} + \begin{bmatrix} k_2 x_2 \\ k_2 y_2 \end{bmatrix} = \begin{bmatrix} k_1 x_1 + k_2 x_2 \\ k_1 y_1 + k_2 y_2 \end{bmatrix}$. As an example, for the specific vectors $V = \begin{bmatrix} \sin t \\ 2 \end{bmatrix}$ and $W = \begin{bmatrix} \cos t \\ e^t \end{bmatrix}$, a linear combination has the form $\begin{bmatrix} k_1 \sin t + k_2 \cos t \\ 2k_1 + k_2 e^t \end{bmatrix}$.

It is important to know that the *associative* and *distributive rules* of algebra hold for matrix addition and the product of a matrix and a vector. For example, if $A$ and $B$ are matrices; $V$ and $W$ are vectors; and $k, k_1$, and $k_2$ are scalars, then

$$A(kV) = k(AV)$$

$$A(V + W) = AV + AW,$$

and

$$A(k_1 V + k_2 W) = A(k_1 V) + A(k_2 W) = k_1(AV) + k_2(AW).$$

These results are discussed further in Section B.3.

Finally, note that the matrix $\begin{bmatrix} 1 & 0 \\ 0 & 1 \end{bmatrix}$ acts as an *identity* for multiplication: $\begin{bmatrix} 1 & 0 \\ 0 & 1 \end{bmatrix}\begin{bmatrix} x \\ y \end{bmatrix} = \begin{bmatrix} x \\ y \end{bmatrix}$ for any vector $\begin{bmatrix} x \\ y \end{bmatrix}$. In the context of two-dimensional systems, the matrix $\begin{bmatrix} 1 & 0 \\ 0 & 1 \end{bmatrix}$ is called the **identity matrix** and is denoted by $I$.

In the next section we will see how matrix notation gives us insight into the nature of a system's solutions. To understand the solutions more fully, we will introduce some additional concepts from linear algebra.

## Exercises 5.1

### A

1. Express each of the following systems of linear equations in matrix terms—that is, in the form $AX = B$, where $A, X$, and $B$ are matrices.

   a. $3x + 4y = -7$
   $-x - 2y = 5$

   b. $\pi a - 3b = 4$
   $5a + 2b = -3$

**c.** $x - y + z = 7$
$-x + 2y - 3z = 9$
$2x - 3y + 5z = 11$

[Think about what would make sense in (c).]

**2.** If $A = \begin{bmatrix} 1 & 2 \\ 3 & 4 \end{bmatrix}$, $B = \begin{bmatrix} 0 & -2 \\ 3 & 1 \end{bmatrix}$, $I = \begin{bmatrix} 1 & 0 \\ 0 & 1 \end{bmatrix}$, $V = \begin{bmatrix} 2 \\ -1 \end{bmatrix}$, and $W = \begin{bmatrix} 3 \\ 2 \end{bmatrix}$, calculate each of the following:

    **a.** $2A - 3B$

    **b.** $AV$

    **c.** $BW$

    **d.** $-2V + 5W$

    **e.** $A(3V - 2W)$

    **f.** $(A - 5I)W$

**3.** If $A = \begin{bmatrix} 1 & 2 \\ 3 & 4 \end{bmatrix}$ and $V = \begin{bmatrix} x \\ y \end{bmatrix}$, solve the equation $AV = \begin{bmatrix} 1 \\ 3 \end{bmatrix}$ for $V$ (i.e., find values for $x$ and $y$).

**4.** Find the derivative of each of the following vectors:

    **a.** $X(t) = \begin{bmatrix} t^3 - 2t^2 + t \\ e^t \sin t \end{bmatrix}$

    **b.** $V(x) = \begin{bmatrix} 2 \cos x \\ -3e^{-2x} \end{bmatrix}$

    **c.** $B(u) = \begin{bmatrix} e^{-u} + e^u \\ 2 \cos u - 5 \sin u \end{bmatrix}$

    **d.** $Y(t) = \begin{bmatrix} (t^2 + 1)e^{-t} \\ t \sin t \end{bmatrix}$

Convert each system of differential equations in Problems 5–10 to the matrix form $\dot{X} = AX$.

**5.** $\dot{x} = 2x + y$
$\dot{y} = 3x + 4y$

**6.** $\dot{x} = x - y$
$\dot{y} = y - 4x$

**7.** $\dot{x} = 2x + y$
$\dot{y} = 4y - x$

**8.** $\dot{x} = x$
$\dot{y} = y$

**9.** $\dot{x} = -2x + y$
$\dot{y} = -2y$

**10.** $\dot{x} - 8y + x = 0$
$\dot{y} - y - x = 0$

**B**

1. Using the technique shown in Section 4.6, write each of the following second-order equations as a system of first-order equations and then express the system in matrix form.

   **a.** $y'' - 3y' + 2y = 0$
   **b.** $5y'' + 3y' - y = 0$
   **c.** $y'' + \omega^2 y = 0$, where $\omega$ is a constant.

2. Show that the origin is the only equilibrium point of the system

   $$\dot{x} = ax + by$$
   $$\dot{y} = cx + dy,$$

   where $a, b, c,$ and $d$ are constants, with $ad - bc \neq 0$.

**C**

1. If $A(t) = \begin{pmatrix} a_{11}(t) & a_{12}(t) \\ a_{21}(t) & a_{22}(t) \end{pmatrix}$ and $B(t) = \begin{pmatrix} b_{11}(t) \\ b_{21}(t) \end{pmatrix}$, both matrices having entries that are differentiable functions of $t$, show that

   $$\frac{d}{dt}[A(t)B(t)] = \frac{dA(t)}{dt}B(t) + A(t)\frac{dB(t)}{dt}.$$

## 5.2 TWO-DIMENSIONAL SYSTEMS OF FIRST-ORDER LINEAR EQUATIONS

### 5.2.1 Eigenvalues and Eigenvectors

To get a handle on linear systems of ordinary differential equations, including their qualitative behavior and their possible closed-form solutions, we will focus on linear two-dimensional systems of the form

$$\dot{x} = ax + by$$
$$\dot{y} = cx + dy, \tag{5.2.1}$$

where $x$ and $y$ both depend on the variable $t$, and $a, b, c,$ and $d$ are constants. Our analysis of such simple (but important) systems will prepare us to understand the treatment of higher-order linear systems in Section 5.7.

First, let's write the System (5.2.1) in matrix form:

$$\begin{bmatrix} \dot{x} \\ \dot{y} \end{bmatrix} = \begin{bmatrix} a & b \\ c & d \end{bmatrix} \begin{bmatrix} x \\ y \end{bmatrix},$$

or

$$\frac{d}{dt}\begin{bmatrix} x \\ y \end{bmatrix} = \begin{bmatrix} a & b \\ c & d \end{bmatrix}\begin{bmatrix} x \\ y \end{bmatrix},$$

or

$$\dot{X} = AX, \tag{5.2.2}$$

where $X = \begin{bmatrix} x \\ y \end{bmatrix}$ and $A = \begin{bmatrix} a & b \\ c & d \end{bmatrix}$. Ignoring the fact that the capital letters represent matrices, what does the form of Equation (5.2.2) remind you of? Have you seen a differential equation of this form before? If we use lowercase letters and write the equation as $\dot{x} = ax$, we get a familiar separable equation representing exponential growth or decay. (See Section 2.1, especially Example 2.1.1.) This observation suggests that the solution of System (5.2.1) or the matrix Equation (5.2.2) may have something to do with exponentials.

Let's make a shrewd guess and then examine the consequences of our guess. (This was Euler's strategy, described in Section 4.1.) In particular, let us assume that $x(t) = c_1 e^{\lambda t}$ and $y(t) = c_2 e^{\lambda t}$ for constants $\lambda$, $c_1$, and $c_2$. (Stating that $\lambda$, the coefficient of $t$, is the same for both $x$ and $y$ is a simplifying assumption.) Substituting $\begin{bmatrix} c_1 e^{\lambda t} \\ c_2 e^{\lambda t} \end{bmatrix}$ for $X$ in (5.2.2), we get

$$\begin{bmatrix} c_1 \lambda e^{\lambda t} \\ c_2 \lambda e^{\lambda t} \end{bmatrix} = \begin{bmatrix} a & b \\ c & d \end{bmatrix}\begin{bmatrix} c_1 e^{\lambda t} \\ c_2 e^{\lambda t} \end{bmatrix} = \begin{bmatrix} a & b \\ c & d \end{bmatrix}e^{\lambda t}\begin{bmatrix} c_1 \\ c_2 \end{bmatrix},$$

or

$$\lambda e^{\lambda t}\begin{bmatrix} c_1 \\ c_2 \end{bmatrix} = \begin{bmatrix} a & b \\ c & d \end{bmatrix}e^{\lambda t}\begin{bmatrix} c_1 \\ c_2 \end{bmatrix},$$

or (dividing out the exponential factor and switching right and left sides)

$$A\tilde{X} = \lambda\tilde{X}, \tag{5.2.3}$$

where $\tilde{X} = \begin{bmatrix} c_1 \\ c_2 \end{bmatrix}$. Note that our reasonable guess about $x$ and $y$ has allowed us to replace our original differential equation problem with a pure algebra problem. Equation (5.2.3) is in matrix terms and has nothing (apparently) to do with differential equations. Given a $2 \times 2$ matrix $A$ and a $2 \times 1$ column matrix $\tilde{X}$, we can try to solve (5.2.3) for the value or values of $\lambda$, each called a **characteristic value** or **eigenvalue** of the matrix $A$. (Remember how we first used this term in Sections 4.1–4.2. The connection between the earlier use of the term *eigenvalue* and the current use will be established shortly.) Eigenvalues will play an important role in solving linear systems and in understanding the qualitative behavior of solutions.

Furthermore, if we have solved Equation (5.2.3) for its eigenvalues $\lambda$, then for each value of $\lambda$ we can solve (5.2.3) for the corresponding vector or vectors $\tilde{X}$. Each such *nonzero* vector $\tilde{X}$ is called an **eigenvector** (or **characteristic vector**) of the system. We see that if both entries of $\tilde{X}$ are zero, then $\tilde{X}$ satisfies (5.2.3) for any value of $\lambda$, but this is the trivial case. **In all the discussion that follows, we will assume that $c_1$ and $c_2$ are not both zero**—that is, at least one of the two constants is not zero.

Before getting involved in symbolism, terminology, and the general problem of solving the matrix equation $A\tilde{X} = \lambda\tilde{X}$, let's look at a specific example in detail.

## ■ Example 5.2.1 Solving a Linear System with Eigenvalues and Eigenvectors

Suppose we have the system

$$\dot{x} = -2x + y$$
$$\dot{y} = -4x + 3y, \tag{*}$$

which we can write as $\dot{X} = \begin{bmatrix} \dot{x} \\ \dot{y} \end{bmatrix} = \begin{bmatrix} -2 & 1 \\ -4 & 3 \end{bmatrix} \begin{bmatrix} x \\ y \end{bmatrix}$. We want to find the general solution of this system.

## Substitution

Assuming, say, that $c_1 \neq 0$, we substitute $x = c_1 e^{\lambda t}$ and $y = c_2 e^{\lambda t}$ into (*) and get $\lambda c_1 e^{\lambda t} = -2c_1 e^{\lambda t} + c_2 e^{\lambda t} = e^{\lambda t}(-2c_1 + c_2)$ and $\lambda c_2 e^{\lambda t} = -4c_1 e^{\lambda t} + 3c_2 e^{\lambda t} = e^{\lambda t}(-4c_1 + 3c_2)$. If we simplify each equation by dividing out the exponential term and moving all terms to the left-hand side, we get

$$\text{(A)} \ (\lambda + 2)c_1 - c_2 = 0$$
$$\text{(B)} \ 4c_1 + (\lambda - 3)c_2 = 0. \tag{**}$$

Now we want to solve (**) for $\lambda$.

## Solving for $\lambda$

If we multiply Equation (A) by $(\lambda - 3)$ and then add the resulting equation to (B), we get $(\lambda - 3)(\lambda + 2)c_1 + 4c_1 = 0$, or $(\lambda^2 - \lambda - 2)c_1 = 0$. Because we have assumed that $c_1$ is not zero, we must have $\lambda^2 - \lambda - 2 = 0$. This means that the eigenvalues of $A$ are $\lambda = 2$ and $\lambda = -1$. (*Go through all the algebra carefully.*) Note that we didn't have to know $c_1$ to find $\lambda$. We just had to know that $c_1$ was not zero. It is important that we could have assumed that $c_2$ was not zero and come to the same conclusion. (*Check this.*)

## Solving the System of ODEs

If we take the eigenvalue $\lambda = 2$, we have $x(t) = c_1 e^{2t}$ and $y(t) = c_2 e^{2t}$, so that $X_1 = \begin{bmatrix} x \\ y \end{bmatrix} = \begin{bmatrix} c_1 e^{2t} \\ c_2 e^{2t} \end{bmatrix} = e^{2t} \begin{bmatrix} c_1 \\ c_2 \end{bmatrix}$. But when $\lambda = 2$, the equations in (**) both represent the single equation $4c_1 - c_2 = 0$, so that we have the relation $c_2 = 4c_1$. Then we can write $X_1 = e^{2t} \begin{bmatrix} c_1 \\ c_2 \end{bmatrix} = e^{2t} \begin{bmatrix} c_1 \\ 4c_1 \end{bmatrix} = c_1 \begin{bmatrix} 1 \\ 4 \end{bmatrix} e^{2t}$, which is a one-parameter family of solutions of the System (*). We're saying that the pair of functions $x(t) = c_1 e^{2t}$ and $y(t) = 4c_1 e^{2t}$ is a nontrivial solution of our system for any nonzero constant $c_1$.

Similarly, if we take the eigenvalue $\lambda = -1$, then the System (**) reduces to the single equation $c_1 - c_2 = 0$ and we can define $X_2 = e^{-t} \begin{bmatrix} c_1 \\ c_2 \end{bmatrix} = e^{-t} \begin{bmatrix} c_1 \\ c_2 \end{bmatrix} = c_1 \begin{bmatrix} 1 \\ 1 \end{bmatrix} e^{-t}$, which is also a one-parameter family of solutions of the system. In other words, the pair of functions $x(t) = c_1 e^{-t}$ and $y(t) = c_1 e^{-t}$ is also a nontrivial solution of our system for any nonzero constant $c_1$.

It is easy to see that the Superposition Principle we have been using since Chapter 2 allows us to conclude that

$$X = k_1 X_1 + k_2 X_2 = k_1 \left[ c_1 \begin{bmatrix} 1 \\ 4 \end{bmatrix} e^{2t} \right] + k_2 \left[ c_1 \begin{bmatrix} 1 \\ 1 \end{bmatrix} e^{-t} \right] = C_1 \begin{bmatrix} 1 \\ 4 \end{bmatrix} e^{2t} + C_2 \begin{bmatrix} 1 \\ 1 \end{bmatrix} e^{-t}$$

is the general solution of the system $\dot{x} = -2x + y, \dot{y} = -4x + 3y$. The constants $C_1$ and $C_2$ can be chosen to match arbitrary initial data. ∎

## 5.2.2 Geometric Interpretation of Eigenvectors

The vector $\begin{bmatrix} 1 \\ 4 \end{bmatrix}$ that appears in the preceding example is called an *eigenvector* (or *characteristic vector*) corresponding to the eigenvalue (or characteristic value) $\lambda = 2$. This vector is a nonzero solution, $\tilde{X}$, of $A\tilde{X} = \lambda \tilde{X}$ when $\lambda = 2$. This means that there are infinitely many eigenvectors corresponding to the eigenvalue $\lambda = 2$—all the vectors $\begin{bmatrix} c_1 \\ c_2 \end{bmatrix}$ such that $4c_1 - c_2 = 0$, or all the nonzero vectors of the form $\begin{bmatrix} c_1 \\ 4c_1 \end{bmatrix}$ are eigenvectors associated with $\lambda = 2$. Choosing $c_1 = 1$ gives us the simple particular vector $V_1 = \begin{bmatrix} 1 \\ 4 \end{bmatrix}$, which can be called the *representative* eigenvector. Graphically, this eigenvector represents a straight line from the origin to the point $(1, 4)$ in the $c_1$-$c_2$ plane. Similarly, the vector $V_2 = \begin{bmatrix} 1 \\ 1 \end{bmatrix}$ is the representative eigenvector corresponding to the eigenvalue $\lambda = -1$ and can be interpreted as a straight line from $(0, 0)$

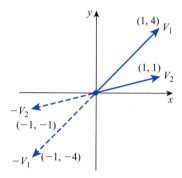

**Figure 5.1**

*Representative eigenvectors* $V_1 = \begin{bmatrix} 1 \\ 4 \end{bmatrix}$ *and* $V_2 = \begin{bmatrix} 1 \\ 1 \end{bmatrix}$

to $(1, 1)$ in the $c_1$-$c_2$ plane. (See the description of vectors in Section B.1.) Figure 5.1 shows $V_1$ and $V_2$ in the $c_1$-$c_2$ plane.

### 5.2.3 The General Problem

Now let's consider the equation $A\tilde{X} = \lambda\tilde{X}$, where $A = \begin{bmatrix} a & b \\ c & d \end{bmatrix}$ and $\tilde{X} = \begin{bmatrix} c_1 \\ c_2 \end{bmatrix}$ and at least one of the numbers $c_1$ and $c_2$ is nonzero. In the discussion that follows, we'll assume that $c_1 \neq 0$. Written out as individual equations, $A\tilde{X} = \lambda\tilde{X}$ has the form

$$ac_1 + bc_2 = \lambda c_1$$
$$cc_1 + dc_2 = \lambda c_2$$

or

$$\text{(A)} \ (a - \lambda)c_1 + bc_2 = 0$$
$$\text{(B)} \ cc_1 + (d - \lambda)c_2 = 0$$

and we want to determine $\lambda$.

We can solve this algebraic system by the method of elimination as follows:

**1.** Multiply Equation (A) by $d - \lambda$ to obtain

$$(d - \lambda)(a - \lambda)c_1 + b(d - \lambda)c_2 = 0.$$

**2.** Multiply Equation (B) by $-b$ to get

$$-bcc_1 - b(d - \lambda)c_2 = 0.$$

**3.** Add the equations found in steps 1 and 2 to get

$$(d - \lambda)(a - \lambda)c_1 - bcc_1 = 0,$$

or

$$\left[\lambda^2 - (a + d)\lambda + (ad - bc)\right]c_1 = 0. \ (Check \ the \ algebra.)$$

**4.** Because we assumed that $c_1 \neq 0$ at the beginning of this discussion, we must have

$$\lambda^2 - (a + d)\lambda + (ad - bc) = 0. \tag{5.2.4}$$

This equation is called the **characteristic equation** of the matrix $A$, and its roots are the eigenvalues of $A$. We'll see the connection between this equation and the characteristic equation we introduced in Section 4.1 shortly.

Using the quadratic formula, we find that

$$\lambda = \frac{(a + d) \pm \sqrt{(a + d)^2 - 4(ad - bc)}}{2}.$$

If we had assumed that $c_2 \neq 0$ at the beginning, we would have found the same solution for $\lambda$. Then for each distinct value of $\lambda$ that we find, we can substitute that value into the system

$$(a - \lambda)c_1 + bc_2 = 0$$
$$cc_1 + (d - \lambda)c_2 = 0$$

and solve for $\tilde{X} = \begin{bmatrix} c_1 \\ c_2 \end{bmatrix}$, the corresponding eigenvector.

There are two things to notice about the characteristic equation of $A$,

$$\lambda^2 - (a + d)\lambda + (ad - bc) = 0$$

and the resulting formula for $\lambda$:

**1.** The expression $a + d$ is just the sum of the *main diagonal* (upper left, lower right) elements of the matrix $A = \begin{bmatrix} a & b \\ c & d \end{bmatrix}$. In linear algebra, this is called the trace of $A$. For example, if $A = \begin{bmatrix} -7 & 2 \\ 0 & 4 \end{bmatrix}$, then the trace of $A$ is $(-7) + 4 = -3$.

**2.** The expression $ad - bc$ is formed from the matrix of coefficients $A = \begin{bmatrix} a & b \\ c & d \end{bmatrix}$ as follows: Multiply the main diagonal elements and then subtract the product of the other diagonal elements (upper right, lower left). The number calculated this way is called the determinant of the coefficient matrix. Symbolically, $\det(A) = \det \begin{bmatrix} a & b \\ c & d \end{bmatrix} = ad - bc$. For example, if $A = \begin{bmatrix} -7 & -3 \\ 2 & 4 \end{bmatrix}$, then $\det(A) = (-7)(4) - (-3)(2) = -28 - (-6) = -28 + 6 = -22$. The determinant of a matrix $A$ is often denoted by the symbol $|A|$, so the rule for calculation in the $2 \times 2$ case can be given as

$$|A| = \begin{vmatrix} a & b \\ c & d \end{vmatrix} = ad - bc.$$

Observations 1 and 2 provide us with an alternative way of viewing the characteristic equation:

$$\lambda^2 - (\text{trace of } A)\lambda + \det(A) = 0. \tag{5.2.5}$$

The roots of the characteristic Equation (5.2.5)—the eigenvalues—lead to eigenvectors and ultimately to the general solution of a linear system. Let's look at an example using this shortcut.

### ■ Example 5.2.2 Solving a Linear System with Eigenvalues and Eigenvectors

The following equations constitute a simple model for detecting diabetes:

$$\frac{dg}{dt} = -2.92g - 4.34h$$

$$\frac{dh}{dt} = 0.208g - 0.780h,$$

where $g(t)$ denotes excess glucose concentration in the bloodstream and $h(t)$ represents excess insulin concentration. "Excess" refers to concentrations above the equilibrium values. We want to determine the solution at any time $t$.

### Eigenvalues

The matrix form of our equations is $\frac{d}{dt}X = \begin{bmatrix} dg/dt \\ dh/dt \end{bmatrix} = \begin{bmatrix} -2.92 & -4.34 \\ 0.208 & -0.780 \end{bmatrix} \begin{bmatrix} g \\ h \end{bmatrix}$, so that the matrix of coefficients is $A = \begin{bmatrix} -2.92 & -4.34 \\ 0.208 & -0.780 \end{bmatrix}$. We see that the *trace* of $A$ is $-2.92 + (-0.780) = -3.7$ and the *determinant* of $A$ is $-2.92(-0.780) - (-4.34)(0.208) = 3.18032$. We see that the characteristic equation is

$$\lambda^2 - (\text{trace of } A)\lambda + \det(A) = \lambda^2 + 3.7\lambda + 3.18032 = 0.$$

Solving this by calculator or CAS, we find that the eigenvalues are $\lambda_1 = -2.34212$ and $\lambda_2 = -1.35788$, rounded to five decimal places.

## Eigenvectors

Now we substitute each eigenvalue in the equations

$$(a - \lambda)c_1 + bc_2 = 0$$
$$cc_1 + (d - \lambda)c_2 = 0$$

and solve for the corresponding eigenvector $\tilde{X} = \begin{bmatrix} c_1 \\ c_2 \end{bmatrix}$. In our problem, we must substitute in the equations

$$(-2.92 - \lambda)c_1 - 4.34c_2 = 0$$
$$0.208c_1 + (-0.780 - \lambda)c_2 = 0.$$

If $\lambda = -2.34212$, then the equations are

$$-0.57788c_1 - 4.34c_2 = 0$$
$$0.208c_1 + 1.56212c_2 = 0.$$

But these two equations are really only one distinct equation, $c_2 = -0.13315c_1$. (Solve each equation for $c_2$ and see for yourself.) Therefore, to ensure that at least one element of the eigenvector is an integer, we can take $c_1 = 1$ and $c_2 = -0.13315$, so that an eigenvector corresponding to the eigenvalue $\lambda = -2.34212$ is

$$\tilde{X}_1 = \begin{bmatrix} 1 \\ -0.13315 \end{bmatrix}.$$

Similarly, if we use the other eigenvalue, $\lambda = -1.35788$, we can take the single equation $(-2.92 - \lambda)c_1 - 4.34c_2 = 0$ and substitute the eigenvalue to get $(-2.92 + 1.35788)c_1 - 4.34c_2 = 0$, so that $c_2 = -0.35994c_1$. If we take $c_1 = 1$, we must have $c_2 = -0.35994$, and an eigenvector corresponding to the eigenvalue $\lambda = -2.34212$ is

$$\tilde{X}_2 = \begin{bmatrix} 1 \\ -0.35994 \end{bmatrix}.$$

## The Solution

The Superposition Principle gives the general solution as

$$\tilde{X} = C_1\tilde{X}_1 + C_2\tilde{X}_2 = C_1 \begin{bmatrix} 1 \\ -0.13315 \end{bmatrix} e^{-2.34212t} + C_2 \begin{bmatrix} 1 \\ -0.35994 \end{bmatrix} e^{-1.35788t}.$$

If we were given initial concentrations of glucose and insulin, we could determine the constants $C_1$ and $C_2$. (See Problem B2 in Exercises 5.2.) ∎

### 5.2.4 The Geometric Behavior of Solutions

In the next few examples, we will get a preview of how the behavior of a two-dimensional system of linear differential equations with constant coefficients depends on the eigenvalues and eigenvectors of its matrix of coefficients. We'll illustrate some typical phase portraits. Then, in Sections 5.3–5.5, we'll give a systematic description of all possible behaviors of such linear systems, using the nature of their eigenvalues and eigenvectors.

### ■ Example 5.2.3 Example 5.2.1 Revisited—A Saddle Point

Let's look again at the system from Example 5.2.1:

$$\dot{x} = -2x + y$$
$$\dot{y} = -4x + 3y.$$

As we saw earlier, the eigenvalues of this system are $\lambda_1 = 2$ and $\lambda_2 = -1$, with corresponding representative eigenvectors $V_1 = \begin{bmatrix} 1 \\ 4 \end{bmatrix}$ and $V_2 = \begin{bmatrix} 1 \\ 1 \end{bmatrix}$. The general solution was given by

$$X = C_1 \begin{bmatrix} 1 \\ 4 \end{bmatrix} e^{2t} + C_2 \begin{bmatrix} 1 \\ 1 \end{bmatrix} e^{-t} = \begin{bmatrix} C_1 e^{2t} + C_2 e^{-t} \\ 4C_1 e^{2t} + C_2 e^{-t} \end{bmatrix}.$$

Figure 5.2 shows some trajectories for this system of linear equations. These are particular solutions of $\frac{dy}{dx} = \frac{dy/dt}{dx/dt} = \frac{-4x+3y}{-2x+y}$. Note in particular that the lines $y = 4x$ and $y = x$ appear as trajectories. These trajectories are actually four *half-lines*: $y = 4x$ for $x > 0$, $y = 4x$ for $x < 0$, $y = x$ for $x > 0$, and $y = x$ for $x < 0$.

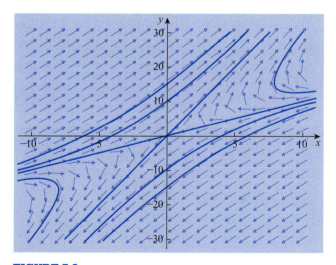

**FIGURE 5.2**
*Phase portrait of the system $\dot{x} = -2x + y$, $\dot{y} = -4x + 3y$*

A little basic algebra shows us that the origin is the only equilibrium point, and it is called a **saddle point** in this situation. A saddle point is the two-dimensional version of the *node* we discussed in Section 2.6. What characterizes a saddle point is that solutions can approach the equilibrium point along one direction (as though it were a *sink*), yet move away from the point in another direction (as though it were a *source*).[1] In particular, it turns out that one trajectory is the half-line $y = 4x$ in the first quadrant, along which the motion is *away* from the origin, and another trajectory is the line $y = x$ also in the first quadrant, along which the movement is *toward* the origin. The straight lines $y = 4x$ and $y = x$ are *asymptotes* for the other trajectories (as $t \to \pm\infty$). You may not be able to see this clearly from the phase portrait that your graphing utility generates unless you play with the range of $t$ and choose initial values carefully, but you can see this and other behavior *analytically* (see Problem B3 in Exercises 5.2). ∎

## ■ Example 5.2.4 A Source

Now let's look at the system of differential equations

$$\dot{x} = 2x + y$$
$$\dot{y} = 3x + 4y.$$

First of all, note that the system's only equilibrium point—where $\dot{x} = 0$ and $\dot{y} = 0$—is the origin of the phase plane, $(x, y) = (0, 0)$. (You should verify this using the ordinary algebra of simultaneous equations.)

Using the formula given by Equation (5.2.5), we see that the characteristic equation of our system is $\lambda^2 - (2 + 4)\lambda + ((2)(4) - (1)(3)) = 0$, or $\lambda^2 - 6\lambda + 5 = 0$, which has the roots $\lambda_1 = 5$ and $\lambda_2 = 1$. To find the eigenvectors corresponding to these eigenvalues, we must solve the matrix equation $A\tilde{X} = \lambda\tilde{X}$, where $A = \begin{bmatrix} 2 & 1 \\ 3 & 4 \end{bmatrix}$, $\lambda = 5$ or 1, and $\tilde{X} = \begin{bmatrix} c_1 \\ c_2 \end{bmatrix}$. This matrix equation is equivalent to the system

$$\begin{aligned} &(1)\ (2 - \lambda)c_1 + c_2 = 0 \\ &(2)\ 3c_1 + (4 - \lambda)c_2 = 0. \end{aligned} \qquad (5.2.6)$$

Substituting the first eigenvalue, $\lambda = 5$, in (5.2.6) gives us

$$\begin{aligned} &(1)\ -3c_1 + c_2 = 0 \\ &(2)\ 3c_1 - c_2 = 0. \end{aligned}$$

---

[1] This terminology is usually seen in a multivariable calculus course: If you look at a horse's saddle in the tail-to-head direction, it appears that the center of the saddle is lower than the front or back, so that the center seems to be a *minimum* point on the saddle's surface. However, if you look *across* the saddle from one side of the horse, it appears that the center is at the peak of a stirrup-to-stirrup curve, so the center seems like a *maximum* point. In fact, a *saddle point* is neither a minimum nor a maximum.

There is really only one equation here, and its solution is given by $c_2 = 3c_1$. Thus, the eigenvectors corresponding to the eigenvalue $\lambda = 2$ have the form $\begin{bmatrix} c_1 \\ 3c_1 \end{bmatrix}$, or $c_1 \begin{bmatrix} 1 \\ 3 \end{bmatrix}$. If we let $c_1 = 1$, we get the "neat" representative eigenvector $V_1 = \begin{bmatrix} 1 \\ 3 \end{bmatrix}$.

When we use the other eigenvalue, $\lambda = 1$, in the System (5.2.6), we find that

$$(1)\ c_1 + c_2 = 0$$
$$(2)\ 3c_1 + 3c_2 = 0,$$

which has the solution $c_2 = -c_1$. Therefore, the eigenvectors in this case have the form $\begin{bmatrix} c_1 \\ -c_1 \end{bmatrix}$, or $c_1 \begin{bmatrix} 1 \\ -1 \end{bmatrix}$. Thus, our representative eigenvector can be $V_2 = \begin{bmatrix} 1 \\ -1 \end{bmatrix}$.

Now let's look at the phase portrait corresponding to the original system, a family of trajectories corresponding to the first-order equation

$$\frac{dy}{dx} = \frac{\frac{dy}{dt}}{\frac{dx}{dt}} = \frac{3x + 4y}{2x + y}.$$

This phase portrait is shown in Figure 5.3. The curves $y = 3x$ and $y = -x$, which are straight-line trajectories, are labeled so that we can see the significance of the eigenvectors.

If you look carefully (or find your own phase portrait), you may notice that the trajectories shown are fleeing the origin (as $t \to \infty$) in such a way that any trajectory is tangent to the line

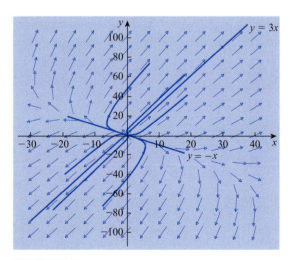

**FIGURE 5.3**
*Phase portrait of the system $\dot{x} = 2x + y$, $\dot{y} = 3x + 4y$*

$y = -x$ at the origin—that is, as $t \to -\infty$. In this situation, the origin is called an **unstable node** (specifically, a **source** or **repeller**). We'll come to a better understanding of this behavior in Section 5.3. ∎

The next example reveals another type of source for a two-dimensional system.

## ■ Example 5.2.5 A Spiral Source

Look at the system

$$\frac{dx}{dt} = x + y$$
$$\frac{dy}{dt} = -4x + y.$$

Note that once again the origin is this system's only equilibrium point. (*Check this for yourself.*) Because the matrix of coefficients has $a = 1, b = 1, c = -4$, and $d = 1$, we use Formula (5.2.4) to determine that the characteristic equation of this system is $\lambda^2 - 2\lambda + 5 = 0$, so the quadratic formula gives us the eigenvalues $\lambda_1 = 1+2i$ and $\lambda_2 = 1-2i$. When we get complex eigenvalues such as this complex conjugate pair, the eigenvectors will turn out to have complex numbers as entries and to have no useful direct geometric significance.[2] We'll deal with this situation in more detail in Section 5.5. The phase portrait for this system is shown in Figure 5.4.

**FIGURE 5.4**

*Phase portrait of the system $\frac{dx}{dt} = x + y$, $\frac{dy}{dt} = -4x + y$*

$(x(0), y(0)) = (-4, 0), (-2, 0), (2, 0), (3, 0); -5 \leq t \leq 0.7$

---

[2] See Section 4.4 of *Applied Linear Algebra* by L. Sadun (Upper Saddle River, NJ: Prentice Hall, 2001).

We can see that the trajectories are spirals that move outward, away from the equilibrium point, in a clockwise direction. In this case, as in the previous example, the equilibrium point is called a *source* (or a *repeller*). Other systems with complex eigenvalues may correspond to spirals that move in a *counterclockwise* direction or to spirals that move in *toward* the equilibrium point (clockwise or counterclockwise). ∎

These examples should convince you that trajectories can behave quite differently near equilibrium points. In the next section, we will examine how the trajectories of a two-dimensional system can be classified.

## Exercises 5.2

### A

1. Calculate the determinant of each of the following matrices by hand:

   a. $\begin{bmatrix} -3 & 5 \\ -4 & 1 \end{bmatrix}$

   b. $\begin{bmatrix} 4 & 2 \\ 10 & 5 \end{bmatrix}$

   c. $\begin{bmatrix} 6t & -4 \\ \sin t & t^3 \end{bmatrix}$

   d. $\begin{bmatrix} \cos\theta & \sin\theta \\ -\sin\theta & \cos\theta \end{bmatrix}$

2. Find the eigenvalues and eigenvectors of matrices (a) and (b) in Problem 1.
3. Find a 2 × 2 matrix with eigenvalues 1 and 3 and corresponding eigenvectors

$$\begin{bmatrix} 1 \\ 1 \end{bmatrix} \text{ and } \begin{bmatrix} 1 \\ 2 \end{bmatrix}.$$

For each system in Problems 4–9, (a) convert to the matrix form $\dot{X} = AX$; (b) find the characteristic equation; (c) find all eigenvalues; (d) describe all eigenvectors corresponding to each eigenvalue found in part (c). Parts (a)–(d) should be done without the aid of a calculator or CAS.

4. $\dot{x} = -x + 4y$
   $\dot{y} = 2x - 3y$
5. $\dot{x} = x - y$
   $\dot{y} = y - 4x$
6. $\dot{x} = -4x + 2y$
   $\dot{y} = 2x - y$
7. $\dot{x} = x$
   $\dot{y} = y$
8. $\dot{x} = -6x + 4y$
   $\dot{y} = -3x + y$

**9.** $\dot{x} = 5x - y$
$\dot{y} = 2x + y$

**B**

**1.** Consider the system

$$ax + by = e$$

$$cx + dy = f,$$

where $a, b, c, d, e,$ and $f$ are constants, with $ad - bc \neq 0$.

**a.** Show that the solution is given by $x = \dfrac{de - bf}{ad - bc}, y = \dfrac{af - ce}{ad - bc}$.

**b.** Express your solution in part (a) in terms of the determinants

$$\begin{vmatrix} a & e \\ c & f \end{vmatrix}, \begin{vmatrix} a & b \\ c & d \end{vmatrix}, \text{ and } \begin{vmatrix} e & b \\ f & d \end{vmatrix}.$$

**2.** In Example 5.2.2, find the solution of the system satisfying the initial conditions $g(0) = g_0$ and $h(0) = 0$. (You may use technology to solve the resulting system of algebraic equations.)

**3.** In Example 5.2.3, the system $\dot{x} = -2x + y, \dot{y} = -4x + 3y$ was shown to have the solution

$$X = \begin{pmatrix} C_1 e^{2t} + C_2 e^{-t} \\ 4C_1 e^{2t} + C_2 e^{-t} \end{pmatrix}.$$

**a.** Substitute for $x(t)$ and $y(t)$ in the right-hand side of the expression

$$\frac{dy}{dx} = \frac{-4x + 3y}{-2x + y}.$$

**b.** Use the result of part (a) to show that the slope of any trajectory not on either of the lines determined by the eigenvectors approaches 4, the slope of the eigenvector corresponding to the larger of the two distinct eigenvalues. [*Hint*: Factor out $e^{2t}$, the dominant term for large positive values of $t$.]

**c.** Use the result of part (a) to show that the slope of any trajectory not on either of the lines determined by the eigenvectors approaches 1, the slope of the eigenvector corresponding to the smaller of the two eigenvalues. [*Hint*: Factor out $e^{-t}$, the dominant term for large negative values of $t$.]

**4.** Use technology to sketch the phase portrait of the system in Problem B3. Then sketch in the eigenvectors (getting them from the answers in the back of the book if necessary) and comment on the behavior of the trajectories with respect to the origin. (Use both positive and negative values of $t$.)

5. Use technology to sketch the phase portrait of the system in Problem B4 . Then sketch in the eigenvectors (using your CAS if necessary) and comment on the behavior of the trajectories with respect to the origin. (Use both positive and negative values of $t$.)

6. A substance $X$ decays into substance $Y$ at rate $k_1 > 0$, and $Y$ in turn decays into another substance at rate $k_2 > 0$. The system

$$\frac{dx}{dt} = -k_1 x$$

$$\frac{dy}{dt} = k_1 x - k_2 y$$

describes the process, where $x(t)$ and $y(t)$ represent the amount of $X$ and $Y$, respectively. Assume that $k_1 \neq k_2$.

a. Find the eigenvalues of the system.

b. Find the eigenvectors corresponding to each of the eigenvalues found in part (a).

c. Solve for $x(t)$ and $y(t)$ and then find $\lim_{t \to \infty} x(t)$ and $\lim_{t \to \infty} y(t)$, interpreting your answers in physical terms.

7. The following system models the exchange of nutrients between mother and fetus in the placenta:

$$\frac{dc_1}{dx} = -\alpha_1 (c_1 - c_2)$$

$$\frac{dc_2}{dx} = -\alpha_2 (c_1 - c_2),$$

where $c_1(x)$ is the concentration of nutrient in the maternal bloodstream at a distance $x$ along the placental membrane and $c_2(x)$ is the concentration of nutrient in the fetal bloodstream at a distance $x$. Here, $\alpha_1$ and $\alpha_2$ are constants, $\alpha_1 \neq \alpha_2$.

a. If $c_1(0) = c_0$ and $c_2(0) = C_0$, use eigenvalues and eigenvectors to solve the system for $c_1(x)$ and $c_2(x)$.

b. Solve for $c_1(x)$ and $c_2(x)$ by converting the system into a single second-order differential equation and using the techniques of Section 4.1.

8. Consider the spring-mass system described by $\ddot{x} + b\dot{x} + kx = 0$.

a. Find all values of $b$ and $k$ for which this system has real, distinct eigenvalues.

b. Find the general solution of the system for the values of $b$ and $k$ found in part (a).

c. Find the solution of the system that satisfies the initial condition $x(0) = 1$.

d. Describe the motion of the mass in the situation described in part (c).

9. The behavior of a *damped pendulum* is described near the lowest point of its trajectory by the linear equation $\ddot{x} = -k\dot{x} - gx/L$, where $k$ is the damping coefficient, $g$ is the acceleration due to gravity, and $L$ is the length of the pendulum.

a. Express the differential equation as a linear system.

b. Find the characteristic equation of the system.

    **c.** Find all eigenvalues of the system.

    **d.** Describe all eigenvectors.

    **e.** If $k^2 > 4g/L$, solve the system and state what happens as $t \to \infty$.

## C

**1.** Consider the system $\dot{x} = Ax$, where $A = \begin{bmatrix} a & b \\ c & d \end{bmatrix}$. Suppose that the trajectories spiral as in Example 5.2.5, possibly in the opposite direction. The polar form (Section B.1) of a spiral trajectory provides the *polar angle* $\theta(t) = \arctan\left(\frac{y(t)}{x(t)}\right)$; and the direction of the spiral will be clockwise if $d\theta/dt < 0$ and counterclockwise if $d\theta/dt > 0$. Show that the direction of a spiral trajectory depends on the sign of $c$, the lower left-hand entry of the matrix $A$, as follows.

    **a.** Show that $\det(A) = ad - bc > 0$.

    **b.** Show that $(\text{trace of } A)^2 < 4(\det(A))$—that is, $(a + d)^2 < 4(ad - bc)$.

    **c.** Show that $d\theta/dt = (x\dot{y} - y\dot{x})/(x^2 + y^2)$, and that the sign of $d\theta/dt$ equals the sign of $x\dot{y} - y\dot{x}$.

    **d.** Show that

$$x\dot{y} - y\dot{x} = c\left[x + \left(\frac{d-a}{2c}\right)y\right]^2 + \frac{y^2}{4c}\left[4(ad - bc) - (a + d)^2\right]$$

and explain why the sign of $d\theta/dt$ equals the sign of $c$.

## 5.3 THE STABILITY OF HOMOGENEOUS LINEAR SYSTEMS: UNEQUAL REAL EIGENVALUES

First of all, we should have guessed by now that a linear system $\dot{X} = AX$ of ordinary differential equations, where $\det(A) \neq 0$, has exactly one equilibrium point, $(0, 0)$. (See Problem B2 of Exercises 5.1.) If $\det(A) = 0$, however, the system may have many other equilibrium solutions. As promised in the preceding section, the *stability* of a system—the behavior of trajectories with respect to the equilibrium point(s)—will be explained completely in terms of the eigenvalues and eigenvectors of the matrix $A$.

Because the characteristic equation of a two-dimensional system is a quadratic equation, we know that there are two eigenvalues, $\lambda_1$ and $\lambda_2$. There are only three possibilities for these eigenvalues: (1) The eigenvalues are both real numbers with $\lambda_1 \neq \lambda_2$; (2) the eigenvalues are real numbers with $\lambda_1 = \lambda_2$; or (3) the eigenvalues are complex numbers: $\lambda_1 = p + qi$ and $\lambda_2 = p - qi$, where $p$ and $q$ are real numbers (called the *real part* and the *imaginary part*, respectively) and $i = \sqrt{-1}$. In situation 3, we say that $\lambda_1$ and $\lambda_2$ are *complex conjugates* of each other. (You may want to review Appendix C, especially Section C.3, for more information about complex numbers.) The nature of the eigenvalues will play an important role in the analysis of systems of linear equations, just as it did for second- and higher-order linear equations with constant coefficients in Sections 4.1–4.3. In this section we will deal with possibility 1 in the foregoing list, leaving situations 2 and 3 for the next two sections.

### 5.3.1 Unequal Real Eigenvalues

First, suppose that the matrix $A$ in the system $\dot{X} = AX$ has two real eigenvalues $\lambda_1$ and $\lambda_2$ with $\lambda_1 \neq \lambda_2$. Let $V_1$ and $V_2$ be the corresponding representative eigenvectors. Then we'll show that the general solution of the system is given by

$$X(t) = c_1 e^{\lambda_1 t} V_1 + c_2 e^{\lambda_2 t} V_2, \qquad (5.3.1)$$

where $c_1$ and $c_2$ are arbitrary constants.

Geometrically, the first term on the right-hand side of (5.3.1) represents a straight-line trajectory parallel[3] to $V_1$, and the second term describes a line parallel to $V_2$ (see Figure 5.5). Note that these trajectories lie in the phase plane (the $x$-$y$ plane).

If both $c_1$ and $c_2$ are nonzero, then the solution $X(t)$ is a linear combination of the two basic terms whose relative contributions change with time. In this situation, the trajectories curve in a way that will be described later.

To see why (5.3.1) is the general solution, first note that each term is itself a solution of the system. If, for example, we consider $X_1(t) = c_1 e^{\lambda_1 t} V_1$, then $\dot{X}_1(t) = c_1 \lambda_1 e^{\lambda_1 t} V_1$ and $AX_1 = A(c_1 e^{\lambda_1 t} V_1) = c_1 e^{\lambda_1 t} (AV_1) = c_1 e^{\lambda_1 t} (\lambda_1 V_1) = \lambda_1 c_1 e^{\lambda_1 t} V_1$ because $V_1$ is an eigenvector corresponding to $\lambda_1$. (See Section 5.1 for properties of matrix multiplication.) Therefore, $\dot{X}_1(t) = AX_1$. Now let's see that if $X_1$ and $X_2$ are any solutions of the system, then the linear

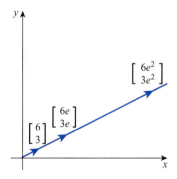

**FIGURE 5.5**
$V = 3e^t \begin{bmatrix} 2 \\ 1 \end{bmatrix}$ for $t = 0, 1,$ and 2

---

[3] Two vectors $V$ and $W$ are **parallel** if $W = cV$ for some nonzero constant $c$. In other words, parallel vectors lie on the same straight line through the origin, pointing in the same direction (if $c > 0$) or in opposite directions (if $c < 0$). See Section B.1.

combination $X = k_1X_1 + k_2X_2$ is also a solution for any constants $k_1$ and $k_2$:

$$\dot{X} = \overbrace{(k_1X_1 + k_2X_2)}^{\bullet} = k_1\dot{X}_1 + k_2\dot{X}_2 = k_1(AX_1) + k_2(AX_2)$$
$$= A(k_1X_1) + A(k_2X_2)$$
$$= A(k_1X_1 + k_2X_2) = AX.$$

These steps follow from the algebraic properties of matrices (Section 5.1) and of derivatives, and this property of solutions of linear systems is another version of the Superposition Principle that we have encountered several times before. (For example, see Section 4.1.)

We can argue (somewhat loosely) that (5.3.1) represents a solution of a two-dimensional system (or its equivalent second-order equation) and has two arbitrary constants and hence is the *general* solution of the system $\dot{X} = AX$. To be rigorous, we can use the fact that any initial condition $X_0 = X(t_0) = \begin{bmatrix} x(t_0) \\ y(t_0) \end{bmatrix} = \begin{bmatrix} x_0 \\ y_0 \end{bmatrix}$ for the system can be written as a linear combination of the eigenvectors—$X_0 = k_1V_1 + k_2V_2$ for some constants $k_1$ and $k_2$—so a solution (5.3.1) can be found to satisfy any initial condition $X(t_0) = X_0$. (You'll be asked to prove these assertions in Problems C2 and C3 in Exercises 5.3.) Finally, the Existence and Uniqueness Theorem of Section 4.6 allows us to say that (5.3.1) is the *only* solution.

### 5.3.2 The Impossibility of Dependent Eigenvectors

If one of the eigenvectors is a scalar multiple of the other—say $V_2$ is a multiple of $V_1$—then the expression in (5.3.1) collapses to a scalar multiple of $V_1$ and there is only one arbitrary constant. This expression can't represent the general solution of a second-order equation.

Fortunately, this collapse can't happen with our current assumption. It is easy to prove that if a $2 \times 2$ matrix $A$ has distinct eigenvalues $\lambda_1$ and $\lambda_2$ with corresponding eigenvectors $V_1$ and $V_2$, then neither eigenvector is a scalar multiple of the other. Suppose that $V_2 = cV_1$, where $c$ is a nonzero scalar. Then $V_2 - cV_1 = \mathbf{0}$, the zero vector, and we must have

$$\mathbf{0} = A(V_2 - cV_1) = AV_2 - c(AV_1) = \lambda_2V_2 - c(\lambda_1V_1)$$
$$= \lambda_2(cV_1) - c(\lambda_1V_1) = c(\lambda_2 - \lambda_1)V_1.$$

But then, because $c \neq 0$ and $V_1$ (as an eigenvector) is nonzero, we must conclude that $(\lambda_2 - \lambda_1) = 0$—contradicting the assumption that we have distinct eigenvalues.

### 5.3.3 Unequal Positive Eigenvalues

In the expression for the general solution, $c_1e^{\lambda_1 t}V_1 + c_2e^{\lambda_2 t}V_2$, suppose that $\lambda_1 > \lambda_2 > 0$. First, note that as $t$ increases, both eigenvector multiples point *away* from the origin so all solutions *grow* with time. (The algebraic signs of the constants $c_1$ and $c_2$ influence the quadrants in which the solutions grow.) To understand the *relative* rates at which the individual terms grow, we can factor out the exponential corresponding to the larger eigenvalue and write $X(t) = e^{\lambda_1 t}(c_1 V_1 + c_2 e^{(\lambda_2 - \lambda_1)t} V_2)$.

Note that $e^{(\lambda_2 - \lambda_1)t} \to 0$ as $t \to +\infty$ because $\lambda_2 - \lambda_1 < 0$. Therefore, $X(t) \approx e^{\lambda_1 t} c_1 V_1$ as $t$ gets larger and larger. Noting that $e^{\lambda_1 t} c_1 V_1$ is parallel to $V_1$, we see that the slope of any trajectory $X(t)$ approaches the slope of the line determined by $V_1$. This says that trajectories will curve *away* from the origin and their slopes will approach the slope of the line determined by the eigenvector $V_1$, corresponding to the larger eigenvalue. In this situation, the equilibrium point $(0, 0)$ is called a *source* (*unstable node, repeller*). (Recall our discussions in Section 2.5.) In "backward time," as $t \to -\infty$, the trajectories will be asymptotic to the line determined by the eigenvector $V_2$ because then the first term in the linear combination $c_1 e^{\lambda_1 t} V_1 + c_2 e^{\lambda_2 t} V_2$ is approaching zero faster than the second term. This says that if we move *backward*, the trajectories *enter* the origin tangent to the line determined by $V_2$.

We are now ready to re-examine an earlier example in light of the preceding two paragraphs.

## ■ Example 5.3.1 Unequal Positive Eigenvalues: A Source

First of all, the system

$$\dot{x} = 2x + y$$
$$\dot{y} = 3x + 4y$$

that we saw in Example 5.2.4 has two positive unequal eigenvalues, $\lambda_1 = 5$ and $\lambda_2 = 1$, with corresponding eigenvectors $V_1 = \begin{bmatrix} 1 \\ 3 \end{bmatrix}$ and $V_2 = \begin{bmatrix} 1 \\ -1 \end{bmatrix}$. Therefore, the general solution is

$$X(t) = c_1 e^{5t} \begin{bmatrix} 1 \\ 3 \end{bmatrix} + c_2 e^t \begin{bmatrix} 1 \\ -1 \end{bmatrix} = \begin{bmatrix} c_1 e^{5t} + c_2 e^t \\ 3c_1 e^{5t} - c_2 e^t \end{bmatrix} = \begin{bmatrix} x(t) \\ y(t) \end{bmatrix}.$$

Figure 5.6 is a more detailed version of Figure 5.3, the phase portrait of our system. The new graph shows several trajectories and the way in which they curve away from the origin, their slopes approaching the slope of the line determined by the eigenvector $V_1 = \begin{bmatrix} 1 \\ 3 \end{bmatrix}$ corresponding to the larger eigenvalue $\lambda = 5$.

Analytically, we can examine the equation $\frac{dy}{dx} = \frac{\frac{dy}{dt}}{\frac{dx}{dt}} = \frac{3x + 4y}{2x + y}$, whose solutions make up the phase portrait—that is, the equation giving the slopes of trajectories in the $x$-$y$ plane. Substituting $x(t) = c_1 e^{5t} + c_2 e^t$ and $y(t) = 3c_1 e^{5t} - c_2 e^t$ from the general solution given previously, we get $\frac{dy}{dx} = \frac{15c_1 e^{5t} - c_2 e^t}{5c_1 e^{5t} + c_2 e^t}$. For large values of $t$, the expression for $\frac{dy}{dx}$ is dominated by the $e^{5t}$ terms, which we can factor out:

$$\frac{dy}{dx} = \frac{e^{5t}(15c_1 - c_2 e^{-4t})}{e^{5t}(5c_1 + c_2 e^{-4t})} = \frac{15c_1 - c_2 e^{-4t}}{5c_1 + c_2 e^{-4t}}.$$

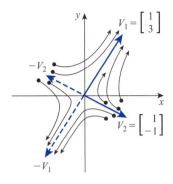

**FIGURE 5.6**

*Trajectories of the system $\dot{x} = 2x + y$, $\dot{y} = 3x + 4y$*
*Bold points ● indicate initial positions ($t = 0$) for trajectories*

The condition $c_1 = 0$ would mean that we are dealing with the straight-line trajectory determined by the eigenvector $V_2 = \begin{bmatrix} 1 \\ -1 \end{bmatrix}$. But if $c_1 \neq 0$, as $t \to \infty$, we see that the slope of any trajectory tends to $\frac{15c_1 - 0}{5c_1 + 0} = 3$, the slope of the line determined by the eigenvector $V_1 = \begin{bmatrix} 1 \\ 3 \end{bmatrix}$.

If we consider large *negative* values of $t$—that is, if we run the trajectories backward in time—then $e^t$ is the dominant term in the expression for $\frac{dy}{dx}$ and we can factor it out:

$$\frac{dy}{dx} = \frac{15c_1 e^{5t} - c_2 e^t}{5c_1 e^{5t} + c_2 e^t} = \frac{e^t\left(15c_1 e^{4t} - c_2\right)}{e^t\left(5c_1 e^{4t} + c_2\right)} = \frac{15c_1 e^{4t} - c_2}{5c_1 e^{4t} + c_2}.$$

The preceding expression tells us that if $c_2 \neq 0$, then as $t \to -\infty$, the slope of any trajectory tends to $\frac{0 - c_2}{0 + c_2} = -1$, the slope of the line determined by the eigenvector $V_2 = \begin{bmatrix} 1 \\ -1 \end{bmatrix}$. If we have $c_2 = 0$, we will be on the straight-line trajectory determined by the eigenvector $V_1 = \begin{bmatrix} 1 \\ 3 \end{bmatrix}$. We conclude that if $c_2 \neq 0$, then any trajectory is tangent to the line $y = -x$ at the origin—that is, as $t \to -\infty$.    ■

### 5.3.4 Unequal Negative Eigenvalues

If both eigenvalues are *negative* (say $\lambda_1 < \lambda_2 < 0$), then both eigenvector multiples point *toward* the origin, and all solutions *decrease* or *decay* with time. To see this, write (5.3.1) in the form

$$X(t) = \begin{bmatrix} c_1 \\ e^{-\lambda_1 t} \end{bmatrix} V_1 + \begin{bmatrix} c_2 \\ e^{-\lambda_2 t} \end{bmatrix} V_2 = \begin{bmatrix} c_1 \\ e^{Kt} \end{bmatrix} V_1 + \begin{bmatrix} c_2 \\ e^{Mt} \end{bmatrix} V_2,$$

where $K = -\lambda_1$ and $M = -\lambda_2$ are *positive* constants. Then clearly, both terms of $X(t)$ approach the origin as $t \to +\infty$. Because $\lambda_1 < \lambda_2$ we have $-\lambda_1 > -\lambda_2$, or $K > M$, so the first term in the expression for $X(t)$ approaches the origin faster than the second term. We will see in the next example that as $t$ increases, trajectories curve *toward* the origin, closer to the eigenvector $V_2$ (or its negative if $c_2 < 0$), corresponding to the larger eigenvalue. Under these circumstances, we say that $(0, 0)$ is a *stable node*, or *sink*.

## ■ Example 5.3.2 Unequal Negative Eigenvalues: A Sink

Suppose we look at the system

$$\dot{x} = -4x + y$$
$$\dot{y} = 3x - 2y.$$

The characteristic equation is $\lambda^2 + 6\lambda + 5 = 0$ and the eigenvalues are negative and unequal: $\lambda_1 = -5$ and $\lambda_2 = -1$. Using the linear algebra capabilities of a CAS, we find that the corresponding representative eigenvectors are $V_1 = \begin{bmatrix} -1 \\ 1 \end{bmatrix}$ and $V_2 = \begin{bmatrix} 1 \\ 3 \end{bmatrix}$. (Don't be disturbed if your CAS produces eigenvectors that are different from the book's—yours should lie on the same line as the ones given here. Your slopes $y/x$ should be $-1$ and 3.)

The general solution of our system is

$$X(t) = c_1 e^{-5t} \begin{bmatrix} -1 \\ 1 \end{bmatrix} + c_2 e^{-t} \begin{bmatrix} 1 \\ 3 \end{bmatrix} = \begin{bmatrix} -c_1 e^{-5t} + c_2 e^{-t} \\ c_1 e^{-5t} + 3c_2 e^{-t} \end{bmatrix}.$$

It is clear from the negative exponents in the expression for $X(t)$ that $X(t) \to \begin{bmatrix} 0 \\ 0 \end{bmatrix}$ as $t \to \infty$, so the origin is a *sink*. Figure 5.7 shows some typical trajectories and seems to indicate that the trajectories are tangent to the line determined by the eigenvector $V_2 = \begin{bmatrix} 1 \\ 3 \end{bmatrix}$.

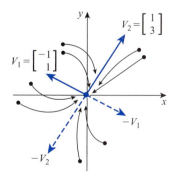

**FIGURE 5.7**

*Trajectories for the system $\dot{x} = -4x + y, \dot{y} = 3x - 2y$*
*Bold points • indicate initial positions ($t = 0$) for trajectories*

Recognizing that $e^{-t}$ is larger than $e^{-5t}$ for large values of $t$, we look at

$$\frac{dy}{dx} = \frac{3x - 2y}{-4x + y} = \frac{-5c_1 e^{-5t} - 3c_2 e^{-t}}{5c_1 e^{-5t} - c_2 e^{-t}}$$

$$= \frac{e^{-t}\left(-5c_1 e^{-4t} - 3c_2\right)}{e^{-t}\left(5c_1 e^{-4t} - c_2\right)} = \frac{-5c_1 e^{-4t} - 3c_2}{5c_1 e^{-4t} - c_2}.$$

If $c_2 \neq 0$, then $\frac{dy}{dx}$ approaches $\frac{-3c_2}{-c_2} = 3$, the slope of the eigenvector $V_2 = \begin{bmatrix} 1 \\ 3 \end{bmatrix}$, as $t \to \infty$. If $c_2 = 0$, then the trajectory is on the straight line determined by the eigenvector $\begin{bmatrix} -1 \\ 1 \end{bmatrix}$. ■

## 5.3.5 Unequal Eigenvalues with Opposite Signs

If the eigenvalues have *opposite* signs (say $\lambda_1 < 0 < \lambda_2$), then look at the general solution $X(t) = c_1 e^{\lambda_1 t} V_1 + c_2 e^{\lambda_2 t} V_2$ to see that the term $c_1 e^{\lambda_1 t} V_1$ (corresponding to the negative eigenvalue $\lambda_1$) points *toward* the origin, whereas $c_2 e^{\lambda_2 t} V$ points *away* from the origin (Figure 5.8).

In this case, trajectories *approach* the origin along one direction and veer *away* from the origin along another. In this situation we describe $(0, 0)$ as a *saddle point*. Look back at Example 5.2.3, especially Figure 5.2.

Let's consider a new example of what happens when the eigenvalues of a system have opposite signs.

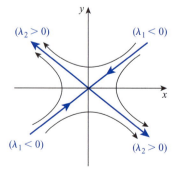

**FIGURE 5.8**

*Typical eigenvectors for the case $\lambda_1 < 0 < \lambda_2$*

# ■ Example 5.3.3 Unequal Eigenvalues with Opposite Signs: A Saddle Point

Let's investigate the system $\frac{dx}{dt} = x+5y$, $\frac{dy}{dt} = x-3y$. The characteristic equation is $\lambda^2 + 2\lambda - 8 = 0$. The eigenvalues and their corresponding eigenvectors are $\lambda_1 = -4$, $V_1 = \begin{bmatrix} 1 \\ -1 \end{bmatrix}$; $\lambda_2 = 2$, $V_2 = \begin{bmatrix} 5 \\ 1 \end{bmatrix}$. The general solution is

$$X(t) = c_1 e^{-4t} \begin{bmatrix} 1 \\ -1 \end{bmatrix} + c_2 e^{2t} \begin{bmatrix} 5 \\ 1 \end{bmatrix} = \begin{bmatrix} c_1 e^{-4t} + 5c_2 e^{2t} \\ -c_1 e^{-4t} + c_2 e^{2t} \end{bmatrix}.$$

We can see that the straight-line trajectory $c_1 e^{-4t} V_1 = c_1 e^{-4t} \begin{bmatrix} 1 \\ -1 \end{bmatrix} = \begin{bmatrix} c_1 e^{-4t} \\ -c_1 e^{-4t} \end{bmatrix}$ approaches the origin as $t \to \infty$. (There are actually *two* half-line trajectories, one for positive $c_1$ and one for negative $c_1$. See Figure 5.9.) But the half-line trajectories corresponding to $c_2 e^{2t} V_2 = c_2 e^{2t} \begin{bmatrix} 5 \\ 1 \end{bmatrix} = \begin{bmatrix} 5c_2 e^{2t} \\ c_2 e^{2t} \end{bmatrix}$ for positive and negative values of $c_2$ are clearly growing *away* from the origin with increasing $t$.

Substituting the expressions for $x(t)$ and $y(t)$ in the formula for $\frac{dy}{dx}$ and factoring out the dominant term for large $t$, we get

$$\frac{dy}{dx} = \frac{x - 3y}{x + 5y} = \frac{4c_1 e^{-4t} + 2c_2 e^{2t}}{-4c_1 e^{-4t} + 10c_2 e^{2t}}$$

$$= \frac{e^{2t} \left( 4c_1 e^{-6t} + 2c_2 \right)}{e^{2t} \left( -4c_1 e^{-6t} + 10c_2 \right)} = \frac{4c_1 e^{-6t} + 2c_2}{-4c_1 e^{-6t} + 10c_2}.$$

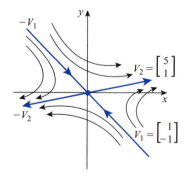

**FIGURE 5.9**

Trajectories for the system $\frac{dx}{dt} = x + 5y$, $\frac{dy}{dt} = x - 3y$

If $c_2 \neq 0$, we see that as $t \to \infty$, $\frac{dy}{dx}$ tends to $\frac{2c_2}{10c_2} = \frac{1}{5}$, the slope of the eigenvector $V_2$. This says that the slopes of trajectories not on the straight lines determined by $V_1$ and $V_2$ approach the slope of $V_2$, the eigenvalue associated with the positive eigenvalue. As $t \to -\infty$, the slopes of these trajectories tend to the slope of $V_1$. Figure 5.9 shows this partial-source/partial-sink behavior with respect to the origin, which is a *saddle point*. ■

### 5.3.6 Unequal Eigenvalues, One Eigenvalue Equal to Zero

Finally, we consider the situation in which we have two unequal eigenvalues, but one of them is 0. Suppose that $\lambda_1 = 0$ and $\lambda_2 \neq 0$. This means that the characteristic equation can be written in the form $0 = (\lambda - 0)(\lambda - \lambda_2) = \lambda^2 - \lambda_2 \lambda$. We know from Section 5.2 that the constant term of the characteristic equation equals $\det(A)$. Clearly, in this case we have $\det(A) = 0$. Therefore, we should not expect the origin to be the only equilibrium point (see Problem B2 of Exercises 5.1). In fact, *every point $(x, 0)$ of the horizontal axis may be an equilibrium point for such a system.* (Problem B9 in Exercises 5.3 asks for a proof of this assertion.) If $V_1$ is the eigenvector associated with $\lambda_1 = 0$, we know that $A(c_1 V_1) = c_1 A(V_1) = c_1 \lambda_1 V_1 = \mathbf{0}$—that is, *each point on the line determined by $V_1$ is an equilibrium point.*

The general solution in this situation has the form

$$X(t) = c_1 e^{(0)t} V_1 + c_2 e^{\lambda_2 t} V_2 = c_1 V_1 + c_2 e^{\lambda_2 t} V_2.$$

Note that if $\lambda_2 > 0$ and $t \to \infty$, then $X(t)$ grows without bound. But if $t \to -\infty$, so that we are traveling backward along a trajectory, then the trajectory approaches $c_1 V_1$, the line determined by $V_1$. Similarly, if $\lambda_2 < 0$ and $t \to \infty$, then $X(t)$ approaches the line determined by $V_1$, whereas if $t \to -\infty$, then $X(t)$ grows without bound. In any case, each trajectory will be a half-line parallel (in the usual plane-geometry sense) to the eigenvector $V_2$, with one endpoint on the line determined by $V_1$. (The constant vector $c_1 V_1$ just shifts $c_2 e^{\lambda_2 t} V_2$ horizontally and vertically.)

The next example should explain the geometry of the trajectories when we have one eigenvalue equal to 0.

### ■ Example 5.3.4 Unequal Eigenvalues, One Eigenvalue Equal to Zero

Figure 5.10 shows the phase portrait for the system $\dot{x} = y$, $\dot{y} = y$, whose eigenvalues are 0 and 1 and whose corresponding eigenvectors are $\begin{bmatrix} 1 \\ 0 \end{bmatrix}$ and $\begin{bmatrix} 1 \\ 1 \end{bmatrix}$, respectively.

Therefore, the equations of the trajectories are $X(t) = c_1 \begin{bmatrix} 1 \\ 0 \end{bmatrix} + c_2 e^t \begin{bmatrix} 1 \\ 1 \end{bmatrix} = \begin{bmatrix} c_1 + c_2 e^t \\ c_2 e^t \end{bmatrix}$. This says (Exercise C4) that any trajectory not on the line determined by $V = \begin{bmatrix} 1 \\ 0 \end{bmatrix}$ has the equation $y(t) = x(t) + k$, so these trajectories form an infinite family of straight lines parallel to $y = x$. Note that the eigenvector $\begin{bmatrix} 1 \\ 0 \end{bmatrix}$ corresponding to the zero eigenvalue determines two half-line trajectories, the positive $x$-axis and the negative $x$-axis. In our example, it is easy to see that

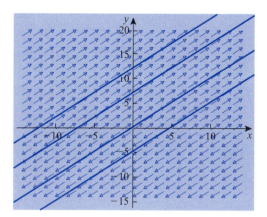

**FIGURE 5.10**

*Phase portrait for the system $\dot{x} = y, \dot{y} = y$*

every point $(x, 0)$ of the horizontal axis is an equilibrium point: $\dot{x} = y = 0$ and $\dot{y} = y = 0$ imply that $y = 0$ and the $x$-coordinate is completely arbitrary. The fact that the nonzero eigenvalue is positive makes the points on the $x$-axis *sources*. (If necessary, review the last full paragraph before this example.) ■

By looking at Examples 5.2.3–5.2.5 and the examples in this section, we notice that a solution starting in a direction different from those of the eigenvectors is curved, representing [as we know from (5.3.1)] a linear combination, $c_1 e^{\lambda_1 t} V_1 + c_2 e^{\lambda_2 t} V_2$, of two exponential solutions that have different rates of change (indicated by the eigenvalues). If we look at enough phase portraits, we may also realize that there is a tendency for the "fast" eigenvector (associated with the larger of two unequal eigenvalues) to have the stronger influence on the solutions. Trajectories curve toward the direction of this eigenvector as $t \to \infty$.

In the next section, we'll investigate what happens when there is a repeated real eigenvalue and when there seems to be only one eigenvector corresponding to two real eigenvalues.

## Exercises 5.3

### A

For each of the Systems 1–11, (a) find the eigenvalues and their corresponding eigenvectors and (b) sketch/plot a few trajectories and show the position(s) of the eigenvector(s). Do part (a) manually, but if the eigenvalues are irrational numbers, you may use technology to find the corresponding eigenvectors.

1. $\dot{x} = 3x, \dot{y} = 2y$
2. $\dot{x} = -x, \dot{y} = -2y$
3. $x' = -3x - y, y' = 4x + 2y$
4. $\dot{r} = 5r + 4s, \dot{s} = -2r - s$

**5.** $\dot{x} = x + 5y, \dot{y} = x - 3y$
**6.** $\dot{x} = 2x + 3y, \dot{y} = x + y$
**7.** $\dot{x} = -3x + y, \dot{y} = 4x - 2y$
**8.** $x' = -4x + 2y, y' = -3x + y$
**9.** $x' = -2x - y, y' = -x + 2y$
**10.** $\dot{x} = 3y, \dot{y} = -3x$
**11.** $\dot{x} = 3x + 6y, \dot{y} = -x - 2y$

**B**

**1.** Consider the system $\dot{x} = 4x - 3y, \dot{y} = 8x - 6y$.

  **a.** Find the eigenvalues of this system.
  **b.** Find the eigenvectors corresponding to the eigenvalues in part (a).
  **c.** Sketch/plot some trajectories and explain what you see.
  **d.** Write the general solution of the system in the form $X(t) = c_1 e^{\lambda_1 t} V_1 + c_2 e^{\lambda_2 t} V_2$, and then re-examine your explanation in part (c).

**2.** Show that if $X$ is an eigenvector of $A$ corresponding to eigenvalue $\lambda$, then any nonzero multiple of $X$ is also an eigenvector of $A$ corresponding to $\lambda$.

**3.** Solve the initial-value problem $X' = \begin{bmatrix} -2 & 1 \\ -5 & 4 \end{bmatrix} X, X(0) = \begin{bmatrix} 1 \\ 3 \end{bmatrix}$ and describe the behavior of the solution as $t \to \infty$. $\left[ \text{Here, } X(t) = \begin{bmatrix} x(t) \\ y(t) \end{bmatrix}. \right]$

**4.** Write a system of first-order linear equations whose trajectories show the following behaviors:

  **a.** $(0, 0)$ is a sink with eigenvalues $\lambda_1 = -3$ and $\lambda_2 = -5$.
  **b.** $(0, 0)$ is a saddle point with eigenvalues $\lambda_1 = -1$ and $\lambda_2 = 4$.
  **c.** $(0, 0)$ is a source with eigenvalues $\lambda_1 = 2$ and $\lambda_2 = 3$.

**5.** Consider the system $\dot{x} = -x + \alpha y, \dot{y} = -2y$, where $\alpha$ is a constant.

  **a.** Show that the origin is a *sink* regardless of the value of $\alpha$.
  **b.** Assume that $X(t)$ is the solution vector of the system satisfying the initial condition $X(0) = \begin{bmatrix} 0 \\ 0.5 \end{bmatrix}$. Sketch the phase portrait for different values of $\alpha$ and describe how the trajectory $X(t)$, for $t \geq 0$, depends on the value of $\alpha$.

**6.** Consider the following circuit.

The current $I$ through the inductor and the voltage $V$ across the capacitor satisfy the system

$$L\frac{dI}{dt} = -R_1 I - V$$

$$C\frac{dV}{dt} = I - \frac{V}{R_2}.$$

**a.** Find the general solution of the system if $R_1 = 1$ ohm, $R_2 = \frac{3}{5}$ ohm, $L = 2$ henrys, and $C = \frac{2}{3}$ farad.

**b.** Show that $I(t) \to 0$ and $V(t) \to 0$ as $t \to \infty$ regardless of the initial values $I(0)$ and $V(0)$.

**7.** Consider the system of differential equations in the preceding problem. Find a condition on $R_1, R_2,$ $C,$ and $L$ that must be satisfied if the eigenvalues of the coefficient matrix are to be real and distinct.

**8.** Two quantities of a chemical solution are separated by a membrane. If $x(t)$ and $y(t)$ represent the amounts of the chemical at time $t$ on each side of the membrane and $V_1$ and $V_2$ represent the (constant) volume of each solution, respectively, then the *diffusion problem* can be modeled by the system

$$\dot{x} = P\left[\frac{y}{V_2} - \frac{x}{V_1}\right]$$

$$\dot{y} = P\left[\frac{x}{V_1} - \frac{y}{V_2}\right],$$

where $P$ is a positive constant called the *permeability* of the membrane. Note that $\frac{x(t)}{V_1}$ and $\frac{y(t)}{V_2}$ represent the *concentrations* of solution on each side.

**a.** Assuming that $x(0) = x_0$ and $y(0) = y_0$, find the solution of the system IVP without using technology.

**b.** Calculate $\lim_{t\to\infty} x(t)$ and $\lim_{t\to\infty} y(t)$.

**c.** Using part (b), interpret the result $\lim_{t\to\infty} [x(t) + y(t)]$ physically.

**d.** Notice that if $\frac{y}{V_2} > \frac{x}{V_1}$, then $\dot{x} > 0$. Does this say that the chemical moves across the membrane from the side with a lower concentration to the side with a higher concentration or vice versa? Confirm your answer by considering what happens if $\frac{x}{V_1} > \frac{y}{V_2}$ in the second equation.

**9.** Consider the system

$$\dot{x} = ax + by$$
$$\dot{y} = cx + dy,$$

where $a, b, c,$ and $d$ are constants. Show that if $ad - bc = 0$, then every point $(x, 0)$ of the horizontal axis is an equilibrium point for the system. [*Hint:* Solve the system $ax + by = 0, cx + dy = 0$ for $x$.]

## C

**1.** Consider the system

$$\dot{r} = -r - s$$
$$\dot{s} = -\beta r - s,$$

where $\beta$ is a parameter.

**a.** Find the general solution of the system when $\beta = 0.5$. Use the eigenvalues of the coefficient matrix to determine what kind of equilibrium the system has at the origin.

**b.** Find the general solution of the system when $\beta = 2$. Use the eigenvalues of the coefficient matrix to determine what kind of equilibrium the system has at the origin.

**c.** The solutions of the system show two rather different types of behavior for the two values of $\beta$ considered in parts (a) and (b). Find a formula for the eigenvalues in terms of $\beta$ and determine the value of $\beta$ between 0.5 and 2 where the transition from one type of behavior to the other occurs. (This critical value of the parameter is called a *bifurcation point*. See Section 2.7.)

2. Suppose that we have the system $\dot{X} = AX$ and that $V_1$ and $V_2$ are eigenvectors of $A$ such that neither $V_1$ nor $V_2$ is a scalar multiple of the other. Show that any initial condition $X_0 = X(0) = \begin{bmatrix} x_0 \\ y_0 \end{bmatrix}$ can be written as a linear combination of $V_1$ and $V_2$. In other words, show that you can always find scalars $c_1$ and $c_2$ so that $X_0 = c_1 V_1 + c_2 V_2$. [*Hint:* Let $V_1 = \begin{bmatrix} x_1 \\ y_1 \end{bmatrix}$ and $V_2 = \begin{bmatrix} x_2 \\ y_2 \end{bmatrix}$ be the eigenvectors, where you assume that $x_1, x_2, y_1$, and $y_2$ are known. Now convert the equation $X_0 = c_1 V_1 + c_2 V_2$ into a system of algebraic linear equations and go from there.]

3. If the system $\dot{X} = AX$ has two real eigenvalues $\lambda_1$ and $\lambda_2$, with $\lambda_1 \neq \lambda_2$, and $V_1$ and $V_2$ are the corresponding (distinct) eigenvectors, show that $X(t) = c_1 e^{\lambda_1 t} V_1 + c_2 e^{\lambda_2 t} V_2$ satisfies the initial condition $X(0) = X_0 = \begin{bmatrix} x_0 \\ y_0 \end{bmatrix} = c_1 V_1 + c_2 V_2$. (See the preceding problem for the justification of this representation of $X_0$ for some scalars $c_1$ and $c_2$.)

4. As indicated in Example 5.3.4, the system $\dot{x} = y, \dot{y} = y$ has the solution $X(t) = \begin{bmatrix} c_1 + c_2 e^t \\ c_2 e^t \end{bmatrix}$. Show that any trajectory not on the line determined by $V = \begin{bmatrix} 1 \\ 0 \end{bmatrix}$ satisfies the equation $y(t) = x(t) + k$ (in the phase plane) for some constant $k$. (This says that the trajectories form an infinite family of lines parallel to $y = x$.)

## 5.4 THE STABILITY OF HOMOGENEOUS LINEAR SYSTEMS: EQUAL REAL EIGENVALUES

Now let's see what happens if both eigenvalues are real and equal. In other words, the characteristic equation has a *repeated root*, or *double root*. (See Section 4.1 for the second-order homogeneous linear equation case.) A full understanding of this situation requires more linear algebra than we want to pursue right now. However, the following discussions and examples should give us a good idea of what's going on.

### 5.4.1 Equal Nonzero Eigenvalues, Two Independent Eigenvectors

First, suppose that $\lambda_1 = \lambda_2 \neq 0$. If we can find distinct representative eigenvectors $V_1$ and $V_2$ that are not scalar multiples of each other, then we can still write the general solution of the system using (5.3.1): $X(t) = c_1 e^{\lambda_1 t} V_1 + c_2 e^{\lambda_2 t} V_2 = c_1 e^{\lambda_1 t} V_1 + c_2 e^{\lambda_1 t} V_2 = e^{\lambda_1 t}(c_1 V_1 + c_2 V_2)$. If we let $t = 0$, we see that $X(0) = e^{\lambda_1 (0)}(c_1 V_1 + c_2 V_2) = c_1 V_1 + c_2 V_2$, so we can write

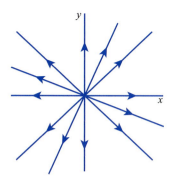

**FIGURE 5.11a**

*Source:* $\lambda > 0$

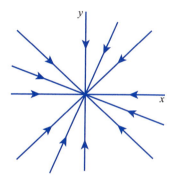

**FIGURE 5.11b**

*Sink:* $\lambda < 0$

$X(t) = e^{\lambda_1 t} X_0$, where $X_0 = X(0)$. (See Problem C2 of Exercises 5.3.) Under these conditions, all trajectories are straight lines through the origin because they are constant multiples of the constant vector $X_0 = c_1 V_1 + c_2 V_2$. The origin is called a **star node** in this case and will be a *source* if $\lambda_1 > 0$ and a *sink* if $\lambda_1 < 0$. Figures 5.11a and 5.11b show possible trajectories for various initial vectors $X_0$.

Let's examine a system for which the origin is a star node.

## ■ Example 5.4.1 The Origin as a Star Node (a Source)

Look at the system $\frac{dx}{dt} = x$, $\frac{dy}{dt} = y$. We can write this in matrix form as $\dot{X} = AX$, where $A = \begin{bmatrix} 1 & 0 \\ 0 & 1 \end{bmatrix}$. It is easy to see that $A$ has eigenvalues $\lambda_1 = 1 = \lambda_2$. (*Check this.*) By the way we defined the product of a matrix and a vector in Section 5.1, we see that our matrix of coefficients $A$ is such that $AV = V = 1 \cdot V = \lambda_1 V$ for *every* vector $V$. In particular, any *nonzero* vector $V$ is an eigenvector corresponding to the eigenvalue 1. *Be sure you understand the preceding*

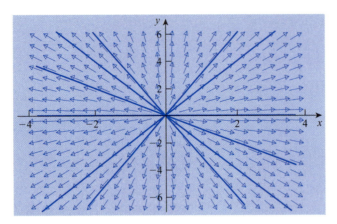

**FIGURE 5.12**

Phase portrait of the system $\frac{dx}{dt} = x$, $\frac{dy}{dt} = y$

*statement.* A particularly simple eigenvector to work with is $V_1 = \begin{bmatrix} 1 \\ 0 \end{bmatrix}$. It is easy to see that the vector $V_2 = \begin{bmatrix} 0 \\ 1 \end{bmatrix}$ is not a multiple of $V_1$ because any scalar multiple of $V_1$ would have the form $\begin{bmatrix} c \\ 0 \end{bmatrix}$, where $c$ is a constant. Therefore, we can write the solution of our system as

$$X(t) = c_1 e^t V_1 + c_2 e^t V_2 = c_1 e^t \begin{bmatrix} 1 \\ 0 \end{bmatrix} + c_2 e^t \begin{bmatrix} 0 \\ 1 \end{bmatrix} = \begin{bmatrix} c_1 e^t \\ c_2 e^t \end{bmatrix}.$$

Of course, because each of our original (separable) differential equations contains only one variable, we could solve each one separately to get the same result in the form $x(t) = c_1 e^t$, $y(t) = c_2 e^t$. As we indicated in the discussion right before this example, the trajectories are straight lines through the origin, and Figure 5.12 shows that the origin, a star node, is a *source*.

■

### 5.4.2 Equal Nonzero Eigenvalues, Only One Independent Eigenvector

Now suppose that $\lambda_1 = \lambda_2 \neq 0$, but our single eigenvalue has *only one distinct representative eigenvector*. What we mean is that all eigenvectors corresponding to the single distinct eigenvalue are scalar multiples of each other. Geometrically, this says that all eigenvectors lie on the same straight line through the origin. Then if we tried to use the solution form (5.3.1), we would get

$$X(t) = c_1 e^{\lambda_1 t} V + c_2 e^{\lambda_1 t} V = (c_1 + c_2) e^{\lambda_1 t} V = k e^{\lambda_1 t} V.$$

But how can the general solution of a two-dimensional system or second-order equation have only one arbitrary constant?

What we have to do here is find another solution of the system that is *independent* of the one solution we found using the single eigenvalue and its representative eigenvector. This is similar to the technique we used in solving a second-order linear equation with a repeated eigenvalue (see Section 4.1). In our situation, an independent solution is one that is not a scalar multiple of the first solution. If we *do* find another eigenvector corresponding to the single eigenvalue, but one that is independent of the original eigenvector, then the solution can still be written in the form $X(t) = c_1 e^{\lambda_1 t} V_1 + c_2 e^{\lambda_2 t} V_2$.

It turns out that we *can* find a substitute for an independent eigenvector. Although we won't go into all the linear algebraic details here, we can at least try to explain the end result. Another (independent) solution of the system must have the form

$$X_2(t) = t e^{\lambda t} V + e^{\lambda t} W, \tag{5.4.1}$$

where $V$ is the original eigenvector corresponding to the single eigenvalue $\lambda$, and where $W$, called a **generalized eigenvector (of order 2)**, is a vector that satisfies the matrix equation

$$(A - \lambda I)W = V. \tag{5.4.2}$$

(See Problem B5 in Exercises 5.4.)

We can easily see that the vector defined by (5.4.1) is a solution of the system. If $X(t) = t e^{\lambda t} V + e^{\lambda t} W$, then $\dot{X}(t) = t(\lambda e^{\lambda t} V) + e^{\lambda t} V + \lambda e^{\lambda t} W = (\lambda t + 1) e^{\lambda t} V + \lambda e^{\lambda t} W$ and, because (5.4.2) implies that $AW = V + \lambda W$,

$$AX = A(t e^{\lambda t} V + e^{\lambda t} W) = t e^{\lambda t}(AV) + e^{\lambda t}(AW) = t e^{\lambda t}(\lambda V) + e^{\lambda t}(V + \lambda W)$$

$$= (\lambda t + 1) e^{\lambda t} V + \lambda e^{\lambda t} W.$$

Thus, $\dot{X} = AX$—that is, (5.4.1) defines a solution of the system.

Next, we must solve Equation (5.4.2) for $W$, and then we can write the general solution of the system as

$$X(t) = c_1 e^{\lambda t} V + c_2 \left[ t e^{\lambda t} V + e^{\lambda t} W \right]. \tag{5.4.3}$$

(The theory of linear algebra shows that we can always solve for $W$ in Equation (5.4.2) if $V$ is an eigenvector of $A$ corresponding to eigenvalue $\lambda$.)

Now let's look at an example in which we have equal nonzero eigenvalues but only one distinct representative eigenvector.

## ■ Example 5.4.2 Equal Nonzero Eigenvalues, Only One Distinct Eigenvector

Consider the system $\dot{x} = -2x + y$, $\dot{y} = -2y$. We can write this in matrix form as $\dot{X} = AX$, where $A = \begin{bmatrix} -2 & 1 \\ 0 & -2 \end{bmatrix}$. The characteristic polynomial of $A$ is $\lambda^2 + 4\lambda + 4 = (\lambda + 2)^2$, so

$\lambda = -2$ is a repeated root. Then the matrix equation $AV = \lambda V = -2V$ is equivalent to the system

$$-2x + y = -2x$$
$$-2y = -2y,$$

or

$$y = 0$$
$$-2y = -2y.$$

From this we see that any eigenvector $\begin{bmatrix} x \\ y \end{bmatrix}$ must have the form $\begin{bmatrix} x \\ 0 \end{bmatrix} = x \begin{bmatrix} 1 \\ 0 \end{bmatrix}$ for arbitrary values of $x$. Therefore, we can take $V = \begin{bmatrix} 1 \\ 0 \end{bmatrix}$ as the only independent eigenvector that corresponds to the eigenvalue $-2$. Now we must find a vector $W = \begin{bmatrix} r \\ s \end{bmatrix}$ satisfying $(A - \lambda I)W = V$.

In our problem, $(A - \lambda I)W = V$ becomes

$$\left\{ \begin{bmatrix} -2 & 1 \\ 0 & -2 \end{bmatrix} - (-2) \begin{bmatrix} 1 & 0 \\ 0 & 1 \end{bmatrix} \right\} \begin{bmatrix} r \\ s \end{bmatrix} = \begin{bmatrix} 1 \\ 0 \end{bmatrix},$$

$$\left\{ \begin{bmatrix} -2 & 1 \\ 0 & -2 \end{bmatrix} + \begin{bmatrix} 2 & 0 \\ 0 & 2 \end{bmatrix} \right\} \begin{bmatrix} r \\ s \end{bmatrix} = \begin{bmatrix} 1 \\ 0 \end{bmatrix},$$

or

$$\begin{bmatrix} 0 & 1 \\ 0 & 0 \end{bmatrix} \begin{bmatrix} r \\ s \end{bmatrix} = \begin{bmatrix} 1 \\ 0 \end{bmatrix},$$

which is equivalent to the algebraic system

$$0 \cdot r + 1 \cdot s = 1$$
$$0 \cdot r + 0 \cdot s = 0.$$

This tells us that $s = 1$ and $r$ is a "free variable"—that is, $r$ is completely arbitrary. For convenience, let $r = 0$ so that our generalized eigenvector is $W = \begin{bmatrix} 0 \\ 1 \end{bmatrix}$. Finally, we can write the general solution of our system in the form (5.4.3):

$$X(t) = c_1 e^{-2t} \begin{bmatrix} 1 \\ 0 \end{bmatrix} + c_2 \left[ te^{-2t} \begin{bmatrix} 1 \\ 0 \end{bmatrix} + e^{-2t} \begin{bmatrix} 0 \\ 1 \end{bmatrix} \right]$$

$$= \begin{bmatrix} c_1 e^{-2t} + c_2 te^{-2t} \\ c_2 e^{-2t} \end{bmatrix}.$$

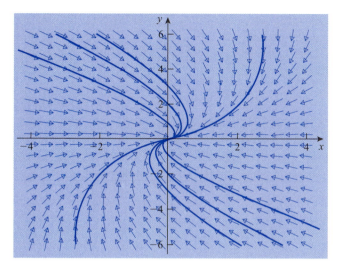

**FIGURE 5.13**

*Trajectories for the system* $\dot{x} = -2x + y, \dot{y} = -2y$

Figure 5.13, generated by a CAS, shows that the trajectories spiral in toward the origin, in such a way that they are tangent to the eigenvector $V = \begin{bmatrix} 1 \\ 0 \end{bmatrix}$ or its negative at the origin. (Note that the vector $V$ is part of the positive $x$-axis.)

Whenever we have a system with equal nonzero eigenvalues but only one distinct eigenvector, the phase portrait will consist of spirals *approaching* the origin when the repeated eigenvalue is *negative*, and the phase portrait will consist of spirals moving *outward* if the eigenvalue is *positive*. A negative eigenvalue makes the origin a **spiral sink**; a positive eigenvalue makes the origin a **spiral source**. Furthermore, if the eigenvalue is negative, the slopes of all trajectories not on the line determined by the one eigenvector approach the slope of this line as $t \to \infty$. A positive eigenvalue indicates that the slopes of all trajectories not on the line determined by the one eigenvector approach the slope of this line as $t \to -\infty$. (Problem B6 in Exercises 5.4 asks for a proof of the preceding two assertions.)

### 5.4.3 Both Eigenvalues Zero

Finally, let's assume that $\lambda_1 = \lambda_2 = 0$. If there are two linearly independent eigenvectors $V_1$ and $V_2$, then the general solution is $X(t) = c_1 e^{0 \cdot t} V_1 + c_2 e^{0 \cdot t} V_2 = c_1 V_1 + c_2 V_2$, a single vector of constants. If there is only one linearly independent eigenvector $V$ corresponding to the eigenvalue 0, then we can find a generalized eigenvector and use Formula (5.4.3):

$$X(t) = c_1 e^{\lambda t} V + c_2 \left[ t e^{\lambda t} V + e^{\lambda t} W \right].$$

For $\lambda = 0$, we get $X(t) = c_1V + c_2[tV + W] = (c_1 + c_2t)V + c_2W$. In Exercise B7 you will investigate a system that has both eigenvalues zero.

## Exercises 5.4

### A

For each of the Systems 1–8, (a) find the eigenvalues and their corresponding linearly independent eigenvectors and (b) sketch/plot a few trajectories and show the position(s) of the eigenvector(s) if they do not have complex entries. Do part (a) manually, but if the eigenvalues are irrational numbers, you may use technology to find the corresponding eigenvectors.

1. $\dot{x} = 3x, \dot{y} = 3y$
2. $\dot{x} = -4x, \dot{y} = x - 4y$
3. $\dot{x} = 2x + y, \dot{y} = 4y - x$
4. $\dot{x} = 3x - y, \dot{y} = 4x - y$
5. $\dot{x} = 2y - 3x, \dot{y} = y - 2x$
6. $\dot{x} = 5x + 3y, \dot{y} = -3x - y$
7. $\dot{x} = -3x - y, \dot{y} = x - y$
8. $\dot{x} = \sqrt{2}x + 5y, \dot{y} = \sqrt{2}y$

### B

1. Given a characteristic polynomial $\lambda^2 + \alpha\lambda + \beta$, what condition on $\alpha$ and $\beta$ guarantees that there is a repeated eigenvalue?

2. Let $A = \begin{bmatrix} a & b \\ c & d \end{bmatrix}$. Show that $A$ has only one eigenvalue if and only if $[\text{trace }(A)]^2 - 4\det(A) = 0$.

3. Write a system of first-order linear equations for which $(0, 0)$ is a sink with eigenvalues $\lambda_1 = -2$ and $\lambda_2 = -2$.

4. Write a system of first-order linear equations for which $(0, 0)$ is a source with eigenvalues $\lambda_1 = 3$ and $\lambda_2 = 3$.

5. Show that if $V$ is an eigenvector of a $2 \times 2$ matrix $A$ corresponding to eigenvalue $\lambda$ and vector $W$ is a solution of $(A - \lambda I)W = V$, then $V$ and $W$ are linearly independent. [See Equations (5.4.2)–(5.4.3).] [*Hint*: Suppose that $W = cV$ for some scalar $c$. Then show that $V$ must be the zero vector.]

6. Suppose that a system $\dot{X} = AX$ has only one eigenvalue $\lambda$, and that every eigenvector is a scalar multiple of one fixed eigenvector, $V$. Then Equation (5.4.3) tells us that any trajectory has the form $X(t) = c_1e^{\lambda t}V + c_2\left[te^{\lambda t}V + e^{\lambda t}W\right] = te^{\lambda t}\left[\frac{1}{t}(c_1V + W) + c_2V\right]$.

   a. If $\lambda < 0$, show that the slope of $X(t)$ approaches the slope of the line determined by $V$ as $t \to \infty$. [*Hint*: $\frac{e^{-\lambda t}}{t}X(t)$, as a scalar multiple of $X(t)$, is parallel to $X(t)$.]
   b. If $\lambda < 0$, show that the slope of $X(t)$ approaches the slope of the line determined by $V$ as $t \to -\infty$.

7. Consider the system $\dot{x} = 6x + 4y, \dot{y} = -9x - 6y$.

   a. Show that the only eigenvalue of the system is 0.
   b. Find the single independent eigenvector $V$ corresponding to $\lambda = 0$.

c. Show that every trajectory of this system is a straight line parallel to $V$, with trajectories on opposite sides of $V$ moving in opposite directions. [*Hint*: First, for any trajectory not on the line determined by $V$, look at its slope, $dy/dx$.]

**C**

1. Prove that $c_1 e^{\lambda t} \begin{bmatrix} 1 \\ 0 \end{bmatrix} + c_2 e^{\lambda t} \begin{bmatrix} t \\ 1 \end{bmatrix}$ is the general solution of $\dot{X} = AX$,

$$\text{where } A = \begin{bmatrix} \lambda & 1 \\ 0 & \lambda \end{bmatrix}.$$

2. Suppose the matrix $A$ has repeated real eigenvalues $\lambda$ and there exists a pair of linearly independent eigenvectors associated with $A$. Prove that $A = \begin{bmatrix} \lambda & 0 \\ 0 & \lambda \end{bmatrix}$.

3. A special case of the *Cayley-Hamilton Theorem* states that if $\lambda^2 + \alpha\lambda + \beta = 0$ is the characteristic equation of a matrix $A$, then $A^2 + \alpha A + \beta I$ is the zero matrix. (We say that a $2 \times 2$ matrix always satisfies its own characteristic equation.) Using this result, show that if a $2 \times 2$ matrix $A$ has a repeated eigenvalue $\lambda$ and $V = \begin{bmatrix} x \\ y \end{bmatrix}$, then either $V$ is an eigenvector of $A$ or else $(A - \lambda I)V$ is an eigenvector of $A$.

## 5.5 THE STABILITY OF HOMOGENEOUS LINEAR SYSTEMS: COMPLEX EIGENVALUES

### 5.5.1 Complex Eigenvalues and Complex Eigenvectors

Now let's examine what occurs when the matrix $A$ in the system $\dot{X} = AX$ has *complex* eigenvalues. As we've already stated, any complex root $\lambda$ of the quadratic characteristic equation $\lambda^2 - (a + d)\lambda + (ad - bc) = 0$ must occur as part of a *complex conjugate pair*: $\lambda = p \pm qi$. As we'll see, the behavior of trajectories in the case of complex eigenvalues depends on the *real part, p*, of the complex eigenvalues. When the eigenvalues of a matrix are complex numbers, the eigenvectors will also have complex entries (see Appendix C), and therefore the algebra of the situation will be slightly more complicated.

The most important point to realize is that when $A$ has complex eigenvalues, the general solution of $\dot{X} = AX$ has the same form as (5.3.1), $X(t) = c_1 e^{\lambda_1 t} V_1 + c_2 e^{\lambda_2 t} V_2$. In other words, the Superposition Principle holds, but we have to deal with the fact that this formula will produce vectors whose elements are complex functions or numbers. For example, in the context of the general solution formula, the phrase *multiplying by a scalar* refers to multiplying vectors (whose entries may be complex numbers) by complex numbers.

Fortunately, there are some useful results that aid us in our work with complex eigenvalues and eigenvectors:

1. A crucial fact to recall is Euler's formula, which we saw in Section 4.1:

$$e^{p+qi} = e^p(\cos(q) + i\sin(q)).$$

This result will be useful in simplifying complex-valued expressions and will show us how to obtain real-valued solutions of $\dot{X} = AX$.

2. Another important fact is that eigenvectors corresponding to complex conjugate *eigenvalues are conjugate to each other.* If the eigenvalue $\lambda_1 = p + qi$ has a corresponding

eigenvector $V_1 = \begin{bmatrix} a_1 + b_1 i \\ a_2 + b_2 i \end{bmatrix} = \begin{bmatrix} a_1 \\ a_2 \end{bmatrix} + i \begin{bmatrix} b_1 \\ b_2 \end{bmatrix} = U + iW$, then $\lambda_2 = \overline{\lambda}_1 = p - qi$ has a

corresponding eigenvector $V_2 = \overline{V}_1 = \begin{bmatrix} \overline{a_1 + b_1 i} \\ \overline{a_2 + b_2 i} \end{bmatrix} = \begin{bmatrix} a_1 - b_1 i \\ a_2 - b_2 i \end{bmatrix} = \begin{bmatrix} a_1 \\ a_2 \end{bmatrix} - i \begin{bmatrix} b_1 \\ b_2 \end{bmatrix} =$

$U - iW$. The proof of this result follows from the properties of the conjugate: Suppose that $AV_1 = \lambda_1 V_1$. Then $\overline{(AV_1)} = \overline{(\lambda_1 V_1)}$, so $\overline{A}\,\overline{V}_1 = \overline{\lambda}_1 \overline{V}_1$, or (because all elements of $A$ are real) $A\overline{V}_1 = \overline{\lambda}_1 \overline{V}_1 = \lambda_2 \overline{V}_1$. That is, $\overline{V}_1$ is an eigenvector corresponding to $\lambda_2 = \overline{\lambda}_1$. To see how valuable results 1 and 2 are, let's suppose that $\lambda = p + qi$ is an eigenvalue of the matrix $A$ and that $V = U + iW$ is a corresponding eigenvector. If we define $X(t) = e^{\lambda t}V$, then $AX = A\left(e^{\lambda t}V\right) = e^{\lambda t}(AV) = e^{\lambda t}(\lambda V) = \lambda e^{\lambda t}V = \dot{X}$, so $X(t)$ is a solution of $\dot{X} = AX$. Using Euler's formula and the properties of complex multiplication (see Section C.1), we have

$$X(t) = e^{\lambda t}V = e^{(p+qi)t}V = e^{pt}(\cos qt + i\sin qt)(U + iW)$$
$$= e^{pt}\left\{(\cos qt)U - (\sin qt)W\right\} + ie^{pt}\left\{(\cos qt)W + (\sin qt)U\right\}.$$

Then the *real part* and the *imaginary part* of $X(t)$ can be considered separately:

$$X_1(t) = \text{Re}\left\{X(t)\right\} = e^{pt}\left\{(\cos qt)U - (\sin qt)W\right\}$$
$$X_2(t) = \text{Im}\left\{X(t)\right\} = e^{pt}\left\{(\cos qt)W + (\sin qt)U\right\}.$$

The important observation here is that $X_1(t)$ *and* $X_2(t)$ *are real-valued linearly inde-pendent solutions of the system* $\dot{X} = AX$. (Problem B3 in Exercises 5.5 asks for a proof that the same two solutions result from taking the real and imaginary parts of $e^{\overline{\lambda}t}\overline{V}$.)

We will justify this observation for the real part of $X(t)$, leaving the proof for the imaginary part as Problem B4 in Exercises 5.5. First, we write $X_R = \text{Re}\left\{X(t)\right\} = \frac{X+\overline{X}}{2}$ (see Section C.1 if necessary). Then

$$AX_R = A\left(\frac{X + \overline{X}}{2}\right) = \tfrac{1}{2}A\left(X + \overline{X}\right) = \tfrac{1}{2}\left(AX + A\overline{X}\right)$$

$$= \tfrac{1}{2}\left(\dot{X} + \overline{(AX)}\right) = \tfrac{1}{2}\left(\dot{X} + \dot{\overline{X}}\right) = \text{Re}(\dot{X}) = (\dot{X})_R = \overset{\cdot}{\overline{(X_R)}}.$$

Now the Superposition Principle tells us that $c_1 X_1(t) + c_2 X_2(t)$ is also a solution—in fact, it is the *general solution* of the system. The proofs of this last fact in Section 5.3 are valid here. We can take the scalars $c_1$ and $c_2$ to be real numbers.

As a first example of working with complex eigenvalues and eigenvectors, let's look at the equation $\frac{d^2\theta}{dt^2} + k^2 \sin\theta = 0$, which describes the motion of an *undamped pendulum*. Here, $\theta$ is the angle the pendulum makes with the vertical, and $k^2 = \frac{g}{L}$, where $g$ is the acceleration due to gravity and $L$ is the length of the pendulum. This famous equation is nonlinear and will be treated fully in Section 7.4, but for small angles $\theta$, $\sin\theta \approx \theta$, so we can consider the *linearized* equation $\frac{d^2\theta}{dt^2} + k^2\theta = 0$. The system form of the linear pendulum equation has complex eigenvalues.

Let's see how to work with the complexities (pun intended) of this situation.

### ■ Example 5.5.1 A System with Complex Eigenvalues

First, we convert the linearized pendulum equation to a system (see Problem C2 of Exercises 4.7 for the nonlinear case). Letting $x = \theta$ and $y = \frac{d\theta}{dt} = \frac{dx}{dt}$, we convert our linear second-order homogeneous equation into the system $\frac{dx}{dt} = y$, $\frac{dy}{dt} = -k^2 x$. (Be sure that you remember how to carry out this conversion.)

In matrix form, we have the system $\frac{d}{dt}\begin{bmatrix} x \\ y \end{bmatrix} = \begin{bmatrix} 0 & 1 \\ -k^2 & 0 \end{bmatrix}\begin{bmatrix} x \\ y \end{bmatrix}$, with characteristic equation $\lambda^2 + k^2 = 0$ and complex conjugate eigenvalues $\lambda_1 = ki$ and $\lambda_2 = -ki$. (*Verify all the statements in the preceding sentence.*) The equation $AV = \lambda_1 V$ has the form $\begin{bmatrix} 0 & 1 \\ -k^2 & 0 \end{bmatrix}\begin{bmatrix} x \\ y \end{bmatrix} = ki\begin{bmatrix} x \\ y \end{bmatrix} = \begin{bmatrix} kix \\ kiy \end{bmatrix}$, which is equivalent to the algebraic system

$$y = kix$$
$$-k^2 x = kiy.$$

Because the second equation is just $ki$ times the first, we see that we can take $x$ as arbitrary and $y = kix$, which gives us the eigenvector $V = \begin{bmatrix} x \\ kix \end{bmatrix} = x\begin{bmatrix} 1 \\ ki \end{bmatrix}$. Letting $x = 1$, we get the representative eigenvector

$$V_1 = \begin{bmatrix} 1 \\ ki \end{bmatrix} = \begin{bmatrix} 1 \\ 0 \end{bmatrix} + i\begin{bmatrix} 0 \\ k \end{bmatrix}.$$

From the discussion preceding this example, we realize that we don't have to worry about the second (conjugate) eigenvalue and its associated eigenvector. The general solution of our original equation and its system version can be obtained from the information we already have. We start with the solution

$$\hat{X}(t) = e^{kit}\left(\begin{bmatrix}1\\0\end{bmatrix} + i\begin{bmatrix}0\\k\end{bmatrix}\right) = (\cos kt + i\sin kt)\left(\begin{bmatrix}1\\0\end{bmatrix} + i\begin{bmatrix}0\\k\end{bmatrix}\right)$$

$$= \left((\cos kt)\begin{bmatrix}1\\0\end{bmatrix} - (\sin kt)\begin{bmatrix}0\\k\end{bmatrix}\right) + i\left((\cos kt)\begin{bmatrix}0\\k\end{bmatrix} + (\sin kt)\begin{bmatrix}1\\0\end{bmatrix}\right).$$

Because the real and imaginary parts of the preceding expression are linearly independent solutions of the system, the general solution is given by

$$X(t) = c_1\left((\cos kt)\begin{bmatrix}1\\0\end{bmatrix} - (\sin kt)\begin{bmatrix}0\\k\end{bmatrix}\right) + c_2\left((\cos kt)\begin{bmatrix}0\\k\end{bmatrix} + (\sin kt)\begin{bmatrix}1\\0\end{bmatrix}\right)$$

$$= c_1\begin{bmatrix}\cos kt\\-k\sin kt\end{bmatrix} + c_2\begin{bmatrix}\sin kt\\k\cos kt\end{bmatrix} = \begin{bmatrix}c_1\cos kt + c_2\sin kt\\-kc_1\sin kt + kc_2\cos kt\end{bmatrix}.$$

Figure 5.14 shows some trajectories for this system when $k = 1$. These curves are circles centered at the origin. We say that the origin is a **center** for the system. You should try to generate your own phase portrait by choosing different values of $k$ and various initial points for each value of $k$.

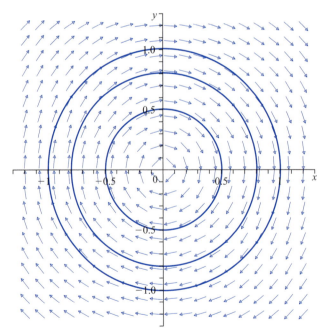

**FIGURE 5.14**

Trajectories for the system $\frac{dx}{dt} = y$, $\frac{dy}{dt} = -x$, $0 \le t \le 7$
Initial points: $(x(0), y(0)) = (1, 0), (0.5, 0), (0, 0.8)$

The next example provides a more challenging problem algebraically.

## ■ Example 5.5.2 A System with Complex Eigenvalues

According to *Kirchhoff's Second Law*, an electric circuit with resistance of 2 ohms, capacitance of 0.5 farad, inductance of 1 henry, and no driving electromotive force can be modeled by the second-order linear equation $\ddot{Q} + 2\dot{Q} + 2Q = 0$, where $Q = Q(t)$ is the charge on the capacitor at time $t$. If $Q(0) = 1$ and $\dot{Q}(0) = 0$, we want to determine the charge on the capacitor at time $t \geq 0$.

We write our second-order equation as a system of first-order equations by introducing new variables: Let $x = Q$ and $y = \dot{x} = \dot{Q}$, so $\dot{y} = \ddot{Q} = -2Q - 2\dot{Q} = -2x - 2y$. Then the original second-order equation is equivalent to the system

$$\dot{x} = y$$
$$\dot{y} = -2x - 2y,$$

which can be written in matrix form as $\dot{X} = \begin{bmatrix} 0 & 1 \\ -2 & -2 \end{bmatrix}\begin{bmatrix} x \\ y \end{bmatrix}$. The matrix of coefficients has characteristic equation $\lambda^2 + 2\lambda + 2 = 0$, with roots $-1 + i$ and $-1 - i$. Working with the first of these eigenvalues, we see that any eigenvector must satisfy the matrix equation

$$\begin{bmatrix} 0 & 1 \\ -2 & -2 \end{bmatrix}\begin{bmatrix} x \\ y \end{bmatrix} = (-1 + i)\begin{bmatrix} x \\ y \end{bmatrix},$$

which is equivalent to the equations

$$y = -x + ix$$
$$-2x - 2y = -y + iy.$$

Substituting the first equation in the second equation, we get

$$-2x - 2[-x + ix] = -[-x + ix] + i[-x + ix]$$
$$-2x + 2x - 2ix = x - ix - ix - x \text{ (remembering that } t^2 = -1)$$
$$-2ix = -2ix.$$

The preceding equation, an identity, says that *any* value of $x$ will be a solution. If we choose $x = 1$ for convenience, then the first equation gives us $y = -1 + i$, so the representative eigenvector is

$$V_1 = \begin{bmatrix} 1 \\ i - 1 \end{bmatrix} = \begin{bmatrix} 1 \\ -1 \end{bmatrix} + i \cdot \begin{bmatrix} 0 \\ 1 \end{bmatrix} = U + iW.$$

As in the previous example, we work with the solution provided by one of the complex conjugate eigenvalues and its representative eigenvector:

$$\hat{X}(t) = e^{(-1+i)t}\left(\begin{bmatrix}1\\-1\end{bmatrix}+i\begin{bmatrix}0\\1\end{bmatrix}\right) = e^{-t}(\cos t + i\sin t)\left(\begin{bmatrix}1\\-1\end{bmatrix}+i\begin{bmatrix}0\\1\end{bmatrix}\right)$$

$$= e^{-t}\left((\cos t)\begin{bmatrix}1\\-1\end{bmatrix}-(\sin t)\begin{bmatrix}0\\1\end{bmatrix}\right) + ie^{-t}\left((\cos t)\begin{bmatrix}0\\1\end{bmatrix}+(\sin t)\begin{bmatrix}1\\-1\end{bmatrix}\right).$$

Extracting the real and imaginary parts of this last complex-valued expression, we express the general solution as

$$X(t) = c_1 e^{-t}\left((\cos t)\begin{bmatrix}1\\-1\end{bmatrix}-(\sin t)\begin{bmatrix}0\\1\end{bmatrix}\right) + c_2 e^{-t}\left((\cos t)\begin{bmatrix}0\\1\end{bmatrix}+(\sin t)\begin{bmatrix}1\\-1\end{bmatrix}\right)$$

$$= c_1\begin{bmatrix}e^{-t}\cos t\\-e^{-t}\cos t - e^{-t}\sin t\end{bmatrix} + c_2\begin{bmatrix}e^{-t}\sin t\\e^{-t}\cos t - e^{-t}\sin t\end{bmatrix}$$

$$= e^{-t}\begin{bmatrix}c_1\cos t + c_2\sin t\\(c_2 - c_1)\cos t - (c_2 + c_1)\sin t\end{bmatrix} = \begin{bmatrix}x(t)\\y(t)\end{bmatrix}.$$

Now, using the initial conditions $x(0) = Q(0) = 1$ and $y(0) = \dot{Q}(0) = 0$ in the general solution just given, we get the condition $\begin{bmatrix}c_1\\c_2 - c_1\end{bmatrix} = \begin{bmatrix}1\\0\end{bmatrix}$, which implies that $c_1 = 1$ and $c_2 = 1$. Thus, the solution of our original initial-value problem is $Q(t) = x(t) = e^{-t}(\cos t + \sin t)$. (See Figure 5.15.) Because the current, $I$, is defined as the rate of change of $Q$, we get a bonus: $I(t) = \dot{Q}(t) = y(t) = -2e^{-t}\sin t$.

As satisfying as this analytical solution may be, a natural question is what the trajectories for this system look like. Figure 5.16 shows five trajectories, corresponding to different initial conditions. The trajectory for the IVP we started with is second from the bottom.

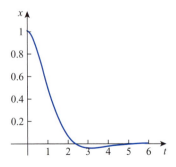

**FIGURE 5.15**
*Graph of $x(t) = e^{-t}(\cos t + \sin t)$, $0 \le t \le 6$*

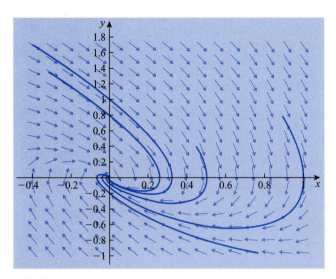

**FIGURE 5.16**

*Trajectories for the system $\dot{x} = y, \dot{y} = -2x - 2y, -0.3 \le t \le 4$*
*Initial conditions: $(x(0), y(0)) = (1, 0), (0.5, 0), (0, 0.8), (0, 1), (0.5, -0.8)$*

Note that in Figure 5.16 the trajectories are *spirals* moving *toward* the equilibrium solution, the origin. We say that the origin is a **spiral sink**. If we examine the general solution, we can see why the trajectories behave this way. First of all, there is no straight-line direction along which the trajectories approach the origin. In the preceding example, the expressions for both $x(t)$ and $y(t)$ have trigonometric terms that contribute oscillations, movements back and forth across the $x$-axis. But in addition, each entry of the general solution has a factor of $e^{-t}$, which *dampens* these oscillations for positive values of $t$. Thus, as $t$ increases in a positive direction, the amplitudes of these oscillations tend to 0. A look at Euler's formula explains the existence of this decaying exponential: *The real part, p, of the eigenvalue pair is negative.* Figure 5.15 shows a plot of $x$ against $t$ for the particular solution with $x(0) = 1$ and $y(0) = 0$.

The graph of $y$ against $t$ is similar. In terms of the spring-mass problems we analyzed in various examples of Section 4.8, we can interpret our problem as representing a system with *damped oscillations*. (See Example 4.8.2, especially Figure 4.19a.)

As we'll see in some of the exercises following this section, if the eigenvalues are $p \pm qi$ and $p > 0$, then we get spirals that wind *away* from $(0, 0)$ as $t$ increases. Here, we say that the origin is a *spiral source*. This corresponds to oscillatory solutions with increasing amplitudes and describes *resonance*. (See Example 4.8.4, especially Figure 4.22.)

The case where $p = 0$, so that we have *pure imaginary eigenvalues*, is interesting. Now the trajectories are *closed, nonintersecting curves that encircle the origin*. This corresponds to the situation in which we have *undamped oscillations*. (See Example 5.5.1 and Example 4.8.1, especially Figure 4.16.)

Table 5.1 Summary of Stability Criteria for Two-Dimensional Linear Systems

| Eigenvalues | Stability | References |
|---|---|---|
| **REAL** | | |
| *Unequal* | | |
| Both > 0 | Unstable node (source, repeller) | Examples 5.2.4 and 5.3.1 |
| Both < 0 | Stable node (sink, attractor) | Examples 5.2.2 and 5.3.2 |
| Different signs | Saddle point | Examples 5.2.1, 5.2.3, and 5.3.3 |
| One = 0, the other ≠ 0 | Whole line of equilibrium points | Example 5.3.4 and Problem B9 of Exercises 5.3 |
| *Equal* | | |
| Both > 0 | Unstable node (source, repeller) | Example 5.4.1 |
| Both < 0 | Stable node (sink, attractor) | Example 5.4.2 |
| Both = 0 | "Algebraically unstable" | Problem B7 of Exercises 5.4 |
| **COMPLEX** | | |
| Real part > 0 | Spiral source (unstable spiral, repeller) | Example 5.2.5 |
| Real part < 0 | Spiral sink (stable spiral) | Example 5.5.2 |
| Real part = 0 | Center (neutral center, stable center) | Example 5.5.1 |

Now let us stand back and summarize all these cases. Table 5.1 categorizes the stability of two-dimensional autonomous systems, referring to relevant examples or exercises.

## Exercises 5.5

### A

For each of the Systems 1–10, (a) find the eigenvalues and their corresponding eigenvectors and (b) sketch/plot a few trajectories and show the position(s) of the eigenvector(s) if they do not have complex entries.

1. $\dot{r} = -r - 2s, \dot{s} = 2r - s$
2. $\dot{x} = 3x - 2y, \dot{y} = 2x + 3y$
3. $\dot{x} = -0.5x - y, \dot{y} = x - 0.5y$
4. $\dot{x} = x + y, \dot{y} = -3x - y$
5. $\dot{x} = 2x + y, \dot{y} = -3x - y$
6. $\dot{x} = -0.5x - y, \dot{y} = x - 0.5y$
7. $\dot{x} = y - 7x, \dot{y} = -2x - 5y$
8. $\dot{x} = x - 3y, \dot{y} = 3x + y$

**9.** $\dot{x} = 6x - y, \dot{y} = 5x + 4y$

**10.** $\dot{x} + x + 5y = 0, \dot{y} - x - y = 0$

## B

**1.** Write systems of first-order linear equations whose trajectories show the following behaviors:

   **a.** $(0, 0)$ is a spiral source with eigenvalues $\lambda_1 = 2 + 2i$ and $\lambda_2 = 2 - 2i$.

   **b.** $(0, 0)$ is a stable center with eigenvalues $\lambda_1 = -3i$ and $\lambda_2 = 3i$.

   **c.** $(0, 0)$ is a spiral sink with eigenvalues $\lambda_1 = -1 + 2i$ and $\lambda_2 = -1 - 2i$.

**2.** Consider the system

$$\dot{x} = y$$
$$\dot{y} = -x - \beta y,$$

where $\beta$ is a parameter.

   **a.** By using technology to draw trajectories, examine the stability of the equilibrium solution for $\beta = -1, -0.1, 0, 0.1$, and $1$.

   **b.** Does there seem to be a *bifurcation point*—that is, a critical value of $\beta$ at which the stability changes its nature? (Read/review Section 2.7.)

   **c.** Find a formula for the eigenvalues of the system, showing their dependence on $\beta$.

   **d.** Relate the information found in part (c) to the stability summary in Table 5.1 and answer the question in part (b) with increased authority.

**3.** If $\lambda$ is a complex eigenvalue of matrix $A$, $V = U + iW$ is a corresponding eigenvector, and $X(t) = e^{\lambda t}V$, then we have seen that

$$X_1(t) = \text{Re}\{X(t)\} = e^{pt}\{(\cos qt)U - (\sin qt)W\}$$

$$X_2(t) = \text{Im}\{X(t)\} = e^{pt}\{(\cos qt)W + (\sin qt)U\}$$

are real-valued linearly independent solutions of the system $\dot{X} = AX$. Show that the same two solutions can be obtained by taking the real and imaginary parts of $e^{\bar{\lambda}t}\bar{V}$. (Thus, the second term of the familiar solution formula $c_1 e^{\lambda_1 t}V_1 + c_2 e^{\lambda_2 t}V_2 = c_1 e^{\lambda_1 t}V_1 + c_2 e^{\bar{\lambda}_1 t}\bar{V}_1$ is unnecessary.)

**4.** Show that if $X(t)$ is a complex-valued solution of the system $\dot{X} = AX$, then so is $X_{\text{I}} = \text{Im}(X) = \frac{X - \bar{X}}{2i}$, the imaginary part of $X(t)$.

**5.** The following two-loop electrical circuit illustration can be modeled by the system

$$\frac{di_1}{dt} = -\left(\frac{R_1 + R_2}{L}\right)i_1 + \frac{R_2}{L}i_2$$

$$\frac{di_2}{dt} = -\left(\frac{R_1 + R_2}{L}\right)i_1 + \left(\frac{R_2}{L} - \frac{1}{R_2 C}\right)i_2.$$

Using eigenvalues and eigenvectors, solve the initial value problem $i_1(0) = 1, i_2(0) = 0$, when $R_1 = R_2 = 1, L = 1$, and $C = 3$. (Use technology to find the eigenvectors.)

## C

1. In Section 5.3, we used the result that the eigenvectors corresponding to distinct eigenvalues are linearly independent. Use this result to show that the real and imaginary parts of complex eigenvectors are linearly independent.

2. The change in the amounts $x$ and $y$ of two substances that enter a certain chemical reaction can be described by the initial-value problem

$$\dot{x} = -3x + \alpha y, \dot{y} = \beta x - 2y; x(0) = y(0) = 1,$$

where $\alpha$ and $\beta$ are two parameters that depend on the conditions of reaction (temperature, humidity, etc.). Are there values of $\alpha$ and $\beta$ for which the solution of the initial-value problem is a periodic function of time?

## 5.6 NONHOMOGENEOUS SYSTEMS

### 5.6.1 The General Solution

The linear systems we have been dealing with so far are called **homogeneous** systems. Basically, this means that they can be expressed in the form $\dot{X} = AX$ with no "leftover" terms. If a linear system has to be written as $\dot{X} = AX + B(t)$, where $B(t)$ is a vector of the form $\begin{bmatrix} b_1(t) \\ b_2(t) \end{bmatrix}$, then we say that the system is **nonhomogeneous**. For example, in matrix terms, the system $\frac{dx}{dt} = x + \sin t, \frac{dy}{dt} = t - y$ must be written as $\begin{bmatrix} \dot{x}(t) \\ \dot{y}(t) \end{bmatrix} = \begin{bmatrix} 1 & 0 \\ 0 & -1 \end{bmatrix} \begin{bmatrix} x \\ y \end{bmatrix} + \begin{bmatrix} \sin t \\ t \end{bmatrix}$ and so is nonhomogeneous.

Don't confuse the distinction between *autonomous* and *nonautonomous* with that between *homogeneous* and *nonhomogeneous*. For example, if $b_1(t)$ and $b_2(t)$ are constant functions (not both zero), then we have a system that is both autonomous and nonhomogeneous. (See, for instance, Example 4.7.3.)

The techniques that were introduced in Section 4.2 for second-order nonhomogeneous equations generalize to systems, but the calculations are more complicated. To get a handle on solving a nonhomogeneous linear system, we need a fundamental fact about linear systems:

The general solution, $X_{\mathrm{GNH}}$, of a linear nonhomogeneous system is obtained by finding a *particular* solution, $X_{\mathrm{PNH}}$, of the *nonhomogeneous* system and adding it to the *general* solution, $X_{\mathrm{GH}}$, of the associated *homogeneous* system.

You should see this as an application of the Superposition Principle and as an extension of the result we saw for single linear differential equations (Section 4.2). Symbolically, we can write $X_{GNH} = X_{GH} + X_{PNH}$. Using the definitions of these terms, we can see that this sum of vectors is a solution of the nonhomogeneous system:

$$\dot{X}_{GNH} = \dot{X}_{GH} + \dot{X}_{PNH} = AX_{GH} + \{AX_{PNH} + B(t)\}$$
$$= A(X_{GH} + X_{PNH}) + B(t) = AX_{GNH} + B(t).$$

(*Be sure you follow this.*) You should see that $X_{GH}$, as a general solution, must contain two arbitrary constants, so the expression for $X_{GNH}$ contains two arbitrary constants.

Let's look at a simple example showing the structure of a nonhomogeneous system's solution.

### ■ Example 5.6.1 The Solution of a Nonhomogeneous System

The system

$$\dot{x} = x + y + 2e^{-t}$$
$$\dot{y} = 4x + y + 4e^{-t}$$

can be written in the form $\dot{X}(t) = \begin{bmatrix} 1 & 1 \\ 4 & 1 \end{bmatrix} X + \begin{bmatrix} 2e^{-t} \\ 4e^{-t} \end{bmatrix} = \begin{bmatrix} 1 & 1 \\ 4 & 1 \end{bmatrix} X + 2e^{-t} \begin{bmatrix} 1 \\ 2 \end{bmatrix}$. The system

has eigenvalues $\lambda_1 = 3$ and $\lambda_2 = -1$, with corresponding eigenvectors $V_1 = \begin{bmatrix} 1 \\ 2 \end{bmatrix}$ and $V_2 = \begin{bmatrix} 1 \\ -2 \end{bmatrix}$. (*Check this.*) Then the general solution of the associated homogeneous system $\dot{X}(t) = \begin{bmatrix} 1 & 1 \\ 4 & 1 \end{bmatrix} X$ is

$$X_{GH} = c_1 e^{3t} \begin{bmatrix} 1 \\ 2 \end{bmatrix} + c_2 e^{-t} \begin{bmatrix} 1 \\ -2 \end{bmatrix}.$$

You should verify that a particular solution of the original nonhomogeneous system is given by $X_{PNH} = e^{-t} \begin{bmatrix} 0 \\ -2 \end{bmatrix} = \begin{bmatrix} 0 \\ -2e^{-t} \end{bmatrix}$. Therefore, the general solution of the nonhomogeneous system is

$$X_{GNH} = X_{GH} + X_{PNH} = c_1 e^{3t} \begin{bmatrix} 1 \\ 2 \end{bmatrix} + c_2 e^{-t} \begin{bmatrix} 1 \\ -2 \end{bmatrix} + \begin{bmatrix} 0 \\ -2e^{-t} \end{bmatrix}$$

$$= \begin{bmatrix} c_1 e^{3t} + c_2 e^{-t} \\ 2c_1 e^{3t} - 2c_2 e^{-t} - 2e^{-t} \end{bmatrix}.$$

Check that this is the general solution of the original nonhomogeneous system. ■

## 5.6.2 The Method of Undetermined Coefficients

The challenge in working with a nonhomogeneous system is to find a particular solution of the nonhomogeneous system. There are various techniques for finding a particular solution. We can use the *variation of parameters* technique of Section 4.4, but for systems the calculations involved are very tedious. Therefore, we'll restrict our attention to the method of undetermined coefficients (Section 4.3), which is not so powerful but is easier to use. As we've seen in Section 4.3 and Section 4.4, this method requires intelligent guessing. We have to ask ourselves what terms are contained in $B(t)$ but not in $X_{GH}$—and then guess at the form of $X_{PNH}$ on the basis of this information.

We should note that this method of undetermined coefficients can be used only when the vector $B(t) = \begin{bmatrix} b_1(t) \\ b_2(t) \end{bmatrix}$ contains terms that are constants, exponential functions, sines, cosines, polynomials, or any sum or product of such terms. For other kinds of functions making up $B(t)$, $X_{PNH}$ must be found using some other technique (for example, variation of parameters).

The next example illustrates the method with its resulting algebraic complexities.

### ■ Example 5.6.2 Using the Method of Undetermined Coefficients

Let's consider the system $\frac{dx}{dt} = x + \sin t$, $\frac{dy}{dt} = t - y$ that we discussed at the beginning of this section. We have $\dot{X} = AX + B(t)$, where $A = \begin{bmatrix} 1 & 0 \\ 0 & -1 \end{bmatrix}$ and $B(t) = \begin{bmatrix} \sin t \\ t \end{bmatrix} = \sin t \begin{bmatrix} 1 \\ 0 \end{bmatrix} + t \begin{bmatrix} 0 \\ 1 \end{bmatrix}$.

The eigenvalues of $A$ are 1 and $-1$, with corresponding eigenvectors

$$\begin{bmatrix} 1 \\ 0 \end{bmatrix} \text{ and } \begin{bmatrix} 0 \\ 1 \end{bmatrix}.$$

Thus, the general solution of the homogeneous system can be written as

$$X_{GH} = c_1 e^t \begin{bmatrix} 1 \\ 0 \end{bmatrix} + c_2 e^{-t} \begin{bmatrix} 0 \\ 1 \end{bmatrix}.$$

(*Verify the statements in this paragraph for yourself.*)

Now we look for a particular solution of the original nonhomogeneous equation. First, we compare the terms of $B(t)$ with the terms of $X_{GH}$ to see whether there is any duplication. In this case, we see that the terms $\sin t$ and $t$ are not terms that can be obtained just from $X_{GH}$. Because our system is equivalent to a single second-order differential equation, we realize that we must find a function that can combine with its own first and second derivatives to yield $B(t)$. We take a guess that $X_{PNH}$ must look like $C \sin t + D \cos t + Et + F$, where $C, D, E,$ and $F$ are *vectors of constants*. Our trial solution for $X_{PNH}$ consists of a linear combination of the functions $\sin t$ and $t$ and their derivatives—a linear combination with *undetermined coefficients*.

Let's substitute our guess into the nonhomogeneous system:

$$
\overbrace{C\cos t - D\sin t + E}^{\dot{X}_{\text{PNH}}} = A\left(\overbrace{C\sin t + D\cos t + Et + F}^{X_{\text{PNH}}}\right) + \overbrace{\sin t\begin{bmatrix}1\\0\end{bmatrix} + t\begin{bmatrix}0\\1\end{bmatrix}}^{B(t)}
$$

$$
= AC\sin t + AD\cos t + AEt + AF + \sin t\begin{bmatrix}1\\0\end{bmatrix} + t\begin{bmatrix}0\\1\end{bmatrix}.
$$

When we collect like terms, matching the coefficients of functions on each side, we get the following system:

(1) $C = AD$          [The coefficients of $\cos t$ must be equal.]

(2) $-D = AC + \begin{bmatrix}1\\0\end{bmatrix}$    [The coefficients of $\sin t$ must be equal.]

(3) $0 = AE + \begin{bmatrix}0\\1\end{bmatrix}$    [The coefficients of $t$ must be equal.]

(4) $E = AF$          [The constant terms must be equal.]

Remembering that $A = \begin{bmatrix}1 & 0\\0 & -1\end{bmatrix}$, we can solve Equation (3) for $E$:

$$
AE = -\begin{bmatrix}0\\1\end{bmatrix} = \begin{bmatrix}0\\-1\end{bmatrix}, \text{ or } \begin{bmatrix}1 & 0\\0 & -1\end{bmatrix}\begin{bmatrix}e_1\\e_2\end{bmatrix} = \begin{bmatrix}0\\-1\end{bmatrix},
$$

so $e_1 = 0$ and $e_2 = 1$. (*Check this.*) Now that we know $E$, we can use Equation (4) to find $F$:

$$
\begin{bmatrix}0\\1\end{bmatrix} = \begin{bmatrix}1 & 0\\0 & -1\end{bmatrix}\begin{bmatrix}f_1\\f_2\end{bmatrix},
$$

so $f_1 = 0$ and $f_2 = -1$.

If we multiply both sides of (1) by $A$, we get $AC = A^2D = D$ (because $A^2 = I$, the $2 \times 2$ identity matrix), which we can substitute into Equation (2): $-D = D + \begin{bmatrix}1\\0\end{bmatrix}$, or $\begin{bmatrix}-1\\0\end{bmatrix} = 2D = \begin{bmatrix}2d_1\\2d_2\end{bmatrix}$, so $d_1 = -\frac{1}{2}$ and $d_2 = 0$. (*Make sure you follow all this.*) Finally, we solve (1) for $C$:

$\begin{bmatrix}c_1\\c_2\end{bmatrix} = \begin{bmatrix}1 & 0\\0 & -1\end{bmatrix}\begin{bmatrix}-1/2\\0\end{bmatrix}$, so $c_1 = -\frac{1}{2}$ and $c_2 = 0$.

We have determined all the coefficients. Putting the pieces together, we have

$$
X_{\text{PNH}} = \begin{bmatrix}-1/2\\0\end{bmatrix}\sin t + \begin{bmatrix}-1/2\\0\end{bmatrix}\cos t + \begin{bmatrix}0\\1\end{bmatrix}t + \begin{bmatrix}0\\-1\end{bmatrix} = \begin{bmatrix}-\frac{1}{2}(\sin t + \cos t)\\t - 1\end{bmatrix},
$$

and we finally obtain

$$X_{GNH} = X_{GH} + X_{PNH}$$

$$= c_1 e^t \begin{bmatrix} 1 \\ 0 \end{bmatrix} + c_2 e^{-t} \begin{bmatrix} 0 \\ 1 \end{bmatrix} + \begin{bmatrix} -\frac{1}{2}(\sin t + \cos t) \\ t - 1 \end{bmatrix} = \begin{bmatrix} c_1 e^t - \frac{1}{2}(\sin t + \cos t) \\ t - 1 + c_2 e^{-t} \end{bmatrix}$$

as the general solution of the original nonhomogeneous equation.

Note that the system in this example is *uncoupled*—that is, each equation contains only one unknown function. Problem B1 in Exercises 5.6 asks you to solve each equation separately to obtain the same answer as the one shown here. ■

Practice in the technique of undetermined coefficients leads to a more systematic way of guessing a possible solution of the nonhomogeneous system. The second column of Table 5.2 indicates the component of $X_{PNH}$ that corresponds to the matching component $b_i(t)$ of $B(t)$. If $b_i(t)$ is a sum of different functions, then it is a consequence of the Superposition Principle that the matching component of $X_{PNH}$ is a sum of trial solutions.

There is an exception to the neatness of the table. If $b_i(t)$ contains terms that duplicate any corresponding parts of $X_{GH}$, then each corresponding trial term must be multiplied by $t^m$, where $m$ is the smallest positive integer that eliminates the duplication.

In Example 5.6.2, we had $b_1(t) = \sin t$ and $b_2(t) = t$—a trigonometric function ($a \cos rt + b \sin rt$, with $a = 0$, $r = 1$, and $b = 1$) and a first-degree polynomial ($P_n(t) = a_n t^n + a_{n-1} t^{n-1} + \cdots + a_1 t + a_0$, where $n = 1$, $a_1 = 1$, and $a_0 = 0$). There was no duplication between $X_{GH}$ and $B(t)$ because the terms making up $X_{GH}$ are exponential functions. Consequently, our educated guess for $X_{PNH}$ consisted of a linear combination of sine and cosine plus a first-degree polynomial.

Let's use the instant wisdom conferred by Table 5.2 to solve the next problem.

**Table 5.2** Trial Particular Solutions for Nonhomogeneous Systems

| $b_i(t)$ | Form of Trial Solution |
|---|---|
| $c \neq 0$, a constant | $K$, a constant |
| $P_n(t) = a_n t^n + a_{n-1} t^{n-1} + \cdots + a_1 t + a_0$ | $Q_n(t) = c_n t^n + c_{n-1} t^{n-1} + \cdots + c_1 t + c_0$ |
| $c e^{at}$ | $K e^{at}$ |
| $a \cos rt + b \sin rt$ | $\alpha \cos rt + \beta \sin rt$ |
| $e^{Rt}(a \cos rt + b \sin rt)$ | $e^{Rt}(\alpha \cos rt + \beta \sin rt)$ |
| $P_n(t) e^{at}$ | $Q_n(t) e^{at}$ |

## ■ Example 5.6.3 Undetermined Coefficients

Suppose we try to solve the system $\frac{dx}{dt} = y$, $\frac{dy}{dt} = 3y - 2x + 2t^2 + 3e^{2t}$. We can write this system as $\dot{X} = AX + B(t)$, where $A = \begin{bmatrix} 0 & 1 \\ -2 & 3 \end{bmatrix}$ and $B(t) = \begin{bmatrix} 0 \\ 2t^2 + 3e^{2t} \end{bmatrix} = (2t^2 + 3e^{2t}) \begin{bmatrix} 0 \\ 1 \end{bmatrix}$. The eigenvalues of $A$ are 1 and 2, with corresponding eigenvectors $\begin{bmatrix} 1 \\ 1 \end{bmatrix}$ and $\begin{bmatrix} 1 \\ 2 \end{bmatrix}$. (*Verify this.*)

We know that the general solution of the homogeneous system is given by

$$X_{GH} = c_1 e^t \begin{bmatrix} 1 \\ 1 \end{bmatrix} + c_2 e^{2t} \begin{bmatrix} 1 \\ 2 \end{bmatrix}.$$

To find a particular solution of the nonhomogeneous system, we compare the terms of $B(t)$ with the terms of $X_{GH}$ to see whether there is any duplication. In this example, ignoring constants, we see that $e^{2t}$ appears in both $X_{GH}$ and $B(t)$. We also recognize that the term $t^2$ in $B(t)$ is *not* found in $X_{GH}$. Using Table 5.2 and the description of how to handle duplicate terms, we guess that $X_{PNH}$ must look like

$$Ct^2 + Dt + E + Fe^{2t} + Gte^{2t},$$

where $C, D, E, F$, and $G$ are vectors of constants. Note that because there is a second-degree term, our trial particular solution contains a full quadratic polynomial and multiplying $e^{2t}$ by $t$ eliminates the duplication.

If we substitute this guess into the nonhomogeneous system, we get

$$2Ct + D + 2Fe^{2t} + Ge^{2t} + 2Gte^{2t} = A\left(Ct^2 + Dt + E + Fe^{2t} + Gte^{2t}\right) + (2t^2 + 3e^{2t}) \begin{bmatrix} 0 \\ 1 \end{bmatrix}$$

$$= ACt^2 + ADt + AE + AFe^{2t} + AGte^{2t} + (2t^2 + 3e^{2t}) \begin{bmatrix} 0 \\ 1 \end{bmatrix}$$

$$= \left(AC + \begin{bmatrix} 0 \\ 2 \end{bmatrix}\right) t^2 + ADt + AE + \left(AF + \begin{bmatrix} 0 \\ 3 \end{bmatrix}\right) e^{2t} + AGte^{2t}.$$

Matching the coefficients of like terms on each side, we get the system

(1) $0 = AC + \begin{bmatrix} 0 \\ 2 \end{bmatrix}$      [The coefficients of $t^2$ must be equal.]

(2) $2C = AD$      [The coefficients of $t$ must be equal.]

(3) $D = AE$      [The constant terms must be equal.]

(4) $2F + G = AF + \begin{bmatrix} 0 \\ 3 \end{bmatrix}$      [The coefficients of $e^{2t}$ must be equal.]

(5) $2G = AG$      [The coefficients of $te^{2t}$ must be equal.]

Working through these equations (see Problem B3 in Exercises 5.6), we find that

$$C = \begin{bmatrix} 1 \\ 0 \end{bmatrix}, D = \begin{bmatrix} 3 \\ 2 \end{bmatrix}, E = \begin{bmatrix} \frac{7}{2} \\ 3 \end{bmatrix}, F = \begin{bmatrix} 0 \\ 3 \end{bmatrix}, \text{ and } G = \begin{bmatrix} 3 \\ 6 \end{bmatrix}.$$

Now that we've determined the coefficients $C, D, E, F,$ and $G$, we can construct the particular solution of the nonhomogeneous equation.

$$X_{\text{PNH}} = \begin{bmatrix} 1 \\ 0 \end{bmatrix} t^2 + \begin{bmatrix} 3 \\ 2 \end{bmatrix} t + \begin{bmatrix} \frac{7}{2} \\ 3 \end{bmatrix} + \begin{bmatrix} 0 \\ 3 \end{bmatrix} e^{2t} + \begin{bmatrix} 3 \\ 6 \end{bmatrix} te^{2t}.$$

Finally, we get the general solution of the nonhomogeneous equation:

$$X_{\text{GNH}} = X_{\text{GH}} + X_{\text{PNH}}$$

$$= c_1 e^t \begin{bmatrix} 1 \\ 1 \end{bmatrix} + c_2 e^{2t} \begin{bmatrix} 1 \\ 2 \end{bmatrix} + \begin{bmatrix} 1 \\ 0 \end{bmatrix} t^2 + \begin{bmatrix} 3 \\ 2 \end{bmatrix} t + \begin{bmatrix} \frac{7}{2} \\ 3 \end{bmatrix} + \begin{bmatrix} 0 \\ 3 \end{bmatrix} e^{2t} + \begin{bmatrix} 3 \\ 6 \end{bmatrix} te^{2t}$$

$$= \begin{bmatrix} c_1 e^t + c_2 e^{2t} + t^2 + 3t + \frac{7}{2} + 3te^{2t} \\ c_1 e^t + 2c_2 e^{2t} + 2t + 3 + 3e^{2t} + 6te^{2t} \end{bmatrix}$$

$$= \begin{bmatrix} c_1 e^t + (c_2 + 3t) e^{2t} + t^2 + 3t + \frac{7}{2} \\ c_1 e^t + (2c_2 + 3 + 6t) e^{2t} + 2t + 3 \end{bmatrix}.$$

Of course, this means that $x(t) = c_1 e^t + (c_2 + 3t)e^{2t} + t^2 + 3t + 7/2$ and $y(t) = c_1 e^t + (2c_2 + 3 + 6t)e^{2t} + 2t + 3$ are the solutions of our system. You should check to see that these functions satisfy our original system. ∎

When the nonhomogeneous system is also *autonomous*—that is, it has the form $\dot{X} = AX + B(t)$, where the entries of $B(t)$ are *constants*—we can analyze the stability of the system's solutions by finding the equilibrium point(s) (no longer the origin) and considering the eigenvalues and eigenvectors of the matrix $A$.

## ■ Example 5.6.4 Stability of an Autonomous Nonhomogeneous System

We return to the system of Example 4.7.3:

$$\dot{x} = 7y - 4x - 13$$
$$\dot{y} = 2x - 5y + 11.$$

To find the equilibrium point(s), we solve the algebraic system

$$-4x + 7y = 13$$
$$2x - 5y = -11$$

to find that $(2, 3)$ is the only equilibrium point. (The details in this example are left as parts of Problem B5 in Exercises 5.6.)

We can write our system of differential equations in the form

$$\dot{X} = AX + B(t) = \begin{bmatrix} -4 & 7 \\ 2 & -5 \end{bmatrix} \begin{bmatrix} x \\ y \end{bmatrix} + \begin{bmatrix} -13 \\ 11 \end{bmatrix}.$$

Because the eigenvalues of $A$ are $\lambda_1 = \left(\sqrt{57} - 9\right)/2$ and $\lambda_2 = -\left(\sqrt{57} + 9\right)/2$, both of which are negative real numbers, Table 5.1 at the end of Section 5.5 tells us that the equilibrium point $(2, 3)$ is a *sink*. (Go back to take another look at Figure 4.9.) ■

Despite its limitations, the method of undetermined coefficients is very useful. In Chapter 6, we'll see another way of solving systems of nonhomogeneous linear equations, by means of the *Laplace transform*. This transform method is particularly useful in solving initial-value problems.

## Exercises 5.6

### A

1. Find the particular solution of the system in Example 5.6.2 that satisfies $x(0) = 0, y(0) = 1$.
2. Find the particular solution of the system in Example 5.6.3 that satisfies $x(0) = -1, y(0) = 2$.

Without using technology, find the general solution of each of the systems in Problems 3–15. You may check your answers using a CAS.

3. $\dot{x} = y + 2e^t, \dot{y} = x + t^2$
4. $\dot{x} = y - 5\cos t, \dot{y} = 2x + y$
5. $\dot{x} = 3x + 2y + 4e^{5t}, \dot{y} = x + 2y$
6. $\dot{x} = 3x - 4y + e^{-2t}, \dot{y} = x - 2y - 3e^{-2t}$
7. $\dot{x} = 4x + y - e^{2t}, \dot{y} = y - 2x$ [*Hint*: Multiples of both $e^{2t}$ and $te^{2t}$ should appear in your guess for $X_{PNH}$.]
8. $\dot{x} = 2y - x + 1, \dot{y} = 3y - 2x$ [*Hint*: Multiples of both $e^t$ and $te^t$ should appear in your guess for $X_{PNH}$.]
9. $\dot{x} = 5x - 3y + 2e^{3t}, \dot{y} = x + y + 5e^{-t}$
10. $\dot{x} = x + y + 1 + e^t, \dot{y} = 3x - y$
11. $\dot{x} = 2x - y, \dot{y} = 2y - x - 5e^t \sin t$
12. $\dot{x} = x + 2y, \dot{y} = x - 5\sin t$
13. $\dot{x} = y, \dot{y} = -2x - 3y + \sin t + e^t$
14. $\dot{x} = -2x + y + 2e^{-t}, \dot{y} = x - 2y + 3t$
15. $\dot{x} = x + y + e^t, \dot{y} = y + e^{-t}$

**B**

1. Consider each equation in Example 5.6.2 as a first-order linear equation and solve each equation separately, confirming that you get the same answer as in the worked-out example. (You may have to review Section 2.2 and the technique of integration by parts.)

2. **a.** Use technology to draw the phase portrait for the system in Example 5.6.2.
   **b.** Draw a graph of $x(t)$ vs. $t$.
   **c.** Draw a graph of $y(t)$ vs. $t$.

3. Assume that you have Equations (1)–(5) in Example 5.6.3. Let $C = \begin{bmatrix} c_1 \\ c_2 \end{bmatrix}$, $D = \begin{bmatrix} d_1 \\ d_2 \end{bmatrix}$, etc. Solve for the vectors $C, D, E, F$, and $G$ in the following order:

   **a.** Use Equation (1) to show that $C = \begin{bmatrix} 1 \\ 0 \end{bmatrix}$.

   **b.** Use Equation (2) to show that $D = \begin{bmatrix} 3 \\ 2 \end{bmatrix}$.

   **c.** Use Equation (3) to show that $E = \begin{bmatrix} 7/2 \\ 3 \end{bmatrix}$.

   **d.** Assuming that $G$ is not the zero vector, use Equation (5) to derive a general form for $G$. (There is an arbitrary constant involved.)

   **e.** Substitute the general form for $G$ found in part (d) into Equation (4) to determine the concrete form of $G$. Then use this information to see that a convenient form for $F$ is $\begin{bmatrix} 0 \\ 3 \end{bmatrix}$.

4. **a.** Use technology to draw the phase portrait for the system in Example 5.6.3.
   **b.** Draw the graph of $x(t)$ vs. $t$, assuming that $x(0) = 50$.
   **c.** Draw the graph of $y(t)$ vs. $t$, assuming that $y(0) = 100$.

5. Look at the system in Example 5.6.4.

   **a.** Show that the only equilibrium point is $(2, 3)$.
   **b.** Show that the eigenvalues of the matrix of coefficients $A$ are

   $$\lambda_1 = \left(\sqrt{57} - 9\right)/2 \quad \text{and} \quad \lambda_2 = -\left(\sqrt{57} + 9\right)/2.$$

   **c.** Find eigenvectors corresponding to $\lambda_1$ and $\lambda_2$.
   **d.** Express the general solution of the homogeneous system in terms of the eigenvalues and eigenvectors found in parts (b) and (c).
   **e.** Find a particular solution of the nonhomogeneous system.
   **f.** Put the answers to parts (d) and (e) together to get the general solution of the nonhomogeneous system. Then determine what happens as $t \to \infty$.

6. Newton's laws of motion give the following system as a model for the motion of an object falling under the influence of gravity:

$$\frac{dy}{dt} = v(t)$$

$$\frac{dv}{dt} = g - cv(t); y(0) = 0, v(0) = 0$$

for $0 \le t \le T$, where $y(T) = H$. Here, $y(t)$ denotes the downward distance from the spot where the object was dropped to the place where the falling object is at time $t$; $v(t)$ is the velocity; $g$ is the gravitational constant; and $c$ is the *drag coefficient*, representing air resistance.

**a.** Without using technology, solve this nonhomogeneous system for $y(t)$ and $v(t)$.

**b.** Find $\lim\limits_{t \to \infty} v(t)$ and interpret your answer in physical terms.

**7.** A cold medication moving through the body can be modeled[4] by the IVP

$$\dot{x} = -k_1 x + I$$

$$\dot{y} = k_1 x - k_2 y; \quad x(0) = 0, y(0) = 0,$$

where $x(t)$ and $y(t)$ are the amounts of medication in the GI tract and the bloodstream, respectively, at time $t$ measured in hours elapsed since the initial dosage. Here, $I > 0$ is the constant dosage rate and $k_1, k_2$ are positive transfer rates (out of the GI tract and bloodstream, respectively).

**a.** Without using technology, solve the nonhomogeneous system for $x(t)$ and $y(t)$.

**b.** Find $\lim\limits_{t \to \infty} x(t)$ and $\lim\limits_{t \to \infty} y(t)$.

**c.** Assume that the decongestant part of a continuous acting capsule (such as Contac®) has $k_1 = 1.386/\text{hr}$ and $k_2 = 0.1386/\text{hr}$ and that the antihistamine portion has $k_1 = 0.6931/\text{hr}$ and $k_2 = 0.0231/\text{hr}$. Also assume that $I = 1/6$ (i.e., one unit per six hours). Use technology to graph $x(t)$ against $t$ and $y(t)$ against $t$ for the decongestant on the same set of axes.

**d.** Assuming the data given in part (c), use technology to graph $x(t)$ and $y(t)$ for the antihistamine on the same set of axes.

**8.** The buying behavior of the public toward a particular product can be modeled by

$$\frac{dB}{dt} = b(M - \beta B)$$

$$\frac{dM}{dt} = a(B - \alpha M) + cA,$$

where $B = B(t)$ is the level of buying, $M = M(t)$ is a measure of the public's motivation or attitude toward the product, and $A = A(t)$ is the advertising policy. The parameters $a, b, c, \alpha$, and $\beta$ are all assumed positive.

**a.** Show that for constant advertising (i. e., $A(t)$ is a constant function), the buying levels tend to a limiting value over time.

**b.** If $\alpha = \beta = 2, a = b = c = 1, B(0) = M(0) = 0$, and

$$A(t) = \begin{cases} 100 \, \text{units} & \text{for } 0 < t < 10 \\ 0 & \text{for } t > 10 \end{cases}$$

determine the complete forecast for the buying behavior—that is, find $B(t)$.

---

[4] This model is based on the work of Edward Spitznagel of Washington University and was first communicated to me by Courtney Coleman, Harvey Mudd College.

**9.** A political race between two candidates can be modeled by the system

$$\frac{dx}{dt} = ax - by + e$$

$$\frac{dy}{dt} = -cx + dy - e,$$

where $x(t)$ and $y(t)$ represent the number of supporters of the two candidates and $e > 0$ denotes a gain in supporters for the first candidate based on contact between those committed to the first candidate and those who are uncommitted. The parameters $a, b, c,$ and $d$ are positive.

**a.** Determine conditions involving $a$, $b$, and $e$ guaranteeing that $dx/dt > 0$.
**b.** Determine conditions involving $c$, $d$, and $e$ guaranteeing that $dy/dt > 0$.
**c.** Determine conditions guaranteeing equilibrium in the system.

**10.** Consider a closed, two-compartment model in which the initial concentrations of a dye are 2 mg/liter in Compartment 1 and 10 mg/liter in Compartment 2. The compartments have constant volumes of 10 and 20 liters, respectively, and are separated by a permeable membrane that allows transfer between the compartments at the rate of 0.25/hour.

**a.** Determine formulas for the concentrations of dye at any time $t$ in each compartment.
**b.** Determine what happens to the concentrations in each compartment as $t \to \infty$.

## C

**1.** During World War I, the English scientist F. W. Lanchester (1868–1946) devised several mathematical models for the new art of aerial combat. These models have since been extended and applied to various modern conflicts. One model, describing the interaction of two conventional armies (as opposed to guerrilla forces or a mixture of conventional and guerrilla forces), is given by

$$\frac{dx}{dt} = -ay + f(t) - c$$

$$\frac{dy}{dt} = -bx + g(t) - d; \ x(0) = \alpha, y(0) = \beta,$$

where $x(t)$ and $y(t)$ represent the strengths of the opposing forces at time $t$; $a$ and $b$ denote nonnegative loss rates; $c$ and $d$ are constant noncombat losses per day; and $f(t)$ and $g(t)$ denote reinforcement rates in number of combatants per day.

**a.** Assuming that $f(t) = k$ and $g(t) = l$ ($k$ and $l$ are constants) during a battle, determine the strengths of each army at time $t$ during the battle.
**b.** If $\alpha > \frac{l-d}{b} > 0$ and $\beta > \frac{k-c}{a} > 0$, determine the conditions under which the $y$-force will be wiped out.
**c.** Assume that $a = 0.006, b = 0.008, c = d = 1000, k = 6000, l = 4000, \alpha = 90{,}000$, and $\beta = 200{,}000$, where $c, d, k,$ and $l$ are measured in men per day. Use technology to graph $x(t)$ and $y(t)$ for $0 \le t \le 50$. Then use the graphs to determine the time $t^*$ when $x(t^*) = y(t^*)$. Which side is winning after 50 days?

**2.** A two-compartment model for cholesterol flow yields the following nonhomogeneous system:

$$\frac{dx}{dt} = -(\alpha + \beta)x + \gamma y + K_1 + K_2$$

$$\frac{dy}{dt} = \beta x - \gamma y + K_3,$$

where $x$ and $y$ denote the amounts of cholesterol in the two compartments, $\beta$ and $\gamma$ represent rates at which cholesterol moves from one compartment to the other, $\alpha$ represents a rate of excretion, and $K_1, K_2, K_3$ denote the rates at which cholesterol flows into the compartments.

**a.** Solve the system, using technology for lengthy algebraic calculations.

**b.** Describe the behavior of the solutions $x(t)$ and $y(t)$ over a long period of time.

## 5.7 GENERALIZATIONS: THE $n \times n$ CASE ($n \geq 3$)

### 5.7.1 Matrix Representation

We are going to extend our previous analysis of systems, first to $3 \times 3$ systems and then to $n$th-order linear systems. We can use matrix notation to represent a third-order system with constant coefficients

$$\dot{x}_1 = a_{11}x_1 + a_{12}x_2 + a_{13}x_3$$

$$\dot{x}_2 = a_{21}x_1 + a_{22}x_2 + a_{23}x_3$$

$$\dot{x}_3 = a_{31}x_1 + a_{32}x_2 + a_{33}x_3$$

symbolically, in the form $\dot{X} = AX$, where

$$X = \begin{bmatrix} x_1 \\ x_2 \\ x_3 \end{bmatrix}, \dot{X} = \begin{bmatrix} \dot{x}_1 \\ \dot{x}_2 \\ \dot{x}_3 \end{bmatrix},$$

and

$$A = \begin{bmatrix} a_{11} & a_{12} & a_{13} \\ a_{21} & a_{22} & a_{23} \\ a_{31} & a_{32} & a_{33} \end{bmatrix}.$$

### ■ Example 5.7.1 Matrix Representation of a $3 \times 3$ System

The system $\dot{x} = -2x + 4y - z, \dot{y} = 5x - y + 3z, \dot{z} = x + z$ can be written first in the usual vertical way

$$\begin{array}{ll} \dot{x} = -2x + 4y - z & \dot{x} = -2x + 4y - z \\ \dot{y} = 5x - y + 3z & \text{or} \quad \dot{y} = 5x - y + 3z \\ \dot{z} = x + z & \dot{z} = x + 0y + z \end{array}$$

and then more compactly as

$$\begin{bmatrix} \dot{x} \\ \dot{y} \\ \dot{z} \end{bmatrix} = \begin{bmatrix} -2 & 4 & -1 \\ 5 & -1 & 3 \\ 1 & 0 & 1 \end{bmatrix} \begin{bmatrix} x \\ y \\ z \end{bmatrix}.$$

■

## 5.7.2 Eigenvalues and Eigenvectors

It is important to understand that the concepts of eigenvalue and eigenvector are valid for any system of $n$ equations in $n$ unknowns ($n \geq 2$). Specifically, given a system $\dot{X} = AX$, where $X$ is a nonzero $3 \times 1$ column matrix (vector) and $A$ is a $3 \times 3$ matrix, then an *eigenvalue* $\lambda$ is a solution of the equation $AX = \lambda X$. Given an eigenvalue $\lambda$, an *eigenvector* associated with $\lambda$ is a nonzero vector $V$ that satisfies the equation $AV = \lambda V$.

The equation $AX = \lambda X$ can be expressed as $AX - \lambda X = 0$, where $0$ denotes the $3 \times 1$ vector consisting entirely of zeros. This matrix equation is equivalent to the homogeneous algebraic system

$$(a_{11} - \lambda) x_1 + a_{12} x_2 + a_{13} x_3 = 0$$
$$a_{21} x_1 + (a_{22} - \lambda) x_2 + a_{23} x_3 = 0$$
$$a_{31} x_1 + a_{32} x_2 + (a_{33} - \lambda) x_3 = 0,$$

or

$$\begin{bmatrix} a_{11} - \lambda & a_{12} & a_{13} \\ a_{21} & a_{22} - \lambda & a_{23} \\ a_{31} & a_{32} & a_{33} - \lambda \end{bmatrix} \begin{bmatrix} x_1 \\ x_2 \\ x_3 \end{bmatrix} = \begin{bmatrix} 0 \\ 0 \\ 0 \end{bmatrix}. \qquad (5.7.1)$$

Now the matrix of coefficients in (5.7.1) can be expressed as

$$\begin{bmatrix} a_{11} & a_{12} & a_{13} \\ a_{21} & a_{22} & a_{23} \\ a_{31} & a_{32} & a_{33} \end{bmatrix} - \begin{bmatrix} \lambda & 0 & 0 \\ 0 & \lambda & 0 \\ 0 & 0 & \lambda \end{bmatrix} = \begin{bmatrix} a_{11} & a_{12} & a_{13} \\ a_{21} & a_{22} & a_{23} \\ a_{31} & a_{32} & a_{33} \end{bmatrix} - \lambda \begin{bmatrix} 1 & 0 & 0 \\ 0 & 1 & 0 \\ 0 & 0 & 1 \end{bmatrix} = A - \lambda I,$$

so the equation $AX - \lambda X = 0$ can be written as $(A - \lambda I)X = 0$, where $I$ is the $3 \times 3$ *identity matrix* consisting of ones down the main diagonal and zeros elsewhere. (The matrix $I$ is such that $IX = X$ for any $3 \times 1$ vector $X$. See Section 5.1 for the $2 \times 2$ case.)

The equation $(A - \lambda I)X = 0$ represents a homogeneous algebraic system of three linear equations in three unknowns, and the theory of linear algebra indicates that there is a number $\Delta$ depending on the matrix of coefficients with the following important property:

The System (5.7.1) has only the zero solution $x_1 = x_2 = x_3 = 0$ if $\Delta \neq 0$. However, if $\Delta = 0$, then there is a solution $x_1, x_2, x_3$ with at least one of the $x_i (i = 1, 2, 3)$ different from zero.

This number $\Delta$ is the *determinant* of the matrix of coefficients in (5.7.1), denoted by $\det(A - \lambda I)$, and it is the extension to three dimensions of the determinant introduced in Section 5.2. (See Problem B2 of Exercises 5.1 and Problem B1 of Exercises 5.2 for the significance of the $2 \times 2$ determinant in the solution of a system of equations.) Therefore, $(A - \lambda I)X = \mathbf{0}$ has a nonzero solution $X$ only if $\det(A - \lambda I) = 0$. An important fact is that $\det(A - \lambda I)$ is a third-degree polynomial in $\lambda$, called the **characteristic polynomial** of $A$, so the *eigenvalues of $A$ are the roots of the characteristic equation* $\det(A - \lambda I) = 0$. There are algorithms for calculating determinants of $3 \times 3$ systems, but they are tedious and any graphing calculator or CAS can evaluate them. In particular, a CAS will provide characteristic polynomials, eigenvalues, and corresponding eigenvectors. Also, there are formulas for solving cubic equations, but these methods are more complicated than the quadratic formula, and it is advisable to use your calculator or computer to solve such equations.

Let's use technology in the next example to calculate determinants, eigenvalues, and eigenvectors for a three-dimensional system.

## ■ Example 5.7.2 Eigenvalues and Eigenvectors via a CAS

Let's look at the matrix of coefficients in Example 5.7.1:

$$A = \begin{bmatrix} -2 & 4 & -1 \\ 5 & -1 & 3 \\ 1 & 0 & 1 \end{bmatrix}.$$

A computer algebra system provides the information that $\det(A) = -7$, the characteristic equation is $\lambda^3 + 2\lambda^2 - 20\lambda + 7 = 0$, and the eigenvalues (rounded to four decimal places) are $\lambda_1 = 3.3485, \lambda_2 = -5.7143$, and $\lambda_3 = 0.3658$. The corresponding representative eigenvectors are

$$V_1 = \begin{bmatrix} 2.3485 \\ 3.3903 \\ 1 \end{bmatrix}, V_2 = \begin{bmatrix} -6.7143 \\ 6.4848 \\ 1 \end{bmatrix}, \text{ and } V_3 = \begin{bmatrix} -0.6342 \\ -0.1251 \\ 1 \end{bmatrix}.$$

Don't be concerned if your CAS or calculator gives you eigenvectors that are different from these. You should check to see that each eigenvector you find is a constant multiple of one of the vectors $V_1, V_2$, and $V_3$ given here.   ■

## 5.7.3 Linear Independence and Linear Dependence

At this point you should be asking yourself, "What do these eigenvalues and eigenvectors tell me about the system?" Just as in the $2 \times 2$ case, we can write the general solution of a $3 \times 3$ system in terms of the eigenvalues and eigenvectors of the matrix of coefficients. To see what's going on, we'll need a few concepts that we have already seen in the $2 \times 2$ case. For example, given a number of vectors $v_1, v_2, \ldots, v_k$, a **linear combination** of these vectors is a vector that has the form $a_1 v_1 + a_2 v_2 + \cdots + a_k v_k$ for some choice of scalars $a_1, a_2, \ldots, a_k$. The collection of vectors is called **linearly independent** if the only way you can have $a_1 v_1 + a_2 v_2 + \cdots + a_k v_k = \mathbf{0}$ (the zero vector) is to have $a_1 = a_2 = \cdots = a_k = 0$. If you *could* find scalars $a_i$, not all zero, so that a linear combination of the vectors $v_i$ was equal to the zero vector, then we say that the collection of vectors is **linearly dependent**. To see what linear dependence means, suppose that $a_1 v_1 + a_2 v_2 + \cdots + a_k v_k = \mathbf{0}$ and one of the scalars, say $a_j$, is not zero. Then we can write

$$a_1 v_1 + a_2 v_2 + \cdots + a_j v_j + \cdots + a_k v_k = \mathbf{0},$$

$$a_j v_j = -a_1 v_1 - a_2 v_2 - \cdots - a_{j-1} v_{j-1} - a_{j+1} v_{j+1} - \cdots - a_k v_k,$$

or

$$v_j = \left(-\frac{a_1}{a_j}\right) v_1 + \left(-\frac{a_2}{a_j}\right) v_2 + \cdots + \left(-\frac{a_{j-1}}{a_j}\right) v_{j-1}$$
$$+ \left(-\frac{a_{j+1}}{a_j}\right) v_{j+1} + \cdots + \left(-\frac{a_k}{a_j}\right) v_k.$$

This last line tells us that if a collection of vectors is linearly dependent, then at least one of the vectors is a linear combination of the others.

Let's see some examples of these concepts.

### ■ Example 5.7.3 Linearly Independent and Linearly Dependent Vectors

The three vectors $\begin{bmatrix} 1 \\ 0 \\ 2 \end{bmatrix}$, $\begin{bmatrix} 0 \\ 1 \\ 2 \end{bmatrix}$, and $\begin{bmatrix} 1 \\ 2 \\ 0 \end{bmatrix}$ are linearly *independent* because the equation

$$a_1 \begin{bmatrix} 1 \\ 0 \\ 2 \end{bmatrix} + a_2 \begin{bmatrix} 0 \\ 1 \\ 2 \end{bmatrix} + a_3 \begin{bmatrix} 1 \\ 2 \\ 0 \end{bmatrix} = \begin{bmatrix} 0 \\ 0 \\ 0 \end{bmatrix}$$

is equivalent to the algebraic system

$$a_1 + a_3 = 0$$
$$a_2 + 2a_3 = 0$$
$$2a_1 + 2a_2 = 0,$$

which you can solve to find that $a_1 = a_2 = a_3 = 0$. (*Do the work!*)

On the other hand, the collection of vectors $\begin{bmatrix} 3 \\ 4 \\ -4 \end{bmatrix}$, $\begin{bmatrix} 0 \\ 1 \\ 2 \end{bmatrix}$, and $\begin{bmatrix} 1 \\ 2 \\ 0 \end{bmatrix}$ is linearly *dependent* because the vector equation

$$a_1 \begin{bmatrix} 3 \\ 4 \\ -4 \end{bmatrix} + a_2 \begin{bmatrix} 0 \\ 1 \\ 2 \end{bmatrix} + a_3 \begin{bmatrix} 1 \\ 2 \\ 0 \end{bmatrix} = \begin{bmatrix} 0 \\ 0 \\ 0 \end{bmatrix}$$

is equivalent to the algebraic system

$$3a_1 + a_3 = 0$$
$$4a_1 + a_2 + 2a_3 = 0$$
$$-4a_1 + 2a_2 = 0,$$

which has infinitely many solutions of the form $a_1 = K$, $a_2 = 2K$, and $a_3 = -3K$. In particular, we can let $K = 1$, so we have the nonzero solution $a_1 = 1$, $a_2 = 2$, and $a_3 = -3$. Note, for example, that we can write

$$\begin{bmatrix} 3 \\ 4 \\ -4 \end{bmatrix} = -2 \begin{bmatrix} 0 \\ 1 \\ 2 \end{bmatrix} + 3 \begin{bmatrix} 1 \\ 2 \\ 0 \end{bmatrix}.$$

◼

Now suppose that we have the system $\dot{X} = AX$, where $X$ is a $3 \times 1$ vector and $A$ is a $3 \times 3$ matrix of constants. If $A$ has three distinct real eigenvalues $\lambda_1, \lambda_2, \lambda_3$, then the theory of linear algebra tells us that the corresponding eigenvectors $V_1, V_2, V_3$ are linearly independent. Furthermore, the vectors $e^{\lambda_1 t}V_1, e^{\lambda_2 t}V_2, e^{\lambda_3 t}V_3$ are linearly independent, and the general solution of $\dot{X} = AX$ is

$$X(t) = c_1 e^{\lambda_1 t}V_1 + c_2 e^{\lambda_2 t}V_2 + c_3 e^{\lambda_3 t}V_3, \tag{5.7.2}$$

where $c_1, c_2$, and $c_3$ are arbitrary constants. Compare this with (5.3.1).

## ◼ Example 5.7.4 Solving a $3 \times 3$ System via Eigenvalues and Eigenvectors

Consider the system

$$\dot{x} = 4x + z$$
$$\dot{y} = -2y$$
$$\dot{z} = -z.$$

The matrix of coefficients is $A = \begin{bmatrix} 4 & 0 & 1 \\ 0 & -2 & 0 \\ 0 & 0 & -1 \end{bmatrix}$, and a CAS calculates the eigenvalues to be $\lambda_1 = 4, \lambda_2 = -1$, and $\lambda_3 = -2$, with corresponding eigenvectors

$$V_1 = \begin{bmatrix} 1 \\ 0 \\ 0 \end{bmatrix}, \ V_2 = \begin{bmatrix} 1 \\ 0 \\ -5 \end{bmatrix}, \text{ and } V_3 = \begin{bmatrix} 0 \\ 1 \\ 0 \end{bmatrix}.$$

Note that these vectors must be linearly independent because the eigenvalues are distinct real numbers. Thus, by (5.7.2), the general solution of our system is given by

$$X(t) = c_1 e^{4t} \begin{bmatrix} 1 \\ 0 \\ 0 \end{bmatrix} + c_2 e^{-t} \begin{bmatrix} 1 \\ 0 \\ -5 \end{bmatrix} + c_3 e^{-2t} \begin{bmatrix} 0 \\ 1 \\ 0 \end{bmatrix}$$

$$= \begin{bmatrix} c_1 e^{4t} + c_2 e^{-t} \\ c_3 e^{-2t} \\ -5 c_2 e^{-t} \end{bmatrix}.$$

In this example, we could have noticed that the second and third differential equations making up our original system were separable. After solving each of these, we could have substituted for $z$ in the first equation, which would then be a simple linear equation in $x$. (*Do this and compare your answer with the one given previously.*)

A trajectory in $x$-$y$-$z$ space (corresponding to the initial conditions $x(0) = 2, y(0) = 5$, and $z(0) = -10$) is shown in Figure 5.17, and the same trajectories in the $t$-$z$ plane and the $y$-$z$ plane are shown in Figures 5.18a and 5.18b, respectively.

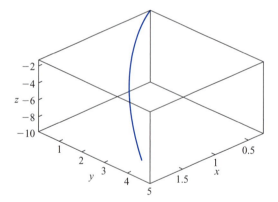

**FIGURE 5.17**
*Solution of $\dot{x} = 4x + z, \dot{y} = -2y, \dot{z} = -z; x(0) = 2, y(0) = 5, z(0) = -10; 0 \leq t \leq 2$*

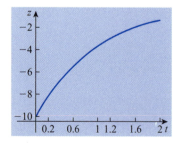

**FIGURE 5.18a**
*Graph of $z(t), 0 \leq t \leq 2$*

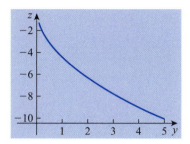

**FIGURE 5.18b**
*Graph of $z(t)$ vs. $y(t)$, $0 \leq t \leq 2$*

Note that the graph of a solution of this system is really four-dimensional, a set of points of the form $(t, x(t), y(t), z(t))$. Therefore, what Figure 5.17 is showing is a *projection* of a four-dimensional curve onto three-dimensional $x$-$y$-$z$ space. ∎

Accepting the fact that a $3 \times 3$ matrix has a cubic characteristic equation, we realize that we can have (1) three distinct real eigenvalues, (2) one distinct real eigenvalue and a different repeated real eigenvalue, (3) one repeated real eigenvalue, or (4) one real eigenvalue and a complex conjugate pair of eigenvalues. Possibilities 1 and 4 are handled easily by Formula (5.7.2). However, when we have repeated eigenvalues, we must find linearly independent eigenvectors, sometimes by calculating one or more generalized eigenvectors. (Go back to Example 5.4.2 and the discussion preceding it. Also see Problem C4 in Exercises 5.7.)

It should be clear how important the theory of linear algebra is to a full understanding of higher-order differential equations and their equivalent systems, but we will not investigate that theory further in this book.

The next example shows how techniques that we developed for two-dimensional systems in Section 5.5 can be extended to three-dimensional systems.

### ■ Example 5.7.5 Solving a $3 \times 3$ System—Complex Eigenvalues

Look at the system $\dot{x} = x, \dot{y} = 2x + y - 2z, \dot{z} = 3x + 2y + z$. The matrix of coefficients is $A = \begin{bmatrix} 1 & 0 & 0 \\ 2 & 1 & -2 \\ 3 & 2 & 1 \end{bmatrix}$. A CAS provides the characteristic equation $\lambda^3 - 3\lambda^2 + 7\lambda - 5 = (\lambda - 1)(\lambda^2 - 2\lambda + 5) = 0$, which has roots $\lambda_1 = 1, \lambda_2 = 1 + 2i$, and $\lambda_3 = 1 - 2i$. A CAS also gives the corresponding eigenvectors $V_1 = \begin{bmatrix} 2 \\ -3 \\ 2 \end{bmatrix}, V_2 = \begin{bmatrix} 0 \\ i \\ 1 \end{bmatrix}$, and $V_3 = \begin{bmatrix} 0 \\ -i \\ 1 \end{bmatrix}$. (Remember that your calculator or CAS may give you eigenvectors that look different from these. Just check to see that yours are multiples of the ones used here. Also, note that $V_2$ and $V_3$ are conjugates of each other.)

Now we can use (5.7.2) to write the general solution in the form

$$X(t) = c_1 e^t \begin{bmatrix} 2 \\ -3 \\ 2 \end{bmatrix} + c_2 e^{(1+2i)t} \begin{bmatrix} 0 \\ i \\ 1 \end{bmatrix} + c_3 e^{(1-2i)t} \begin{bmatrix} 0 \\ -i \\ 1 \end{bmatrix}.$$

However, we realize that $X_1(t) = e^t \begin{bmatrix} 2 \\ -3 \\ 2 \end{bmatrix}$ is a solution of the system by itself.

Furthermore, extending what we saw in Section 5.5, we know that we need work only with the *first* complex eigenvalue-eigenvector pair, because the other eigenvalue and eigenvector are conjugates that produce the same solutions (see Problem B3 of Exercises 5.5). Therefore, we consider only

$$\tilde{X}(t) = e^{(1+2i)t} \begin{bmatrix} 0 \\ i \\ 1 \end{bmatrix} = e^t(\cos(2t) + i\sin(2t)) \begin{bmatrix} 0 \\ i \\ 1 \end{bmatrix}$$

$$= e^t \begin{bmatrix} 0 \\ -\sin(2t) + i\cos(2t) \\ \cos(2t) + i\sin(2t) \end{bmatrix} = e^t \begin{bmatrix} 0 \\ -\sin(2t) \\ \cos(2t) \end{bmatrix} + ie^t \begin{bmatrix} 0 \\ \cos(2t) \\ \sin(2t) \end{bmatrix}.$$

From the preceding expression, we derive two linearly independent real-valued solutions of our system:

$$X_2(t) = \text{Re}\left\{\tilde{X}(t)\right\} = e^t \begin{bmatrix} 0 \\ -\sin(2t) \\ \cos(2t) \end{bmatrix} \text{ and } X_3(t) = \text{Im}\left\{\tilde{X}(t)\right\} = e^t \begin{bmatrix} 0 \\ \cos(2t) \\ \sin(2t) \end{bmatrix}.$$

Finally, the Superposition Principle tells us that

$$X(t) = c_1 X_1 + c_2 X_2 + c_3 X_3$$

$$= c_1 e^t \begin{bmatrix} 2 \\ -3 \\ 2 \end{bmatrix} + c_2 e^t \begin{bmatrix} 0 \\ -\sin(2t) \\ \cos(2t) \end{bmatrix} + c_3 e^t \begin{bmatrix} 0 \\ \cos(2t) \\ \sin(2t) \end{bmatrix}$$

$$= \begin{bmatrix} 2c_1 e^t \\ -3c_1 e^t - c_2 e^t \sin(2t) + c_3 e^t \cos(2t) \\ 2c_1 e^t + c_2 e^t \cos(2t) + c_3 e^t \sin(2t) \end{bmatrix}$$

$$= e^t \begin{bmatrix} 2c_1 \\ -3c_1 - c_2 \sin(2t) + c_3 \cos(2t) \\ 2c_1 + c_2 \cos(2t) + c_3 \sin(2t) \end{bmatrix}$$

is the real-valued general solution of the original system. If you use technology to solve this problem, be aware that your CAS may express the solution functions in a different but equivalent way. ∎

## 5.7.4 Nonhomogeneous Systems

It is important to realize that we can also handle larger *nonhomogeneous* systems in this way, using the relationship explored in Section 5.6: $X_{GNH} = X_{GH} + X_{PNH}$. The method of undetermined coefficients becomes algebraically messier as the size of the system increases, and in Chapter 6 we'll examine a better way of handling such systems.

## 5.7.5 Generalization to $n \times n$ Systems

Everything we've done with $2 \times 2$ and $3 \times 3$ systems of equations can be generalized to $n \times n$ systems. We can express a homogeneous $n$th-order linear system with constant coefficients

$$\dot{x}_1 = a_{11}x_1 + a_{12}x_2 + a_{13}x_3 + \cdots + a_{1n}x_n$$
$$\dot{x}_2 = a_{21}x_1 + a_{22}x_2 + a_{23}x_3 + \cdots + a_{2n}x_n$$
$$\dot{x}_3 = a_{31}x_1 + a_{32}x_2 + a_{33}x_3 + \cdots + a_{3n}x_n$$
$$\vdots \qquad\qquad \vdots \qquad\quad \vdots$$
$$\dot{x}_n = a_{n1}x_1 + a_{n2}x_2 + a_{n3}x_3 + \cdots + a_{nn}x_n$$

as $\dot{X} = AX$, where

$$X = \begin{bmatrix} x_1 \\ x_2 \\ x_3 \\ \vdots \\ x_n \end{bmatrix}, \quad \dot{X} = \begin{bmatrix} \dot{x}_1 \\ \dot{x}_2 \\ \dot{x}_3 \\ \vdots \\ \dot{x}_n \end{bmatrix}, \quad \text{and} \quad A = \begin{bmatrix} a_{11} & a_{12} & a_{13} & \cdots & a_{1n} \\ a_{21} & a_{22} & a_{23} & \cdots & a_{2n} \\ a_{31} & a_{32} & a_{33} & \cdots & a_{3n} \\ \vdots & \vdots & \vdots & \vdots & \vdots \\ a_{n1} & a_{n2} & a_{n3} & \cdots & a_{nn} \end{bmatrix}.$$

## ∎ Example 5.7.6 Matrix Form of a Four-Dimensional System

A compartmental analysis (see Section 2.3) of the processes involved in protein synthesis in animals and humans uses radioactive isotopes as tracers. A particular four-compartment model of this situation could lead to a system such as

$$\dot{x}_1 = -2x_1 + x_2$$
$$\dot{x}_2 = x_1 - 2x_2$$
$$\dot{x}_3 = x_1 + x_2 - x_3$$
$$\dot{x}_4 = x_3,$$

where $x_i(t)$ denotes the fraction of the total administered radioactivity attached to the material (albumen) in compartment $i(i = 1, 2, 3, 4)$. The coefficients indicate flow rates of the radioactive material from compartment to compartment.

In matrix terms, we can express this system as

$$\begin{bmatrix} \dot{x}_1 \\ \dot{x}_2 \\ \dot{x}_3 \\ \dot{x}_4 \end{bmatrix} = \begin{bmatrix} -2 & 1 & 0 & 0 \\ 1 & -2 & 0 & 0 \\ 1 & 1 & -1 & 0 \\ 0 & 0 & 1 & 0 \end{bmatrix} \begin{bmatrix} x_1 \\ x_2 \\ x_3 \\ x_4 \end{bmatrix}.$$

■

Extending the theory underlying the solution of algebraic systems of three linear equations in three unknowns to systems of $n$ equations in $n$ unknowns, we state that any $n \times n$ matrix has a determinant and that eigenvalues and eigenvectors can be defined for such square matrices. Specifically, given a system $\dot{X} = AX$, where $X$ is an $n \times 1$ column matrix (vector) and $A$ is an $n \times n$ matrix, an eigenvalue $\lambda$ is a solution of the equation $\det(A - \lambda I) = 0$, where $I$ is the $n \times n$ identity matrix consisting of ones down the main diagonal and zeros elsewhere. Given an eigenvalue $\lambda$, an eigenvector associated with $\lambda$ is a nonzero vector $V$ satisfying the equation $AV = \lambda V$.

The characteristic equation of an $n \times n$ matrix is an $n$th-degree polynomial. However, once a polynomial has degree greater than or equal to 5, there is no longer a general formula that gives the zeros. In general, the only way to tackle such equations is to use *approximation* methods. A CAS—or even a graphing calculator—has various algorithms to do this.

Now suppose that we have the system $\dot{X} = AX$, where $X$ is an $n \times 1$ vector and $A$ is an $n \times n$ matrix of constants. If $A$ has $n$ distinct real eigenvalues $\lambda_1, \lambda_2, \ldots, \lambda_n$, then the theory of linear algebra guarantees that the corresponding eigenvectors $V_1, V_2, \ldots, V_n$ are linearly independent. Furthermore, the vectors $e^{\lambda_1 t}V_1, e^{\lambda_2 t}V_2, \ldots, e^{\lambda_n t}V_n$ are linearly independent, and the general solution of $\dot{X} = AX$ is

$$X(t) = c_1 e^{\lambda_1 t}V_1 + c_2 e^{\lambda_2 t}V_2 + \cdots + c_n e^{\lambda_n t}V_n, \tag{5.7.3}$$

*where $c_1, c_2, \ldots, c_n$ are arbitrary constants.* You should expect the usual complications when there are repeated real roots, complex conjugate pairs of roots, and so forth.

If we investigate a mechanical system (Figure 5.19) consisting of two springs attached to two movable masses, the physics of the situation gives us a pair of second-order linear differential equations. In turn, this system of two equations can be expressed as a system of four first-order linear equations.

The following example assumes that we start from the equilibrium position by giving one mass an initial velocity. Most of the computational work will be done by a CAS.

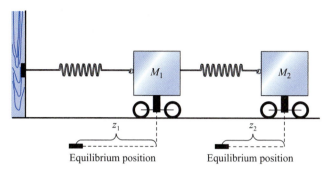

**FIGURE 5.19**
*The spring-mass system for Example 5.7.7*

## ■ Example 5.7.7 A Four-Dimensional System from Mechanics

Let's consider the system

$$\frac{d^2 z_1}{dt^2} = -11z_1 + 6z_2$$

$$\frac{d^2 z_2}{dt^2} = -6z_2 + 6z_1,$$

where $z_1$ is the distance of mass 1 from its equilibrium position and $z_2$ is the distance of mass 2 from equilibrium. We'll assume the initial conditions $z_1(0) = 0$, $z_1'(0) = 0$, $z_2(0) = 0$, and $z_2'(0) = 1$.

## Representation as a First-Order System

Introducing the new variables $x_1 = z_1, x_2 = \frac{dz_1}{dt}, x_3 = z_2$, and $x_4 = \frac{dz_2}{dt}$, we convert our pair of second-order equations into the four-dimensional system of first-order equations

$$\frac{dx_1}{dt} = x_2$$

$$\frac{dx_2}{dt} = -11x_1 + 6x_3$$

$$\frac{dx_3}{dt} = x_4$$

$$\frac{dx_4}{dt} = -6x_3 + 6x_1,$$

with $x_1(0) = x_2(0) = x_3(0) = 0$ and $x_3'(0) = x_4(0) = 1$.

## Matrix Representation, Eigenvalues, Eigenvectors

We can express the last system as $\frac{d}{dt}X = \begin{bmatrix} 0 & 1 & 0 & 0 \\ -11 & 0 & 6 & 0 \\ 0 & 0 & 0 & 1 \\ 6 & 0 & -6 & 0 \end{bmatrix} \begin{bmatrix} x_1 \\ x_2 \\ x_3 \\ x_4 \end{bmatrix} = AX$. A CAS gives the

characteristic equation of matrix $A$ as $\lambda^4 + 17\lambda^2 + 30 = 0$, which we can factor as $(\lambda^2 + 15)(\lambda^2 + 2) = 0$, so the eigenvalues are $\lambda_1 = \sqrt{15}i, \lambda_2 = -\sqrt{15}i, \lambda_3 = \sqrt{2}i$, and $\lambda_4 = -\sqrt{2}i$. The corresponding eigenvectors are

$$V_1 = \begin{bmatrix} 3 \\ 3\sqrt{15}i \\ -2 \\ -2\sqrt{15}i \end{bmatrix}, V_2 = \begin{bmatrix} 3 \\ -3\sqrt{15}i \\ -2 \\ 2\sqrt{15}i \end{bmatrix}, V_3 = \begin{bmatrix} 2 \\ 2\sqrt{2}i \\ 3 \\ 3\sqrt{2}i \end{bmatrix}, \text{ and } V_4 = \begin{bmatrix} 2 \\ -2\sqrt{2}i \\ 3 \\ -3\sqrt{2}i \end{bmatrix}.$$

If you check this with a CAS, remember that you may get a different (but equivalent) form for the eigenvectors.

## The General Solution

On the basis of our previous experience with complex conjugate pairs of eigenvalues and eigenvectors, we can just work with the pairs $\lambda_1, V_1$ and $\lambda_3, V_3$. First, we know that

$$\hat{X}(t) = e^{\lambda_1 t}V_1 = e^{\sqrt{15}it} \begin{bmatrix} 3 \\ 3\sqrt{15}i \\ -2 \\ -2\sqrt{15}i \end{bmatrix} = \left(\cos\left(\sqrt{15}t\right) + i\sin\left(\sqrt{15}t\right)\right) \begin{bmatrix} 3 \\ 3\sqrt{15}i \\ -2 \\ -2\sqrt{15}i \end{bmatrix}$$

$$= \begin{bmatrix} 3\cos\left(\sqrt{15}t\right) \\ -3\sqrt{15}\sin\left(\sqrt{15}t\right) \\ -2\cos\left(\sqrt{15}t\right) \\ 2\sqrt{15}\sin\left(\sqrt{15}t\right) \end{bmatrix} + i \begin{bmatrix} 3\sin\left(\sqrt{15}t\right) \\ 3\sqrt{15}\cos\left(\sqrt{15}t\right) \\ -2\sin\left(\sqrt{15}t\right) \\ -2\sqrt{15}\cos\left(\sqrt{15}t\right) \end{bmatrix} = X_1(t) + iX_2(t),$$

where both $X_1(t)$ and $X_2(t)$ are linearly independent real-valued solutions of the system. Then we have

$$\tilde{X}(t) = e^{\lambda_3 t}V_3 = e^{\sqrt{2}it} \begin{bmatrix} 2 \\ 2\sqrt{2}i \\ 3 \\ 3\sqrt{2}i \end{bmatrix} = \left(\cos\left(\sqrt{2}t\right) + i\sin\left(\sqrt{2}t\right)\right) \begin{bmatrix} 2 \\ 2\sqrt{2}i \\ 3 \\ 3\sqrt{2}i \end{bmatrix}$$

$$= \begin{bmatrix} 2\cos\left(\sqrt{2}t\right) \\ -2\sqrt{2}\sin\left(\sqrt{2}t\right) \\ 3\cos\left(\sqrt{2}t\right) \\ -3\sqrt{2}\sin\left(\sqrt{2}t\right) \end{bmatrix} + i \begin{bmatrix} 2\sin\left(\sqrt{2}t\right) \\ 2\sqrt{2}\cos\left(\sqrt{2}t\right) \\ 3\sin\left(\sqrt{2}t\right) \\ 3\sqrt{2}\cos\left(\sqrt{2}t\right) \end{bmatrix} = X_3(t) + iX_4(t),$$

where $X_3(t)$ and $X_4(t)$ are linearly independent real-valued solutions of the system. The general solution is

$$X(t) = c_1 X_1 + c_2 X_2 + c_3 X_3 + c_4 X_4$$

$$= c_1 \begin{bmatrix} 3\cos\left(\sqrt{15}t\right) \\ -3\sqrt{15}\sin\left(\sqrt{15}t\right) \\ -2\cos\left(\sqrt{15}t\right) \\ 2\sqrt{15}\sin\left(\sqrt{15}t\right) \end{bmatrix} + c_2 \begin{bmatrix} 3\sin\left(\sqrt{15}t\right) \\ 3\sqrt{15}\cos\left(\sqrt{15}t\right) \\ -2\sin\left(\sqrt{15}t\right) \\ -2\sqrt{15}\cos\left(\sqrt{15}t\right) \end{bmatrix}$$

$$+ c_3 \begin{bmatrix} 2\cos\left(\sqrt{2}t\right) \\ -2\sqrt{2}\sin\left(\sqrt{2}t\right) \\ 3\cos\left(\sqrt{2}t\right) \\ -3\sqrt{2}\sin\left(\sqrt{2}t\right) \end{bmatrix} + c_4 \begin{bmatrix} 2\sin\left(\sqrt{2}t\right) \\ 2\sqrt{2}\cos\left(\sqrt{2}t\right) \\ 3\sin\left(\sqrt{2}t\right) \\ 3\sqrt{2}\cos\left(\sqrt{2}t\right) \end{bmatrix}.$$

## Particular Solutions

The initial conditions $x_1(0) = x_2(0) = x_3(0) = 0$ and $x_3'(0) = x_4(0) = 1$ imply (Problem B1 in Exercises 5.7) that $c_1 = c_3 = 0$ and $c_2 = -2\sqrt{15}/195, c_4 = 3\sqrt{2}/26$. Therefore,

$$z_1(t) = x_1(t) = \frac{3\sqrt{2}}{13}\sin\left(\sqrt{2}t\right) - \frac{2\sqrt{15}}{65}\sin\left(\sqrt{15}t\right)$$

and

$$z_2(t) = x_3(t) = \frac{9\sqrt{2}}{26}\sin\left(\sqrt{2}t\right) + \frac{4\sqrt{15}}{195}\sin\left(\sqrt{15}t\right).$$

■

If we are interested in the *stability* of an $n \times n$ system rather than its exact solution, we can give a simplified version of the results we have seen for $2 \times 2$ systems (see Table 5.1 in Section 5.5): If $A$ is an $n \times n$ matrix of constants, then the equilibrium solution $X = 0$ for the system $\dot{X} = AX$ is asymptotically stable (that is, it is a *sink*) if every eigenvalue of $A$ has a negative real part and is unstable if $A$ has at least one eigenvalue with a positive real part. Furthermore, if *all* eigenvalues have positive real parts, the $n$-dimensional origin $X = 0$ is a *source*; and if some eigenvalues have positive real parts and others have negative real parts, the equilibrium point is called a *saddle point*.

In the next chapter we'll learn another way to handle initial-value problems involving systems of linear equations. The method of *Laplace transforms*, especially when implemented by a CAS, is a powerful tool for solving various applied problems.

## Exercises 5.7

**A**

For each of the systems in Problems 1–6, (a) write the system in the form $\dot{X} = AX$; (b) use technology to find eigenvalues and representative eigenvectors; and (c) express the general solution as a single real-valued vector of functions.

1. $\dfrac{dx}{dt} = x - y + z$

   $\dfrac{dy}{dt} = x + y - z$

   $\dfrac{dz}{dt} = 2x - y$

2. $\dfrac{dx}{dt} = x - 2y - z$

   $\dfrac{dy}{dt} = -x + y + z$

   $\dfrac{dz}{dt} = x - z$

3. $\dfrac{dx}{dt} = 3x - y + z$

   $\dfrac{dy}{dx} = x + y + z$

   $\dfrac{dz}{dt} = 4x - y + 4z$

4. $\dfrac{dx}{dt} = 2x + y$

   $\dfrac{dy}{dt} = x + 3y - z$

   $\dfrac{dz}{dt} = 2y + 3z - x$

5. $\dfrac{dx}{dt} = 2x - y + z$

   $\dfrac{dy}{dt} = x + 2y - z$

   $\dfrac{dz}{dt} = x - y + 2z$

**6.**  $\dfrac{dx}{dt} = 2x + 2z - y$

$\dfrac{dy}{dt} = x + 2z$

$\dfrac{dz}{dt} = y - 2x - z$

**7.** For the system in Problem 1, use your CAS to plot the $x$-$y$-$z$ space trajectory passing through the point $(0, 1, 0)$ when $t = 0$.

**8.** For the system in Problem 4, use your CAS to plot the $x$-$y$-$z$ space trajectory passing through the point $(1, 1, -1)$ when $t = 0$.

## B

**1.** In Example 5.7.7, use the initial conditions to show that $c_1 = c_3 = 0$ and $c_2 = -2\sqrt{15}/195, c_4 = 3\sqrt{2}/26$.

**2. a.** Solve the initial value problem

$$\frac{dx}{dt} = z + y - x$$

$$\frac{dy}{dt} = z + x - y$$

$$\frac{dz}{dt} = x + y + z,$$

$$x(0) = 1, y(0) = -1/3, z(0) = 0.$$

**b.** Use the explicit solution found in part (a) to calculate $x(0.5)$, $y(0.5)$, and $z(0.5)$.

**c.** Use two or more numerical methods found in your CAS to approximate $x(0.5)$, $y(0.5)$, and $z(0.5)$. Compare the answers to each other and to the exact answers in part (a).

**3.** Solve the initial value problem

$$\frac{dx}{dt} = y + z$$

$$\frac{dy}{dt} = z + x$$

$$\frac{dz}{dt} = x + y$$

$$x(0) = -1, y(0) = 1, z(0) = 0.$$

**4.** Consider the system

$$\ddot{x} = 2x + \dot{x} + y$$

$$\dot{y} = 4x + 2y.$$

**a.** Convert this system to a system of three first-order equations, $\dot{Y} = AY$.

**b.** Use technology to find the eigenvalues of the matrix $A$ in part (a).

c. Use technology to find two linearly independent eigenvectors corresponding to the eigenvalues found in (b).

d. Take $W = \begin{bmatrix} t \\ 1 \\ -1 - 2t \end{bmatrix}$ as a third eigenvector that is independent of the two found in part (c) and give the general solution of $\dot{Y} = AY$.

e. Find the general solution $x(t), y(t)$ of the original system.

5. Consider the two-mass, three-spring system shown here.

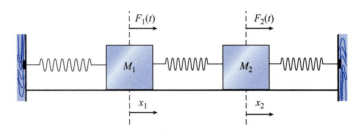

If there are no external forces, and if the masses and spring constants are equal and of unit magnitude, then the equations of motion are

$$x_1'' = -2x_1 + x_2, \ x_2'' = x_1 - 2x_2.$$

a. Transform the system of equations into a system of four first-order equations by letting $y_1 = x_1, y_2 = x_1', y_3 = x_2$, and $y_4 = x_2'$.

b. Find the eigenvalues of the matrix of coefficients for the system in part (a).

c. Solve the system in part (a) subject to the initial conditions $y_1(0) = 2, y_2(0) = 1, y_3(0) = 2$, $y_4(0) = 1$. Describe the physical motion of the spring-mass system corresponding to this solution.

d. Solve the system in part (a) subject to the initial conditions $y_1(0) = 2, y_2(0) = \sqrt{3}, y_3(0) = -2$, $y_4(0) = -\sqrt{3}$. Describe the physical motion of the spring-mass system corresponding to this solution.

e. Observe that the spring-mass system has two natural modes of oscillation in this problem. How are the natural frequencies related to the eigenvalues of the coefficient matrix? Do you think that there might be a third natural mode of oscillation with a different frequency?

## C

1. There are three tanks (see the following figure) that pump fluid back and forth in the following way: Tank A pumps fluid into tank B at a rate of 1% of its volume per hour and also back into itself at a rate of 1% of the volume per hour. Tank B pumps into itself, tank A, and tank C, all at a rate of 2% of its volume per hour. Tank C pumps into tank B at a rate of 2% of its volume per hour and back into itself at the rate of 3% of its volume per hour. Assuming that the initial volumes in tanks A, B, and C are 23,000, 1000, and 1000 liters, respectively, describe the changes in volume of fluid in

each tank as functions of time. (Use technology only to find the eigenvalues and corresponding eigenvectors.)

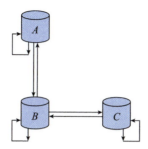

2. Suppose that you have a *double pendulum*—that is, one pendulum suspended from another, as shown in the following figure. The laws of physics, after a simplifying change of variables, give us the following system as a model for small oscillations about the equilibrium position:

$$\dot{x} = y$$
$$\dot{y} = -x + \alpha u$$
$$\dot{u} = v$$
$$\dot{v} = x - u.$$

Here, $\alpha = (m_2/m_1)(1 + m_2/m_1)^{-1}$, $x = \theta_1$, $u = \theta_2$ and $y$ and $v$ are the angular velocities $\dot{\theta}_1$ and $\dot{\theta}_2$, respectively. For this problem, let $\alpha = 0.3$.

a. Express the system in matrix form.
b. Use technology to find the eigenvalues of the system.
c. Use technology to find eigenvectors corresponding to the eigenvalues found in part (b).
d. Find the general real-valued solution of the system.

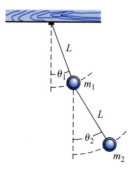

3. Consider the pair of 50 gallon tanks shown here. Initially, tank I is full of compound B and tank II is full of compound C. Start to introduce compound A into each tank at the rates shown in the figure.

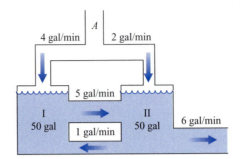

a. Let $x_1(t)$ and $x_2(t)$ denote the amount of compound A in tanks I and II, respectively. Similarly, define $y_1(t), y_2(t), z_1(t)$, and $z_2(t)$ for the amounts of compounds B and C in tanks I and II. Now write a system of six nonhomogeneous differential equations describing the flow of the various substances into and out of tanks I and II, expressing any fractions in decimal form. Be sure to write initial conditions.

b. Use technology to solve the IVP expressed in part (a).

c. Use technology to graph $x_1(t), y_1(t)$, and $z_1(t)$ against $t$, all on the same set of axes.

d. Use technology to graph $x_2(t), y_2(t)$, and $z_2(t)$ against $t$, all on the same set of axes.

4. Suppose you have a system $\dot{X} = AX$, where $A$ is a $3 \times 3$ matrix that has an eigenvalue $\lambda$ of multiplicity 3 and corresponding eigenvector $V$. Then it can be shown that the general solution of the system can be written as $c_1 X_1 + c_2 X_2 + c_3 X_3$, where $X_1 = e^{\lambda t} V, X_2 = e^{\lambda t}(W + tV)$, $X_3 = e^{\lambda t}(U + tW + \frac{t^2}{2}V)$, $W$ satisfies $(A - \lambda I)W = V$, and $U$ satisfies $(A - \lambda I)U = W$.

a. Find the repeated eigenvalue and representative eigenvector for the system

$$x' = x + y + z$$
$$y' = 2x + y - z$$
$$z' = -3x + 2y + 4z.$$

b. Use the method described above and technology to write the general solution of this system.

5. Find the general solution of the nonautonomous, nonhomogeneous system

$$\frac{dx}{dt} = 2t$$
$$\frac{dy}{dt} = 3x + 2t$$
$$\frac{dz}{dt} = x + 4y + t$$

a. by using ideas from Problem C4, followed by the technique of *undetermined coefficients*. (See Section 5.6.)

**b.** by solving the first equation and then substituting this solution in the second equation, and so forth.

## SUMMARY

By using matrices and their properties, we can write any $n \times n$ autonomous system of linear equations with constant coefficients

$$\dot{x}_1 = a_{11}x_1 + a_{12}x_2 + a_{13}x_3 + \cdots + a_{1n}x_n$$
$$\dot{x}_2 = a_{21}x_1 + a_{22}x_2 + a_{23}x_3 + \cdots + a_{2n}x_n$$
$$\dot{x}_3 = a_{31}x_1 + a_{32}x_2 + a_{33}x_3 + \cdots + a_{3n}x_n$$
$$\vdots \qquad \qquad \vdots \qquad \qquad \vdots$$
$$\dot{x}_n = a_{n1}x_1 + a_{n2}x_2 + a_{n3}x_3 + \cdots + a_{nn}x_n$$

in the compact form $\dot{X} = AX$, where

$$X = \begin{bmatrix} x_1 \\ x_2 \\ x_3 \\ \vdots \\ x_n \end{bmatrix}, \quad \dot{X} = \begin{bmatrix} \dot{x}_1 \\ \dot{x}_2 \\ \dot{x}_3 \\ \vdots \\ \dot{x}_n \end{bmatrix}, \quad \text{and} \quad A = \begin{bmatrix} a_{11} & a_{12} & a_{13} & \cdots & a_{1n} \\ a_{21} & a_{22} & a_{23} & \cdots & a_{2n} \\ a_{31} & a_{32} & a_{33} & \cdots & a_{3n} \\ \vdots & \vdots & \vdots & \vdots & \vdots \\ a_{n1} & a_{n2} & a_{n3} & \cdots & a_{nn} \end{bmatrix}.$$

For two-dimensional systems in which the matrix of coefficients $A$ has a nonzero determinant, the origin is the only equilibrium point. The qualitative behavior (stability) of such a linear system is completely determined by the eigenvalues and eigenvectors of $A$. If the system has two real eigenvalues $l_1$ and $l_2$, with $\lambda_1 \neq \lambda_2$, and $V_1$ and $V_2$ are the corresponding (linearly independent) eigenvectors, then the general solution of the system is given by $X(t) = c_1 e^{\lambda_1 t} V_1 + c_2 e^{\lambda_2 t} V_2$. If the system has two real and equal eigenvalues, we can try to find two linearly independent eigenvectors corresponding to the single eigenvalue. If we can't find two such eigenvectors, we can start with one eigenvector $V$ and calculate a *generalized eigenvector* $W$ so that the general solution of the system can be written in the form $X(t) = c_1 e^{\lambda t} V + c_2 \left[ te^{\lambda t} V + e^{\lambda t} W \right]$. Finally, if the system has a pair of complex conjugate eigenvalues, we can still write the solution as $X(t) = c_1 e^{\lambda_1 t} V_1 + c_2 e^{\lambda_2 t} V_2$, but we have to use *Euler's formula*, $e^{p+qi} = e^p (\cos(q) + i \sin(q))$, to simplify this expression and wind up with *real-valued* solutions. Specifically, first we get a solution of the form $X(t) = X_1(t) + iX_2(t)$, where $X_1(t)$ and $X_2(t)$ are real-valued matrix (vector) functions called the *real part* and the *imaginary part*, respectively, of $X(t)$. Then the real-valued general solution is $X(t) = C_1 X_1(t) + C_2 X_2(t)$, where $C_1$ and $C_2$ are real numbers. Using these forms for the general solution of our system, we can analyze the stability of the system qualitatively in terms of eigenvalues and eigenvectors. These results were summarized in Table 5.1 at the end of Section 5.5.

We may have a *nonhomogeneous* system

$$\dot{x}_1 = a_{11}x_1 + a_{12}x_2 + a_{13}x_3 + \cdots + a_{1n}x_n + b_1(t)$$
$$\dot{x}_2 = a_{21}x_1 + a_{22}x_2 + a_{23}x_3 + \cdots + a_{2n}x_n + b_2(t)$$
$$\dot{x}_3 = a_{31}x_1 + a_{32}x_2 + a_{33}x_3 + \cdots + a_{3n}x_n + b_3(t)$$
$$\vdots \qquad\qquad \vdots \qquad\qquad \vdots$$
$$\dot{x}_n = a_{n1}x_1 + a_{n2}x_2 + a_{n3}x_3 + \cdots + a_{nn}x_n + b_n(t),$$

which can be written as $\dot{X} = AX + B(t)$, with

$$X = \begin{bmatrix} x_1 \\ x_2 \\ x_3 \\ \vdots \\ x_n \end{bmatrix}, \ \dot{X} = \begin{bmatrix} \dot{x}_1 \\ \dot{x}_2 \\ \dot{x}_3 \\ \vdots \\ \dot{x}_n \end{bmatrix},$$

$$A = \begin{bmatrix} a_{11} & a_{12} & a_{13} & \cdots & a_{1n} \\ a_{21} & a_{22} & a_{23} & \cdots & a_{2n} \\ a_{31} & a_{32} & a_{33} & \cdots & a_{3n} \\ \vdots & \vdots & \vdots & \vdots & \vdots \\ a_{n1} & a_{n2} & a_{n3} & \cdots & a_{nn} \end{bmatrix}, \quad \text{and} \quad B(t) = \begin{bmatrix} b_1(t) \\ b_2(t) \\ b_3(t) \\ \vdots \\ b_n(t) \end{bmatrix}.$$

In this situation, we know that the general solution, $X_{\text{GNH}}$, of a linear nonhomogeneous system is obtained by finding a particular solution, $X_{\text{PNH}}$, of the nonhomogeneous system and adding it to the general solution, $X_{\text{GH}}$, of the homogeneous system: $X_{\text{GNH}} = X_{\text{GH}} + X_{\text{PNH}}$. The method of undetermined coefficients can be used to make an intelligent guess about the particular solution if the entries of vector $B(t)$ contain terms that are constants, exponential functions, sines, cosines, polynomials, or any sum or product of such terms. For other kinds of functions making up $B(t)$, $X_{\text{PNH}}$ must be found using some other technique (for example, variation of parameters).

Although we started with a thorough analysis of the equilibrium points and the stability of the system near these points for *two*-dimensional systems of equations with constant coefficients, we saw eventually that the concepts of eigenvalue and eigenvector were meaningful for systems of $n$ equations. Specifically, given a system $\dot{X} = AX$, where $X$ is an $n \times 1$ column matrix (vector) and $A$ is an $n \times n$ matrix, an eigenvalue $\lambda$ is a solution of the equation $\det(A - \lambda I) = 0$, where $I$ is the $n \times n$ identity matrix consisting of ones down the main diagonal and zeros elsewhere. We know that $\det(A - \lambda I)$ is an $n$th-degree polynomial in $\lambda$. Given an eigenvalue $\lambda$, an eigenvector associated with $\lambda$ is a nonzero vector $V$ satisfying the equation $AV = \lambda V$.

For values of $n$ greater than 3, we lose the ordinary intuitive geometric interpretation of our results. Also, when $n$ is greater than or equal to 5, there is no general procedure we can follow to solve the characteristic equations. We must use approximation methods, and technology becomes crucial here. The question of the multiplicity of eigenvalues leads to complicated linear-algebra considerations, and the general vector form of the solution of a system is difficult to describe without delving more deeply into linear algebra.

## PROJECT 5-1

### A Vicious Circle

There are three species of omnivores on an island: xaccoons, yadgers, and zoyotes. Xaccoons eat zoyotes, zoyotes prey on yadgers, and yadgers find xaccoons delicious. Given the species' individual birth rates and predation rates, we can set up the following system:

$$\dot{x} = 21x - 9y$$
$$\dot{y} = 2y - 4z$$
$$\dot{z} = 6z - 7x.$$

Here, $x(t), y(t),$ and $z(t)$ denote the populations of the three species (in an obvious way) at time $t$, where $t$ is in centuries.

Long before human beings arrived on the island, the three species lived in a state of equilibrium. At time $t = 0$, shortly after humans discovered the island and disrupted the equilibrium by hunting, chopping down trees, and so on, there were 300 xaccoons, 598 yadgers, and 323 zoyotes. (This is no longer the equilibrium state.)

   **a.** Write the system in matrix terms, and use technology to find the eigenvalues and eigenvectors of the system.

   **b.** Solve the system IVP manually, using the results of part (a). (You may use technology to solve an algebraic system of three equations in three unknowns.)

   **c.** What were the equilibrium populations before humans arrived on the scene? [Look at the solution in part (b) as $t \to -\infty$.]

   **d.** Substitute the answers found in part (c) into the three differential equations making up the system. For each population, what does the result say about its birth rate compared to its loss by predation during the period of equilibrium?

   **e.** Graph $x(t)$ against $t, 0 \le t \le 0.1$. Graph $y(t)$ against $t$, for $0 \le t \le 0.1$ and then for $0 \le t \le 0.2$. Graph $z(t)$ against $t$ for $0 \le t \le 0.16$. (Use technology to obtain these graphs.)

   **f.** Which is the most acutely endangered species? After how many years will it become extinct? At the time of this species' extinction, what will be the populations of the surviving species?

## PROJECT 5-2

## Go with the Flow

The setup for a complicated mixing system is shown in the accompanying figure. We have three tanks interconnected as indicated by pipes whose capacity of flow is 5 gal/min. Initially, tank I contains 20 gallons of red paint, tank II contains 30 gallons of yellow paint, and tank III contains 40 gallons of blue paint. What will be the mixture of paints in each tank at the end of 5 minutes? *Use technology to solve this problem.*

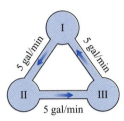

*Comments*: Let $x_i$ = the amount of red paint in tank $i$ at time $t$, $y_i$ = the amount of yellow paint in tank $i$ at time $t$, and $z_i$ = the amount of blue paint in tank $i$ at time $t$. Then the initial conditions say that at $t = 0, x_1 = 20, x_2 = 0, x_3 = 0; y_1 = 0, y_2 = 30, y_3 = 0;$ and $z_1 = 0, z_2 = 0, z_3 = 40$. Furthermore, the flow pattern in the diagram, together with the observation that the total volume in each tank is constant—20, 30, and 40 gallons, respectively—leads to a system of nine differential equations. *However, the system of three equations containing the variables $x_1, x_2,$ and $x_3$ is exactly the same as the system containing the $y_i$ and the system containing the $z_i$. The only difference lies in the initial conditions. Physically, the obvious symmetry of the tanks and the flow pattern explain this. Therefore, you need solve only one of the three-dimensional systems and use it to find all three sets of variables by varying the initial conditions.*

# CHAPTER 6

# The Laplace Transform

## INTRODUCTION

The idea of a *transform*, or *transformation*, is a very important one in mathematics and problem solving in general. When you are faced with a difficult problem, it is often a good idea to change it in some way into an easier problem, solve that easier problem, and then take your solution and apply it to your original problem. One of the first examples of this process that you have seen involves the idea of a *logarithm*. When John Napier and others developed logarithms in the early 1600s, they served as an aid to calculation. Given a difficult multiplication problem, you could transform it into an addition problem, perform the addition, and then transform the answer back into the answer to the original problem. For example, if you wanted to multiply 8743 by 2591, you could apply the natural logarithm to this product, getting the sum $\ln(8743) + \ln(2591) = 9.07600865918\ldots + 7.85979918056\ldots = 16.9358078397\ldots$. Then you would reverse the process by determining the number whose natural logarithm is $16.9358078397\ldots$. That number, 22,653,113, should be the original product. The process of going from the sum of logs back to the original product is called an *inverse* transformation. Of course, we recognize that the inverse of the logarithmic transformation is the exponential transformation:

In elementary calculus, you evaluated certain integrals by changing variables—transforming an integral in terms of $x$, say, to what you hope is a simpler integral in another variable, say $u$. Back in Section 2.2 you encountered a transformation when you introduced an *integrating factor* into a linear equation. By multiplying the equation by the appropriate exponential factor, you transformed the left-hand side into an exact derivative, which could then be integrated to yield the unknown function. You solved the equation by changing it into an equivalent form that was easier to deal with. The entire philosophy of using transformations in solving problems can be stated simply: I. TRANSFORM; II. SOLVE; and III. INVERT.

The important mathematical tool known as the *Laplace transform* is named for the great French mathematician Pierre-Simon de Laplace (1749–1827) who studied its properties, but it was probably used earlier by Euler. This transformation will be useful to us because it removes derivatives from differential equations and replaces them with *algebraic* expressions. In this way, differential equations are replaced by algebraic equations. This transformation turns out to be particularly powerful when we are dealing with initial-value problems, nonhomogeneous equations with discontinuous forcing terms, and systems of differential equations. The downside is that the use of the Laplace transform is restricted to the solution of *linear* differential equations and *linear* systems of differential equations.

## 6.1  THE LAPLACE TRANSFORM OF SOME IMPORTANT FUNCTIONS

We start by assuming that $f(t)$ is a function that is defined for $t \geq 0$. The **Laplace transform** of this function, $\mathcal{L}[f(t)]$, is defined as

$$\mathcal{L}[f(t)] = \int_0^\infty f(t)e^{-st}dt, \tag{6.1.1}$$

when this improper integral exists. Note that after you've integrated with respect to $t$, the result will have the parameter $s$ in it—that is, this integral is a *function* of the parameter $s$, so we can write $\mathcal{L}[f(t)] = F(s)$. We have transformed our function in $t$ to a function in $s$.

Before we give some examples, let's just examine the integral in (6.1.1). From basic calculus, we know that the improper integral is defined as $\lim_{b \to \infty} \int_0^b f(t)e^{-st}dt$ when this limit exists.

There are two important requirements here. First, the ordinary Riemann integral $\int_0^b f(t)e^{-st}dt$ must exist for every $b > 0$; and then the limit must exist as $b \to \infty$. Both requirements are taken care of if we stick to *continuous* or *piecewise continuous* functions $f(t)$ for which there exist positive constants $M$ and $K$ such that $|f(t)| < e^{Mt}$ for all $t \geq K$. This says that the function $f$ doesn't grow faster than an exponential function, so the integrand $f(t)e^{-st}$ in (6.1.1) behaves like the function $e^{Mt} \cdot e^{-st} = e^{-(s-M)t}$ for values of $s$ greater than $M$ and for $t$ large enough. The improper integral of this kind of function converges. (See Section A.6 for basic definitions and examples.)

Now suppose that $f(t) \equiv 1$. Then

$$F(s) = \mathcal{L}[1] \int_0^\infty 1 \cdot e^{-st}dt = \lim_{b \to \infty} \int_0^b e^{-st}dt = \lim_{b \to \infty} \left. \frac{e^{-st}}{-s} \right|_0^b$$

$$= \frac{1}{-s}\left(\lim_{b \to \infty}\left(e^{-sb} - 1\right)\right) = \frac{1}{s}.$$

From this you can see that the Laplace transform of a constant function $f(t) \equiv C$ is $\frac{C}{s}$ for $s > 0$. (*Yes?*)

Next, we can find $\mathcal{L}[t]$ by using integration by parts and the value of $\mathcal{L}[1]$. For $s > 0$,

$$\mathcal{L}[t] = \int_0^\infty te^{-st}\,dt = \lim_{b\to\infty}\int_0^b te^{-st}\,dt$$

$$= \lim_{b\to\infty}\left\{ \left.\frac{te^{-st}}{-s}\right|_0^b - \int_0^b \frac{e^{-st}}{-s}\,dt \right\} = 0 + \frac{1}{s}\mathcal{L}[1] = \frac{1}{s^2}.$$

Similarly, we can show that $\mathcal{L}[t^2] = \frac{2}{s^3}$ and $\mathcal{L}[t^3] = \frac{6}{s^4}$ for $s > 0$. (See Problems A1 and A3 in Exercises 6.1.) In general, for all integers $n \geq 0$, it can be shown that

$$\mathcal{L}[t^n] = \frac{n!}{s^{n+1}} \quad (s > 0), \tag{6.1.2}$$

where $0!$ is defined to be 1.

From the basic properties of integrals, we can see that

$$\mathcal{L}[c \cdot f(t)] = c\,\mathcal{L}[f(t)],$$

where $c$ is any real constant, and that

$$\mathcal{L}[f(t) + g(t)] = \mathcal{L}[f(t)] + \mathcal{L}[g(t)],$$

whenever the Laplace transforms of both $f$ and $g$ exist. Any transformation that satisfies the last two properties is called a **linear operator** or a **linear transformation**. (See Section 2.2.) If $c_1$ and $c_2$ are constants, then we can combine the two properties to write

$$\mathcal{L}[c_1 f(t) + c_2 g(t)] = c_1 \mathcal{L}[f(t)] + c_2 \mathcal{L}[g(t)]. \tag{6.1.3}$$

Extending (6.1.3), we can see how to calculate the Laplace transform of any *polynomial function*:

$$\mathcal{L}\left[c_0 + c_1 t + c_2 t^2 + \cdots + c_k t^k + \cdots + c_n t^n\right]$$

$$= \mathcal{L}[c_0] + \mathcal{L}[c_1 t] + \mathcal{L}[c_2 t^2] + \cdots + \mathcal{L}\left[c_k t^k\right] + \cdots + \mathcal{L}[c_n t^n]$$

$$= c_0 \mathcal{L}[1] + c_1 \mathcal{L}[t] + c_2 \mathcal{L}[t^2] + \cdots + c_k \mathcal{L}\left[t^k\right] + \cdots + c_n \mathcal{L}[t^n]$$

$$= \frac{c_0}{s} + \frac{c_1}{s^2} + \frac{2c_2}{s^3} + \frac{6c_3}{s^4} + \cdots + \frac{k!c_k}{s^{k+1}} + \cdots + \frac{n!c_n}{s^{n+1}} \quad (s > 0).$$

If $a$ is a real number, let us find the Laplace transform of $f(t) = e^{at}$, an important function for us because of its frequent appearance in differential equations. By definition,

$$\mathcal{L}\left[e^{at}\right] = \int_0^\infty e^{at} e^{-st} dt = \lim_{b\to\infty} \int_0^b e^{(a-s)t} dt$$

$$= \lim_{b\to\infty} \frac{e^{(a-s)t}}{(a-s)}\bigg|_0^b = \frac{1}{a-s}\left(\lim_{b\to\infty}\left(e^{(a-s)b} - 1\right)\right) = -\frac{1}{a-s}$$

$$= \frac{1}{s-a} \quad \text{for } s > a.$$

Why is this assumption about $s$ needed? In what step is the assumption crucial?

To have tools with which to handle a variety of differential equations, we have to stock our warehouse with different Laplace transforms. Another basic function we should deal with is $\sin at$, where $a$ is a real number. This transform requires two integrations by parts:

For $s > 0$, $\mathcal{L}[\sin at] =$

$$\int_0^\infty \sin at\, e^{-st} dt = \lim_{b\to\infty} \sin at\, \frac{e^{-st}}{-s}\bigg|_0^b - \int_0^\infty a\cos at\,\frac{e^{-st}}{-s} dt = \frac{a}{s}\int_0^\infty \cos at\, e^{-st} dt$$

$$= \frac{a}{s}\left(\lim_{b\to\infty}\cos at\,\frac{e^{-st}}{-s}\bigg|_0^b - \int_0^\infty -a\sin at\,\frac{e^{-st}}{-s} dt\right)$$

$$= \frac{a}{s}\left(\frac{1}{s} - \frac{a}{s}L[\sin at]\right),$$

so that

$$\left(1 + \frac{a^2}{s^2}\right)\mathcal{L}[\sin at] = \frac{a}{s^2} \quad \text{and} \quad \mathcal{L}[\sin at] = \frac{a}{s^2 + a^2}.$$

Using one of the steps from this result, we can easily show that $\mathcal{L}[\cos at] = \frac{s}{s^2+a^2}$. (See Problem A4 in Exercises 6.1.)

To help set the stage for a type of applied differential equation problem that can be handled neatly by using the Laplace transform, let's find $\mathcal{L}[f(t)]$ for the piecewise continuous function defined as follows:

$$f(t) = \begin{cases} t & \text{for } 0 \le t \le 2 \\ 4 - t & \text{for } 2 \le t \le 4 \\ 0 & \text{for } t \ge 4. \end{cases}$$

You should sketch the graph of this function. All we have to do is split the integral in Definition (6.1.1) into three pieces, one corresponding to the interval $[0, 2]$, another corresponding to

[2, 4], and the last matching $[4, \infty)$:

$$\mathcal{L}[f(t)] = \int_0^2 te^{-st}dt + \int_2^4 (4 - t)e^{-st}dt + \int_4^\infty 0 \cdot e^{-st}dt$$

$$= \frac{1 - e^{-2s} - 2se^{-2s}}{s^2} + \frac{e^{-4s} - e^{-2s} + 2se^{-2s}}{s^2} = \frac{1 + e^{-4s} - 2e^{-2s}}{s^2}$$

for $s > 0$. (Carry out all the integrations yourself!)

Finally, before we can apply Laplace transforms to the solution of differential equations, we have to know the transforms of $f', f''$, and higher-order derivatives. So suppose that $F(s) = \mathcal{L}[f(t)]$ exists for $s > c$. Then we have, for $s > c$,

$$\mathcal{L}[f'(t)] = \int_0^\infty f'(t)e^{-st}dt$$

$$= \lim_{b \to \infty} f(t)e^{-st}\Big|_0^b + \int_0^\infty sf(t)e^{-st}dt = -f(0) + s\mathcal{L}[f(t)],$$

which we can write as

$$\mathcal{L}[f'(t)] = s\mathcal{L}[f(t)] - f(0). \tag{6.1.4}$$

Note that in this derivation, we are assuming that $f(b)e^{-sb} \to 0$ as $b \to \infty$.

Now if we assume that $f(be^{-sb})$ also tends to 0 as $b \to \infty$, we can apply Formula (6.1.4) twice, first with $f$ replaced by $f'$, to get

$$\mathcal{L}[f''(t)] = -f'(0) + s\mathcal{L}[f'(t)] = -f'(0) + s[s\mathcal{L}[f(t)] - f(0)]$$

$$= -f'(0) + s^2\mathcal{L}[f(t)] - sf(0),$$

so we can write

$$\mathcal{L}[f''(t)] = s^2\mathcal{L}[f(t)] - sf(0) - f'(0) \quad (\text{for } s > c). \tag{6.1.5}$$

In general, for any positive integer $n$, if the $n$th derivative is continuous (or piecewise continuous), and all the lower-order derivatives are continuous and have the proper growth rate, then

$$\mathcal{L}[f^{(n)}(t)] = s^n\mathcal{L}[f(t)] - \sum_{i=1}^n s^{n-i}f^{(i-1)}(0)$$

$$= s^n\mathcal{L}[f(t)] - s^{n-1}f(0) - s^{n-2}f'(0) - \cdots - sf^{(n-2)}(0) - f^{(n-1)}(0). \tag{6.1.6}$$

It is important to note that this last formula implies that a linear differential equation with constant coefficients will be transformed into a purely algebraic equation—that is, an equation without derivatives. (Recognize that $f^{(k)}(0)$ is a number for $k \geq 0$.)

As a hint of what we'll be doing in the next few sections, we'll convert a differential equation into an algebraic expression by using the Laplace transform.

### ■ Example 6.1.1 The Laplace Transform of a Differential Equation

Let's look at the initial-value problem

$$x'' + 3x' + 2x = 12e^{2t};\; x(0) = 1,\; x'(0) = -1.$$

We're going to apply the Laplace transform to both sides of the equation and substitute the initial conditions where appropriate:

$$\mathcal{L}\left[x'' + 3x' + 2x\right] = \mathcal{L}\left[12e^{2t}\right],$$

or, using the linearity of the transform—(6.1.3),

$$\mathcal{L}[x''] + 3\,\mathcal{L}[x'] + 2\,\mathcal{L}[x] = 12\,\mathcal{L}\left[e^{2t}\right].$$

We have already calculated the Laplace transform of an exponential function. This, together with Formulas (6.1.4) and (6.1.5), allows us to write

$$\left\{s^2\mathcal{L}[x(t)] - sx(0) - x'(0)\right\} + 3\left\{s\,\mathcal{L}[x(t)] - x(0)\right\} + 2\,\mathcal{L}[x(t)] = \frac{12}{s - 2}.$$

Now we substitute the given initial conditions to get

$$\left\{s^2\mathcal{L}[x(t)] - s + 1\right\} + 3\left\{s\,\mathcal{L}[x(t)] - 1\right\} + 2\,\mathcal{L}[x(t)] = \frac{12}{s - 2}.$$

Finally, collecting like terms, we find that

$$\left(s^2 + 3s + 2\right)\mathcal{L}[x(t)] = \frac{12}{s - 2} + s + 2 = \frac{s^2 + 8}{s - 2},$$

so we can solve for $\mathcal{L}[x(t)]$:

$$\mathcal{L}[x(t)] = \frac{s^2 + 8}{s - 2} \cdot \frac{1}{s^2 + 3s + 2} = \frac{s^2 + 8}{(s - 2)\left(s^2 + 3s + 2\right)}$$

$$= \frac{s^2 + 8}{(s - 2)(s + 2)(s + 1)}.$$

*Now what?* We have an unknown function, the solution of an IVP, whose Laplace transform is known. If we can *reverse* the process and figure out what function has this Laplace transform, we can solve our original initial-value problem. This is what we'll focus on in the next section. ∎

## Exercises 6.1

### A

Use the definition and properties of the Laplace transform to find the Laplace transform of the functions in Problems 1–16 and specify the values of $s$ for which each transform exists.

**1.** $f(t) = t^2$

**2.** $g(t) = t^2 - t$

**3.** $f(t) = t^3$

**4.** $h(t) = \cos at$, where $a$ is a real number

**5.** $F(t) = t\,e^{at}$, where $a$ is a real number

**6.** $s(t) = 2\cos 3t$

**7.** $u(t) = 10 + 100\,e^{2\,t}$

**8.** $G(t) = \dfrac{e^{at} - e^{bt}}{a - b}$, where $a$ and $b$ are real numbers, $a \neq b$

**9.** $H(t) = 2t^3 - 7t^2 + 5t - 17$

**10.** $r(t) = 3\sin 5t - 4\cos 5t$

**11.** $U(t) = 2e^t - 3e^{-t} + 4t^2$

**12.** $S(t) = 3 - 5e^{2t} + 4\sin t - 7\cos 3t$

**13.** $F(t) = \begin{cases} t & \text{for } 0 < t < 4 \\ 0 & \text{for } t > 4 \end{cases}$

**14.** $f_a(t) = \begin{cases} t/a & \text{for } t < a \\ 1 & \text{for } t \geq a \end{cases}$, where $a \geq 0$

**15.** $A(t) = \begin{cases} 1 & \text{for } 0 \leq t \leq 1 \\ 2 - t & \text{for } 1 \leq t < 2 \\ 0 & \text{for } 2 \leq t \end{cases}$

**16.** $B(t) = \begin{cases} 1 & \text{for } 0 \leq t \leq 2\pi \\ \cos t & \text{for } 2\pi \leq t \leq 7\pi/2 \\ 0 & \text{for } 7\pi/2 \leq t < \infty \end{cases}$

In Problems 17–25, find the Laplace transform of the solution of each initial value problem, assuming that the Laplace transform exists in each case. (Do not try to solve the IVPs.)

**17.** $y' - y = 0;\quad y(0) = 1$

**18.** $y' + y = e^{-x};\quad y(0) = 1$

**19.** $y' = -y + e^{-2t};\quad y(0) = 2$

**20.** $y' = -y + t^2;\quad y(0) = 1$

**21.** $y'' + y = 0;\quad y(0) = 1, y'(0) = 0$

**22.** $y'' + 4y' + 4y = 0;\quad y(0) = 1, y'(0) = 1$

**23.** $y'' - y' - 2y = 5\sin x;\quad y(0) = 1, y'(0) = -1$

**24.** $2y'' + 3y' - 2y = 1; \quad y(0) = 0, y'(0) = 1/2$

**25.** $y''' - 2y'' + y' = 2e^x + 2x; \quad y(0) = 0, y'(0) = 0, y''(0) = 0$

## B

**1.** Recall that the *hyperbolic sine* and *hyperbolic cosine* are defined as $\sinh(at) = (e^{at} - e^{-at})/2$ and $\cosh(at) = (e^{at} + e^{-at})/2$, respectively. Find the Laplace transform of these two functions and give the values of $s$ for which the transforms exist.

**2.** Consider the IVP $y'' + 3y = w(t); \; y(0) = 2, \; y'(0) = 0$, where

$$w(t) = \begin{cases} t & \text{for } 0 \le t < 1 \\ 1 & \text{for } t \ge 1. \end{cases}$$

    **a.** Find $\mathcal{L}[w(t)]$.

    **b.** Find $\mathcal{L}[y(t)]$.

**3.** Find the Laplace transform of the following periodic function.

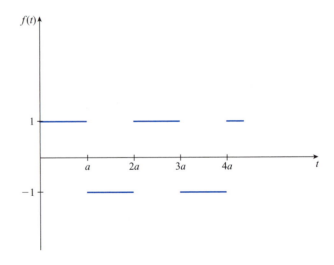

**4.** Determine $\mathcal{L}[\sin at]$ using the fact that $\sin at$ satisfies the differential equation $y'' + a^2y = 0$. Do the same for $\mathcal{L}[\cos at]$.

**5.** Apply Formula (6.1.5) to the function $f''(t)$ to show that

$$\mathcal{L}[f'''(t)] = s^3 \mathcal{L}[f(t)] - s^2 f(0) - s f'(0) - f''(0).$$

**6.** If $\mathcal{L}[f(t)]$ exists for $s = \alpha$, prove that it exists for all $s > \alpha$.

**7.** If $f(t) = e^{t^2}$ for $t \ge 0$, show that there are no constants $M$ and $K$ such that $\left| f(t) \right| < e^{Mt}$ for all $t \ge K$. Thus, show that the Laplace transform of $f(t)$ doesn't exist. [*Hint:* $e^{t^2} < e^{Mt}$ implies that $t^2 < Mt$ for $t$ large enough.]

8. Prove that $\mathcal{L}\left[e^{\sqrt[3]{t}}\right]$ exists, but $\mathcal{L}\left[e^{e^t}\right]$ does not exist.

9. Show that $f(t) = 1$ for $t > 0$ and $g(t) = \begin{cases} 5 & \text{for } t = 3 \\ 1 & \text{for } t \neq 3 \end{cases}$ have the same Laplace transforms, namely $1/s$ for $s > 0$. Can you think of other functions with the same Laplace transform? Explain your answer.

10. Define

$$f(t) = \begin{cases} 1 & \text{for } t = 0 \\ t & \text{for } t > 0. \end{cases}$$

   a. Find $\mathcal{L}[f(t)]$ and $\mathcal{L}[f'(t)]$.
   b. Is it true for this function that $\mathcal{L}[f'(t)] = s\,\mathcal{L}[f(t)] - f(0)$? Explain.

## C

1. Use Definition (6.1.1) and the fact that $\int_0^\infty e^{-x^2}\,dx = \frac{\sqrt{\pi}}{2}$ to find the Laplace transform of $f(t) = \frac{1}{\sqrt{t}} = t^{-\frac{1}{2}}$. [*Hint*: Make the substitution $t = \frac{u^2}{s}$.]

2. Suppose $f(t)$ is a periodic function with period $T$—that is, $f(t + T) = f(t)$ for all $t$—such that $\mathcal{L}[f(t)]$ exists.

   a. Show that

$$\mathcal{L}[f(t)] = \frac{1}{1 - e^{-sT}} \int_0^T f(t)\,e^{-sT}\,dt.$$

   [*Hint*: Make a substitution and recall geometric series.]
   b. Use the result of part (a) to find the Laplace transform of the function given in Problem B3.

3. Suppose that $a$ is any real number and that $F(s) = \mathcal{L}[f(t)]$. Show that

$$\mathcal{L}\left[e^{at}f(t)\right] = F(s - a) \text{ for } s > a.$$

[This is usually called the *First Shift Formula*. See Formulas (6.3.1a) and (6.3.1b) for the *Second Shift Formula*.]

# 6.2 THE INVERSE TRANSFORM AND THE CONVOLUTION

## 6.2.1 The Inverse Laplace Transform

Recall that in Example 6.1.1 we took an initial-value problem, applied the Laplace transform, and then wound up with the Laplace transform of the solution of the IVP. Now we would like to *reverse* the transformation process so that, given $\mathcal{L}[f(t)]$ as a function of the parameter $s$, we can find $f(t)$. This involves the idea of the **inverse Laplace transform**, $\mathcal{L}^{-1}$.

Now think back to the concept of inverse of a *function*. When you first encountered the inverse in calculus or precalculus, you may have worked with both the formal definition and the graphical interpretation in terms of a "horizontal line test." In any case, the important idea is that to have an inverse function $f^{-1}$ we must guarantee that for any element in the *range* of the original function $f$, there is one and only one corresponding element in the *domain* of $f$. Another way of saying this is that *a function has an inverse if and only if it is a one-to-one function*.

For our purposes, the important fact is that *if the Laplace transforms of the continuous functions f and g exist and are equal for $s \geq c$ (c,a constant), then $f(t) = g(t)$ for all $t \geq 0$*. This says that a continuous function can be uniquely recovered from its Laplace transform. (Compare Problem B9 in Exercises 6.1.) Letting $\mathcal{L}[f(t)] = F(s)$, we can express the definition of the inverse Laplace transform as follows:

$$\mathcal{L}^{-1}[F] = f \text{ if and only if } \mathcal{L}[f] = F. \tag{6.2.1}$$

We can easily verify (see Problem B1 in Exercises 6.1) that the inverse Laplace transform is a linear transformation:

$$\mathcal{L}^{-1}\left[c_1 F(t) + c_2 G(t)\right] = c_1 \mathcal{L}^{-1}[F(t)] + c_2 \mathcal{L}^{-1}[G(t)]. \tag{6.2.2}$$

Now how do we find the inverse Laplace transform in practice? It turns out that the relationship between calculating a Laplace transform and determining its inverse is similar to that between differentiation and antidifferentiation.

This means that in calculus, the indefinite integral of a function $f$ answers the question "What is a function whose derivative is $f$?" (Note that in calculus the answer to this question is not unique.) Just as a list of differentiation formulas helps us to construct a list of antidifferentiation formulas (indefinite integrals), so will a table of Laplace transforms aid us in finding inverses. In the examples that follow, we will use the information in Table 6.1. Some of these transforms were derived in Section 6.1; others were given as exercises.

Now let's return to Example 6.1.1 and solve the initial-value problem using Laplace transforms and the inverse Laplace transform.

### ■ Example 6.2.1 Solving an IVP Using the Inverse Laplace Transform

The IVP was $x'' + 3x' + 2x = 12e^{2t}, x(0) = 1, x'(0) = -1$, and we found that

$$\mathcal{L}[x(t)] = \frac{s^2 + 8}{(s - 2)(s + 2)(s + 1)}.$$

If we try to work with the given expression for the transform (the single rational expression in $s$), we would have a tough time figuring out what function $x(t)$ might have this as its Laplace transform. This expression doesn't seem to correspond to any of the forms in the second column of Table 6.1.

**Table 6.1** Some Laplace Transforms

|     | $f(t)$ | $F(s) = \mathcal{L}[f(t)]$ |
| --- | --- | --- |
| 1 | $t^n (n = 0, 1, 2, \ldots)$ | $\dfrac{n!}{s^{n+1}}, s > 0$ |
| 2 | $e^{at}$ | $\dfrac{1}{s - a}, s > a$ |
| 3 | $\sin at$ | $\dfrac{a}{s^2 + a^2}, s > 0$ |
| 4 | $\cos at$ | $\dfrac{s}{s^2 + a^2}, s > 0$ |
| 5 | $e^{at} \sin bt$ | $\dfrac{b}{(s - a)^2 + b^2}, s > a$ |
| 6 | $e^{at} \cos bt$ | $\dfrac{s - a}{(s - a)^2 + b^2}, s > a$ |
| 7 | $t \sin at$ | $\dfrac{2as}{(s^2 + a^2)^2}, s > 0$ |
| 8 | $t \cos at$ | $\dfrac{s^2 - a^2}{(s^2 + a^2)^2}, s > 0$ |
| 9 | $f'(t)$ | $sF(s) - f(0)$ |
| 10 | $f''(t)$ | $s^2 F(s) - sf(0) - f'(0)$ |
| 11 | $e^{at} f(t)$ | $F(s - a), s > a$ |

However, we can use the partial-fractions technique to express the transform as the sum of three simpler terms, each of which matches an entry in the table:

$$\frac{s^2 + 8}{(s - 2)(s + 2)(s + 1)} = \frac{1}{s - 2} + \frac{3}{s + 2} - \frac{3}{s + 1}.$$

You should be able to see, for example, that the term $\frac{3}{s+2} \left( = \frac{3}{s-(-2)} \right)$ is the Laplace transform of $3e^{-2t}$. Applying the inverse transform to each side of

$$\mathcal{L}[x(t)] = \frac{1}{s - 2} + \frac{3}{s + 2} - \frac{3}{s + 1}$$

and using (6.2.1) and the linearity of $\mathcal{L}^{-1}$, we see that

$$x(t) = \mathcal{L}^{-1}[\mathcal{L}[x(t)]] = \mathcal{L}^{-1}\left[\frac{1}{s-2} + \frac{3}{s+2} - \frac{3}{s+1}\right]$$

$$= \mathcal{L}^{-1}\left[\frac{1}{s-2}\right] + 3\mathcal{L}^{-1}\left[\frac{1}{s+2}\right] - 3\mathcal{L}^{-1}\left[\frac{1}{s+1}\right]$$

$$= e^{2t} + 3e^{-2t} - 3e^{-t},$$

where we have used formula 2 from Table 6.1 three times.  ∎

The alternative to using the Laplace transform to solve the initial-value problem in the preceding example is to go back to the method first explained in Section 4.2: First find the general solution of the homogeneous equation $x'' + 3x' + 2x = 0$; then find a particular solution of the nonhomogeneous equation; $x'' + 3x' + 2x = 12e^{2t}$; finally, add these two solutions together to get the general solution of the original *nonhomogeneous* equation. And even then we would not be finished, because we would have to use the initial conditions to determine the two arbitrary constants in the general solution. Note that the Laplace transform method enables us to handle the nonhomogeneous equation and initial conditions all at once.

Now let's see what the Laplace transform method does in an important applied problem that we first saw as Example 2.2.5. (Also see Problems B9–B11 in Exercises 2.2.)

### ■ Example 6.2.2 Solving a Circuit Problem via the Laplace Transform

The current $I$ flowing in a particular electrical circuit can be described by the initial-value problem $L\frac{dI}{dt} + RI = v_0 \sin(\omega t), I(0) = 0$. Here, $L$, $R$, $v_0$, and $\omega$ are positive constants.

First, we apply the Laplace transform to each side of the differential equation:

$$\mathcal{L}\left[L\frac{dI}{dt} + RI\right] = \mathcal{L}[v_0 \sin(\omega t)]$$

$$L\mathcal{L}\left[\frac{dI}{dt}\right] + R\mathcal{L}[I(t)] = v_0\,\mathcal{L}[\sin(\omega t)]$$

$$sL\mathcal{L}[I(t)] - LI(0) + R\mathcal{L}[I(t)] = v_0\left(\frac{\omega}{s^2 + \omega^2}\right)$$

$$(Ls + R)\mathcal{L}[I(t)] - LI(0) = v_0\left(\frac{\omega}{s^2 + \omega^2}\right)$$

$$L\left(s + \frac{R}{L}\right)\mathcal{L}[I(t)] = v_0\left(\frac{\omega}{s^2 + \omega^2}\right)$$

so that we have $\mathcal{L}[I(t)] = \left(\frac{v_0}{L}\right) \cdot \omega \cdot \frac{1}{\left(s+\frac{R}{L}\right)(s^2+\omega^2)}$. To find the inverse Laplace transform, we have to use the method of partial fractions on the right-hand side:

$$\frac{1}{\left(s + \frac{R}{L}\right)(s^2 + \omega^2)} = \frac{A}{\left(s + \frac{R}{L}\right)} + \frac{Bs + C}{s^2 + \omega^2}.$$

With a little effort, we find that

$$\frac{1}{\left(s + \frac{R}{L}\right)(s^2 + \omega^2)} = \frac{\frac{1}{\left(\frac{R^2}{L^2}+\omega^2\right)}}{s + \frac{R}{L}} + \frac{-\frac{1}{\left(\frac{R^2}{L^2}+\omega^2\right)}s + \frac{\left(\frac{R}{L}\right)}{\left(\frac{R^2}{L^2}+\omega^2\right)}}{s^2 + \omega^2}$$

and

$$\mathcal{L}[I(t)] = \left(\frac{v_0}{L}\right) \cdot \omega \cdot \frac{1}{\left(s + \frac{R}{L}\right)(s^2 + \omega^2)}$$

$$= \left(\frac{v_0}{L}\right) \cdot \frac{\omega}{\left(\frac{R^2}{L^2} + \omega^2\right)} \left\{\frac{1}{s + \frac{R}{L}} + \frac{-s}{s^2 + \omega^2} + \frac{\frac{R}{L}}{s^2 + \omega^2}\right\}.$$

*Check the last three equalities.* The final step is to apply the inverse Laplace transform to both sides of this last equation and then use formulas 2, 3, and 4 from Table 6.1.

$$I(t) = \left(\frac{v_0}{L}\right) \frac{\omega}{\left(\frac{R^2}{L^2} + \omega^2\right)} \left\{\mathcal{L}^{-1}\left[\frac{1}{s + \frac{R}{L}}\right] - \mathcal{L}^{-1}\left[\frac{s}{s^2 + \omega^2}\right] + \frac{R}{L}\mathcal{L}^{-1}\left[\frac{1}{s^2 + \omega^2}\right]\right\}$$

$$= \left(\frac{v_0}{L}\right) \frac{\omega}{\left(\frac{R^2}{L^2} + \omega^2\right)} \left\{e^{-\frac{R}{L}t} - \cos(\omega t) + \frac{R}{L}\frac{1}{\omega}\sin(\omega t)\right\}$$

$$= \left(\frac{v_0}{L}\right) \frac{1}{\left(\frac{R^2}{L^2} + \omega^2\right)} \left\{\omega e^{-\frac{R}{L}t} - \omega \cos(\omega t) + \frac{R}{L}\sin(\omega t)\right\}.$$

Compare this solution to the one obtained in Example 2.2.5. ■

## ■ Example 6.2.3 Solving an IVP Using the Inverse Laplace Transform

Let's look at the initial-value problem $\ddot{x} - 2\dot{x} = e^t\,(t - 3)$, $x(0) = 2 = \dot{x}(0)$. As before, we take Laplace transforms of both sides and use the table. Letting $\mathcal{L}[x(t)] = X(s)$, we get

$$\{s^2 X(s) - sx(0) - \dot{x}(0)\} - 2\{sX(s) - x(0)\} = \mathcal{L}\left[e^t(t - 3)\right]. \tag{#}$$

To evaluate the right-hand side, first note that

$$\mathcal{L}[t - 3] = \frac{1}{s^2} - \frac{3}{s} = F(s),$$

so if we use entry 11 in Table 6.1 (with $a = 1$), we get

$$\mathcal{L}\left[e^t(t - 3)\right] = F(s - 1) = \frac{1}{(s - 1)^2} - \frac{3}{s - 1} = \frac{4 - 3s}{(s - 1)^2}.$$

If we return to (#) and put in our initial conditions, we get

$$s(s - 2)X(s) = 2s - 2 + \frac{4 - 3s}{(s - 1)^2}$$

$$= \frac{2(s - 1)^3 + (4 - 3s)}{(s - 1)^2} = \frac{(s - 2)\left(2s^2 - 2s - 1\right)}{(s - 1)^2}.$$

Therefore, we conclude that

$$X(s) = \frac{2s^2 - 2s - 1}{s(s - 1)^2} = \frac{3}{s - 1} - \frac{1}{(s - 1)^2} - \frac{1}{s}.$$

We go to the table to find the function $x(t) = \mathcal{L}^{-1}[X(t)]$, where we have to use entries 1 and 11 for the second term. The solution of the IVP is $x(t) = 3e^t - te^t - 1$. (*Check that this is the solution.*) ∎

## 6.2.2 The Convolution

In each of the preceding three examples, applying the Laplace transform yielded an expression for $\mathcal{L}[f(t)]$ that seemed to involve the product of two or more transforms. Because we didn't know any way to find the inverse transform of such products, we had to resort to the messiness of a partial-fraction decomposition. This, at least, enabled us to use the linearity of the inverse transform.

There is, however, a way to deal with this problem—a method that involves a special product of functions.

> The **convolution** of two functions $f$ and $g$ is the integral
> $$(f * g)(t) = \int_0^t f(r)g(t - r)dr,$$
> provided that the integral exists for $t > 0$.

For example, the convolution of $\cos t$ and $t$ is

$$(\cos t) * t = \int_0^t (\cos r)(t - r)dr = \int_0^t t \cos r \, dr - \int_0^t r \cos r \, dr$$

$$= t \int_0^t \cos r \, dr - \int_0^t r \cos r \, dr = 1 - \cos t$$

after using integration by parts for the last integral. For this example, you should verify that $(\cos t) * (t) = (t) * (\cos t)$.

Convolution has important algebraic properties (see Problem B5 in Exercises 6.2), but the most significant property for us right now is that the Laplace transform of a convolution of two functions is equal to the product of the Laplace transforms of these two functions. More precisely, suppose that $f$ and $g$ are two functions whose Laplace transforms exist. Let $F(s) = \mathcal{L}[f(t)]$ and $G(s) = \mathcal{L}[g(t)]$. Then the **Convolution Theorem** says that

$$\mathcal{L}\left[(f * g)(t)\right] = \mathcal{L}\left[\int_0^t f(r)g(t-r)dr\right] = \mathcal{L}[f(t)] \cdot \mathcal{L}[g(t)] = F(s) \cdot G(s).$$

Now let's revisit part of Example 6.2.2 to see how the convolution property helps us.

### ■ Example 6.2.4 (Example 6.2.2 Revisited)

How can we find $\mathcal{L}^{-1}\left[\dfrac{1}{\left(s+\frac{R}{L}\right)(s^2+\omega^2)}\right]$?

The expression inside the brackets is the product of two transforms $F$ and $G$:

$$F(s)G(s) = \left[\frac{1}{\left(s + \frac{R}{L}\right)\left(s^2 + \omega^2\right)}\right],$$

where $F(s) = \dfrac{1}{\left(s+\frac{R}{L}\right)}$ and $G(s) = \dfrac{1}{(s^2+\omega^2)}$. Entries 2 and 3 of Table 6.1 tell us that $f(t) = \mathcal{L}^{-1}[F(s)] = e^{-\frac{R}{L}t}$ and $g(t) = \mathcal{L}^{-1}[G(s)] = \frac{1}{\omega}\sin(\omega t)$. Then the Convolution Theorem enables us to conclude that

$$\mathcal{L}^{-1}\left[F(s)G(s)\right] = \int_0^t f(r)g(t-r)dr,$$

or

$$\mathcal{L}^{-1}\left[\frac{1}{\left(s + \frac{R}{L}\right)\left(s^2 + \omega^2\right)}\right] = \int_0^t e^{-\frac{R}{L}r}\frac{1}{\omega}\sin\omega(t-r)dr$$

$$= \frac{1}{\omega}\int_0^t e^{-\frac{R}{L}r}\sin\omega(t-r)dr.$$

A computer algebra system evaluates this integral as

$$\frac{L\left(\omega L e^{\left(-\frac{LR}{t}\right)} - \omega L\cos(\omega t) + R\sin(\omega t)\right)}{\left(R^2 + \omega^2 L^2\right)\omega}.$$

A bit of algebra will show you that this corresponds to the inverse transform found in Example 6.2.2 via partial fractions. (*Do the work.*)  ∎

### 6.2.3 Integral Equations and Integro-Differential Equations

The Convolution Theorem is certainly useful in solving differential equations, but it can also help us solve **integral equations**—equations involving an integral of the unknown function— and **integro-differential equations**—those involving both a derivative and an integral of the unknown function.

### ∎ Example 6.2.5 The Convolution Theorem and an Integral Equation

A store manager finds that the proportion of merchandise that remains unsold at time $t$ after she has bought the merchandise is given by $f(t) = e^{-1.5t}$. She wants to find the rate at which she should purchase the merchandise so that the stock in the store remains constant.

Suppose that the store starts off by buying an amount $A$ of the merchandise at time $t = 0$ and buys at a rate $r(t)$ subsequently. Over a short time interval $u \leq t \leq u + \Delta u$, an amount $r(t) \cdot \Delta u$ is bought by the store, and at time $t$ the portion of this remaining unsold is $e^{-1.5(t-u)}r(u)\Delta u$. Then the amount of previously purchased merchandise remaining unsold at time $t$ is given by

$$Ae^{-1.5t} + \int_0^t e^{-1.5(t-u)}r(u)du.$$

Because this is the total stock of the store and the store manager wants it to remain constant at its initial value, we must have

$$A = Ae^{-1.5t} + \int_0^t e^{-1.5(t-u)}r(u)du,$$

and the required restocking rate $r(t)$ is the solution of this integral equation.

If we look carefully at the integral on the right-hand side of this last equation, we should recognize something familiar about its form: It looks like a convolution—in fact, it is $e^{-1.5t} * r(t)$. Now we can rewrite the integral equation in the form

$$A = Ae^{-1.5t} + \left(e^{-1.5t} * r(t)\right).$$

Taking the Laplace transform of each side and letting $R(s) = \mathcal{L}[r(t)]$, we get

$$\mathcal{L}[A] = A \mathcal{L}\left[e^{-1.5t}\right] + \mathcal{L}\left[e^{-1.5t} * r(t)\right] = \frac{A}{s+1.5} + \frac{1}{s+1.5} \cdot R(s),$$

$$\frac{A}{s} = \frac{A}{s+1.5} + \frac{1}{s+1.5} \cdot R(s),$$

$$(s + 1.5) \left( \frac{A}{s} - \frac{A}{s + 1.5} \right) = R(s),$$

$$\frac{1.5A}{s} = R(s).$$

Applying the inverse Laplace transform to each side, we find that $r(t) = 1.5A$. That is, the restocking rate should be a constant one-and-a-half times the original amount bought. (*Check that this is a solution of our original integral equation.*)  ■

## ■ Example 6.2.6 An Integro-Differential Equation

The following integro-differential equation can also be solved using the properties of the Laplace transform:

$$\frac{dx}{dt} + x(t) - \int_0^t x(r) \sin(t - r) dr = -\sin t, x(0) = 1.$$

As in the preceding example, we recognize that the integral in our equation represents a convolution, this time $(x * \sin)(t)$. Therefore, taking the Laplace transform of each side of the equation, we get

$$\mathcal{L}[dx/dt] + \mathcal{L}[x(t)] - \mathcal{L}[(x * \sin)(t)] = \mathcal{L}[-\sin t],$$

or, using formula 10 in Table 6.1 and the Convolution Theorem,

$$[s \, \mathcal{L}[x(t)] - x(0)] + \mathcal{L}[x(t)] - \mathcal{L}[x(t)] \cdot \mathcal{L}[\sin t] = -\frac{1}{s^2 + 1},$$

which becomes

$$[s \, \mathcal{L}[x(t)] - 1] + \mathcal{L}[x(t)] - \mathcal{L}[x(t)] \cdot \frac{1}{s^2 + 1} = -\frac{1}{s^2 + 1}.$$

This simplifies to

$$\left( \frac{s^3 + s^2 + s}{s^2 + 1} \right) \mathcal{L}[x(t)] = \frac{s^2}{s^2 + 1},$$

so we wind up with $\mathcal{L}[x(t)] = \dfrac{s^2}{s^3 + s^2 + s} = \dfrac{s}{s^2 + s + 1}.$

A bit of clever algebra shows us that

$$\frac{s}{s^2 + s + 1} = \frac{s + \frac{1}{2}}{\left(s + \frac{1}{2}\right)^2 + \left(\frac{\sqrt{3}}{2}\right)^2} - \frac{\frac{1}{2}}{\left(s + \frac{1}{2}\right)^2 + \left(\frac{\sqrt{3}}{2}\right)^2}$$

$$= \frac{s - \left(-\frac{1}{2}\right)}{\left(s - \left(-\frac{1}{2}\right)\right)^2 + \left(\frac{\sqrt{3}}{2}\right)^2} - \frac{1}{\sqrt{3}} \frac{\frac{\sqrt{3}}{2}}{\left(s - \left(-\frac{1}{2}\right)\right)^2 + \left(\frac{\sqrt{3}}{2}\right)^2}.$$

Using formulas 5 and 6 of Table 6.1 to invert this transform, we find that

$$x(t) = e^{-t/2} \cos\left(\frac{\sqrt{3}t}{2}\right) - \frac{1}{\sqrt{3}} e^{-t/2} \sin\left(\frac{\sqrt{3}t}{2}\right).$$

(*Checking that this is the solution involves a bit of work, but try it!*) ■

### 6.2.4 The Laplace Transform and Technology

Most computer algebra systems have built-in Laplace transform and inverse transform capabilities. In particular, some systems (for example, *Maple*) have sophisticated differential equation solvers with a "laplace" option for initial-value problems. If you have such an option at your command, learn to use it. However, realize that all the machinery is under the covers, so you have to develop an understanding of what the system is really doing.

Be aware that some computer algebra systems (such as *Mathematica* and MATLAB) can find Laplace transforms and their inverses but have no direct way of solving a linear IVP with these tools. In this case, you have to apply the Laplace transform to the differential equation, solve for the transform $\mathcal{L}[x(t)]$ of the solution algebraically (via a *solve* command or by hand), use technology to find the inverse transform $\mathcal{L}^{-1}[\mathcal{L}[x(t)]]$, and finally substitute the initial conditions.

Determine what your options are in using technology to solve IVPs via the Laplace transform. Some of the exercises that follow will help you do this.

### Exercises 6.2

**A**

1. Find the inverse Laplace transform of $\frac{1}{s^2+9}$.

2. Find the inverse Laplace transform of $\frac{s}{s^2-a^2}$.

3. Find the inverse Laplace transform of $\frac{s}{s^2+2}$.

4. Find the inverse Laplace transform of $\frac{a}{s^2(s^2+a^2)}$.

5. Find the inverse Laplace transform of $\frac{1}{s(s^2+2s+2)}$.

6. Find the inverse Laplace transform of $\frac{2s-10}{s^2-4s+20}$.

7. Find the inverse Laplace transform of $\frac{4}{s-2} - \frac{3s}{s^2+16} + \frac{5}{s^2+4}$.

8. Find the inverse Laplace transform of $\frac{3s+7}{s^2-2s-3}$.

9. Find the inverse Laplace transform of $\frac{2s^2-4}{(s+1)(s-2)(s-3)}$.

**B**

1. If $F(t)$ and $G(t)$ are the Laplace transforms of $f(t)$ and $g(t)$, respectively, and $c_1$ and $c_2$ are constants, show that

$$\mathcal{L}^{-1}[c_1 F(t) + c_2 G(t)] = c_1 \mathcal{L}^{-1}[F(t)] + c_2 \mathcal{L}^{-1}[G(t)].$$

2. Suppose $y_1(t)$ is the solution of the IVP
   $y'' + ay' + by = f_1(t);\ y(0) = y'(0) = 0$, where $a$ and $b$ are constants.

   **a.** Compute an expression for $\mathcal{L}[y_1]$.
   **b.** Suppose $y_2(t)$ is the solution of the IVP

   $$y'' + ay' + by = f_2(t);\ y(0) = y'(0) = 0$$

   for a different forcing function $f_2(t)$. Show that

   $$\frac{\mathcal{L}[f_2]}{\mathcal{L}[y_2]} = \frac{\mathcal{L}[f_1]}{\mathcal{L}[y_1]}.$$

   **c.** Show that $\mathcal{L}[y_2] = \mathcal{L}[f_2] \cdot \frac{\mathcal{L}[y_1]}{\mathcal{L}[f_1]}$. (This says that we can use the solution with any forcing function and zero initial conditions to compute solutions for other forcing functions.)

3. **a.** Show that the Laplace transform of $t^n f(t)$ is $(-1)^n F^{(n)}(s)$, where $F(s) = \mathcal{L}[f(t)]$.
   **b.** Use the result of part (a) and the derivative of the function $F(s) = \ln(2 + \frac{3}{s})$, $s > 0$, to find its inverse Laplace transform.

4. Find the convolution $f * g$ of each of the following pairs of functions:

   **a.** $f(t) = t^2, g(t) = 1$
   **b.** $f(t) = t, g(t) = e^{-t}$ for $t \geq 0$
   **c.** $f(t) = t^2, g(t) = (t^2 + 1)$ for $t \geq 0$
   **d.** $f(t) = e^{-at}, g(t) = e^{-bt}$ ($a, b$ constants)
   **e.** $f(t) = \cos t, g(t) = \cos t$

   [*Hint*: For part (e) you need some trigonometric identities.]

5. Prove the following properties of the convolution of functions:

   **a.** $f * g = g * f$ [Commutativity]
   **b.** $(f * g) * h = f * (g * h)$ [Associativity]
   **c.** $f * (g + h) = f * g + f * h$ [Distributivity]
   **d.** $f * 0 = 0$, but $f * 1 \neq f$ and $f * f \neq f^2$ in general. (In particular, $1 * 1 \neq 1$.)

6. **a.** Using property (b) of Problem B5, find $1 * 1 * 1$.
   **b.** Find $1 * t * t^2$.

**7.** Use the Convolution Theorem to find the Laplace transform of

$$f(t) = \int_0^t \cos(t - r) \sin r \, dr.$$

**8.** Find the Laplace transform of

$$h(t) = \int_0^t e^{t-v} \sin v \, dv.$$

**9.** Find the solution of the IVP $y'' + 3y' + 2y = 4t^2$ ; $y(0) = 0$, $y'(0) = 0$.

**10.** Solve the IVP $y'' + 4y' + 4y = e^{-2x}$ ; $y(0) = 0$, $y'(0) = 1$.

**11.** Solve the IVP $y''' - 2y'' + y' = 2e^x + 2x$; $y(0) = 0$, $y'(0) = 0$, $y''(0) = 0$.

**12.** Solve the IVP $y'' + 6y' + 9y = H(x)$; $y(0) = 0$, $y'(0) = 0$, where $H(x)$ is a known function of $x$. [*Hint*: Use the Convolution Theorem.]

**13.** An electrical circuit that is initially unforced but is plugged into a particular alternating voltage source at time $t = \pi$ can be modeled by the IVP

$$Q'' + 2Q' + 2Q = \begin{cases} 0 & \text{for } 0 \le t < \pi \\ -\sin t & \text{for } t \ge \pi, \end{cases}$$

with $Q(0) = 0$ and $Q'(0) = 1$. Solve the IVP for $Q(t)$, the charge on the capacitor at time $t$.

**14.** The equation $v\frac{dv}{ds} = v\cos(2s) - v^2$ describes the velocity $v$ of a piston moving into an oil-filled cylinder under a variable force. Here, $s$ is the distance moved in time $t$.

   **a.** Rewrite the given equation as a linear equation with constant coefficients.
   **b.** Assuming that $v(0) = 0$, use the Laplace transform to solve for $v$ as a function of $s$. Is there a singular solution?
   **c.** Use technology to graph the nontrivial solution found in part (b) for $0 \le s \le 20$.

**15.** Solve the integral equation for $f$ : $f(t) = 4t + \int_0^t f(t - r) \sin r \, dr$.

**16.** Solve for $g$ : $g(t) - t = -\int_0^t (t - r) g(r) \, dr$.

**C**

**1.** Solve for $x$ : $\dot{x}(t) = 1 - \int_0^t x(t - r) e^{-2r} \, dr$, $x(0) = 1$.

**2.** Solve the integro-differential equation

$$\dot{y} + y + \int_0^t y(u) \, du = 1, \text{ with } y(0) = 0.$$

**3.** Solve the equation $\dot{x} - 4x + 4 \int_0^t x(u) du = t^3 e^{2t}$, with $x(0) = 0$.

**4.** Solve the equation $f''(x) + \int_0^x e^{2(x-y)} f'(y) dy = 1$; $f(0) = 0, f'(0) = 0$.

## 6.3 TRANSFORMS OF DISCONTINUOUS FUNCTIONS

Differential equations are often used to model complex systems. In some situations, models have to deal with abrupt changes in these systems. In the circuit problem described in Example 2.2.5 (or Example 6.2.2), we have the equation $L\frac{dI}{dt} + RI = v_0 \sin(\omega t)$, where the right-hand side (the forcing term) represents a continuous alternating-current source. Now suppose that the voltage $E(t)$ were applied for only a short period of time and then discontinued. Mathematically, this means that the forcing term would have the form

$$f(t) = \begin{cases} E(t) & \text{for } 0 \le t \le a \\ 0 & \text{for } t > a. \end{cases}$$

Perhaps we have a switch that we can open and close so that the voltage is applied, removed, and then applied again:

$$g(t) = \begin{cases} E(t) & \text{for } 0 \le t \le a \\ 0 & \text{for } a < t < b \\ E(t) & \text{for } t \ge b. \end{cases}$$

Problem B12 in Exercises 2.2, in which advertising expenditure is terminated after a certain period of time, is another illustration of this kind of behavior. The common element here is abrupt change. In mathematical terms, we are dealing with *piecewise continuous functions*.

### 6.3.1 The Heaviside (Unit Step) Function

In Section 6.1 we saw a simple example of the Laplace transform applied to a piecewise continuous function. We computed the transform directly from the definition, breaking the integral up into two parts. This can be tedious if there are several intervals involved in the definition of the function. Now we are going to see how these kinds of functions can be expressed in such a way that the Laplace transform method doesn't have to consider separate intervals.

We start with the **unit step function** $U$ defined by

$$U(t) = \begin{cases} 0 & \text{if } t < 0 \\ 1 & \text{if } t \ge 0. \end{cases}$$

(This is sometimes called the **Heaviside (unit step) function** for the English electrical engineer Oliver Heaviside (1850–1925), who developed many of the applications of Laplace transforms we will see.) We can say that the function $U$ is "off" ($= 0$) for negative values of $t$ and "on" ($= 1$) for values of $t$ greater than or equal to 0. This "switching" aspect makes $U$ an important building block in modeling abrupt changes.

It follows that

$$U(t - a) = \begin{cases} 0 & \text{if } t < a \\ 1 & \text{if } t \ge a. \end{cases}$$

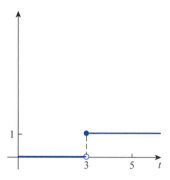

**FIGURE 6.1**

*Graph of $U(t - 3)$, $t \geq 0$*

The function has a jump discontinuity at $t = a$. Figure 6.1 shows $U(t - 3)$ for $t \geq 0$.

The nice thing is that these step functions can be used to express a piecewise continuous function in terms of a single formula. For example, if

$$f(t) = \begin{cases} A(t) & \text{for } t < a \\ B(t) & \text{for } t \geq a, \end{cases}$$

then we can see that $f(t) = A(t) + U(t-a)[B(t)-A(t)]$: If $t < a$, then $U(t-a) = 0$, so $f(t) = A(t)$; whereas if $t \geq a$, then we have $U(t - a) = 1$, so $f(t) = A(t) + [B(t) - A(t)] = B(t)$. (*Okay?*)

This technique can be extended to functions such as

$$g(t) = \begin{cases} A(t) & \text{for } a \leq t < b \\ B(t) & \text{for } b \leq t < c \\ C(t) & \text{for } c \leq t < d. \end{cases}$$

We can write $g(t) = U(t - a)A(t) + U(t - b)[B(t) - A(t)] + U(t - c)[C(t) - B(t)]$. You should be sure that you see how this works.

When we are solving differential equations that model abrupt changes, the following result comes in handy:

$$\boxed{\begin{array}{l} \text{If } \mathcal{L}[f(t)] \text{ exists for } s > c \text{ and if } a > 0, \text{ then} \\ \mathcal{L}[f(t - a)U(t - a)] = e^{-as}\mathcal{L}[f(t)] \text{ for } s > c. \end{array}} \tag{6.3.1a}$$

This result is usually called the *Second Shift Formula*—see Problem C3 in Exercises 6.1 for the *First Shift Formula*.

Alternatively, we can write (6.3.1a) as

$$f(t - a)U(t - a) = \mathcal{L}^{-1}\left[e^{-as}\mathcal{L}[f(t)]\right]. \tag{6.3.1b}$$

Formula (6.3.1a) follows from a straightforward calculation:

$$\mathcal{L}[f(t-a)U(t-a)] = \int_0^\infty f(t-a)U(t-a)e^{-st}dt = \int_a^\infty f(t-a)e^{-st}dt$$

$$= \int_0^\infty f(u)e^{-s(u+a)}du = e^{-sa}\,\mathcal{L}[f(t)],$$

where we have made the substitution $t - a = u$ in the last integral.

The next example shows how to use the Laplace transform of a unit step function to solve an initial-value problem.

### ■ Example 6.3.1 An IVP with a Discontinuous Forcing Term

Let's look at the initial-value problem

$$x'(t) + x = \begin{cases} t & \text{for } 0 \le t < 4 \\ 1 & \text{for } 4 \le t \end{cases} \qquad x(0) = 1.$$

Using the unit step function, we can write the differential equation as

$$x'(t) + x = t + (1-t)U(t-4) = t - (t-4)U(t-4) - 3U(t-4).$$

[Note that in order to use Formula (6.3.1b) later, we have to use algebra to convert the term $(1-t)U(t-4)$ into the form $f(t-a)U(t-a)$.] Now we apply the Laplace transform to both sides of the equation to get

$$\mathcal{L}[x'(t)] + \mathcal{L}[x(t)] = \mathcal{L}[t] - \mathcal{L}\left[(t-4)U(t-4)\right] - 3\,\mathcal{L}[U(t-4)],$$

or, using (6.1.4), entry 1 in Table 6.1, and then Formula (6.3.1a) twice,

$$s\,\mathcal{L}[x(t)] - 1 + \mathcal{L}[x(t)] = \frac{1}{s^2} - e^{-4s}\mathcal{L}[t] - 3e^{-4s}\mathcal{L}[1],$$

so

$$(s+1)\mathcal{L}[x(t)] = 1 + \frac{1}{s^2} - e^{-4s}\left(\frac{1}{s^2} + \frac{3}{s}\right).$$

Therefore,

$$\mathcal{L}[x(t)] = \frac{1}{s+1} + \frac{1}{s^2(s+1)} - e^{-4s}\left(\frac{3s+1}{s^2(s+1)}\right)$$

$$= [\text{by partial fractions}]\ \frac{2}{s+1} - \frac{1}{s} + \frac{1}{s^2} - e^{-4s}\left(\frac{2}{s} + \frac{1}{s^2} - \frac{2}{s+1}\right).$$

Finally, applying the inverse transform to both sides and using (6.3.1b), we get

$$x(t) = 2e^{-t} - 1 + t - \mathcal{L}^{-1}\left[e^{-4s}\left(\frac{2}{s} + \frac{1}{s^2} - \frac{2}{s+1}\right)\right]$$

$$= 2e^{-t} - 1 + t - \left[U(t)\mathcal{L}^{-1}\left(\frac{2}{s} + \frac{1}{s^2} - \frac{2}{s+1}\right)\right]$$

(where $t$ must be replaced by $t-4$ within the brackets before we're finished)

$$= 2e^{-t} - 1 + t - \left[U(t)\left(2 + t - 2e^{-t}\right)\right]$$

$$= 2e^{-t} - 1 + t - U(t-4)\left(t - 2 - 2e^{-t+4}\right)$$

$$= \begin{cases} 2e^{-t} + t - 1 & \text{for } 0 \le t < 4 \\ 2e^{-t} + 2e^{-t+4} + 1 & \text{for } 4 \le t. \end{cases}$$

∎

The next example shows the application of the Laplace transform and the unit step function to an important type of applied problem.

### ∎ Example 6.3.2 A Cantilever Beam Problem

A wooden beam, the ends of which are considered to be at $x = 0$ and $x = L$ on a horizontal axis, will "give" (that is, bend) when a vertical load, given by $W(x)$ per unit length, acts on the beam (Figure 6.2). (Compare Problem C2 in Exercises 1.3.)

It is known that $y(x)$, the amount of bending, or deflection, in the direction of the load force at the point $x$, satisfies the differential equation $\dfrac{d^4y}{dx^4} = \dfrac{W(x)}{EI}$ for $0 < x < L$.

Here, $E$ and $I$ are constants that describe characteristics of the beam. The graph of $y(x)$ is called the *deflection curve* or *elastic curve*.

Now suppose that we have a *cantilever beam* (such as a diving board)—one that is clamped at the end $x = 0$ and free at the end $x = L$—and that this beam carries a load per unit length

**FIGURE 6.2**
*A loaded beam*

given by

$$W(x) = \begin{cases} W_0 & \text{for } 0 < x < \dfrac{L}{2} \\ 0 & \text{for } \dfrac{L}{2} < x < L. \end{cases}$$

Then engineering mechanics shows that finding the deflection amounts to solving the boundary-value problem

$$\frac{d^4 y}{dx^4} = \frac{W(x)}{EI} \ (\text{for } 0 < x < L); \quad y(0) = 0, y'(0) = 0, y''(L) = 0, y'''(L) = 0.$$

(In physics terms, the quantities $y''(L)$ and $y'''(L)$ are called the *bending moment* and the *shear force*, respectively.)

First of all, note that up to now we have applied the technique of Laplace transforms only to initial-value problems, not to boundary-value problems (BVPs). Second, to use the Laplace transform, we must assume that $y(x)$ and $W(x)$ are defined on the interval $(0, \infty)$ rather than just on $(0, L)$. This means that we should extend the definition of $W(x)$ as follows:

$$W(x) = \begin{cases} W_0 & \text{for } 0 < x < \dfrac{L}{2} \\ 0 & \text{for } x > \dfrac{L}{2}. \end{cases}$$

We can write this function in terms of the unit step function as

$$W(x) = W_0 \left\{ U(x) - U\left(x - \frac{L}{2}\right) \right\}.$$

Now take the Laplace transform of each side of our fourth-order equation, letting $Y = Y(s) = \mathcal{L}[y(x)]$ for convenience. Using (6.1.6), we find that

$$s^4 Y - s^3 y(0) - s^2 y'(0) - s y''(0) - y'''(0) = \frac{W_0}{EI} \left\{ \frac{1 - e^{-sL/2}}{s} \right\}.$$

Note that in Formula (6.1.6) the second and third derivatives of $y$ are evaluated at 0. However, our BVP gives us the values of these derivatives at $L$. Letting $y''(0) = C_1$ and $y'''(0) = C_2$, we can use all the boundary conditions as given and solve the last equation for $Y : Y = \dfrac{C_1}{s^3} + \dfrac{C_2}{s^4} + \dfrac{W_0}{EIs^5}\{1 - e^{2sL/2}\}$. Using the inverse transform, we find that

$$y(x) = \frac{C_1 x^2}{2!} + \frac{C_2 x^3}{3!} + \frac{W_0}{EI}\frac{x^4}{4!} - \frac{W_0}{EI}\frac{\left(x - \frac{L}{2}\right)^4}{4!} U\left(x - \frac{L}{2}\right),$$

which is equivalent to

$$
y(x) =
\begin{cases}
\dfrac{C_1 x^2}{2} + \dfrac{C_2 x^3}{6} + \dfrac{W_0}{24EI} x^4 & \text{for } 0 \le x < \dfrac{L}{2} \\[3mm]
\dfrac{C_1 x^2}{2} + \dfrac{C_2 x^3}{6} + \dfrac{W_0}{24EI} x^4 - \dfrac{W_0}{24EI}\left(x - \dfrac{L}{2}\right)^4 & \text{for } x \ge \dfrac{L}{2}.
\end{cases}
$$

Now we use the conditions $y''(L) = 0$ and $y'''(L) = 0$ to find that $C_1 = \dfrac{W_0 L^2}{8EI}$ and $C_2 = -\dfrac{W_0 L}{2EI}$. (*Be sure to go through the calculations for yourself.*)

Finally, we can write our deflection function as

$$
y(x) =
\begin{cases}
\dfrac{W_0 L^2}{16EI} x^2 - \dfrac{W_0 L}{12EI} x^3 + \dfrac{W_0}{24EI} x^4 & \text{for } 0 \le x < \dfrac{L}{2} \\[3mm]
\dfrac{W_0 L^2}{16EI} x^2 - \dfrac{W_0 L}{12EI} x^3 + \dfrac{W_0}{24EI} x^4 - \dfrac{W_0}{24EI}\left(x - \dfrac{L}{2}\right)^4 & \text{for } \dfrac{L}{2} < x < L.
\end{cases}
$$

∎

## Exercises 6.3

**A**

In Problems 1–6, (a) sketch the graph of each function $f(t)$ and (b) write each function as a sum of multiples of the unit step function $U(t)$.

**1.** $f(t) = \begin{cases} 1 & \text{for } 1 \le t < 2 \\ 0 & \text{elsewhere} \end{cases}$

**2.** $f(t) = \begin{cases} t^2 & \text{for } 0 < t < 2 \\ 4t & \text{for } t > 2 \end{cases}$

**3.** $f(t) = \begin{cases} 1 & \text{for } 1 \le t < 2 \\ -2 & \text{for } 2 \le t < 3 \\ 0 & \text{elsewhere} \end{cases}$

**4.** $f(t) = \begin{cases} t & \text{for } 0 \le t < 2 \\ 4 - t & \text{for } 2 \le t < 4 \\ 0 & \text{elsewhere} \end{cases}$

**5.** $f(t) = \begin{cases} t & \text{for } 0 \le t < 2 \\ t - 2 & \text{for } 2 \le t < 4 \\ 0 & \text{elsewhere} \end{cases}$

**6.** $f(t) = \begin{cases} \sin t & \text{for } 0 < t < \pi \\ \sin 2t & \text{for } \pi < t < 2\pi \\ \sin 3t & \text{for } t > 2\pi \end{cases}$

**7.** Show that $\mathcal{L}\,[t\,U(t - a)] = (1 + a s)\,s^{-2} e^{-as}$ for $a > 0$.

**8.** Calculate $\mathcal{L}\left[t^2 U(t-1)\right]$.

**9.** Show that $\mathcal{L}\left[t^2\,U(t-2)\right] = \frac{2}{s^3} - \frac{2e^{-2s}}{s^3}\left(1+2s+2s^2\right), s > 0.$

**10.** Use Formula (6.3.1a-b) to compute the Laplace transform of the function in Problem 1.

**11.** Use Formula (6.3.1a-b) to compute the Laplace transform of the function in Problem 2.

**12.** Use Formula (6.3.1a-b) to compute the Laplace transform of the function in Problem 3.

**13.** Use Formula (6.3.1a-b) to compute the Laplace transform of the function in Problem 4.

**14.** Consider the function

$$F(t) = \begin{cases} e^{-t} & \text{for } 0 < t < 3 \\ 0 & \text{for } t > 3. \end{cases}$$

    **a.** Show that $F(t)$ can be written as $e^{-t}\left[1 - U(t-3)\right]$.

    **b.** Use Formula (6.3.1a) to find $\mathcal{L}[F(t)]$.

**B**

**1.** Solve the IVP

$$y'' + 4y = U(t-\pi) - U(t-3\pi);\ y(0) = 0,\ y'(0) = 0.$$

**2.** Solve the IVP

$$y^{(4)} + 5y'' + 4y = 1 - U(t-\pi);\ y(0) = y'(0) = y''(0) = y'''(0) = 0.$$

Solve each of the IVPs in Problems 3–10 by writing each discontinuous forcing function as a linear combination of unit step functions and then using the Laplace transform.

**3.** $4y' - 5y = \begin{cases} 0 & \text{for } t < 0 \\ -30t & \text{for } 0 \le t < 1;\ y(0) = 2 \\ 0 & \text{for } t \ge 1 \end{cases}$

**4.** $4y' + 5y = \begin{cases} 0 & \text{for } t < 0 \\ \sin 8t & \text{for } 0 \le t \le 2;\ y(0) = 1 \\ 0 & \text{for } t > 2 \end{cases}$

**5.** $y'' + y = \begin{cases} 1 & \text{for } 0 \le t < \pi/2 \\ 0 & \text{for } t \ge \pi/2 \end{cases};\ y(0) = 0,\ y'(0) = 1$

**6.** $y'' + y = \begin{cases} t/2 & \text{for } 0 \le t < 6 \\ 3 & \text{for } t \ge 6 \end{cases};\ y(0) = 0,\ y'(0) = 1$

**7.** $y'' + 5y' + 2y = \begin{cases} 0 & \text{for } t < 0 \\ 8 & \text{for } 0 \le t \le 1;\ y(0) = 0,\ y'(0) = 0 \\ 0 & \text{for } t > 1 \end{cases}$

**8.** $3y'' + 3y' + 2y = \begin{cases} 0 & \text{for } t < 0 \\ 5 & \text{for } 0 \le t \le 5;\ y(0) = 0,\ y'(0) = 0 \\ 0 & \text{for } t > 5 \end{cases}$

9. $y' - 3y = f(t);\ y(0) = 0$, where the graph of $f(t)$ is

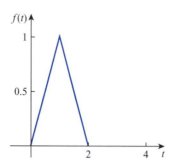

10. $y' + y = g(t);\ y(0) = 0$, where the graph of $g(t)$ is

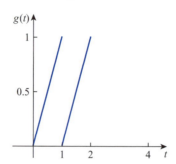

11. Suppose that the fish population in a large lake is growing too rapidly and the local authorities decide to give out fishing licenses that allow a total of $h$ fish to be caught per day over a 30-day period. A model for such a situation could be

$$P'(t) = kP(t) - \begin{cases} h & \text{for } 0 \le t \le 30 \\ 0 & \text{for } t > 30 \end{cases}$$

where $P(t)$ denotes the number of fish in the lake at time $t$ (in days) and $k$ is a positive constant describing the natural growth rate of the fish population.

  **a.** Use technology and the Laplace transform to find an expression for $P(t)$ if $P(0) = A$.

  **b.** Find a relation among $A$, $h$, and $k$ that guarantees that exactly 330 days after the end of the 30-day fishing season, the fish population will once more be at the level $A$.

12. Problem B1 of Exercises 2.3 concerns the population of Botswana from 1975 to 1990 under certain basic assumptions. Now consider the situation that occurs if we start with a population of 0.755 million people in 1975 ($t = 0$) and assume that births and deaths, immigration and emigration balance each other until 1977 ($t = 2$). In 1977, an emigration pattern begins in such a way that the population $P(t)$ can be described by the equation

$$P' - kP = \begin{cases} 0 & \text{if } 0 < t < 2 \\ -a(t - 2) & \text{if } t \ge 2 \end{cases}$$

with $P(0) = 0.755, k = 0.0355$, and $a = 1.60625 \times 10^{-3}$.

**a.** Express the discontinuous function on the right-hand side of the equation in terms of the unit step function.

**b.** Use technology and the Laplace transform to solve for $P(t)$, expressing the answer as a step function.

**c.** Graph the solution on the interval $0 \le t \le 35$ and explain what the graph means in terms of the population of Botswana.

## C

**1.** The IVP $y'' + 3y' + 2y = W(t)$; $y(0) = 0, y'(0) = 0$ represents a damped spring-mass system subjected to a square wave forcing term given by

$$W(t) = U(t-1) - U(t-2).$$

**a.** Graph $W(t)$.

**b.** Without using technology, solve the IVP when $W(t)$ is not present in the system. (That is, make the right-hand side of the differential equation zero.)

**c.** Without using technology, solve the given IVP (that is, with $W(t)$ as the forcing term).

**d.** Use technology to graph the solutions to parts (b) and (c) on the same set of axes. What difference does the forcing term make?

## 6.4 TRANSFORMS OF IMPULSE FUNCTIONS—THE DIRAC DELTA FUNCTION

In the preceding section we dealt with situations in which some abrupt change occurred. To describe it in general terms, we were dealing with systems acted on by some external force that was applied suddenly. But although the change was sudden, force was assumed to have been applied for some measurable period of time. Now we want to examine problems in which there is an external force of large magnitude applied suddenly for a very short period of time. For example, think about a baseball being hit by a major-league player. The time of contact of ball with bat is very brief, but enough force can be applied to send that horsehide soaring into the stands. More dramatic instances of this phenomenon include an electrical surge caused by a power line that is suddenly struck by lightning and a population that is growing at a certain rate until some sudden disaster strikes the community.

Mathematically, we can start to approach this idea by considering a piecewise continuous function that looks like

$$\delta_b(t) = \begin{cases} \frac{1}{b} & \text{for } 0 \le t \le b \\ 0 & \text{for } t > b. \end{cases}$$

Here, we must assume that $\delta_b(t)$, which is pronounced "delta sub $b$ of $t$," does not exist if $b = 0$. This function can represent a force of magnitude $1/b$ applied for a time period of length $b$ (see Figure 6.3).

**FIGURE 6.3**

*The graph of $\delta_b(t)$*

First of all, note that $\int_0^\infty \delta_b(t)dt = \int_0^b \frac{1}{b}dt = 1$ for all values of $b > 0$. Now look at what happens as we allow the value of $b$ to get smaller and smaller. This situation describes a force whose magnitude $1/b$ is getting larger and larger over a shorter and shorter interval of time $(0, b)$. *Can you see what's going on?* More precisely, the unusual nature of this discontinuous function led various physicists, mathematicians, and engineers to consider the limiting behavior of $\delta_b(t)$ as $b \to 0$. In particular, they defined $\delta(t)$ as follows:

$$\delta(t) = \lim_{b \to 0} \delta_b(t) = \begin{cases} \infty & \text{for } t = 0 \\ 0 & \text{for } t \neq 0. \end{cases}$$

This "function" $\delta$ is called the **unit impulse function** or the **Dirac delta function** [named for the English-Belgian theoretical physicist Paul A. M. Dirac (1902–1984), who won the Nobel Prize in 1933 with E. Schrödinger for his work on quantum theory]. More generally, we can define

$$\delta(t - a) = \lim_{b \to 0} \delta_b(t - a) = \begin{cases} \infty & \text{for } t = a \\ 0 & \text{for } t \neq a. \end{cases}$$

In the mathematically precise sense of the word, this limit does not exist and so does not define a function. However, such *generalized functions*, or *distributions*, can be put on a firm mathematical foundation and are very useful in modern physics and engineering theory. Before we look at examples of the delta function's use in solving differential equations, we should try to calculate its Laplace transform. The only reasonable way to do this is to make the formal assumption that

$$\mathcal{L}\left[\delta(t - a)\right] = \lim_{b \to 0} \mathcal{L}\left[\delta_b(t - a)\right].$$

(Mathematically, this raises an important theoretical question of whether

$$\lim_{b \to 0} \mathcal{L}\left[\delta_b(t - a)\right] = \mathcal{L}\left[\lim_{b \to 0} \delta_b(t - a)\right].$$

This question is beyond the scope of this course and will be ignored.)

Now let's write $\delta_b(t - a)$ in terms of the unit step function, as we did for functions in Section 6.3:

$$\delta_b(t - a) = \frac{1}{b}\left[U(t - a) - U(t - (a + b))\right].$$

If we use the linearity of the Laplace transform together with Formula (6.3.1a)—taking $f(t - a) \equiv 1$—we get

$$\mathcal{L}\left[\delta(t - a)\right] = \lim_{b \to 0} \mathcal{L}\left[\delta_b(t - a)\right]$$

$$= \lim_{b \to 0} \frac{1}{b}\left\{\frac{e^{-sa}}{s} - \frac{e^{-s(a+b)}}{s}\right\} = \lim_{b \to 0} e^{-sa}\left\{\frac{1 - e^{-sb}}{bs}\right\}$$

$$= e^{-sa} \lim_{b \to 0}\left\{\frac{1 - e^{-sb}}{bs}\right\} = e^{-sa},$$

where we have used L'Hôpital's Rule to evaluate the indeterminate form in this last limit. (Alternatively, we could have used the series expansion of $(1 - e^{-sb})/bs$ about the point $b = 0$.)

Because we have shown that

$$\mathcal{L}\left[\delta(t - a)\right] = e^{-sa}, \tag{6.4.1a}$$

it seems reasonable to take $a = 0$ and conclude that

$$\mathcal{L}\left[\delta(t)\right] = 1. \tag{6.4.1b}$$

Now let's see how to solve a differential equation involving an impulse function. You may want to review the spring-mass problems in Section 4.8.

## ■ Example 6.4.1 Solving an ODE That Involves the Dirac Delta Function

A mass attached to a spring is released from rest 1 meter below the equilibrium position for the spring-mass system and begins to move up and down. After 3 seconds, the mass is struck by a hammer in a downward direction. We suppose the undamped system is governed by the IVP

$$\frac{d^2x}{dt^2} + 9x = 3\delta(t - 3); \quad x(0) = 1, \quad \frac{dx}{dt}(0) = 0,$$

where $x(t)$ denotes the displacement from equilibrium at time $t$, and we want to determine a formula for $x(t)$. (Note that the impulse force applied at $t = 3$ has magnitude 3.)

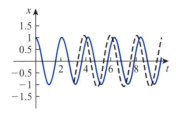

**FIGURE 6.4**

Graph of $x(t) = \begin{cases} \cos 3t & \text{for } t < 3 \\ \cos 3t + \sin 3(t-3) & \text{for } 3 \leq t \end{cases}$

Let $X = X(s) = \mathcal{L}[x(t)]$. Then, taking the Laplace transform of both sides of our ODE and using (6.1.5) and (6.4.1a) with $a = 3$, we get

$$s^2X - s + 9X = 3e^{-3s},$$

so we can solve for $X$:

$$X = \frac{s}{s^2 + 9} + e^{-3s}\frac{3}{s^2 + 9}.$$

Applying the inverse transform yields

$$x(t) = \cos 3t + \sin 3(t-3)U(t-3)$$

$$= \begin{cases} \cos 3t & \text{for } t < 3 \\ \cos 3t + \sin 3(t-3) & \text{for } 3 \leq t. \end{cases}$$

$\blacksquare$

Figure 6.4 is the graph of $x(t)$, where the solid curve shows the displacement of the mass if the hammer had not hit it.

## Exercises 6.4

**A**

**1.** Evaluate the integral $\int_{-\infty}^{\infty} \delta(t - 3\pi/2) \cos 2t \ dt$.

**2.** Evaluate the integral $\int_{0}^{1} t^3 \delta\left(t + \frac{1}{3}\right) dt$.

**3.** Evaluate $\mathcal{L}\left[\delta(t - \pi) \cos t^3\right]$.

Solve the IVPs in Problems 4–13:

**4.** $y'' = \delta(t - a); \ y(0) = 0, y'(0) = 0$

**5.** $y' + 8y = \delta(t - 1) + \delta(t - 2); \ y(0) = 0$

**6.** $y'' + y = \delta(t - 2); \ y(0) = 0, \ y'(0) = 0$

**7.** $y'' + 2y' - 8y = \delta(t); \ y(0) = 0, \ y'(0) = 0$

**8.** $2y'' + y' + 2y = \delta(t - 5); \ y(0) = 0, y'(0) = 0$

**9.** $y'' + 2y' + y = 2\delta(t - 1)$; $y(0) = 1$, $y'(0) = 1$

**10.** $y'' + 6y' + 109y = \delta(t - 1) - \delta(t - 7)$; $y(0) = 0$, $y'(0) = 0$

**11.** $y'' + y = 1 + \delta(t - 2\pi)$; $y(0) = 1, y'(0) = 0$

**12.** $y'' + y = 4\delta\left(t - \frac{3}{2}\pi\right)$; $y(0) = 0$, $y'(0) = 1$

**13.** $y^{(iv)} - y = \delta(t - 1)$; $y(0) = 0$, $y'(0) = 0$, $y''(0) = 0$, $y'''(0) = 0$

## B

**1.** A uniform beam of length $L$ carries a load $W$ concentrated at $x = L/2$. The beam is embedded at its left end and is free at its right end. The deflection $y(x)$ is governed by the equation $E I \frac{d^4 y}{dx^4} = W\delta$ $\left(x - \frac{L}{2}\right)$, where $y(0) = 0$, $y'(0) = 0$, $y''(L) = 0$, and $y'''(L) = 0$. Use the Laplace transform to determine the deflection $y(x)$.

**2.** If $a, b$, and $c$ are constants, show that the solution $x(t)$ of the linear IVP

$$x''(t) + ax'(t) + bx(t) = \delta(t - c); \ x(0) = 0, x'(0) = 0$$

is $x(t) = k(t - c)U(t - c)$, where $k(t) = \mathcal{L}^{-1}\left[\frac{1}{s^2 + as + b}\right]$.

**3.** Suppose we have the equation $y'' + ay' + by = f(t)$, where $a$ and $b$ are constants and $f$ is a piecewise continuous function whose Laplace transform exists. Show that the effect of replacing $f(t)$ by $f(t) + c\,\delta(t)$, where $c$ is a constant, is the same as increasing the initial value of $y'(0)$ by the constant $c$.

**4. a.** Show that $L[\delta(t - a)f(t)] = e^{-as}f(a)$.

   **b.** Use the result in part (a) and the result of Problem C2 to solve the IVP

$$y'' + 2y' + y = \delta(t - 1)\,t; \ y(0) = 0, y'(0) = 0.$$

## C

**1.** If, at time $t = a$, the upper end of an undamped spring-mass system is jerked upward suddenly and returned to its original position, the equation modeling the situation is $mx'' + kx = kH\,\delta(t - a)$; $x(0) = x_0$, $x'(0) = x_1$, where $m$ is the mass, $k$ is the spring constant, and $H$ is a constant.

   **a.** Solve the IVP manually, with $x(0) = 0 = x'(0)$.

   **b.** Use the solution found in part (a) to explain the significance of the constant $H$.

   **c.** Choose a value for $H$ so that the mass achieves a prescribed displacement from equilibrium $A$ for $t \geq a$.

**2.** If the function $g(t)$ is continuous at $a$, show that $\int_0^\infty \delta(t - a)g(t)\,dt = g(a)$. [*Hint*: Use the Mean Value Theorem for integrals.]

**3.** Consider the IVP $y'' + 2y = \sum_{n=1}^\infty \delta(t - n)$; $y(0) = 0$, $y'(0) = 0$.

   **a.** Find the Laplace transform of the solution of the IVP.

   **b.** Solve the IVP.

   **c.** What happens to the solution of the IVP as $t \to \infty$?

## 6.5 TRANSFORMS OF SYSTEMS OF LINEAR DIFFERENTIAL EQUATIONS

We have seen what the Laplace transform does to a single linear equation with constant coefficients. It should be easy to see that when initial conditions are given, the Laplace transform converts a system of linear differential equations with constant coefficients into a system of simultaneous algebraic equations. Then we can solve the algebraic equations for the *transformed* solution functions. Finally, applying the inverse transform to these functions gives us the solutions of the original system of linear ODEs.

Conceptually, this process is easy. The algebraic details, however, may be something else. Problems of this kind make us appreciate the availability of technology.

### ■ Example 6.5.1 Solving a Linear System via the Laplace Transform

Let's start with the system

$$\frac{dx}{dt} = -3x + y$$

$$\frac{dy}{dt} = x - 3y,$$

where we want the solutions $x(t)$ and $y(t)$ that satisfy $x(0) = 2$ and $y(0) = 3$. (This system was discussed briefly in Example 1.3.5.)

Applying the Laplace transform to each side of each equation gives us the system

$$s\mathcal{L}[x(t)] - x(0) = -3\mathcal{L}[x(t)] + \mathcal{L}[y(t)]$$

$$s\mathcal{L}[y(t)] - y(0) = \mathcal{L}[x(t)] - 3\mathcal{L}[y(t)].$$

Inserting the initial conditions and simplifying the resulting equations, we get the system

$$(s+3)\mathcal{L}[x(t)] - \mathcal{L}[y(t)] = 2$$

$$(s+3)\mathcal{L}[y(t)] - \mathcal{L}[x(t)] = 3.$$

Now we solve the preceding system for $\mathcal{L}[x(t)]$ and $\mathcal{L}[y(t)]$ just as we would solve any algebraic system of two equations in two unknowns. (To simplify things, you could let $\mathcal{L}[x(t)] = X$ and $\mathcal{L}[y(t)] = Y$.) For instance, we can eliminate the variable $\mathcal{L}[y(t)]$ by multiplying the first equation by $(s+3)$ and then adding the result to the second equation. When the dust settles, we get

$$\left\{(s+3)^2 - 1\right\}\mathcal{L}[x(t)] = 2(s+3) + 3,$$

so we find that

$$\mathcal{L}[x(t)] = \frac{2s+9}{(s+3)^2 - 1} = \frac{2s+9}{[(s+3)+1][(s+3)-1]}$$

$$= \frac{2s + 9}{(s + 4)(s + 2)} = \frac{-\frac{1}{2}}{s + 4} + \frac{\frac{5}{2}}{s + 2} = \frac{-\frac{1}{2}}{s - (-4)} + \frac{\frac{5}{2}}{s - (-2)}$$

and

$$x(t) = -\frac{1}{2}\mathcal{L}^{-1}\left[\frac{1}{s - (-4)}\right] + \frac{5}{2}\mathcal{L}^{-1}\left[\frac{1}{s - (-2)}\right]$$

$$= -\frac{1}{2}e^{-4t} + \frac{5}{2}e^{-2t}.$$

Now we could go through this process again to eliminate $\mathcal{L}[x(t)]$ and solve for $y(t)$ this time (see Problem A1 in Exercises 6.5)—or we could just substitute our solution for $x(t)$ in the first equation of our original system and solve for $y$:

$$y(t) = \frac{dx}{dt} + 3x = \frac{d}{dx}\left(-\frac{1}{2}e^{-4t} + \frac{5}{2}e^{-2t}\right) + 3\left(-\frac{1}{2}e^{-4t} + \frac{5}{2}e^{-2t}\right)$$

$$= 2e^{-4t} - 5e^{-2t} - \frac{3}{2}e^{-4t} + \frac{15}{2}e^{-2t} = \frac{1}{2}e^{-4t} + \frac{5}{2}e^{-2t}.$$

■

The next example shows how we can handle a system of two second-order linear equations. In particular, note that we don't have to write this as a system of four first-order equations. The Laplace transform technique works directly on higher-order derivatives via Formula (6.1.6) or, in this case, by (6.1.5).

## ■ Example 6.5.2 A System of Second-Order Equations

The system IVP we want to solve is

$$\frac{d^2x}{dt^2} - 4x + \frac{dy}{dt} = 0$$

$$-4\frac{dx}{dt} + \frac{d^2y}{dt^2} + 2y = 0,$$

with $x(0) = 0, x'(0) = 1, y(0) = -1$, and $y'(0) = 2$.

Applying the Laplace transform to each side of each equation, we get

$$\mathcal{L}\left[x''(t)\right] - 4\mathcal{L}\left[x(t)\right] + \mathcal{L}\left[y'(t)\right] = 0$$

$$-4\mathcal{L}\left[x'(t)\right] + \mathcal{L}\left[y''(t)\right] + 2\mathcal{L}\left[y(t)\right] = 0.$$

Using (6.1.4) and (6.1.5), we can write the preceding system as

$$s^2\mathcal{L}[x(t)] - x'(0) - sx(0) - 4\mathcal{L}[x(t)] + s\mathcal{L}[y(t)] - y(0) = 0$$

$$-4s\mathcal{L}[x(t)] + 4x(0) + s^2\mathcal{L}[y(t)] - y'(0) - sy(0) + 2\mathcal{L}[y(t)] = 0.$$

Now we insert the initial conditions and simplify the resulting equations to get

$$\left(s^2 - 4\right) \mathcal{L}[x(t)] + s\mathcal{L}[y(t)] = 0$$
$$-4s\mathcal{L}[x(t)] + \left(s^2 + 2\right) \mathcal{L}[y(t)] = 2 - s. \tag{*}$$

As in the previous example, we can solve these equations by realizing that they constitute a system of ordinary algebraic equations in the unknowns $\mathcal{L}[x(t)]$ and $\mathcal{L}[y(t)]$. If we multiply the first equation of (*) by $4s$, multiply the second by $s^2 - 4$, and then add the resulting equations, we obtain

$$\left(s^4 + 2s^2 - 8\right) \mathcal{L}[y(t)] = -s^3 + 2s^2 + 4s - 8,$$

so

$$\mathcal{L}[y(t)] = \frac{-s^3 + 2s^2 + 4s - 8}{s^4 + 2s^2 - 8} = \frac{-s^3 + 2s^2 + 4s - 8}{\left(s^2 + 4\right)\left(s^2 - 2\right)}$$

$$= \frac{-s^3 + 2s^2 + 4s - 8}{\left(s^2 + 4\right)\left(s + \sqrt{2}\right)\left(s - \sqrt{2}\right)}$$

$$= \frac{1}{6}\left[\frac{1 + \sqrt{2}}{s + \sqrt{2}} + \frac{1 - \sqrt{2}}{s - \sqrt{2}} - \frac{8(s - 2)}{s^2 + 4}\right] \tag{**}$$

$$= \frac{1}{6}\left[\frac{1 + \sqrt{2}}{s + \sqrt{2}} + \frac{1 - \sqrt{2}}{s - \sqrt{2}} - 8\frac{s}{s^2 + 2^2} + 8\frac{2}{s^2 + 2^2}\right].$$

Using entries 2, 3, and 4 of the table of transforms (Table 6.1), we see that

$$y(t) = \frac{1}{6}\left[\left(1 + \sqrt{2}\right)e^{-\sqrt{2}t} + \left(1 - \sqrt{2}\right)e^{\sqrt{2}t} - 8\cos 2t + 8\sin 2t\right].$$

To find $\mathcal{L}[x(t)]$, we can go back to System (*) and eliminate $\mathcal{L}[y(t)]$, or we can substitute expression (**) for $\mathcal{L}[y(t)]$ in either equation of (*) and solve for $\mathcal{L}[x(t)]$. Let's try the latter method.

Using (**) and the first equation in (*), we find that

$$\left(s^2 - 4\right) \mathcal{L}[x(t)] + s\left(\frac{-s^3 + 2s^2 + 4s - 8}{\left(s^2 + 4\right)\left(s^2 - 2\right)}\right) = 0.$$

Solving for $\mathcal{L}[x(t)]$, we get

$$\mathcal{L}[x(t)] = -s\left(\frac{-s^3 + 2s^2 + 4s - 8}{\left(s^2 - 4\right)\left(s^2 + 4\right)\left(s^2 - 2\right)}\right) = \frac{s(s - 2)^2(s + 2)}{(s - 2)(s + 2)\left(s^2 + 4\right)\left(s^2 - 2\right)}$$

$$= \frac{s(s-2)}{\left(s^2+4\right)\left(s+\sqrt{2}\right)\left(s-\sqrt{2}\right)}$$

$$= -\frac{1}{12}\left[\frac{2+\sqrt{2}}{s+\sqrt{2}} + \frac{2-\sqrt{2}}{s-\sqrt{2}} - 4\left(\frac{s+2}{s^2+4}\right)\right]$$

$$= -\frac{1}{12}\left[\frac{2+\sqrt{2}}{s+\sqrt{2}} + \frac{2-\sqrt{2}}{s-\sqrt{2}} - 4\left(\frac{s}{s^2+2^2} + \frac{2}{s^2+2^2}\right)\right].$$

Formulas 2, 3, and 4 from Table 6.1 tell us that

$$x(t) = -\frac{1}{12}\left[\left(2+\sqrt{2}\right)e^{-\sqrt{2}t} + \left(2-\sqrt{2}\right)e^{\sqrt{2}t} - 4\cos 2t - 4\sin 2t\right].$$

■

You should confirm that these are the solutions to the original IVP.

## Exercises 6.5

### A

1. Eliminate $\mathcal{L}[x(t)]$ from the algebraic system

$$(s+3)\mathcal{L}[x(t)] - \mathcal{L}[y(t)] = 2$$
$$(s+3)\mathcal{L}[y(t)] - \mathcal{L}[x(t)] = 3$$

and then solve for $y(t)$. (See Example 6.5.1.)
Solve the IVPs in Problems 2–13 by using the Laplace transform.

2. $\{x' = y,\ y' = -x\};\quad x(0) = 2,\ y(0) = -1$
3. $\{x' = 2x - 3y,\ y' = y - 2x\};\quad x(0) = 8,\ y(0) = 3$
4. $\{x' = 12x + 5y,\ y' = -6x + y\};\quad x(0) = 0,\ y(0) = 1$
5. $\{x' = -2x + y,\ y' = -9x + 4y\};\quad x(0) = 5,\ y(0) = -3$
6. $\{x' = -6x + 2y,\ y' = -7x + 3y\};\quad x(0) = 1,\ y(0) = 0$
7. $\{x' = x + y,\ y' = -4x + y\};\quad x(0) = 1,\ y(0) = 3$
8. $\{x' = y' + 6y,\ y' = \frac{3}{2}x - \frac{1}{2}x'\};\quad x(0) = 2,\ y(0) = 3$
9. $\{x' + x - 5y = 0,\ y' + 4x + 5y = 0\};\quad x(0) = -1,\ y(0) = 2$
10. $\{x' + y' = -3x - 2y + e^{-2t},\ 2x' + y' = -2x - y + 1\};\quad x(0) = 0,\ y(0) = 0$
11. $\{x' = x - y - e^{-t},\ y' = 2x + 3y + e^{-t}\};\quad x(0) = 1,\ y(0) = 0$
12. $\{x' + y' = x,\ y' + z' = x,\ z' + x' = x\};\quad x(0) = 1,\ y(0) = 1,\ z(0) = 1$
13. $\{x' - 3x - 6y = 27t^2,\ x' + y' - 3y = 5e^t\};\quad x(0) = 5,\ y(0) = -1$

**B**

1. Solve the system IVP $\{x'' + y' = 4x,\ 4x' - y'' = 9y\}$; $x(0) = 0, x'(0) = 1,\ y(0) = -1, y'(0) = 2$ by using the Laplace transform.

2. Solve the system IVP $\{x'' - y' = -t + 1,\ x' - x + 2y' = 4e^t\}$; $x(0) = 0, x'(0) = 1, y(0) = 0$ by using the Laplace transform.

3. The system

$$mx'' = -k_1(x - a\theta) - k_2(x + a\theta)$$
$$mr^2\theta'' = k_1 a(x - a\theta) - k_2\, a\, (x + a\theta)$$

models the motion of a slab of mass $m$ mounted on two springs, as shown in the following figure. Here, $x$ is the vertical displacement of the center of mass and $\theta$ is the angle shown. The constant $r$ represents the radius of gyration of the slab about the appropriate axis through the center of mass. Use the Laplace transform and technology to solve the system for $x$ and $\theta$ if $m = 1, k_1 = 1, k_2 = 2$, $a = 1, r = 1, x(0) = 1, x'(0) = 0, \theta(0) = 0.1$, and $\theta'(0) = 0$.

4. The system consisting of two pendulums connected by a spring (see the following figure) has its motion approximated by the system of equations

$$mx'' + m\omega_0^2 x = -k(x - y)$$
$$my'' + m\omega_0^2 y = -k(y - x),$$

where $L$ is the length of each pendulum, $g$ is the gravitational constant, and $\omega_0^2 = g/L$. Use the Laplace transform and technology to solve this system with $m = 1, L = 5, g = 32, k = 2$, and the initial conditions $x(0) = 0, x'(0) = 2, y(0) = 0$, and $y'(0) = 2$.

5. The following circuit is described by the system

$$L_1\dot{I}_1 + R_1\,(I_1 - I_2) = v(t)$$
$$L_2\dot{I}_2 + R_2 I_2 + R_1\,(I_2 - I_1) = 0.$$

Determine $I_1$ and $I_2$ when the switch is closed if $L_1 = L_2 = 2$ henrys, $R_1 = 3$ ohms, $R_2 = 8$ ohms, and $v(t) = 6$ volts. Assume that $I_1(0) = I_2(0) = 0$.

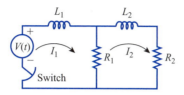

**C**

**1.** Solve the system IVP $\left\{ \frac{d^2x}{dt^2} = y + 4e^{-2t}, \frac{d^2y}{dt^2} = x - e^{-2t} \right\}$;

$$x(0) = y(0) = x'(0) = y'(0) = 0.$$

**2.** In determining the concentration of a chemical in a system consisting of two compartments separated by a membrane, we get the system of equations

$$\dot{x} = ay - bx$$
$$\dot{y} = bx - ay - \beta y,$$

subject to the conditions $x(0) = x^*$ and $y(0) = y^*$, where $x^*$ and $y^*$ are constants. (Here, $x$ and $y$ represent the masses of the chemical in compartments 1 and 2, respectively, at any time $t$, and the constants $a, b,$ and $\beta$ are positive constants of proportionality related to the rate of flow of the chemical from one compartment to another.)

**a.** Solve this system of equations using Laplace transforms.

**b.** Letting $p = \frac{1}{2}(b + a + \beta)$ and $q = \frac{1}{2}\sqrt{(b + a + \beta)^2 - 4\beta b}$, show that $q$ is a (positive) *real* number and that $p > q$.

**c.** Using the solution found in part (a) and the results of part (b), show that the chemical masses $x$ and $y$ approach zero steadily.

# 6.6 A QUALITATIVE ANALYSIS VIA THE LAPLACE TRANSFORM

In Chapter 5 (specifically, in Sections 5.2–5.5), we analyzed autonomous two-dimensional systems of linear equations and their equivalent single second-order homogeneous equations by means of eigenvalues and eigenvectors. This qualitative analysis, which was very neat and very satisfying, depended on the roots of polynomial equations (the *characteristic equations*). However, the eigenvalue method didn't stretch quite far enough to handle general *nonhomogeneous* systems (Section 5.6).

### 6.6.1 Homogeneous Equations

Despite the emphasis on the Laplace transform as a tool for obtaining exact, closed-form solutions, it turns out that the transform can provide insight into the *qualitative* nature of a solution as well. In fact, because the Laplace transform treats nonhomogeneous equations in essentially the same way as it treats homogeneous equations (with slightly messier algebra), the Laplace transform in effect gives us an extension of eigenvalue analysis to nonhomogeneous equations. Let's examine this by analyzing nonhomogeneous second-order equations of the form

$$c_2 x'' + c_1 x' + c_0 x = f(t), \tag{6.6.1}$$

where $c_2$, $c_1$, and $c_0$ are constants and $c_2 \neq 0$. For the sake of simplicity and clarity, we'll start with $f(t) \equiv 0$, the *homogeneous* case.

Taking the Laplace transform of both sides of (6.6.1)—with $f(t) \equiv 0$—we get

$$c_2 \mathcal{L}[x''(t)] + c_1 \mathcal{L}[x'(t)] + c_0 \mathcal{L}[x(t)] = 0,$$

which becomes

$$c_2 \left\{ s^2 \mathcal{L}[x(t)] - sx(0) - x'(0) \right\} + c_1 \left\{ s\mathcal{L}[x(t)] - x(0) \right\} + c_0 \mathcal{L}[x(t)] = 0,$$

or, after simplifying,

$$\left( c_2 s^2 + c_1 s + c_0 \right) \mathcal{L}[x(t)] - (c_2 s + c_1) x(0) - c_2 x'(0) = 0.$$

Solving, we find that

$$\mathcal{L}[x(t)] = \frac{(c_2 s + c_1) x(0)}{c_2 s^2 + c_1 s + c_0} + \frac{c_2 x'(0)}{c_2 s^2 + c_1 s + c_0}. \tag{6.6.2}$$

We should note something significant about the denominator, $c_2 s^2 + c_1 s + c_0$, of the Laplace transform of the solution. It is the *characteristic polynomial* corresponding to the second-order differential equation $c_2 x'' + c_1 x' + c_0 x = 0$ or to the equivalent system

$$x_1' = x_2$$

$$x_2' = \left( -\frac{c_0}{c_2} \right) x_1 - \left( \frac{c_1}{c_2} \right) x_2.$$

*Interesting!* If the right-hand side of Equation (6.6.2) is expressed as a single fraction (rational function) with no common factors in the numerator and denominator, then the zeros of the characteristic polynomial—the values of $s$ that make the denominator zero—are called the **poles**, or **singularities**, of the transform $\mathcal{L}[x(t)]$.

Now suppose that $\lambda_1$ and $\lambda_2$ are the zeros of the characteristic polynomial—the *eigenvalues* of the system. Let's see what happens if both zeros are *real* numbers. For convenience, also assume that $\lambda_1 \neq \lambda_2$.

First, we can divide through by $c_2$ and write $c_2 s^2 + c_1 s + c_0 = 0$ in the equivalent form $s^2 + \left(\frac{c_1}{c_2}\right)s + \left(\frac{c_0}{c_2}\right) = 0$; then we can write $s^2 + \left(\frac{c_1}{c_2}\right)s + \left(\frac{c_0}{c_2}\right) = (s - \lambda_1)(s - \lambda_2)$. Now, returning to (6.6.2), we can apply the inverse transform to each side to get

$$x(t) = \mathcal{L}^{-1}\left[\frac{(c_2 s + c_1)\,x(0)}{c_2 s^2 + c_1 s + c_0}\right] + \mathcal{L}^{-1}\left[\frac{c_2 x'(0)}{c_2 s^2 + c_1 s + c_0}\right]$$

$$= \mathcal{L}^{-1}\left[\frac{(c_2 s + c_1)\,x(0)}{c_2\left(s^2 + \left(\frac{c_1}{c_2}\right)s + \left(\frac{c_0}{c_2}\right)\right)}\right] + \mathcal{L}^{-1}\left[\frac{c_2 x'(0)}{c_2\left(s^2 + \left(\frac{c_1}{c_2}\right)s + \left(\frac{c_0}{c_2}\right)\right)}\right]$$

$$= \mathcal{L}^{-1}\left[\frac{\frac{1}{c_2}(c_2 s + c_1)\,x(0)}{\left(s^2 + \left(\frac{c_1}{c_2}\right)s + \left(\frac{c_0}{c_2}\right)\right)}\right] + \mathcal{L}^{-1}\left[\frac{\frac{1}{c_2}c_2 x'(0)}{\left(s^2 + \left(\frac{c_1}{c_2}\right)s + \left(\frac{c_0}{c_2}\right)\right)}\right]$$

$$= \mathcal{L}^{-1}\left[\frac{\left(s + \frac{c_1}{c_2}\right)x(0)}{(s - \lambda_1)(s - \lambda_2)}\right] + \mathcal{L}^{-1}\left[\frac{x'(0)}{(s - \lambda_1)(s - \lambda_2)}\right]$$

$$= x(0)\mathcal{L}^{-1}\left[\frac{\left(s + \frac{c_1}{c_2}\right)}{(s - \lambda_1)(s - \lambda_2)}\right] + x'(0)\mathcal{L}^{-1}\left[\frac{1}{(s - \lambda_1)(s - \lambda_2)}\right]$$

$$= x(0)\mathcal{L}^{-1}\left[\frac{\frac{c_1 + c_2 \lambda_1}{c_2(\lambda_1 - \lambda_2)}}{s - \lambda_1} - \frac{\frac{c_1 + c_2 \lambda_2}{c_2(\lambda_1 - \lambda_2)}}{s - \lambda_2}\right] + x'(0)\mathcal{L}^{-1}\left[\frac{1}{\lambda_1 - \lambda_2}\left(\frac{1}{s - \lambda_1} - \frac{1}{s - \lambda_2}\right)\right]$$

$$= \frac{x(0)}{\lambda_1 - \lambda_2}\mathcal{L}^{-1}\left[\frac{\lambda_1 + \frac{c_1}{c_2}}{s - \lambda_1} - \frac{\lambda_2 + \frac{c_1}{c_2}}{s - \lambda_2}\right] + \frac{x'(0)}{\lambda_1 - \lambda_2}\mathcal{L}^{-1}\left[\frac{1}{s - \lambda_1} - \frac{1}{s - \lambda_2}\right]$$

$$= A\mathcal{L}^{-1}\left[\frac{1}{s - \lambda_1}\right] - B\mathcal{L}^{-1}\left[\frac{1}{s - \lambda_2}\right] + C\mathcal{L}^{-1}\left[\frac{1}{s - \lambda_1}\right] - D\mathcal{L}^{-1}\left[\frac{1}{s - \lambda_2}\right]$$

$$= K_1\mathcal{L}^{-1}\left[\frac{1}{s - \lambda_1}\right] + K_2\mathcal{L}^{-1}\left[\frac{1}{s - \lambda_2}\right] = K_1 e^{\lambda_1 t} + K_2 e^{\lambda_2 t},$$

where $A, B, C, D, K_1$, and $K_2$ are constants. (*Check the last few lines carefully.*)

### Stability

Imitating the qualitative analysis we did in Chapter 5 for real eigenvalues (see Table 5.1 at the end of Section 5.5 for a summary), we can see that if $\lambda_1$ and $\lambda_2$ are unequal and *positive*, then the origin is a *source*. If $\lambda_1$ and $\lambda_2$ are unequal and *negative*, then the origin is a *sink*. If $\lambda_1$ and $\lambda_2$ have different signs, then the origin is a *saddle point*.

Now suppose that the zeros of our characteristic polynomial are complex numbers: $\lambda_1 = p + qi$, $\lambda_2 = p - qi$. (Remember that complex roots of a quadratic equation occur in complex conjugate pairs.) Then we can write

$$s^2 + \left(\frac{c_1}{c_2}\right)s + \left(\frac{c_0}{c_2}\right) = (s - \lambda_1)(s - \lambda_2) = [s - (p + qi)][s - (p - qi)]$$

$$= [(s - p) - qi][(s - p) + qi] = (s - p)^2 + q^2.$$

Now when we express our solution $x(t)$ in terms of the inverse Laplace transform of functions, we will have $(s - p)^2 + q^2$ in all the denominators and either constants or constant multiples of $s$ in the numerators of these functions. Looking at entries 5 and 6 in the transform table (Table 6.1, in Section 6.2), we realize that the inverse Laplace transforms we will get are constant multiples of either $e^{pt} \sin qt$ or $e^{pt} \cos qt$.

The next example will help us understand these ideas better.

### ■ Example 6.6.1 A Qualitative Analysis via the Laplace Transform

Let's look at the equation $x'' + 3x' + 5x = 0$. The Laplace transform of this equation is

$$\mathcal{L}[x''] + 3\mathcal{L}[x'] + 5\mathcal{L}[x] = 0$$

$$\left\{s^2 \mathcal{L}[x] - sx(0) - x'(0)\right\} + 3\{s\mathcal{L}[x] - x(0)\} + 5\mathcal{L}[x] = 0$$

$$\left(s^2 + 3s + 5\right)\mathcal{L}[x] - (s + 3)x(0) - x'(0) = 0,$$

so

$$\mathcal{L}[x] = \frac{(s + 3)x(0) + x'(0)}{s^2 + 3s + 5} = \frac{(s + 3)x(0)}{s^2 + 3s + 5} + \frac{x'(0)}{s^2 + 3s + 5}.$$

The characteristic polynomial $s^2 + 3s + 5$ has complex conjugate zeros $-\frac{3}{2} + \frac{\sqrt{11}}{2}i$ and $-\frac{3}{2} - \frac{\sqrt{11}}{2}i$. Because the real part is negative, we expect the solution to oscillate with decreasing amplitude. Figure 6.5 shows $x$ against $t$, with $x(0) = 3$ and $x'(0) = 20$.    ■

### 6.6.2 Nonhomogeneous Equations

When we look at the nonhomogeneous version of (6.6.1), we find that

$$X(s) = \mathcal{L}[x(t)] = \frac{\mathcal{L}[f(t)]}{P(s)} + \frac{Q(s)}{P(s)} = \frac{F(s)}{P(s)} + \frac{Q(s)}{P(s)}, \tag{6.6.3}$$

where $P(s)$ is the characteristic polynomial $c_2 s^2 + c_1 s + c_0$ and $Q(s)$ is the linear polynomial $\{c_2 x(0)\}s + \{c_1 x(0) + c_2 x'(0)\}$. (*Verify this.*)

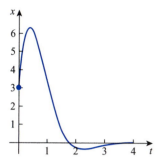

**FIGURE 6.5**
Graph of $x(t) = e^{-3t/2}\left(3\cos\left(\frac{\sqrt{11}}{2}t\right) + \frac{49\sqrt{11}}{11}\sin\left(\frac{\sqrt{11}}{2}t\right)\right), 0 \le t \le 4$

This observation can be expanded to the case of the general $n$th-order linear equation with constant coefficients. In this situation, $P(s)$ is the $n$th-degree characteristic polynomial and $Q(s)$ is a polynomial in $s$ of degree $n - 1$. The coefficients of $Q(s)$ consist of combinations of products of the coefficients in the equation and the $n$ initial conditions.

If we let $W(s) = \frac{1}{P(s)}$, we can write (6.6.3) as

$$X(s) = W(s)F(s) + W(s)Q(s). \tag{6.6.4}$$

Applying the inverse Laplace transform to each side of (6.6.4), we see that

$$x(t) = \mathcal{L}^{-1}[W(s)F(s)] + \mathcal{L}^{-1}[W(s)Q(s)],$$

which expresses the output $x(t)$ of the system as a superposition of two outputs—the first due to the input $f(t)$ and the second due to the initial conditions.

Let's look at a problem we've seen before, as Examples 6.1.1 and 6.2.1.

## ■ Example 6.6.2 Qualitative Analysis of a Nonhomogeneous Equation

Consider the nonhomogeneous initial-value problem

$$x'' + 3x' + 2x = 12e^{2t}; \quad x(0) = x_0, \, x'(0) = x_1.$$

The Laplace transform of this equation is

$$\mathcal{L}[x''] + 3\mathcal{L}[x'] + 2\mathcal{L}[x] = 12\mathcal{L}\left[e^{2t}\right]$$

$$\left\{s^2\mathcal{L}[x] - sx(0) - x'(0)\right\} + 3\left\{s\mathcal{L}[x] - x(0)\right\} + 2\mathcal{L}[x] = \frac{12}{s-2}$$

$$\left(s^2 + 3s + 2\right)\mathcal{L}[x] - (s+3)x_0 - x_1 = \frac{12}{s-2},$$

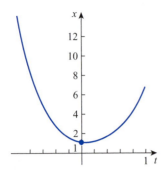

**FIGURE 6.6**

*Graph of $x(t) = e^{2t} + 3e^{-2t} - 3e^{-t}, -1 \le t \le 1$*

so

$$\mathcal{L}[x] = \frac{\frac{12}{s-2}}{s^2 + 3s + 2} + \frac{(s+3)x_0 + x_1}{s^2 + 3s + 2},$$

the form indicated in (6.6.3).

The characteristic polynomial $s^2 + 3s + 2$ has negative real zeros $-1$ and $-2$, so the second term in the preceding equation contributes a part of the solution that decays (tends to 0) as $t \to \infty$. (*Why?*) The first term has an additional pole, at 2. If we imagine the partial-fractions version of the Laplace transform of the solution, written with denominators $s - 2$ and $s^2 + 3s + 2$, we realize that the only terms that appear in the solution are $e^{2t}$, $e^{-t}$, and $e^{-2t}$. Therefore, for increasing values of $t$, the solution is dominated by the term containing $e^{2t}$. Figure 6.6 shows $x$ against $t$, with $x(0) = x0 = 1$ and $x'(0) = x_1 = -1$, as in Examples 6.1.1 and 6.2.1.  ■

### Transfer Functions and Impulse Response Functions

If we consider the nonhomogeneous second-order Equation (6.6.1) with initial conditions $x(0) = 0$, $x'(0) = 0$, then Equation (6.6.3) becomes

$$\mathcal{L}[x(t)] = \frac{1}{c_2 s^2 + c_1 s + c_0} \mathcal{L}[f(t)],$$

or

$$\frac{X(s)}{F(s)} = \frac{1}{c_2 s^2 + c_1 s + c_0}, \tag{6.6.5}$$

where $F(s) = \mathcal{L}[f(t)]$ and $X(s) = \mathcal{L}[x(t)]$. A system with these initial conditions is sometimes described as "relaxed" or at rest until $t = 0$.

In certain areas of engineering (for example, those dealing with feedback and control systems), this ratio (6.6.5) of the Laplace transform of the *output* to the Laplace transform of the *input* is called the **transfer function** of the system modeled by Equation (6.6.1) with all initial values zero. The inverse Laplace transform of the transfer function is called the **impulse response function** for the system, because in physical terms, it describes (for example) the solution when a spring-mass system is struck by a hammer. (See Example 6.4.1, for instance, and Problem B3 of Exercises 6.6.) The analysis of this transfer function provides a picture of what can be called the **response** of the system. The values of $s$ that make the denominator of (6.6.5) zero are called **poles** or **singularities** of the transfer function. On the basis of our analysis of eigenvalues in Chapter 5 and our discussion in this section, you should see that the nature of the poles (real, complex, positive, and so on) determines the behavior of the system. For example, in this second-order situation, the system could be undamped, overdamped, or underdamped, or the response of the system could grow without bound.

## Exercises 6.6

### A

Suppose $X(s) = \mathcal{L}[x(t)]$ is the Laplace transform of the solution of a linear differential equation. For each transform in Problems 1–10, determine the qualitative behavior of $x(t)$ for large values of $t$ without finding the inverse of the transform. (That is, determine if $x(t)$ oscillates, goes to zero, or becomes unbounded as $t$ becomes large.) Note that "oscillates" and "goes to zero" are not mutually exclusive answers, nor are the choices "oscillates" and "becomes unbounded."

**1.** $X(s) = \dfrac{2}{3s + 5}$

**2.** $X(s) = \dfrac{4}{s^2 - 1}$

**3.** $X(s) = \dfrac{s + 1}{s^2 + 1}$

**4.** $X(s) = \dfrac{1}{s^2 + 2s + 10}$

**5.** $X(s) = \dfrac{s - 2}{s^2 - 2s + 1}$

**6.** $X(s) = \dfrac{s + 2}{s^2 + 4}$

**7.** $X(s) = \dfrac{2s + 6}{s^2 + 6s + 18}$

**8.** $X(s) = \dfrac{2s + 5}{s^2 + 3s + 2}$

**9.** $X(s) = \dfrac{s}{s^3 - 1} - 1$

**10.** $X(s) = \dfrac{s}{s^4 + 5s^2 + 4}$

In Problems 11–14, (a) compute the Laplace transform of each solution, (b) find the poles of the Laplace transform of the solution, and (c) discuss the behavior of the solution (oscillatory, unbounded, etc.) without solving the equation.

**11.** $x'' - x = 0;$   $x(0) = 0, x'(0) = 1$

**12.** $\ddot{x} + 2\dot{x} + 2x = e^{-t/10};$   $x(0) = 4, \dot{x}(0) = 1$

**13.** $x'' + 2x' + 2x = e^{-2t} \sin 4t;$   $x(0) = 2, x'(0) = -2$

**14.** $2\ddot{x} + 7\dot{x} + 3x = 2\cos t;$   $x(0) = 1, x'(0) = 0$

## B

1. Suppose $f(t)$ and $x(t)$ are the input and output, respectively, of a linear second-order ODE with constant coefficients.

   **a.** Find the ODE if its transfer function is given by $P(s) = 1/(s^2 + s + 1)$.

   **b.** Find $x(t)$ if $f(t) = [\sin 2(t - 1)]U(t - 1)$.

2. Find the output of the ODE with transfer function $1/(s + 1)$, given that the input $f(t)$ has Laplace transform $F(s) = \frac{se^{-2s}}{s^2+4}$.

3. Consider the initial value problem

$$c_2 x'' + c_1 x' + c_0 x = \delta(t); \quad x(0) = x'(0) = 0,$$

where $\delta(t)$ denotes the unit impulse function (Section 6.4). Show that the transfer function of the system is $X(s) = \frac{1}{c_2 s^2 + c_1 s + c_0}$.

4. Suppose that a linear system is described by the equation

$$\ddot{x} + 2\dot{x} + 5x = f(t); \quad x(0) = 2, \dot{x}(0) = -2.$$

   **a.** Find the transfer function for the system.

   **b.** Find the impulse response function.

   **c.** Give a formula for the solution of the IVP. (Use the result of Problem C1 below. Your answer should contain an integral.)

5. If a linear system is governed by the initial value problem

$$y'' - y' - 6y = g(t); \quad y(0) = 1, y'(0) = 8,$$

   **a.** Find the transfer function for the system.

   **b.** Find the impulse response function.

   **c.** Find a formula for the solution of the IVP. (Use the result of Problem C1 below. Your answer should contain an integral.)

6. Consider the initial value problem

$$y'' + 2y' + 2y = \sin(\alpha t); \quad y(0) = 0, y'(0) = 0.$$

   **a.** Find the transfer function for the system.

   **b.** Find the impulse response function.

   **c.** Find a formula for the solution of the IVP. (Use the result of Problem C1 below. Your answer should contain an integral.)

**7.** Consider the first-order system $a_1 x' + a_0 x = f(t)$, where $a_1$ and $a_0$ are constants, $a_1 \neq 0$.

   **a.** Find the transfer function of this system.

   **b.** Show that the transfer function of a constant-coefficient first-order system can be written as $W(s) = \frac{c}{1+Ts}$, where $c$ is a constant and $T$ is a constant related to the exponential function component of the solution.

## C

**1.** If $I$ is an interval containing the origin and $f$ is continuous on $I$, show that the unique solution to the IVP

$$c_2 x'' + c_1 x' + c_0 x = f(t); \quad x(0) = x_0, x'(0) = x_1$$

is given by $(r * f)(t) + x_H(t)$, where $R = \frac{X(s)}{F(s)}$, $r = \mathcal{L}^{-1}[R](t)$ is the response function, and $x_H(t)$ is the unique solution of the homogeneous equation $c_2 x'' + c_1 x' + c_0 x = 0; x(0) = x_0, x'(0) = x_1$. (Of course, $*$ denotes *convolution*.)

**2.** The *Volterra integral equation* is given by

$$x(t) = g(t) + \int_0^t k(t - \tau) x(\tau) d\tau,$$

where $g$ and $k$ are known functions with Laplace transforms $G = \mathcal{L}[g]$ and $K = \mathcal{L}[k]$. Show that $x(t) = \mathcal{L}^{-1}[\frac{G(s)}{1-K(s)}]$.

## SUMMARY

Transformation methods are important examples of how we can change difficult problems into problems that can be handled more easily. If $f(t)$ is a function that is integrable for $t \geq 0$, then the **Laplace transform** of $f$ is defined by $\mathcal{L}[f(t)] = \int_0^\infty f(t) e^{-st} dt$, when this improper integral exists. The integral will exist if we stick to *continuous* or *piecewise continuous* functions $f(t)$ for which there exist positive constants $M$ and $K$ such that $|f(t)| < e^{Mt}$ for all $t \geq K$. Note that this integral is a function of the parameter $s$, so we can write $\mathcal{L}[f(t)] = F(s)$.

Using basic properties of integrals, we can see that $\mathcal{L}[c \cdot f(t)] = c \cdot \mathcal{L}[f(t)]$, where $c$ is any real constant, and that $\mathcal{L}[f(t) + g(t)] = \mathcal{L}[f(t)] + \mathcal{L}[g(t)]$, whenever the Laplace transforms of both $f$ and $g$ exist. Any transformation that satisfies the last two properties is called a *linear transformation*. If $c_1$ and $c_2$ are constants, we can combine the two properties to write $\mathcal{L}[c_1 f(t) + c_2 g(t)] = c_1 \mathcal{L}[f(t)] + c_2 \mathcal{L}[g(t)]$.

Table 6.1 in Section 6.2 gives the Laplace transform of some important classes of functions, including power functions, exponentials, trigonometric functions, and multiples of these. There are also important formulas for the Laplace transforms of $f'$, $f''$, and higher derivatives. The Laplace transform method enables us to handle a linear nonhomogeneous equation with initial conditions all at once.

Once we have calculated the Laplace transform of a function—in particular, once we have transformed a differential equation into an algebraic equation—we have to be able to reverse the process to gain information about the original problem. An important fact is that *if the Laplace transforms of the continuous functions f and g exist and are equal for $s \geq c$ (c a constant), then $f(t) = g(t)$ for all $t \geq 0$.* This says that a continuous function can be recovered uniquely from its Laplace transform. Letting $\mathcal{L}[f(t)] = F(s)$, we can express the definition of the **inverse Laplace transform** as follows:

$$\mathcal{L}^{-1}[F] = f \quad \text{if and only if} \quad \mathcal{L}[f] = F.$$

It can be shown that the inverse Laplace transform is a linear transformation:

$$\mathcal{L}^{-1}\left[c_1 F(t) + c_2 G(t)\right] = c_1 \mathcal{L}^{-1}[F(t)] + c_2 \mathcal{L}^{-1}[G(t)].$$

In trying to find the inverse transform of an expression that is the product of two or more transforms, we encounter the idea of the *convolution* of two functions. The **convolution** of two functions $f$ and $g$ is the integral $(f * g)(t) = \int_0^t f(r)g(t - r)dr$, provided that the integral exists for $t > 0$. This product has important algebraic properties, and one of the most useful is that the Laplace transform of a convolution of two functions is equal to the product of the Laplace transforms of these two functions. More precisely, suppose that $f$ and $g$ are two functions whose Laplace transforms exist. Let $F(s) = \mathcal{L}[f(t)]$ and $G(s) = \mathcal{L}[g(t)]$. Then the **Convolution Theorem** says that

$$\mathcal{L}[(f * g)(t)] = \mathcal{L}\left[\int_0^t f(r)g(t - r)dr\right] = \mathcal{L}[f(t)] \cdot \mathcal{L}[g(t)] = F(s) \cdot G(s).$$

By using the unit step function (or Heaviside function) $U$, defined by

$$U(t) = \begin{cases} 0 & \text{for } t < 0 \\ 1 & \text{for } t \geq 0, \end{cases}$$

we can model systems in which there are abrupt changes. Mathematically, this means that we can express *piecewise continuous functions* in a simple way, using $U(t)$ as a basic building block.

When we are solving differential equations that model abrupt changes, the following result comes in handy. If $\mathcal{L}[f(t)]$ exists for $s > c$ and if $a > 0$, then

$$\mathcal{L}[f(t - a)U(t - a)] = e^{-as}\mathcal{L}[f(t)] \quad \text{for } s > c.$$

Alternatively, we can write the preceding formula as

$$f(t - a)U(t - a) = \mathcal{L}^{-1}\left[e^{-as}\mathcal{L}[f(t)]\right].$$

If we want to consider problems in which there is an external force of large magnitude applied suddenly for a very short period of time, we need the idea of the **unit impulse function**, or

**Dirac delta function**, defined as

$$\delta(t) = \lim_{b \to 0} \delta_b(t) = \begin{cases} \infty & \text{for } t = 0 \\ 0 & \text{for } t \neq 0, \end{cases}$$

where

$$\delta_b(t) = \begin{cases} \dfrac{1}{b} & \text{for } 0 \leq t \leq b \\ 0 & \text{for } t > b. \end{cases}$$

We can show that $\mathcal{L}[\delta(t-a)] = e^{-sa}$. In particular, $\mathcal{L}[\delta(t)] = 1$.

When initial conditions are given, the Laplace transform converts a system of linear differential equations with constant coefficients to a system of simultaneous algebraic equations. Then we can solve the algebraic equations for the transformed solution functions. Finally, applying the inverse transform to these functions gives us the solutions of the original system of linear ODEs. However neat this sounds conceptually, the algebraic details are often quite messy, and technology comes in handy.

Despite the emphasis on the Laplace transform as a tool for obtaining exact, closed-form solutions, it turns out that the transform can provide insight into the *qualitative* nature of a solution as well. In certain applied areas, when we are considering the important second-order equation $c_2 x'' + c_1 x' + c_0 x = f(t)$, the ratio $\frac{X(s)}{F(s)} = \frac{1}{c_2 s^2 + c_1 s + c_0}$, where $F(s) = \mathcal{L}[f(t)]$ and $X(s) = \mathcal{L}[x(t)]$, is called the **transfer function** of the system modeled by the equation with all initial values zero. The inverse Laplace transform of the transfer function is called the **impulse response function** for the system and describes the solution (and the system) in a way that is similar to the qualitative techniques used in Chapter 5. The analysis of this transfer function provides a picture of what can be called the **response** of the system. The values of $s$ that make the denominator of the transfer function zero are called **poles** or **singularities** of the transfer function. On the basis of the analysis of eigenvalues in Chapter 5 and the discussion in this section, we see that the algebraic nature of the poles determines the behavior of the system.

# PROJECT 6-1

## Residential Segregation[1]

Suppose that two distinct cultural groups, the Yahoos and the Houyhnhnms (pronounced 'hwinems), live in the same city. Their cultural differences are such that one group may annoy the other group so much that the one group may start moving out of the city, which results

---

[1] This project is based on T. P. Dreyer's treatment of a model developed by M. E. Gurtin in "Some Mathematical Models for Population Dynamics that Lead to Segregation," *Quarterly of Applied Mathematics* **32** (1974): 1–9.

in segregation. This scenario leads to the initial-value problem

$$\frac{dY}{dt} = a\,Y(t) + b\,H(t) \quad Y(0) = \alpha$$

$$\frac{dH}{dt} = c\,Y(t) + d\,H(t) \quad H(0) = \beta,$$

$(*)$

where $Y(t)$ and $H(t)$ denote the Yahoo and Houyhnhnm populations, respectively, at time $t$ and $a, b, c, d, \alpha$, and $\beta$ are constants.

The constants $a$ and $d$ denote the net growth rate per population, whereas $b$ and $c$ indicate, respectively, the effect of the presence of the other group.

a. Given the assumption that for both groups the presence of the other group is irritating and encourages emigration, what does this say about the signs of $b$ and $c$?

b. Determine the Laplace transforms of $Y(t)$ and $H(t)$, assuming that these transforms exist.

c. Making the substitutions $\gamma = \frac{1}{2}(a + d)\delta = \frac{1}{2}(a - d)$, and $\omega = \sqrt{\delta^2 + bc}$, rewrite the transforms found in part (b) and then find the partial fraction decompositions of $\mathcal{L}[Y(t)]$ and $\mathcal{L}[H(t)]$.

d. Use the inverse Laplace transform and the given initial conditions to find $Y(t)$ and $H(t)$. (Assuming that $Y(t)$ and $H(t)$ have continuous first derivatives, the second form of the Existence and Uniqueness Theorem in Section 4.9 guarantees that this solution is unique.)

e. Complete segregation occurs when either $Y(t)$ or $H(t)$ is zero at some finite time $t^*$. Explain why such a finite time exists for $Y(t)$ if $\alpha/\beta < -b/(\delta + \omega)$ and such a time exists for $H(t)$ if $\alpha/\beta > -b/(\delta + \omega)$.

f. If $\alpha/\beta = -b/(\delta + \omega)$ (called the tipping ratio), why do the expressions for $Y(t)$ and $H(t)$ found in part (d) indicate that complete segregation cannot occur? What is the tipping ratio if $a = d$?

g. If $b = 0$ and $c < 0$ in System $(*)$, investigate how many Yahoos and Houyhnhnms will be living in the city eventually—that is, as $t \to \infty$.

# CHAPTER 7

# Systems of Nonlinear Differential Equations

## INTRODUCTION

We have discussed various nonlinear equations throughout previous chapters, especially in Chapters 2 and 3, treating them numerically, graphically, and analytically. In general, we can't expect to find the explicit (closed-form) solution of a nonlinear equation, so we are forced to rely on qualitative and computational methods rather than on purely analytical techniques. This complexity is magnified when we address *systems* of nonlinear equations.

In Chapter 5 we analyzed the *stability* of systems of linear differential equations—that is, the behavior of such systems near equilibrium points—and saw that this stability could be described completely in terms of the eigenvalues and eigenvectors of the system. This kind of analysis can be done for nonlinear systems, but it is not quite so satisfactory and complete. One way of carrying out this study is to examine how closely we can approximate (in some sense) a nonlinear system by a linear system and then apply the linear theory.

The modern qualitative theory of stability discussed in Chapter 5 and in this chapter originated in the late 1800s with the work of the French mathematician Henri Poincaré (1854–1912), who was studying nothing less than whether the solar system was a stable system. The equations involved in Poincaré's study of celestial mechanics could not be solved explicitly, so he and others developed implicit (qualitative) methods to deal with the complicated problems of planetary motion. [An excellent account of this work and its consequences is *Celestial Encounters: The Origins of Chaos and Stability* by F. Diacu and P. Holmes (Princeton: Princeton University Press, 1996).]

## 7.1 EQUILIBRIA OF NONLINEAR SYSTEMS

Recall that an *equilibrium point* of a differential equation or a system of differential equations is a constant solution. If we look at the two (somewhat similar) equations (1) $y' = -y$ and (2) $y' = -y(1 - y)$, we will see some important differences between linear and nonlinear equations.

353

Equation (1) is linear but, more fundamentally, *separable*, so it is easy to find the general solution: $y = Ce^{-t}$, where $C$ is an arbitrary constant. (We recognize that $C = y(0)$, the initial state of the system being modeled by the equation.)

Now Equation (2) is nonlinear and separable, and its general solution is $y = \frac{Ce^{-t}}{1-C+Ce^{-t}}$, where $C = y(0)$. (*Verify the solutions to both equations.*)

Let's examine some typical solution curves for Equation (1). Figure 7.1 shows that there is only one equilibrium solution, $y \equiv 0$, and this is a *sink*. (Review Section 2.6 if necessary.) If an object described by the equation starts off at zero (that is, if $C = 0$), it remains at zero for all time. If the object's initial state is not zero, then the object will approach the solution $y \equiv 0$ as its asymptotically stable solution (or sink).

On the other hand, Figure 7.2 shows the same kind of information for Equation (2). For such a nonlinear equation there can be more than one equilibrium solution, in this case $y \equiv 0$ and $y \equiv 1$. Also note that some solutions of a nonlinear equation may "blow up in finite time"—that is, become unbounded as $t$ approaches some finite value.

(*Where does the denominator of the general solution to Equation (2) vanish?*) In contrast, all solutions of a linear equation or a system of linear equations are defined for all values of the independent variable. Finally, looking closely at the behavior of solutions of Equation (2) with different initial values, we see that the solutions starting off above 1 behave differently

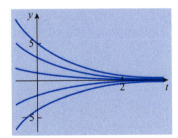

**FIGURE 7.1**
*Solutions of $y' = -y$; $y(0) = 5, 3, 1, -2, -4$*

**FIGURE 7.2**
*Solutions of $y' = -y(1 - y)$*

from those solutions with initial values less than 1. The equilibrium solution $y \equiv 0$ is a *sink* if $y(0) < 1$ and $y \equiv 1$ is a *source* if $y(0) > 1$. Furthermore, for solutions with initial values $C$ greater than 1, the line $t = \ln\left(\frac{C}{C-1}\right)$ is a vertical asymptote. The last three types of behavior cannot occur when we are dealing with a linear equation. You should expect that the situation with nonlinear *systems* is appropriately complicated.

Let's look at an example of a nonlinear system and its behavior near its equilibrium points.

## ■ Example 7.1.1 Stability of a Nonlinear System

The nonlinear system

$$x' = x - x^2 - xy$$
$$y' = -y - y^2 + 2xy$$

represents two populations interacting in a predator-prey relationship. This is essentially a Lotka-Volterra system (see Section 4.7, especially Example 4.7.4) with "crowding" terms (the squared terms) added for both species.

To calculate the equilibrium points of this system, we solve the system $\{x' = 0, y' = 0\}$, which is the same as the nonlinear algebraic system

$$\text{(A)} \quad x(1 - x - y) = 0$$
$$\text{(B)} \quad y(-1 - y + 2x) = 0.$$

Clearly, the origin, $x = y = 0$, is an equilibrium point. Logically, there are only three other cases to examine: (1) $x = 0, y \neq 0$; (2) $x \neq 0, y = 0$; and (3) $x \neq 0, y \neq 0$. Assuming case 1, we can eliminate Equation (A) and examine (B), which becomes $y(-1 - y) = 0$. Because $y \neq 0$, we conclude that $-1 - y = 0$, or $y = -1$. Thus, our second equilibrium point is $(0, -1)$. Moving to case 2, we can ignore Equation (B) and focus on (A), which now looks like $x(1 - x) = 0$. Because we are assuming in case 2 that $x \neq 0$, we can see that $x = 1$, which gives us the third equilibrium point $(1, 0)$. Finally, if $x \neq 0$ and $y \neq 0$, our system of algebraic equations becomes

$$\text{(A2)} \quad x + y = 1$$
$$\text{(B2)} \quad y - 2x = -1.$$

(We have divided out $x$ and $y$ in (A) and (B) and then rearranged the terms of each equation.) Subtracting (B2) from (A2) gives us $3x = 2$, or $x = \frac{2}{3}$. Substituting this value of $x$ in (A2) yields $y = \frac{1}{3}$. Therefore, the last equilibrium point is $\left(\frac{2}{3}, \frac{1}{3}\right)$.

In terms of a population problem, the only interesting equilibrium point is the last one we found. (*Why is this so?*) If we look at a slope field for the original system of nonlinear differential equations near the point $\left(\frac{2}{3}, \frac{1}{3}\right)$, we see some interesting behavior (Figure 7.3a).

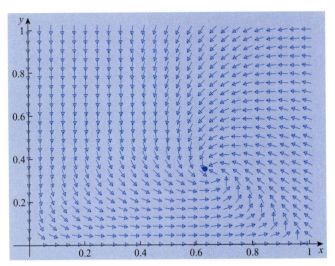

**FIGURE 7.3a**

*Slope field for $x' = x - x^2 - xy, y' = -y - y^2 + 2xy$ near $\left(\frac{2}{3}, \frac{1}{3}\right)$*

**FIGURE 7.3b**

*Phase portrait for $x' = x - x^2 - xy, y' = -y - y^2 + 2xy$ near $\left(\frac{2}{3}, \frac{1}{3}\right)$*
$(x(0), y(0)) = (0.2, 1), (0.8, 0.8), (0.8, 0.5), (1, 0.7), (1, 0.2), (0.5, 1)$

The apparent spiraling of solutions into the equilibrium point can be seen more clearly if we show some (numerically generated) solution curves (Figure 7.3b). Figure 7.3b represents a predator-prey population that is stabilizing. If the units are thousands of creatures, then the $X$ population is heading for a steady population of about 667, whereas the $Y$ population has 333 as its stable value.

Mathematically, however, we should look at the entire phase portrait to understand the complex behavior of nonlinear systems. We'll return for a detailed analysis in Example 7.3.1.

■

## Exercises 7.1

### A

Find all equilibrium points for each of the systems in Problems 1–14, using technology if necessary.

1. $x' = -x + xy, y' = -y + 2xy$
2. $x' = x - xy, y' = y - xy$
3. $x' = x^2 - y^2, y' = x - xy$
4. $x' = 1 - y^2, y' = 1 - x^2$
5. $x' = x + y + 2xy, y' = -2x + y + y^3$
6. $x' = y\left(1 - x^2\right), y' = -x\left(1 - y^2\right)$
7. $x' = x - x^2 - xy, y' = 3y - xy - 2y^2$
8. $x' = 1 - y, y' = x^2 - y^2$
9. $x' = (1 + x)\sin y, y' = 1 - x - \cos y$ [*Hint*: Graph the two equations on the same axes.]
10. $x' = 3y - e^x, y' = 2x - y$ [*Hint*: There are two equilibrium points. Use your CAS to approximate these points.]
11. $x' = y^2 - x^2, y' = x - 1$
12. $x' = x^2 - y, y' = y^2 - x$
13. $x' = xy(1 - x), y' = y\left(1 - \frac{y}{x}\right)$
14. $x' = y, y' = -\sin x - 3y$

### B

1. Use technology to find all equilibrium points of the system

$$x' = -y, y' = \left(x^4 + 4x^3 - x^2 - 4x + y\right)/8.$$

2. A two-mode laser produces two different kinds of photons, whose numbers are $n_1$ and $n_2$. The equations governing the rates of photon production are

$$\dot{n}_1 = G_1 N n_1 - k_1 n_1$$
$$\dot{n}_2 = G_2 N n_2 - k_2 n_2,$$

where $N(t) = N_0 - a_1 n_1 - a_2 n_2$ is the number of excited atoms. The parameters $G_1, G_2, k_1, k_2$, $a_1, a_2$, and $N_0$ are all positive. Use a CAS "solve" command to find all equilibrium points of the system.

3. A *chemostat* is a device for growing and studying bacteria by supplying nutrients and maintaining convenient levels of the bacteria in a culture. (See Project 2-2 at the end of

Chapter 2.) One model of a chemostat is the nonlinear system

$$\frac{dN}{dt} = a_1 \left( \frac{C}{1+C} \right) N - N$$

$$\frac{dC}{dt} = - \left( \frac{C}{1+C} \right) N - C + a_2,$$

where $N(t)$ denotes the bacterial density at time $t$, $C(t)$ denotes the concentration of nutrient, and $a_1, a_2$ are positive parameters. Use technology to find all equilibrium solutions ($N^*, C^*$) of the system.

4. In the absence of damping and any external force, the motion of a pendulum is described by the equation $\frac{d^2\theta}{dt^2} + \frac{g}{L} \sin\theta = 0$, where $\theta$ is the angle between the pendulum and the downward vertical, $g$ is the acceleration due to gravity, and $L$ is the length of the pendulum.

   a. Write this equation as a system of two first-order equations.
   b. Describe all equilibrium points of the system.

**C**

1. Use technology to find all equilibrium solutions of the system

$$x' = x - y^2 + a + bxy, \quad y' = 0.2y - x + x^3,$$

where $a = 1.28$ and $b = 1.4$. (Round to the nearest thousandth.)

## 7.2 LINEAR APPROXIMATION AT EQUILIBRIUM POINTS

One important aspect of *linear* systems is that the behavior of solutions near an equilibrium point ("local" behavior) tells you the behavior of solutions in the entire phase plane. However, even though most of the "nice" properties of linear systems are not present when we analyze nonlinear systems, we may be able to understand the local behavior of nonlinear systems by a process of *linearization* or *linear approximation*. This means that we try to replace the original nonlinear system by a linear system that is "close" or near an equilibrium point. Remember that in Section 3.1 we first discussed *Euler's method*, which involved approximating solution curves by tangent lines.

To see how this might work, let's go back to the nonlinear equation $y' = -y(1-y) = -y+y^2$ discussed in Section 7.1. We know that $y = 0$ is an equilibrium point. Now note that for values of $y$ close to zero, $y^2$ is smaller than $y$. For example, if $y = 0.00001$, then $y^2 = 0.0000000001$. Then, dropping the squared (nonlinear) terms, we can guess that the linear equation $y' = -y$ is a good approximation for the original equation and that the behavior of this last equation near $y = 0$ should tell us how $y' = -y(1-y)$ behaves near $y = 0$. A comparison of the solution curves near $y = 0$ in Figure 7.1 and Figure 7.2 shows us that this is true. However, it should also be clear that we would be wrong to base our analysis of $y' = -y(1-y)$ on $y' = -y$ for *all* initial values.

If we want to analyze the behavior of $y' = -y(1 - y)$ near its other equilibrium point, $y = 1$, we can use a simple change of variable: Let $y = 1 + z$, so that studying the behavior of $y' = -y(1 - y)$ near $y = 1$ is the same as analyzing the behavior of the equation $y = 1 + z$ near $z = 0$. (*Make sure you see this.*) With this change of variable, we get the new equation $z' = -y(1 - y) = (-1 - z)(-z) = z + z^2$. Using the same reasoning as before, we can take $z' = z$ as a good linear approximation near $z = 0$. This last equation has the general solution $z = Ce^t$, so solutions of $z' = z + z^2$ move *away from* $z = 0$ as $t$ increases. But because $y = 1 + z$, solutions of $y' = -y(1 - y)$ near $y = 1$ curve away from $y = 1$, behavior we can verify by looking at Figure 7.2.

As another example, take the second-order nonlinear equation $\frac{d^2x}{dt^2} + \frac{g}{L} \sin x = 0$, which describes the swinging of a pendulum (where $x$ is the angle the pendulum makes with the vertical, $g$ is the acceleration due to gravity, and $L$ is the pendulum's length). This equation is not easy to deal with analytically, so often the nonlinearity is removed by a substitution. For small values of $x$ (that is, for an oscillation of small amplitude), $\sin x \approx x$, so we can replace our original nonlinear equation by the *linear* equation that approximates it: $\frac{d^2x}{dt^2} + \frac{g}{L} x = 0$. This approximate pendulum model has the same mathematical behavior as the undamped spring-mass system; see Equation (4.8.1) in Section 4.8. Despite our success in approximating a nonlinear equation by one that is linear, this is a limited victory. For example, the analysis of the linear approximations implies that all solutions are defined for all values of $t$, but this is clearly not the case for the nonlinear equation. The next example illustrates the failure of linearization more dramatically.

### ■ Example 7.2.1 Linearization Can Mislead

Let's look at the system

$$\dot{x} = y + ax(x^2 + y^2)$$
$$\dot{y} = -x + ay(x^2 + y^2),$$

where $a$ is a given real number.

Clearly, the origin $(x, y) = (0, 0)$ is an equilibrium point regardless of the value of the parameter $a$. In our example the obvious linearized system is

$$\dot{x} = y$$
$$\dot{y} = -x.$$

(Look back at the spring-mass system analyzed in Example 4.8.1.) This can be written in the form $\dot{X} = AX$, where $X = \begin{bmatrix} x \\ y \end{bmatrix}$ and $A = \begin{bmatrix} 0 & 1 \\ -1 & 0 \end{bmatrix}$. The characteristic polynomial is $\lambda^2 + 1 = 0$, so that the eigenvalues of $A$ are purely imaginary: $i$ and $-i$. From Table 5.1 in Section 5.5, we conclude that the origin is a *stable center* of the linearized system. *But this is the wrong conclusion with respect to the original nonlinear system*, as the phase portrait

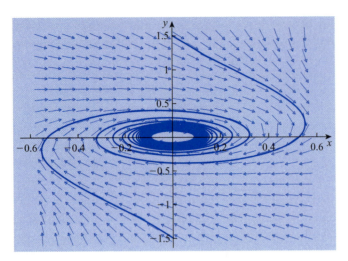

**FIGURE 7.4**
*Trajectories of $\dot{x} = y - x\left(x^2 + y^2\right), \dot{y} = -x - y\left(x^2 + y^2\right), 0 \le t \le 60$*
*$(x(0), y(0)) = (0, 1.5), (0, -1.5)$*

of the original nonlinear system near $(0, 0)$ shows. This portrait (Figure 7.4) corresponds to $a = -1$.

It seems that the trajectories spiral in toward the equilibrium point, indicating that the origin is actually a *spiral sink* for the nonlinear system. However, appearances can be deceiving, and Problem C1 in Exercises 7.2 suggests a way of proving this claim about the origin.

You should suspect that the stability of the original system depends on the value of the parameter $a$. If $a = 0$, for example, then the nonlinear portion of the system disappears, leaving us with a purely linear system—in fact, the same system analyzed in Example 4.8.1 (with $\beta = 1$). As we've said, the origin is a stable center for this linear system, every trajectory closing perfectly after one cycle. Problem C1 asks you to explore these ideas further. ∎

In summary, we can look at this last example as a linear system "perturbed" (disturbed or knocked off kilter) by a nonlinear component. We can write this system as $\dot{X} = \begin{bmatrix} 0 & 1 \\ -1 & 0 \end{bmatrix} X + \begin{bmatrix} f(x, y) \\ g(x, y) \end{bmatrix}$, where $f$ and $g$ are nonlinear functions of $x$ and $y$. As we'll see later, if the nonlinear perturbation is "nice" enough, the behavior of the whole system can be predicted from the behavior of its linear portion near equilibrium points.

## 7.2.1 Almost Linear Systems

If we want to make this discussion of linear approximation mathematically sound, we have to remind ourselves of some basic calculus facts. (See Section A.1 for additional information.) Back in Section 3.1 we discussed *local linearity*, the idea that if we "zoom" in on a point on a curve $y = f(x)$, the curve looks like a straight line—in fact, like a piece of the tangent line drawn to the curve at that point. More precisely, for values of the independent variable $x$ close to $x = a$, we can write $f(x) \approx f(a) + f'(a)(x - a)$. You should recognize that this expression consists of the first two terms of an $n$th-degree ($n \geq 1$) *Taylor polynomial approximation* of $f$ near $x = a$—or, equivalently, the first two terms of the *Taylor series* expansion of $f$ in a neighborhood of $x = a$:

$$f(x) = f(a) + f'(a)(x - a) + \frac{f''(a)}{2!}(x - a)^2 + \frac{f'''(a)}{3!}(x - a)^3 + \cdots$$

$$+ \frac{f^{(n)}}{n!}(x - a)^n + \cdots$$

$$= f(a) + f'(a)(x - a) + (x - a)^2 \left\{ \frac{f''(a)}{2!} + \frac{f'''(a)}{3!}(x - a) + \cdots \right.$$

$$\left. + \frac{f^{(n)}}{n!}(x - a)^{n-2} + \cdots \right\}.$$

We can write this last result as $f(x) \approx f(a) + f'(a)(x - a) + O((x - a)^2)$, where the notation $O((x - a)^2)$ represents the fact that if $x$ is close to $a$ (so that $x - a$ is very small), then the sum of all terms past the second will be bounded by some multiple of $(x - a)^2$. (The series in braces, $\{ \cdots \}$, converges to some constant value.)

Now assume that we have a general nonlinear autonomous system of the form

$$\dot{x} = F(x, y)$$
$$\dot{y} = G(x, y) \tag{7.2.1}$$

which has the origin as an equilibrium point—that is, $F(0, 0) = 0$ and $G(0, 0) = 0$. This last assumption is just for convenience as we develop some methodology. If we can write $F$ as $ax + by + f(x, y)$ and $G$ as $cx + dy + g(x, y)$, where $f$ and $g$ are nonlinear functions, then we can express the system in the form

$$\dot{X} = \begin{bmatrix} a & b \\ c & d \end{bmatrix} X + \begin{bmatrix} f(x, y) \\ g(x, y) \end{bmatrix}.$$

If the nonlinear functions $f$ and $g$ are "small enough" (in a sense to be explained later) that their effect on the system is negligible, then we can call our system "almost linear." Near the origin, our nonlinear system behaves essentially like the linear system $\dot{X} = \begin{bmatrix} a & b \\ c & d \end{bmatrix} X$—that is, like the system

$$\dot{x} = ax + by$$
$$\dot{y} = cx + dy.$$

Earlier we recalled that the tangent line $y = f(a) + f'(a)(x - a)$ gives the best linear approximation of a single-variable function $f$ near $x = a$. For $F(x, y)$, a function of *two* variables, the best approximation near a point $(a, b)$ is provided by the *tangent plane* given by the approximation formula

$$F(x, y) \approx F(a, b) + \frac{\partial F}{\partial x}(a, b)(x - a) + \frac{\partial F}{\partial y}(a, b)(y - b), \tag{7.2.2}$$

where $\frac{\partial F}{\partial x}(a, b)$ and $\frac{\partial F}{\partial y}(a, b)$ denote the partial derivatives of $F$ evaluated at the point $(a, b)$. (See Section A.8.) For example, if we want to approximate $F(x, y) = x^3 + y^3$ near the point $(1, 1)$, we calculate

$$F(1, 1) = 1^3 + 1^3 = 2$$

$$\frac{\partial F}{\partial x} = 3x^2, \frac{\partial F}{\partial x}(1, 1) = 3(1)^2 = 3$$

$$\frac{\partial F}{\partial y} = 3y^2, \frac{\partial F}{\partial y}(1, 1) = 3(1)^2 = 3,$$

so the equation of the tangent plane is $z = 2 + 3(x - 1) + 3(y - 1)$.

You should think of the right-hand side of Equation (7.2.2) as the first-degree Taylor polynomial approximation of $F$, the linear terms in $x$ and $y$ of the two-variable Taylor series expansion of $F$. This approximation ignores the rest of the series consisting of the terms in $x$ and $y$ of the second degree and higher, which we can denote by $f(x, y)$. Thus, in our last example, we can write $x^3 + y^3 \approx 2 + 3(x - 1) + 3(y - 1)$ near $(1, 1)$ or $x^3 + y^3 = 2 + 3(x - 1) + 3(y - 1) + f(x, y)$ near $(1, 1)$. (See Section A.8 for more information on this issue.)

If we choose the point $(a, b)$ to be the origin, then we can rewrite (7.2.1) as

$$\dot{x} = F(0, 0) + \frac{\partial F}{\partial x}(0, 0)x + \frac{\partial F}{\partial y}(0, 0)y + f(x, y)$$

$$\dot{y} = G(0, 0) + \frac{\partial G}{\partial x}(0, 0)x + \frac{\partial G}{\partial y}(0, 0)y + g(x, y)$$

or (remembering that we have assumed $F(0, 0) = G(0, 0) = 0$), as

$$\dot{x} = ax + by + f(x, y)$$
$$\dot{y} = cx + dy + g(x, y), \tag{7.2.3}$$

where $a = \frac{\partial F}{\partial x}(0, 0), b = \frac{\partial F}{\partial y}(0, 0), c = \frac{\partial G}{\partial x}(0, 0)$, and $d = \frac{\partial G}{\partial y}(0, 0)$.

The technical definition of the "smallness" of $f$ and $g$ near the origin is that

$$\lim_{(x,y) \to (0,0)} \frac{f(x, y)}{\sqrt{x^2 + y^2}} = \lim_{(x,y) \to (0,0)} \frac{g(x, y)}{\sqrt{x^2 + y^2}} = 0. \tag{7.2.4}$$

The limits in (7.2.4) just say that near the origin, $f$ and $g$ are small in comparison to $r = \sqrt{x^2 + y^2}$, which is the radial distance of the point $(x, y)$ from the origin.

We define an **almost linear system** as a nonlinear System (7.2.3) that satisfies (7.2.4). In this situation, the linear part

$$\dot{x} = ax + by$$
$$\dot{y} = cx + dy \tag{7.2.5}$$

is called the **associated linear system** or **linear approximation** about the equilibrium point $(0, 0)$.

## ■ Example 7.2.2 A Linear Approximation

Let's examine the behavior of the following system near the origin:

$$\dot{x} = x + 2y + x \cos y$$
$$\dot{y} = -y - \sin y.$$

First of all, we can see that $(0, 0)$ is an equilibrium point for the system. Now we must find the associated linear system, which is not obvious because $x \cos y$ and $-\sin y$ actually contain linear terms that must be combined with the linear terms already visible in the original system.

Substituting the Taylor (or Maclaurin) expansions for $\cos y$ and $\sin y$ in the given equations and collecting terms, we have

$$\dot{x} = x + 2y + x\left(1 - \frac{y^2}{2!} + \frac{y^4}{4!} - \cdots\right) = 2x + 2y + x\left(-\frac{y^2}{2!} + \frac{y^4}{4!} - \cdots\right)$$

$$\dot{y} = -y - \left(y - \frac{y^3}{3!} + \frac{y^5}{5!} - \cdots\right) = -2y - \left(-\frac{y^3}{3!} + \frac{y^5}{5!} - \cdots\right).$$

Thus, the associated linear system is

$$\dot{x} = 2x + 2y$$
$$\dot{y} = -2y,$$

or

$$\dot{X} = \begin{bmatrix} 2 & 2 \\ 0 & -2 \end{bmatrix} X = AX.$$

The characteristic equation of this linear system is given by $\lambda^2 - 4 = 0$, so the eigenvalues are $\lambda = -2$ and $\lambda = 2$.

Table 5.1 in Section 5.5 tells us that two real eigenvalues opposite in sign indicate that we have a *saddle point*. Figure 7.5a shows the slope field for the system; Figure 7.5b shows some

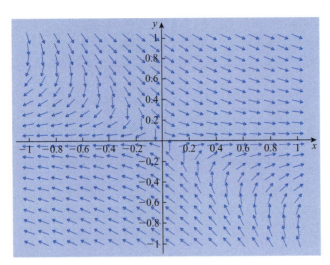

**FIGURE 7.5a**

*Slope field for $\dot{x} = x + 2y + x\cos y$, $\dot{y} = -y - \sin y$*

**FIGURE 7.5b**

*Trajectories for $\dot{x} = x + 2y + x\cos y$, $\dot{y} = -y - \sin y$*

trajectories for the nonlinear system around the origin. Figure 7.5c shows trajectories around the origin for the associated linear system.

We can see from these phase portraits that the linear approximation captures the behavior of the nonlinear system near the origin.

**FIGURE 7.5c**

*Trajectories for* $\dot{x} = 2x + 2y$, $\dot{y} = -2y$

In the next section, we'll generalize our work with almost linear systems in the form of a famous theorem.

## Exercises 7.2

### A

In Problems 1–15, (a) verify that $(0, 0)$ is an equilibrium point of each system and (b) show that each system is almost linear.

1. $x' = 3x + y + xy$, $y' = 2x + 2y - 2xy^2$
2. $x' = x - y + x^2$, $y' = x + y$
3. $x' = x - xy - 8x^2$, $y' = -y + xy$
4. $x' = -4x + y - xy^3$, $y' = x - 2y + 3x^2$
5. $x' = 3\sin x + y$, $y' = 4x + \cos y - 1$
6. $x' = x - y$, $y' = 1 - e^x$
7. $x' = -3x - y - xy$, $y' = 5x + y + xy^3$
8. $x' = y(1 - x^2)$, $y' = -x(1 - y^2)$
9. $x' = -x + x^3$, $y' = -2y$
10. $x' = -2x + 3y + xy$, $y' = -x + y - 2xy^2$
11. $x' = (x - 2y)(y + 4)$, $y' = 2x - y$
12. $x' = (x - 2)(y - 3)$, $y' = (x + 2y)(y - 1)$
13. $x' = 5x - 14y + xy$, $y' = 3x - 8y + x^2 + y^2$
14. $x' = 9x + 5y + xy$, $y' = -7x - 3y + x^2$
15. $x' = \dfrac{1}{2}\left(1 - \dfrac{1}{2}x - \dfrac{1}{2}y\right)x$, $y' = \dfrac{1}{4}\left(1 - \dfrac{1}{3}x - \dfrac{2}{3}y\right)y$

## B

**1.** A woman rows a boat across a river $a$ units wide occupying the strip $0 \le x \le a$ in the $x$-$y$ plane, always rowing toward a fixed point on one bank, say $(0, 0)$. She rows at a constant speed $u$ relative to the water, and the river flows at a constant speed $v$. The situation can be modeled by the equations

$$\dot{x} = -\frac{ux}{\sqrt{x^2 + y^2}}, \ \dot{y} = v - \frac{uy}{\sqrt{x^2 + y^2}},$$

where $(x, y)$ are the coordinates of the boat.

**a.** Use technology to sketch the phase portrait of the system for $u > v$. (Pick some reasonable values of $u$ and $v$.) What happens to the boat over time?

**b.** Use technology to sketch the phase portrait of the system for $u < v$. [Just reverse the values of $u$ and $v$ used in part (a).] What happens to the boat now?

**2.** Show that the system

$$x' = -x + 3y + y \cos \sqrt{x^2 + y^2}$$

$$y' = -x - 5y + x \cos \sqrt{x^2 + y^2}$$

is *not* almost linear.

**3.** Consider the second-order nonlinear equation $\ddot{x} + x - 0.25x^2 = 0$.

**a.** Convert this equation to a nonlinear system of two first-order equations.

**b.** Determine if $(0, 0)$ is an equilibrium solution of the system found in part (a).

**c.** Determine the associated linear system.

## C

**1.** Let's return to the system in Example 7.2.1:

$$\dot{x} = y + ax\left(x^2 + y^2\right)$$

$$\dot{y} = -x + ay\left(x^2 + y^2\right).$$

**a.** Introduce polar coordinates defined by $x = r(t) \cos \theta(t)$, $y = r(t) \sin \theta(t)$. Note that $x^2 + y^2 = r^2$ and use the Chain Rule to show that $x\dot{x} + y\dot{y} = r\dot{r}$.

**b.** In the expression for $r\dot{r}$ found in part (a), substitute for $\dot{x}$ and $\dot{y}$ using the equations in the system and show that $\dot{r} = ar^3$ for $r > 0$.

**c.** Show that $\theta = \arctan(y/x)$ and that $\dot{\theta} = \frac{x\dot{y} - y\dot{x}}{r^2}$. Substitute for $\dot{x}$ and $\dot{y}$ in this last formula to see that $\dot{\theta} = 1$.

**d.** The results of parts (b) and (c) show that our original system is equivalent to the system $\{\dot{r} = ar^3, \dot{\theta} = 1\}$. The second equation says that all trajectories rotate around the origin with constant angular velocity 1. Recognizing that the first equation describes the radial distance from the origin to a point on the trajectory (see the function $d(t)$ introduced in Example 7.2.4), examine what happens to $r(t)$ as $t \to \infty$ in the three cases $a < 0$, $a = 0$, and $a > 0$. What does this say about the nature of the equilibrium point at the origin? Sketch a trajectory (in the $x - y$ plane) for each of the three cases.

2. In Problem B1 above, show that the trajectories in the phase plane are given by $y + \sqrt{x^2 + y^2} = C\, x^{1-\alpha}$, where $\alpha = v/u$ and $C$ is an arbitrary constant.

## 7.3 THE POINCARÉ-LYAPUNOV THEOREM

In this section, we will expand (pun intended) our view of linear approximation as an aid in understanding the stability of nonlinear systems. More generally, suppose that $(a, b)$ is an equilibrium point for the system

$$\dot{x} = F(x, y)$$
$$\dot{y} = G(x, y),$$

which means that $F(a, b) = 0 = G(a, b)$. Using the tangent plane approximation Formula (7.2.2), we can rewrite this system as

$$\dot{x} = F(a, b) + \frac{\partial F}{\partial x}(a, b)(x - a) + \frac{\partial F}{\partial y}(a, b)(y - b) + f(x, y)$$

$$\dot{y} = G(a, b) + \frac{\partial G}{\partial x}(a, b)(x - a) + \frac{\partial G}{\partial y}(a, b)(y - b) + g(x, y)$$

or (because $F(a, b) = 0 = G(a, b)$) as

$$\dot{x} = A(x - a) + B(y - b) + f(x,\ y)$$
$$\dot{y} = C(x - a) + D(y - b) + g(x, y), \tag{7.3.1}$$

where $A = \frac{\partial F}{\partial x}(a, b)$, $B = \frac{\partial F}{\partial y}(a, b)$, $C = \frac{\partial G}{\partial x}(a, b)$, and $D = \frac{\partial G}{\partial y}(a, b)$. Another way to look at this general situation is to realize that we are translating the equilibrium point $(a, b)$ to the origin by using the change of variables $u = x - a$ and $v = y - b$. Of course, this means that $x = u + a$ and $y = v + b$, so that we can rewrite (7.3.1) as

$$\dot{u} = Au + Bv + f(u, v)$$
$$\dot{v} = Cu + Dv + g(u, v),$$

which has $(0, 0)$ as an equilibrium point. Note that this says that any equilibrium point $(a^*, b^*) \neq (0, 0)$ can be transformed to the origin for the purpose of analyzing the stability of the system. Therefore, we can state an important stability result for nonlinear systems in terms of an equilibrium point at the origin.

Suppose we have the nonlinear autonomous system

$$\dot{x} = ax + by + f(x, y)$$
$$\dot{y} = cx + dy + g(x, y), \tag{7.3.2}$$

where $ad - bc \neq 0$, $\displaystyle\lim_{(x,y)\to(0,0)} \frac{f(x,y)}{\sqrt{x^2+y^2}} = \lim_{(x,y)\to(0,0)} \frac{g(x,y)}{\sqrt{x^2+y^2}} = 0$, and the origin is an equilibrium point. If $\lambda_1$ and $\lambda_2$ are the eigenvalues of the associated linear system

$$\dot{y} = cx + dy, \tag{7.3.3}$$

then the equilibrium points of the two systems, (7.3.2) and (7.3.3), are related as follows:

**a.** If the eigenvalues $\lambda_1$ and $\lambda_2$ are **not** equal real numbers or are **not** pure imaginary numbers, then the trajectories of the almost linear System (7.3.2) near the equilibrium point $(0, 0)$ behave the same way as the trajectories of the associated linear System (7.3.3) near the origin. That is, we can use the appropriate table entries given in Section 5.5 to determine whether the origin is a node, a saddle point, or a spiral point of both systems.

**b.** If $\lambda_1$ and $\lambda_2$ are real and equal, then the origin is either a node or a spiral point of both systems. Furthermore, if $\lambda_1 = \lambda_2 < 0$, then the origin is asymptotically stable; and if $\lambda_1 = \lambda_2 > 0$, then the origin is an unstable equilibrium point.

**c.** If $\lambda_1$ and $\lambda_2$ are pure imaginary numbers, then the equilibrium point $(0, 0)$ is either a center or a spiral point of the nonlinear system. Also, this spiral point may be asymptotically stable, stable, or unstable.

This important result was discovered by Poincaré and the Russian mathematician A. M. Lyapunov (1857–1918). The next example shows how to use the Poincaré-Lyapunov theorem.

### ■ Example 7.3.1 An Application of the Poincaré-Lyapunov Theorem

Let's return to the system in Example 7.1.1:

$$x' = x - x^2 - xy$$
$$y' = -y - y^2 + 2xy.$$

We saw that there were four equilibrium points: $(0, 0)$, $(0, -1)$, $(1, 0)$, and $\left(\frac{2}{3}, \frac{1}{3}\right)$.

Near the origin, because the terms $x^2, y^2$, and $xy$ are smaller than the terms $x$ and $y$, we can replace the nonlinear system by its associated linear system

$$x' = x$$
$$y' = -y.$$

The eigenvalues of this linear system are $-1$ and $1$. According to part (a) of the Poincaré-Lyapunov result, the trajectories of the nonlinear system should behave the same way as the trajectories of this associated linear system. Table 5.1 in Section 5.5 tells us that the origin is a *saddle point* for both systems.

If we want to examine what happens near the equilibrium point $(0, -1)$, we make the change of variables $u = x - 0 = x$ and $v = y - (-1) = y + 1$ so that we can rewrite the original system as

$$u' = x' = u - u^2 - u(v - 1) = 2u - u^2 - uv$$
$$v' = y' = -(v - 1) - (v - 1)^2 + 2u(v - 1) = -2u + v - v^2 + 2uv.$$

Then the associated linear system is

$$u' = 2u$$
$$v' = -2u + v,$$

with eigenvalues 1 and 2. (*Check this.*) Now result (a) and the table in Section 5.5 tell us that the equilibrium point $(0, -1)$ is a *source* for the nonlinear system.

The equilibrium point $(1, 0)$ leads us to make the change of variables $u = x - 1$ and $v = y - 0 = y$, so that the nonlinear system is transformed into

$$u' = -u - v - u^2 - uv$$
$$v' = v - v^2 + 2uv,$$

with associated linear system

$$u' = -u - v$$
$$v' = v.$$

The eigenvalues for this last system are $-1$ and $1$, so that $(1, 0)$ is a *saddle point* for both the nonlinear system and its associated linear system.

Finally, we look at the equilibrium point $\left(\frac{2}{3}, \frac{1}{3}\right)$. The transformation $u = x - \frac{2}{3}, v = y - \frac{1}{3}$ leads to the system

$$u' = \frac{-2u}{3} - \frac{2v}{3} - u^2 - uv$$
$$v' = \frac{2u}{3} - \frac{v}{3} - v^2 + 2uv.$$

The linear approximation is given by

$$u' = \frac{-2u}{3} - \frac{2v}{3}$$
$$v' = \frac{2u}{3} - \frac{v}{3},$$

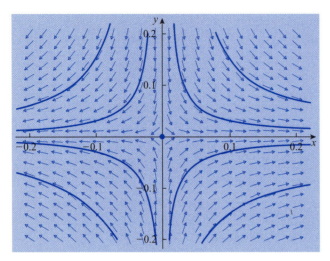

**FIGURE 7.6a**

*Trajectories of $x' = x - x^2 - xy$, $y' = -y - y^2 + 2xy$ near the origin*

**FIGURE 7.6b**

*Trajectories of $x' = x - x^2 - xy$, $y' = -y - y^2 + 2xy$ near $(0, -1)$*

which has eigenvalues $-\frac{1}{2} + \frac{\sqrt{15}}{6}i$ and $-\frac{1}{2} - \frac{\sqrt{15}}{6}i$. Therefore, from result (a) and Table 5.1, we know that $\left(\frac{2}{3}, \frac{1}{3}\right)$ is a *spiral sink*. Look back at Figures 7.3a and 7.3b to see this clearly. Figure 7.6a shows some trajectories near the origin, a saddle point. Figure 7.6b illustrates the behavior of the system near the equilibrium point $(0, -1)$, a source. Finally, Figure 7.6c makes it clear that $(1, 0)$ is indeed a saddle point. ∎

**FIGURE 7.6c**
*Trajectories of $x' = x - x^2 - xy$, $y' = -y - y^2 + 2xy$ near $(1, 0)$*

Now let's examine a system whose stability is not so clear.

## ■ Example 7.3.2 Another Application of Poincaré-Lyapunov

The system we'll investigate is

$$\dot{x} = -x^3 - y$$

$$\dot{y} = x - y^3.$$

Set $\dot{x} = 0$ and $\dot{y} = 0$ and then substitute $y = -x^3$ from the first equation into the second equation. We get $x + x^9 = 0$, or $x(1 + x^8) = 0$, so $x = 0$. It follows that $(0, 0)$ is the only equilibrium point of this system.

The linearized system is

$$\dot{x} = -y$$

$$\dot{y} = x,$$

with characteristic equation $\lambda^2 + 1 = 0$ and eigenvalues $-i$ and $i$. Because the eigenvalues are pure imaginary numbers, case (c) of the Poincaré-Lyapunov result tells us that the origin is either a center or a spiral point of the original nonlinear system. (Note that the origin is a *center* of the associated linear system.) Figure 7.7 shows a typical trajectory, in this case with initial state $(x(0), y(0)) = (-0.5, 0)$ and $t$ running from $-9$ to $100$.

**FIGURE 7.7**
*Trajectories of $\dot{x} = -x^3 - y$, $\dot{y} = x - y^3$ near the origin*

From this, we can see that the trajectory appears to spiral in toward the origin—that is, the equilibrium point is *asymptotically stable*. We could have seen this analytically by defining the function

$$d(t) = \sqrt{x^2(t) + y^2(t)},$$

which gives the distance from any point $(x(t), y(t))$ on a trajectory to $(0, 0)$. Differentiating this function and then substituting from our original equations, we get

$$\dot{d}(t) = \frac{1}{2}\left[x^2(t) + y^2(t)\right]^{-1/2}\left(2x(t)\dot{x}(t) + 2y(t)\dot{y}(t)\right)$$

$$= \frac{x(t)\left(-x^3(t) - y(t)\right) + y(t)\left(x(t) - y^3(t)\right)}{\sqrt{x^2(t) + y^2(t)}} = -\frac{x^4(t) + y^4(t)}{\sqrt{x^2(t) + y^2(t)}} < 0.$$

This says that the distance between points on the trajectory and the origin is *decreasing* with time—that is, the trajectory is always moving closer and closer to the origin. ■

The next example shows another type of behavior.

## ■ Example 7.3.3 Yet Another Application of Poincaré-Lyapunov

The system

$$\dot{x} = 2x - 6x^2 y$$

$$\dot{y} = 2y + x$$

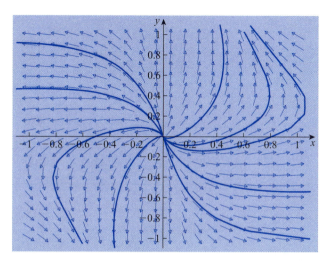

**FIGURE 7.8**
*Trajectories of $\dot{x} = 2x - 6x^2y, \; \dot{y} = 2y + x$ near the origin*

has the origin as its only equilibrium point. *Check this for yourself.* The linearization of this system is

$$\dot{x} = 2x$$
$$\dot{y} = 2y + x,$$

which has the characteristic equation $\lambda^2 - 4\lambda + 4 = (\lambda - 2)^2 = 0$. Because the eigenvalues are positive and equal, we use result (b) to conclude that the origin is an *unstable equilibrium point, a source.* Figure 7.8 shows this. ■

## Exercises 7.3

### A

Each of the almost linear systems in Problems 1–10 has $(0, 0)$ as an equilibrium point. Discuss the type and stability of the origin by examining the associated linear system in each case.

1. $x' = x - y + x^2, y' = x + y$
2. $x' = x - xy - 8x^2, y' = -y + xy$
3. $x' = -4x + y - xy^3, y' = x - 2y + 3x^2$
4. $x' = 3\sin x + y, y' = 4x + \cos y - 1$
5. $x' = x - y, y' = 1 - e^x$
6. $x' = -3x - y - xy, y' = 5x + y + xy^3$
7. $x' = y(1 - x^2), y' = -x(1 - y^2)$

8. $x' = -x + x^3, y' = -2y$

9. $x' = -2x + 3y + xy, y' = -x + y - 2xy^2$

10. $x' = 5x - 14y + xy, y' = 3x - 8y + x^2 + y^2$

**B**

1. Consider the nonlinear system

$$\dot{x} = -x + xy, \quad \dot{y} = 2y - xy + 0.5x.$$

   **a.** Find all equilibrium points.

   **b.** Describe the type and stability of each equilibrium point found in part (a) by examining the associated linear system in each case.

2. Consider the nonlinear system

$$\dot{x} = x - y + 5, \quad \dot{y} = x^2 + 6x + 8.$$

   **a.** Find all equilibrium points.

   **b.** Describe the type and stability of each equilibrium point found in part (a) by examining the associated linear system in each case.

3. Consider the nonlinear system

$$\dot{x} = x(8 - 4x - y), \quad \dot{y} = y(3 - 3x - y),$$

   which describes the populations $x(t)$ and $y(t)$ of two species that are competing for the same resources.

   **a.** Find all equilibrium points of this system.

   **b.** By linearizing about each equilibrium point found in part (a), determine the type and stability of each equilibrium point.

**C**

1. The *Brusselator* is a simple model of a hypothetical chemical oscillator that first appeared in a 1968 paper by Belgian scientists I. Prigogine (a Nobel laureate) and R. Lefever and was named for the capital of their home country. One version of the model is

$$\dot{x} = 1 - (a + 1)x + bx^2 y$$
$$\dot{y} = ax - bx^2 y,$$

   where $x$ and $y$ are concentrations of chemicals and $a, b$ are positive parameters.

   **a.** Use technology, if necessary, to find the only equilibrium solution of this system.

   **b.** Linearize the system about the equilibrium point found in part (a).

   **c.** Find the eigenvalues of the associated linear system. (Technology could be useful here.)

**d.** Using your answers from part (c) and the Poincaré-Lyapunov theorem, discuss the nature of the equilibrium solutions for each of the following cases:

$$(1)\, a = 3, b = 1;$$

$$(2)\, a = 2, b = 7;$$

$$(3)\, a = 1, b = 4.$$

# 7.4  TWO IMPORTANT EXAMPLES OF NONLINEAR EQUATIONS AND SYSTEMS

Now that we know something about the behavior of nonlinear systems, we can apply this knowledge to the analysis of some important nonlinear equations and systems of nonlinear equations.

## 7.4.1  The Lotka-Volterra Equations

As we stated in the discussion preceding Example 4.7.4, the nonlinear *Lotka-Volterra equations* describe a wide class of problems in mathematical ecology and cannot in general be solved in closed form. Now we will look at this system from the Poincaré-Lyapunov point of view.

### ■ Example 7.4.1 The Lotka-Volterra Equations Revisited

The Lotka-Volterra equations are

$$\dot{x} = ax - bxy$$
$$\dot{y} = -cy + dxy,$$

where $a, b, c,$ and $d$ are positive constants. The equilibrium points for this system are solutions of the algebraic system

$$ax - bxy = x(a - by) = 0$$
$$-cy + dxy = y(-c + dx) = 0.$$

Clearly, $x = y = 0$ is a solution—that is, the origin $(0, 0)$ is an equilibrium point. It should also be clear from these last equations that if either $x$ or $y$ is zero, then the other variable must also be zero. Therefore, if there are any other equilibrium points, we must have $x \neq 0$ and $y \neq 0$. In the first algebraic equation, if $x \neq 0$, then we must have $a - by = 0$, so $y = a/b$. From the second equation, we see that if $y \neq 0$, then $-c + dx = 0$, so $x = c/d$. Thus, the only equilibrium points for the Lotka-Volterra system are $(0, 0)$ and $(c/d, a/b)$.

Near the origin, we can replace our original system by the associated linear system

$$\dot{x} = ax$$

$$\dot{y} = -cy,$$

which can be written in matrix form as $\dot{X} = AX$, where $A = \begin{bmatrix} a & 0 \\ 0 & -c \end{bmatrix}$ and $X = \begin{bmatrix} x \\ y \end{bmatrix}$. Now the characteristic equation of $A$ is $\lambda^2 + (c - a)\lambda - ac = 0$, so the eigenvalues are $a$ and $-c$. Because the eigenvalues are real and of opposite signs, Table 5.1 in Section 5.5 indicates that the origin is a *saddle point* for the linearized system. The Poincaré-Lyapunov theorem tells us that $(0, 0)$ is also a saddle point for our original nonlinear system.

To study the behavior of the system near the equilibrium point $(c/d, a/b)$, we transform the system by defining $u = x - c/d$ and $v = y - a/b$. Then our original system becomes

$$\dot{u} = a\left(u + \frac{c}{d}\right) - b\left(u + \frac{c}{d}\right)\left(v + \frac{a}{b}\right)$$

$$\dot{v} = -c\left(v + \frac{a}{b}\right) + d\left(u + \frac{c}{d}\right)\left(v + \frac{a}{b}\right),$$

which simplifies to

$$\dot{u} = \left(-\frac{bc}{d}\right)v - buv$$

$$\dot{v} = \left(\frac{ad}{b}\right)u + duv.$$

The associated linear system is given by $\dot{X} = AX$, where $A = \begin{bmatrix} 0 & -\frac{bc}{d} \\ \frac{ad}{b} & 0 \end{bmatrix}$. The characteristic polynomial here is $\lambda^2 + ac = 0$, so the eigenvalues are $\lambda_1 = \sqrt{ac}\,i$ and $\lambda_2 = -\sqrt{ac}\,i$. Because we have pure imaginary eigenvalues, part (C) of the Poincaré-Lyapunov result tells us that $(c/d, a/b)$ is either a *center* or a *spiral point* for the nonlinear system. (The table in Section 5.5 indicates that $(c/d, a/b)$ is a stable center for the associated linear system, but this doesn't have to be true for our nonlinear system.) Let $a = b = c = d = 1$. Then Figure 7.9a shows the slope field for the nonlinear system near the equilibrium point $(c/d, a/b) = (1, 1)$, and Figure 7.9b depicts some trajectories near $(1, 1)$.

These figures suggest (but do not *prove*) that the equilibrium point $(1, 1)$ is a *stable center* for the nonlinear system. (Problem B1 in Exercises 7.4 proposes some investigations in this direction.)

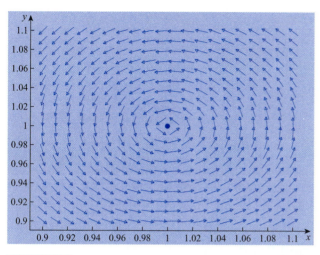

**FIGURE 7.9a**
*Slope field of $\dot{x} = x - xy$, $\dot{y} = -y + xy$ near $(1, 1)$*

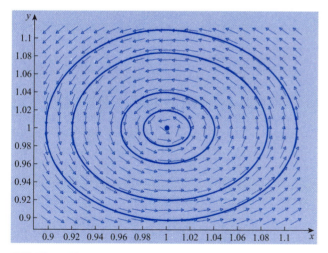

**FIGURE 7.9b**
*Trajectories of $\dot{x} = x - xy$, $\dot{y} = -y + xy$ near $(1, 1)$*

## 7.4.2 The Undamped Pendulum

After our ecological field trip, let's return to the world of physics and look at the motion of a simple pendulum. In Section 7.2, we saw that the second-order nonlinear equation $\frac{d^2\theta}{dt^2} + \frac{g}{L}\sin\theta = 0$ describes the motion of an **undamped pendulum**—that is, a pendulum under the influence of gravity with no friction or air resistance impeding its movement. Here, $\theta$ is the angle the pendulum makes with the vertical, $g$ is the acceleration due to gravity, and $L$ is the pendulum's length (Figure 7.10).

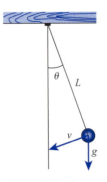

**FIGURE 7.10**
*The undamped pendulum*

## ■ Example 7.4.2 The Undamped Pendulum: A Poincaré-Lyapunov Analysis

Letting $x = \theta$ and $y = \dot{\theta} = \dot{x}$, we can express the single equation $\frac{d^2\theta}{dt^2} + \frac{g}{L}\sin\theta = 0$ as the nonlinear system

$$\dot{x} = y$$

$$\dot{y} = -\frac{g}{L}\sin x.$$

The first thing we have to do is find the equilibrium points of this system. (This was Problem C2 in Exercises 4.7.) Clearly, any equilibrium point $(x, y)$ must have $y = 0$. The equation $-\frac{g}{L}\sin x = 0$ has solutions $x = n\pi, n = 0, \pm 1, \pm 2, \ldots$. Thus, all points of the form $(n\pi, 0)$ for $n = 0, \pm 1, \pm 2, \ldots$ are equilibrium points for the system describing the pendulum's swing. Because the sine function has period $2\pi$—that is, $\sin(x + 2k\pi) = \sin x$ for any integer $k$—the second equation in the system remains the same for angles differing by integer multiples of $2\pi$. Thus, there is no physical difference in the system for such angles. (*Think about this in physical terms.*) Now all the equilibrium point first coordinates that are *even* multiples of $\pi$ differ from 0 by multiples of $2\pi$, so we can just study what happens near $(0, 0)$. (For example, the point $(-8\pi, 0)$ is the same as $(0 + (-4) \cdot 2\pi, 0)$.) Similarly, all the equilibrium point first coordinates that are *odd* multiples of $\pi$ differ from $\pi$ by multiples of $2\pi$, so we can just see what happens to the system near $(\pi, 0)$. (For example, $(17\pi, 0)$ is the same as $(\pi + (8) \cdot 2\pi, 0)$.) Therefore, by analyzing the behavior of the system near the points $(0, 0)$ and $(\pi, 0)$, we can understand the behavior near *any* of the infinite number of equilibrium points.

Near the origin, we can replace $\sin x$ by its Taylor series expansion, so our system can be written as

$$\dot{x} = y$$

$$\dot{y} = -\frac{g}{L}\sin x = -\frac{g}{L}\left(x - \frac{x^3}{3!} + \frac{x^5}{5!} - \frac{x^7}{7!} + \cdots\right)$$

and we see that the linearization of our system is given by

$$\dot{x} = y$$
$$\dot{y} = -\frac{g}{L}x.$$

In matrix form, this becomes $\dot{X} = \begin{bmatrix} 0 & 1 \\ -\frac{g}{L} & 0 \end{bmatrix} X$, with characteristic equation $\lambda^2 + \frac{g}{L} = 0$ and pure imaginary eigenvalues $\lambda = \pm\sqrt{g/L}i$. Once again, part (c) of the Poincaré-Lyapunov theorem points to either a *center* or a *spiral point*. Intuitively, we should realize that this is like the situation with the undamped spring-mass system: In the absence of any kind of resistance, the object will continue to move periodically about its equilibrium state. In our case, we would expect the pendulum to swing back and forth indefinitely. (Compare the pendulum's associated linear system with System (4.8.2) in Example 4.8.1.) Figure 7.11 shows the phase portrait of the nonlinear system with $g = L$ near the origin.

Note what the figure tells us. If the pendulum starts with $x_0 = \theta_0$ anywhere between 0 and $\pi$ and we release the weight at the end (the *bob*), then the pendulum will swing in a clockwise (*negative*) direction toward the vertical position and go past the vertical ($x = \theta = 0$) until it makes the same initial angle on the other side. At this point in time ($x = x_0 = -\theta_0$), the pendulum starts its journey back to the vertical position and then goes past it until $x = \theta_0$ once more. The variable $y$ represents the *angular velocity*, which is zero as we release the pendulum; becomes negative as the velocity increases in a negative (clockwise) direction; attains its maximum as the pendulum swings through the vertical position; and then decreases as the pendulum approaches $x = x_0 = -\theta_0$. At this point, the pendulum begins its swing back toward

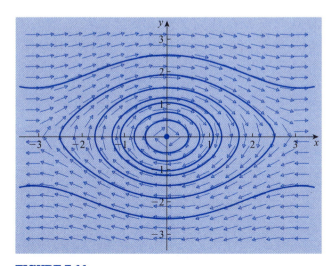

**FIGURE 7.11**

*Trajectories of $\dot{x} = y$, $\dot{y} = -\sin x$ near the origin*

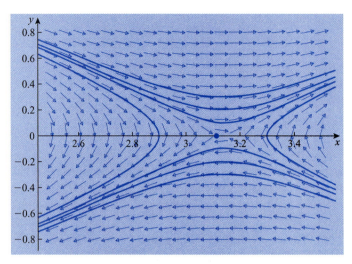

**FIGURE 7.12**

*Trajectories of $\dot{x} = y$, $\dot{y} = -\sin x$ near $(\pi, 0)$*

the center and ultimately back to its initial position, its velocity increasing and decreasing appropriately. (We'll deal with the curves at the top and bottom of Figure 7.11 shortly.)

Now let's examine the pendulum's behavior near the equilibrium point $(\pi, 0)$. The transformation $u = x - \pi$, $v = y - 0$ results in the nonlinear system

$$\dot{u} = v$$

$$\dot{v} = -\frac{g}{L}\sin(u + \pi) = -\frac{g}{L}(-\sin u) = \frac{g}{L}\sin u,$$

with the associated linear system

$$\dot{u} = v$$

$$\dot{v} = \frac{g}{L}u.$$

*Make sure you understand how we arrived here.* This linear system has the characteristic equation $\lambda^2 - \frac{g}{L} = 0$ and eigenvalues $\pm\sqrt{\frac{g}{L}}$. We look to part (a) of our stability theorem (and Table 5.1 in Section 5.5) to see that the equilibrium point $(\pi, 0)$ is a *saddle point*. Figure 7.12 (again with $g = L$) focuses on the system's behavior near this point.

As neat as this analysis seems to be, we've brushed past something we haven't explained yet: the strange curves at the top and bottom of Figure 7.11. If we step back and look at the entire phase portrait (Figure 7.13), this strangeness becomes more evident.

Clearly, if the initial velocity imparted to the undamped pendulum is low enough, the pendulum swings indefinitely back and forth about its equilibrium point $(0, 0)$. Physically, this

**FIGURE 7.13**

*Phase portrait of $\dot{x} = y$, $\dot{y} = -\sin x$*

equilibrium position corresponds to the pendulum at rest ($y = \dot{\theta} = 0$) and hanging straight down ($x = \theta = 0$). If we give the pendulum a high enough initial velocity, it will whirl up and over the top—over and over again in the absence of any friction or air resistance. Its velocity will vary periodically, attaining its minimum at *odd* multiples of $\pi$ (when its position is straight up) and its maximum at *even* multiples of $\pi$ (when it is moving through the straight-down position).

The curves joining the saddle points (odd multiples of $\pi$ on the $x$-axis) need careful explanation. Figure 7.14 focuses on the curves connecting the saddle points $(-\pi, 0)$ and

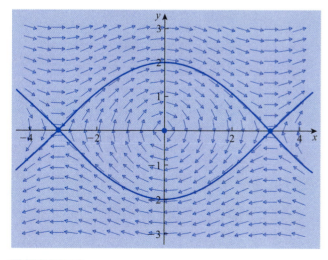

**FIGURE 7.14**

*Separatrices connecting $(-\pi, 0)$ and $(\pi, 0)$*

$(\pi, 0)$. These are called **separatrices** (the plural of **separatrix**); they separate the regions of "normal" behavior from each other. (More technically, they are called **heteroclinic trajectories** or **saddle connections**.) As we've indicated before, the saddle points represent a pendulum pointed straight up and at rest. Physically, then, these heteroclinic trajectories describe the fact that the pendulum slows down just as it approaches the upside-down position.

Problems C1 and C2 in Exercises 7.4 suggest a more analytic way of understanding the undamped pendulum's behavior.  ■

## Exercises 7.4

### A

Find the nontrivial equilibrium point for each of the Lotka-Volterra systems in Problems 1–6. You may need technology.

1. $\dot{x} = 3x - 2xy, \dot{y} = -y + 4xy$
2. $\dot{x} = 0.1x - 0.2xy, \dot{y} = -0.5y + 0.3xy$
3. $\dot{x} = 0.005x - 0.02xy, \dot{y} = -0.3y + 0.4xy$
4. $\dot{x} = x - 2xy, \dot{y} = -3y + 4xy$
5. $\dot{x} = 0.2x - 0.2y, \dot{y} = -3y + xy$
6. $\dot{x} = 3x - 2xy, \dot{y} = -y + \frac{1}{2}xy$
7. Consider the Lotka-Volterra equations (Example 7.4.1) for $a = b = c = d = 1$. To develop some confidence in the power of numerical methods, use whatever Runge-Kutta algorithm and step size your instructor suggests to approximate the solution to the initial value problem with $x(0) = 1$ and $y(0) = 2$ over the interval $[0, 1]$.

### B

1. Consider the Lotka-Volterra equations for $a = b = c = d = 1$. Figure 7.9b shows some trajectories corresponding to this situation. Without relying on the graph, we want to show that the trajectories are closed curves—i.e., that the equilibrium point $(1, 1)$ is a *stable center*.
   a. Show that the slope field for $\frac{dy}{dx}$ is symmetric about the line $y = x$. [*Hint*: Look at what happens if you interchange $x$ and $y$ in the slope equation.]
   b. Argue that if you start at some point $P = (x, y)$ on the line $y = x$ and travel along the trajectory once around the point $(1, 1)$, you wind up back at the same point $P$, so that the curve is closed.
2. Consider the system first examined in Example 4.7.4:

$$\dot{x} = 0.2x - 0.002xy$$

$$\dot{y} = -0.1y + 0.001xy.$$

**a.** Find the equilibrium points for the system.

**b.** Plot the trajectory corresponding to the initial conditions $x(0) = 100$ and $y(0) = 300$. Interpret these initial values and the shape of the trajectory in terms of the predator and prey populations. (Choose the interval $[0, 55]$ for your independent variable $t$.)

**c.** Use the graph of the trajectory found in part (b) to estimate the maximum and minimum values of the populations $x$ and $y$.

**d.** Find the slope equation $\frac{dy}{dx}$ and solve it (implicitly) using the initial conditions given in part (b).

**e.** Use technology to plot the solution found in part (d), using ranges for $x$ and $y$ consistent with your answers to part (c).

**3.** Recall that the Lotka-Volterra system (Example 7.4.1) has the nontrivial equilibrium point $(c/d, a/b)$. To understand the direction of any trajectory for the Lotka-Volterra equations without relying on a graph provided by technology, divide the first quadrant of the $x$-$y$ plane into four subquadrants via the lines $x = c/d$ and $y = a/b$. (Sketch this situation.)

**a.** Show that for $x > c/d$ and $y > a/b$, you have $\dot{x} < 0$ and $\dot{y} > 0$.

**b.** Show that for $x < c/d$ and $y > a/b$, you have $\dot{x} < 0$ and $\dot{y} < 0$.

**c.** Show that for $x < c/d$ and $y < a/b$, you have $\dot{x} > 0$ and $\dot{y} < 0$.

**d.** Show that for $x > c/d$ and $y < a/b$, you have $\dot{x} > 0$ and $\dot{y} > 0$.

**e.** From the results of parts (a)–(d), conclude that any point $(x(t), y(t))$ on a trajectory for the Lotka-Volterra equations moves in a *counterclockwise* direction.

**4.** Recall that in Example 7.4.1 the Lotka-Volterra equations $\dot{x} = ax - bxy, \dot{y} = -cy + dxy$ were linearized to $\dot{u} = (-bc/d)v, \dot{v} = (ad/b)u$ near the equilibrium point $(c/d, a/b)$.

**a.** Find the slope equation $\frac{du}{dv}$ and conclude that the linear system has a solution satisfying $ad^2u^2 + b^2cv^2 = K$, where $K$ is a positive constant.

**b.** Rewrite the solution in part (a) in terms of the original variables $x$ and $y$ and show that you get the equation of an ellipse with center at $(c/d, a/b)$ and with axes parallel to the axes of the $x$-$y$ plane.

**c.** Compute the derivative of each equation of the linearized system to get the equations $\ddot{u} = -acu, \ddot{v} = -acv$—uncoupled second-order linear equations of the form $\ddot{w} = -Rw$.

**d.** Show that the solution of the linearized system is a pair of functions $(u(t), v(t))$ with the same period $2\pi/\sqrt{ac}$.

**5.** Focus on the equation for the *predator* population, $\frac{dy}{dt} = -cy + dxy$.

**a.** Divide the equation by $y$ and integrate between the initial time $t_0$ and some arbitrary time $t$.

**b.** Assuming that the predator population is periodic (see Figure 4.10 or 4.11b, for example) with period $T$, let $t = t_1$ in part (a), so that $t_1 - t_0 = T$ and $y(t_1) = y(t_0)$. Show that the average value of the prey population is $c/d$, the same as the equilibrium population of the prey. (Recall that the *average value* of a function $f$ on the interval $[a, b]$ is defined as $\frac{1}{b-a} \int_a^b f(r)\,dr$.) [*Hint*: Note that $\dot{y}/y = -c + dx$ and integrate from 0 to $T$, using the periodicity of $\ln|y(t)|$.]

**6.** Assuming the result of Problem B5b and that the average value of the predator population $y(t)$ is $a/b$, the equilibrium population of the predator, and also assuming that both the predator and prey populations have the same period $T$, show that the average value of $x(t)y(t)$ equals the average value of $x(t)$ times the average value of $y(t)$. [*Hint*: $(\dot{y} + cy)/d = xy$.]

## C

**1.** Consider the simplified pendulum equation used in Figure 7.11, Example 7.4.2: $\frac{d^2\theta}{dt^2} + \sin\theta = 0$. You're going to show (analytically) that this equation has periodic solutions—i.e., there are closed trajectories in the phase plane corresponding to the system version of the equation.

  **a.** Show that this equation is equivalent to the system

$$\frac{dx}{dt} = y$$

$$\frac{dy}{dt} = -\sin x.$$

  **b.** Show that any trajectory in the phase plane is a solution of $\frac{dy}{dx} = -\frac{\sin x}{y}$.
  **c.** Solve the equation in part (b).
  **d.** Show that there are closed trajectories in the $x$-$y$ plane and hence that the undamped pendulum problem has periodic solutions. [*Hint*: Find suitable values of the constant of integration you get in part (c).]

**2.** When you find the general solution of the equation $\frac{dy}{dx} = -\frac{\sin x}{y}$ [as in part (b) of the preceding problem], you have an arbitrary constant $C$.

  **a.** What values of $C$ give the wavy trajectories at the top and bottom of Figure 7.13?
  **b.** What values of $C$ give the separatrices, as in Figure 7.14?

**3.** For small values of $\theta$, $\sin\theta \approx \theta$, so that the linearized equation of the undamped pendulum is $\ddot{\theta} + \frac{g}{L}\theta = 0$. Work with this equation and the initial conditions $\theta(0) = 0$, $\dot{\theta}(0) = 2$.

  **a.** Find $\theta(t)$ if the length of the pendulum is 8 feet. (Take $g = 32$ ft/sec$^2$.)
  **b.** What is the period of the function found in part (a)?
  **c.** If the pendulum is part of a clock that ticks once for each time the pendulum makes a complete swing, how many ticks does the clock make in one minute?
  **d.** How is the motion of the pendulum affected if the length is changed to $L = 4$?

**4.** The equation $\ddot{\theta} + k\dot{\theta} + \sin\theta = 0$ describes a particular *damped* pendulum—i.e., a pendulum with friction or air resistance. Here, $k$ is a positive constant, the coefficient of friction.

  **a.** Convert this second-order equation to a system of first-order equations.
  **b.** Use technology to produce the phase portrait when $k = 0.1$.
  **c.** Use technology to produce the phase portrait when $k = 0.5$.
  **d.** Compare the phase portraits in parts (b) and (c) and give a physical interpretation of what you see.

## *7.5  VAN DER POL'S EQUATION AND LIMIT CYCLES

### 7.5.1  Van Der Pol's Equation

The next example deals with a famous equation that arose when radios were first developed. The original context was the study of certain electrical circuits containing a vacuum tube ("triode generator"), but the work also has had significant biological applications. The pioneering experiments and the first theoretical analysis were conducted by Dutch electrical engineer Balthasar van der Pol (1889–1959) and others in the 1920s.

### ■ Example 7.5.1 The Van Der Pol Equation

The van der Pol equation (or van der Pol oscillator)

$$x'' + \varepsilon \left( x^2 - 1 \right) x' + x = 0, \tag{7.5.1}$$

where $\varepsilon$ is a positive parameter, can also be interpreted in terms of a spring-mass system with nonlinear resistance. (See Problem A1 in Exercises 7.5.) Equation (7.5.1) can be written as the equivalent system

$$x_1' = x_2$$
$$x_2' = -x_1 + \varepsilon x_2 \left( 1 - x_1^2 \right). \tag{7.5.2}$$

The first thing we have to do is find the equilibrium points of (7.5.2). To get a sense of how this system behaves, let's assume that $\varepsilon = 1$. (See Problems A2 and A3 of Exercises 7.5, which ask you to consider other values of $\varepsilon$.) The linearized version of the nonlinear System (7.5.2) is then

$$x_1' = x_2$$
$$x_2' = -x_1 + x_2$$

with characteristic equation $\lambda^2 - \lambda + 1 = 0$ and eigenvalues $(1 \pm \sqrt{3}i)/2$. This implies that both the nonlinear System (7.5.1) and its linear Approximation (7.5.2) have a *spiral source* at the origin. (*Why?*) However, this particular system exhibits some new, characteristically nonlinear behavior. Figure 7.15a shows the phase portrait of the nonlinear System (7.5.1) near $(0, 0)$.

What is happening here is that several paths starting near the origin spiral *outward* from the origin (as expected) toward a particular closed curve, whereas other trajectories starting farther away from $(0, 0)$ also seem to be approaching this closed curve asymptotically (that is, as $t \to \infty$). Reasonably enough, such a closed trajectory is called a **stable limit cycle**. A stable limit cycle can also be described as a periodic trajectory that attracts other nearby trajectories,

---

* Denotes an optional section.

whereas an **unstable limit cycle** *repels* nearby trajectories. It is important to note that *linear systems never have limit cycles.* (See the discussion of limit cycles following this example.) Note that the phase portrait of the linearized system (Figure 7.15b) shows no limit cycle, only the spiraling away from the origin.

Figure 7.15c shows a plot of $x$ against $t$ with the initial conditions $x(0) = 0.5, x'(0) = -0.5$. This graph reflects the eventual periodicity of the solution and the fact that the spirals work

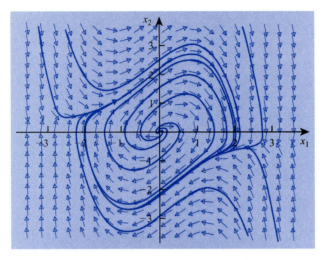

**FIGURE 7.15a**
*Phase portrait of $x'_1 = x_2$, $x'_2 = -x_1 + x_2 \left(1 - x_1^2\right)$ near the origin*

**FIGURE 7.15b**
*Phase portrait of $x'_1 = x_2$, $x'_2 = -x_1 + x_2$ near the origin*

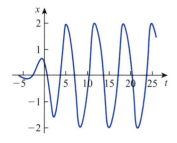

**FIGURE 7.15c**
*Plot of $x(t)$ against $t$, $x(0) = x'(0) = 0.5$, $-6 \leq t \leq 26$*

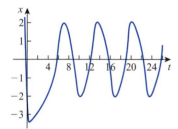

**FIGURE 7.15d**
*Plot of $x(t)$ against $t$, $x(0) = -3$, $x'(0) = -5$, $-0.6 \leq t \leq 26$*

their way *outward* (through increasing values of $t$) to the limit cycle. The solution shows *transient* behavior (temporary or short-lived behavior) at the beginning, before settling into its periodic pattern.

On the other hand, if we choose an initial point ($x(0) = -3, x'(0) = -5$) in a region that appears to be *outside* the limit cycle shown in Figure 7.15a, we see the solution behavior shown by Figure 7.15d. This illustrates how a spiral finds its way *inward* to the limit cycle.

Again, we can see that the solution eventually becomes periodic, after an initial transient stage.    ∎

## 7.5.2 Limit Cycles

The Lotka-Volterra equations (Examples 4.7.4 and 7.4.1) and the undamped spring-mass system (Example 4.8.1) show that autonomous systems sometimes have periodic solutions whose trajectories are closed curves in the phase plane. As we have seen, the van der Pol oscillator, which can be described as a negatively damped nonlinear oscillator (look at the form of the equation), has solutions whose *limiting* behavior (as $t \to \infty$) is that of a finite periodic solution. Such a nontrivial isolated closed trajectory is called a **limit cycle**. Here, "nontrivial" means that the solution curve is not a single point, and "isolated" refers to the fact that no trajectory sufficiently near the limit cycle is also closed.

In general, a *linear* system $\dot{X} = AX$ may have closed trajectories, but they won't be *isolated*: If $X(t)$ is a periodic solution, then so is $cX(t)$ for any nonzero constant $c$. (*Can you show this?*) Therefore, for instance, by choosing $c = (1 - 1/k)(k = 1, 2, 3, \ldots)$, we see that $X(t)$ is being crowded by a one-parameter family of closed trajectories. (In this way the closed trajectories shown in Figure 7.9b of Section 7.3 are not isolated and so could not possibly be limit cycles. You can get trajectories as close to each other as you wish.)

Every trajectory that begins sufficiently near a limit cycle approaches it either for $t \to \infty$ or for $t \to -\infty$. Graphically, this means that such a trajectory either winds itself around the limit cycle or unwinds *from* it. A limit cycle is called **semistable** if trajectories approach one side of it while pulling away from the other side.

As one author has put it,

> The stable limit cycle is the basic model for all self-sustained oscillators—those which return, or recover, to some fundamental periodic orbit when perturbed from it. The stable oscillations, "beating" of the human heart (which returns to some normal rate after we raise it by sprinting), cycles of predator-prey systems, and various electrical circuits are three among myriad examples. Business cycles and certain periodic outbreaks of social unrest ... are, quite possibly, others.[1]

Let's look at other examples of this important phenomenon, the limit cycle. Van der Pol's equation exhibited a *stable* limit cycle, but the next example shows a different type of behavior.

### ■ Example 7.5.2 An Unstable Limit Cycle

Let's examine the autonomous system

$$\dot{x} = -y + x\left(x^2 + y^2 - 1\right)$$
$$\dot{y} = x + y\left(x^2 + y^2 - 1\right).$$

The presence of the algebraic form $x^2 + y^2$, with its suggestion of circularity (rotation), tips us off that we may be able to see things more clearly if we switch to *polar coordinates* (see Section B.1). Making the substitutions $r \cos\theta = r(t)\cos\theta, y = r\sin\theta = r(t)\sin\theta(t)$, and $\theta = \arctan(y/x)$, we have $x^2 + y^2 = r^2$. (You may have seen this sort of substitution in evaluating certain integrals in calculus class—or in Problem C1 of Exercises 7.2.) A few algebraic manipulations (see Problem B1 of Exercises 7.5) give us the polar coordinate form of the system we started with:

$$\dot{r} = \left(r^2 - 1\right)r \quad (r \geq 0)$$
$$\dot{\theta} = 1.$$

---

[1] J. M. Epstein, *Nonlinear Dynamics, Mathematical Biology, and Social Science* (Addison-Wesley, 1997): 121.

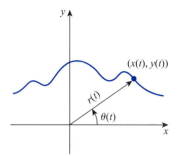

**FIGURE 7.16**

*Motion described in terms of radial distance and angular velocity*

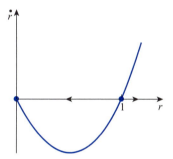

**FIGURE 7.17**

*Phase portrait of $\dot{r} = (r^2 - 1)r$, $r \geq 0$*

This system describes the motion of an object in terms of its radial distance $r = r(t)$ from the origin and its (constant) angular velocity $\dot{\theta}$ in a counterclockwise direction. Figure 7.16 illustrates this in general.

Because the equations are independent (or *uncoupled*), each involving only one dependent variable, we can analyze them separately. We can look at the first equation as a first-order nonlinear equation and consider its phase portrait (Figure 7.17) in the manner of Section 2.5. Recalling that $r$ is nonnegative, we see that the only equilibrium solutions are $r \equiv 0$ and $r \equiv 1$. Note that the first equation tells us that if $r < 1$, then $\dot{r} < 0$, so the trajectory's distance from the origin is decreasing—that is, the trajectory is approaching the origin and moving away from the unit circle ($r \equiv 1, 0 \leq \theta \leq 2\pi$); whereas if $r > 1$, we have $\dot{r} > 0$, so trajectories are also repelled by the unit circle.

From this phase portrait we can see that $r \equiv 0$ is a *sink* and $r \equiv 1$ is a *source*. We could have used the *Derivative Test* of Section 2.6 to see this. (Also see Problem B2 in Exercises 7.5.) In particular, the origin is a *sink* for the system in its original rectangular coordinate form. Figure 7.18 shows the phase portrait in *x-y* space.

From this we see that the unit circle is an *unstable limit cycle*. ∎

**FIGURE 7.18**

*Phase portrait of $\dot{r} = (r^2 - 1)r$, $r \geq 0$, in the x-y phase plane*

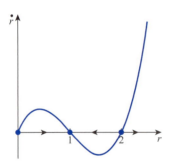

**FIGURE 7.19**

*Phase portrait of $\dot{r} = r(r - 1)(r - 2)$, $r \geq 0$*

Now we're ready for something a bit more complicated, but rewarding.

## ■ Example 7.5.3 A System with Two Limit Cycles

Let's look at the system

$$\dot{r} = r(r - 1)(r - 2), r > 0$$
$$\dot{\theta} = 1.$$

As in the previous example, the system describes the motion of an object in terms of its radial distance $r = r(t)$ from the origin and its (constant) angular velocity $\dot{\theta}$ in a counterclockwise direction. Let's look at the phase portrait of the first equation (Figure 7.19), whose equilibrium solutions are $r \equiv 0$, $r \equiv 1$, and $r \equiv 2$.

As we can see, $r \equiv 0$ is a source, $r \equiv 1$ is a sink, and $r \equiv 2$ is a source. This tells us that the system has two circular limit cycles: one stable ($r \equiv 1$) and one unstable ($r \equiv 2$). Trajectories

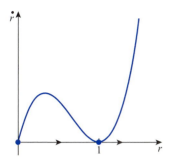

**FIGURE 7.20**
*Phase portrait of $\dot{r} = r(r-1)^2$, $r \geq 0$*

starting out inside the unit circle approach the unit circle as $t \to \infty$, as do trajectories with initial points inside the ring formed by the two circles $r \equiv 1$ and $r \equiv 2$. Any trajectory starting outside the circle of radius 2 moves farther away as $t \to \infty$. ■

The next example illustrates a third kind of limit cycle.

## ■ Example 7.5.4 A Semistable Limit Cycle

What kind of behavior is shown by the following system?

$$\dot{r} = r(r-1)^2, r > 0$$
$$\dot{\theta} = 1.$$

The phase portrait for the first equation (Figure 7.20) tells the story.

The equilibrium point $r \equiv 0$ is a source, whereas $r \equiv 1$ is a node because $\dot{r} > 0$ for $0 < r < 1$ and for $r > 1$ as well. The graphical interpretation of this fact is that the unit circle described by $r \equiv 1$ is a *semistable limit cycle*. Trajectories approach the unit circle from inside it, whereas trajectories that start outside escape the unit circle. ■

Of course, a nonlinear equation or system may have *no* limit cycles (isolated nonconstant periodic solutions). Because nonlinear equations and systems are usually too difficult to solve, other methods—qualitative methods—have been developed to determine the existence or nonexistence of limit cycles. These methods involve advanced mathematical ideas that we won't discuss in this book. Problems B6–B11 of Exercises 7.5 illustrate a *negative* criterion due to the Swedish mathematician Ivar Bendixson (1861–1935).

## EXERCISES 7.5

### A

1. In the discussion of the van der Pol Equation (7.5.1), the comment was made that it can be interpreted as a spring-mass system with nonlinear resistance. Specifically, the term $\varepsilon(x^2 - 1)$ represents a *variable* damping coefficient.

**a.** Explain why $\varepsilon(x^2 - 1) < 0$ when $-1 < x < 1$, so that damping is *negative* for the small oscillations corresponding to $-1 < x < 1$. (This means that small-amplitude oscillations are *amplified* if they become too small.)

**b.** Explain why $\varepsilon(x^2 - 1) > 0$ when $|x| > 1$, so that damping is *positive* for the large oscillations corresponding to $|x| > 1$. (This means that large-amplitude oscillations are made to *decay* if they become too large.)

**2.** Use technology to draw phase portraits of the van der Pol equation for

$$\varepsilon = \frac{1}{4}, \frac{3}{2}, \text{ and } 3.$$

**3.** Consider the van der Pol equation in the system form (7.5.2), where $x_1(0) = 1$ and $x_2(0) = 0$.

**a.** For $\varepsilon = \frac{1}{4}$, graph the trajectory in the $x_1$-$x_2$ plane. Then graph $x_1(t)$ against $t$ and $x_2(t)$ against $t$ on different sets of axes. Use technology.

**b.** For $\varepsilon = 4$, graph the trajectory in the $x_1$-$x_2$ plane. Then graph $x_1(t)$ against $t$ and $x_2(t)$ against $t$ on different sets of axes. Use technology.

**c.** Comment on the differences between the graphs in part (a) and the graphs in part (b).

**B**

**1.** Go back to Example 7.5.2 and look at the trigonometric substitutions suggested there. You're going to verify the polar coordinate form of the system equations.

**a.** Use the Chain Rule to show that $r\dot{r} = x\dot{x} + y\dot{y}$.

**b.** Show that $\dot{\theta} = -\frac{1}{x^2+y^2}(y\dot{x} - x\dot{y})$, or $-r^2\dot{\theta} = (y\dot{x} - x\dot{y})$.

**c.** Show that $x\dot{x} + y\dot{y} = (x^2 + y^2)(x^2 + y^2 - 1) = r^2(r^2 - 1)$. [*Hint*: Multiply the first equation in the system by $x$ and the second equation by $y$ and then add the results.]

**d.** Use part (a) and part (c) to conclude that $\dot{r} = r(r^2 - 1)$.

**e.** Use part (b) and the general method in part (c) to show that $\dot{\theta} = 1$.

**2.** Reconsider the uncoupled system (polar coordinate form) in Example 7.5.2.

**a.** Solve for $r(t)$.

**b.** Solve for $\theta(t)$.

**c.** Use your answers to part (a) and part (b) to construct $x(t)$ and $y(t)$.

**3.** Follow the directions given in the preceding problem for the system in Example 7.5.3.

**4.** Consider the system $\{\dot{r} = r(1 - r^2), \dot{\theta} = 1\}$.

**a.** Show that this is equivalent to the system

$$\dot{x} = x - y - x\left(x^2 + y^2\right)$$
$$\dot{y} = x + y - y\left(x^2 + y^2\right),$$

where $x = r(t)\cos\theta(t)$ and $y = r(t)\sin\theta(t)$.

**b.** Use either form of the system to determine its unique limit cycle.

5. Consider the system $\{\dot{r} = r\left(4 - r^2\right), \dot{\theta} = 1\}$, where $x(t) = r(t)\cos\theta(t)$ and $y = r(t)\sin\theta(t)$. Given the initial conditions $x(0) = 0.1, y(0) = 0$, sketch the graph of $x(t)$ without finding an explicit expression for $x(t)$. [*Hint*: Study Example 7.5.2 carefully.] Suppose we have an autonomous system $\{\dot{x} = f(x, y), \dot{y} = g(x, y)\}$, where $f$ and $g$ have continuous first partial derivatives in some region $R$ of the phase plane that doesn't have any "holes." Then Bendixson's Theorem (or Negative Criterion) states that if $\frac{\partial f}{\partial x} + \frac{\partial g}{\partial y}$ is always positive or always negative at points of $R$, then the system has no periodic solutions in $R$. For example, the system $\{\dot{x} = xy^2, \dot{y} = x^2 + 8y\}$ has no limit cycles anywhere because $\frac{\partial(xy^2)}{\partial x} + \frac{\partial(x^2+8y)}{\partial y} = y^2 + 8 > 0$ for all values of $x$ and $y$ in the plane.

**Use Bendixson's criterion to show that the systems in Problems B6–B9 have no limit cycles in the phase plane.**

6. $\{\dot{x} = x + 2xy + x^3, \ \dot{y} = -y^2 + x^2y\}$
7. $\{\dot{x} = x^3 + x + 7y, \ \dot{y} = x^2y\}$
8. $\{\dot{x} = -2x - x\sin y, \ \dot{y} = -x^2y^3\}$
9. $\{\dot{x} = 2x^3y^4 - 3, \ \dot{y} = 2ye^x + x\}$
10. Show that the system

$$\dot{x} = 12x + 10y + x^2y + y\sin y - x^3$$
$$\dot{y} = x + 14y - xy^2 - y^3$$

has no periodic solution in the disk $x^2 + y^2 \le 8$.

11. A mechanical system with variable damping can be modeled by the equation

$$\ddot{x} + a(x)\dot{x} + b(x) = 0,$$

where $a(x)$ is a positive function.

a. Write this equation in system form.
b. Use the Bendixson criterion shown previously to show that this mechanical system has no nonconstant periodic solution.

## C

1. The system

$$\dot{x} = -y - y^2$$
$$\dot{y} = \frac{1}{2}x - \frac{1}{5}y + xy - \frac{6}{5}y^2$$

was discovered by the Chinese mathematician Tung Chin Chu in the late 1950s in his investigation of a famous unsolved problem on limit cycles.

a. Find the equilibrium point(s) of this system.
b. Use technology to draw a phase portrait for each equilibrium point, focusing on the region around that point. (It's a bit tricky to get a good phase portrait for this problem. Be patient.)

        **c.** Using the phase portrait(s), identify and describe any limit cycle(s) you see with the term *stable* or *unstable*.

**2.** Find all limit cycles of the system

$$\dot{r} = r(r-1)(r-2)^2(r-3)$$

$$\dot{\theta} = 1$$

and identify them as stable, unstable, or semistable.

## SUMMARY

Nonlinear differential equations and systems of nonlinear equations are rarely handled satisfactorily by finding closed-form solutions. In particular, we can't analyze the *stability* of systems of nonlinear equations as easily as we analyzed the stability of linear systems in Chapter 5. The modern study of nonlinear phenomena relies heavily on the qualitative methods pioneered by H. Poincaré and A. Lyapunov at the end of the nineteenth century and in the beginning of the twentieth century. Current technology implements the power of these qualitative techniques.

One of the differences between linear and nonlinear equations is that a nonlinear equation may have more than one equilibrium solution. Another difference is that a solution of a nonlinear equation may "blow up in finite time"—that is, become unbounded as $t$ approaches some finite value. A third difference is that a nonlinear equation or system may be extremely sensitive to initial conditions. A slight change in an initial value may lead to drastic changes in the behavior of the solution or solutions.

A point $(a^*, b^*)$ is an *equilibrium point* of the general nonlinear autonomous system

$$\dot{x} = F(x, y)$$

$$\dot{y} = G(x, y)$$

if $F(a^*, b^*) = 0 = G(a^*, b^*)$. If the origin is an equilibrium point, and the functions $F$ and $G$ are "nice" enough, we may be able to write our system in the form

$$\dot{x} = ax + by + f(x, y)$$

$$\dot{y} = cx + dy + g(x, y),$$

where $f$ and $g$ are nonlinear functions and $a = \frac{\partial F}{\partial x}(0, 0), b = \frac{\partial F}{\partial y}(0, 0), c = \frac{\partial G}{\partial x}(0, 0)$, and $d = \frac{\partial G}{\partial y}(0, 0)$. More generally, if $(a, b) \neq (0, 0)$ is an equilibrium point for the system, we can rewrite the system as

$$\dot{x} = A(x - a) + B(y - b) + f(x, y)$$

$$\dot{y} = C(x - a) + D(y - b) + g(x, y),$$

where $f$ and $g$ are nonlinear and $A = \frac{\partial F}{\partial x}(a, b), B = \frac{\partial F}{\partial y}(a, b), C = \frac{\partial G}{\partial x}(a, b)$, and $D = \frac{\partial G}{\partial y}(a, b)$.

Another way to look at this general situation is to realize that we are translating the equilibrium point $(a, b)$ to the origin by using the change of variables $u = x - a$ and $v = y - b$. Of course, this means that $x = u + a$ and $y = v + b$, so we can rewrite the last system as

$$\dot{u} = Au + Bv + f(u, v)$$
$$\dot{v} = Cu + Dv + g(u, v),$$

which has $(0, 0)$ as an equilibrium point. Note that this says that any equilibrium point $(a, b) \neq (0, 0)$ can be transformed to the origin for the purpose of analyzing the stability of the system.

A nonlinear autonomous system

$$\dot{x} = ax + by + f(x, y)$$
$$\dot{y} = cx + dy + g(x, y),$$

where $ad - bc \neq 0$, $\displaystyle\lim_{(x,y)\to(0,0)} \left( \frac{f(x,y)}{\sqrt{x^2+y^2}} \right) = \lim_{(x,y)\to(0,0)} \left( \frac{g(x,y)}{\sqrt{x^2+y^2}} \right) = 0$, and the origin is an equilibrium point, is called an **almost linear system**, and the reduced system

$$\dot{x} = ax + by$$
$$\dot{y} = cx + dy$$

is called the **associated linear system** (or linear approximation) about the origin.

An important qualitative result discovered by Poincaré and Lyapunov states that if $\lambda_1$ and $\lambda_2$ are the eigenvalues of the associated linear system, then the equilibrium points of the two systems are related as follows:

a. If the eigenvalues $\lambda_1$ and $\lambda_2$ are *not* equal real numbers or are *not* pure imaginary numbers, then the trajectories of the almost linear system near the equilibrium point $(0, 0)$ behave the same way as the trajectories of the associated linear system near the origin. That is, we can use the appropriate entries given in Table 5.1 (Section 5.5) to determine whether the origin is a node, a saddle point, or a spiral point of both systems.

b. If $\lambda_1$ and $\lambda_2$ are real and equal, then the origin is either a node or a spiral point of both systems. Furthermore, if $\lambda_1 = \lambda_2 < 0$, then the origin is asymptotically stable; and if $\lambda_1 = \lambda_2 > 0$, then the origin is an unstable equilibrium point.

c. If $\lambda_1$ and $\lambda_2$ are pure imaginary numbers, then the equilibrium point $(0, 0)$ is either a center or a spiral point of the nonlinear system. Also, this spiral point may be asymptotically stable, stable, or unstable.

In situations (b) and (c), further analysis is necessary to determine the nature of the equilibrium points.

The **Lotka-Volterra equations**, the **undamped pendulum**, and the **van der Pol equation** provide important examples of nonlinear systems and their analyses. In particular, the van der Pol oscillator exhibits uniquely nonlinear behavior in having a **stable limit cycle**, an isolated closed trajectory that (in this case) serves as an asymptotic limit for all other trajectories as $t \to \infty$. Some limit cycles, called **unstable limit cycles**, repel nearby trajectories. Finally, if trajectories near a limit cycle approach it from one side while being repelled from the other side, the cycle is called **semistable**.

## PROJECT 7-1

### Butterflies in Space

In 1963, E. N. Lorenz, an MIT meteorology professor, published a report[2] on the nonlinear system

$$\dot{x} = -\sigma x + \sigma y$$

$$\dot{y} = rx - y - xz$$

$$\dot{z} = xy - bz,$$

where $\sigma, r$, and $b$ are positive parameters.

The equations arose from a model of a layer in the earth's atmosphere, heated from below by the ground that has absorbed sunlight, and cooled from above as it loses heat into space.

**a.** Show that if $0 < r \leq 1$, then the only equilibrium point is $(0, 0, 0)$.

**b.** Show that if $r > 1$, then there are three equilibrium points: $(0, 0, 0)$,

$$\left(\sqrt{b(r-1)}, \sqrt{b(r-1)}, r-1\right), \text{ and } \left(-\sqrt{b(r-1)}, -\sqrt{b(r-1)}, r-1\right).$$

**c.** Let $b = \frac{8}{3}, r = 28$, and $\sigma = 10$ (values used by Lorenz in his initial experiments). For these values of $b, r$, and $\sigma$, linearize the system about the equilibrium points

$$\left(\pm\sqrt{b(r-1)}, \pm\sqrt{b(r-1)}, r-1\right).$$

**d.** Use technology to show that the characteristic equation of the linearized system found in part (c) is $\lambda^3 + (b + \sigma + 1)\lambda^2 + b(r + \sigma)\lambda + 2\sigma b(r-1) = 0$. Show that the characteristic polynomial has a negative real root $\lambda_1 \approx -13.85$ and complex conjugate roots with positive real parts $\approx 0.09$.

---

[2] E. N. Lorenz, "Deterministic Nonperiodic Flow," *J. Atmos. Sci.* **20** (1963): 130–141.

**e.** Use part (d) and the table at the end of Section 5.5 to conclude that the two nonzero equilibrium points given in part (c) are *saddle points* of the system.

**f.** With $b = \frac{8}{3}, r = 28$, and $\sigma = 10$, use technology to plot $x(t)$ against $t$, $y(t)$ against $t$, and $z(t)$ against $t$ for $0 \leq t \leq 10$.

**g.** With $b = \frac{8}{3}, r = 28$, and $\sigma = 10$, use technology to plot $y$ against $x$, $z$ against $y$, and $z$ against $x$.

# Some Calculus Concepts and Results

Appendix A is intended to offer either a brief review of, or an introduction to, selected key ideas of calculus.

## A.1 LOCAL LINEARITY: THE TANGENT LINE APPROXIMATION

The concept of **local linearity** says that if the function $f$ is differentiable (that is, if it has a derivative) at $x = a$ and we "zoom in" on the point $(a, f(a))$ on the graph of $y = f(x)$, then the portion of the curve that surrounds the point looks very much like a straight line—at least to the naked eye. Another way of saying this is to say that the tangent line at the point $(a, f(a))$ is a good approximation to the curve for values of $x$ close to $a$. Figure A.1 illustrates this.

As $x$ takes on values farther away from $a$, we expect the **absolute error** $|E(x)| = |f(x) - f(a) - f'(a)(x - a)|$ to become larger.

Using the point-slope formula from algebra, we can write the equation of this tangent line as $y = f(a) + f'(a)(x - a)$, so we can express this **tangent line approximation** as $f(x) \approx f(a) + f'(a)(x - a)$ for $x$ close to $a$.

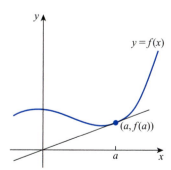

**FIGURE A.1**
*The tangent line approximation*

For example, the equation of the tangent line drawn to the sine curve at the origin is $y = \sin(0) + \cos(0)(x - 0) = x$. This says that near the origin, $\sin x \approx x$. One consequence of this is that $\frac{\sin x}{x} \approx 1$ for values of $x$ near (but not equal to) zero, so we get the famous result $\lim_{x \to 0} \frac{\sin x}{x} = 1$.

## A.2 THE CHAIN RULE

You should know the rules for finding the derivatives of power functions, polynomials, exponential functions, logarithms, and trigonometric and inverse trigonometric functions. You may also have learned about differentiating certain combinations of exponential functions called *hyperbolic functions*. You should know the *Product Rule* and the *Quotient Rule* for differentiation, as well as how to deal with *implicit functions*.

The Chain Rule applies to *composite functions*. Suppose, for example, that a quantity $z$ depends on a quantity $y$ and that the quantity $y$ depends on the value of quantity $x$. Using function notation, we can write this as follows: $z = f(y), y = g(x)$, so $z = f(g(x))$. This says that, ultimately, $z$ depends on (is a function of) $x$. The **Chain Rule** tells us how a change in the value of $x$ affects the value of $z$. In Leibniz notation,

$$\frac{dz}{dx} = \frac{dz}{dy} \cdot \frac{dy}{dx}.$$

This form is useful in many applied problems and in Chapter 4, where the *phase plane* is introduced.

### ■ Example
If $z = y^{57}$ and $y = \sin x$, then

$$\frac{dz}{dx} = \frac{dz}{dy} \cdot \frac{dy}{dx} = \left(57y^{56}\right) \cdot \cos x = 57 \sin^{56} x \cos x.$$

You may have learned another way to see the Chain Rule: If $z = f(g(x))$, then $z' = f'(g(x)) \cdot g'(x)$. This alternative point of view uses the idea of an "inside" function and an "outside" function. Try this on the preceding example, where the 57th-power function is outside and the sine function is inside. ■

## A.3 THE TAYLOR POLYNOMIAL/TAYLOR SERIES

To extend the idea of the tangent line approximation, we look for a polynomial $P_n$ of degree $n$ that approximates a function $f$ as closely as possible on an interval about a point $x = a$. What this means mathematically is that we want the polynomial to satisfy the following closeness conditions: $P_n(a) = f(a), P'_n(a) = f'(a), P''_n(a) = f''(a), P'''_n(a) = f'''(a), \ldots$, and $P_n^{(n)}(a) = f^{(n)}(a)$. For a given function $f$, a point $x = a$, and degree $n$, the polynomial that satisfies all these conditions is given by the formula

$$P_n(x) = f(a) + f'(a)(x-a) + \frac{f''(a)}{2!}(x-a)^2$$

$$+ \frac{f'''(a)}{3!}(x-a)^3 + \cdots + \frac{f^{(n)}(a)}{n!}(x-a)^n.$$

This is called the **Taylor polynomial of degree $n$** about $x = a$, and we can write $f(x) \approx P_n(x)$ for $x$ close to $a$. The closeness of the approximation depends on both the value of $x$ and the value of $n$. In general, the closer the value of $x$ is to the value $a$ and the higher the degree $n$, the better the approximation.

If we consider what happens to a Taylor polynomial as we let $n$ get larger and larger, we arrive at the idea of the (infinite) **Taylor series**:

$$P(x) = \lim_{n\to\infty} P_n(x) = \lim_{n\to\infty} \sum_{k=0}^{n} \frac{f^{(k)}(a)}{k!}(x-a)^k$$

$$= f(a) + f'(a)(x-a) + \frac{f''(a)}{2!}(x-a)^2$$

$$+ \frac{f'''(a)}{3!}(x-a)^3 + \cdots + \frac{f^{(n)}(a)}{n!}(x-a)^n + \cdots.$$

More precisely, suppose that $f$ is a function with derivatives of all order in some interval $(a-r, a+r)$. Then the Taylor series given previously represents the function $f$ on the interval $(a-r, a+r)$ if and only if $\lim_{n\to\infty} R_n(x) = 0$, where $R_n(x)$ is the remainder in Taylor's formula:

$$R_n(x) = f(x) - \left( f(a) + f'(a)(x-a) + \frac{f''(a)}{2!}(x-a)^2 \right.$$

$$\left. + \frac{f'''(a)}{3!}(x-a)^3 + \cdots + \frac{f^{(n)}(a)}{n!}(x-a)^n \right)$$

$$= \frac{f^{(n+1)}(c)}{(n+1)!}(x-a)^{n+1} \quad \text{for some point } c \text{ in } (a-r, a+r).$$

As we saw in Section A.1, any approximation process is subject to error. For example, when working with $\pi$, which has an infinite, nonrepeating decimal representation, we lose accuracy by using 3.14159 or even 3.14159265359 as its value. This, in turn, leads to what is called *propagated error*, the accumulated error resulting from many calculations with rounded values. If each item of data is inaccurate because of rounding of some sort, then the various steps in a calculation process can compound the error. A useful approximation method guarantees that the smaller the round-off error at each stage, the smaller the cumulative round-off error.

Of course, it turns out that sometimes round-off errors cancel each other out to a certain extent—approximate values that are too high may be balanced by values that are too low.

**Truncation error** occurs when we stop (or truncate) an approximation process after a certain number of steps. For example, when we approximate the values of $\sin x$ near $x = 0$ by using the first seven nonzero terms of its (infinite) Taylor series,

$$x - \frac{x^3}{3!} + \frac{x^5}{5!} - \frac{x^7}{7!} + \frac{x^9}{9!} - \frac{x^{11}}{11!} + \frac{x^{13}}{13!}$$

we are introducing truncation error. If we write $\sin x = T_{13}(x) + R_{13}(x)$, where $T_{13}(x)$ is the 13th-degree polynomial just given, a formula we know from calculus gives an upper bound for the absolute truncation error

$$|\sin x - T_{13}(x)| = |R_{13}(x)| = \frac{|\sin c|}{14!}|x|^{14} \leq \frac{|x|^{14}}{14!},$$

where $c$ is a positive number less than $x$. Even if we use the 1001st-degree Taylor polynomial, we are still only approximating and will therefore have truncation error.

Here are some Taylor series that occur often in applications:

$$e^x = 1 + x + \frac{x^2}{2!} + \frac{x^3}{3!} + \frac{x^4}{4!} + \cdots + \frac{x^n}{n!} + \cdots$$

$$\sin x = x - \frac{x^3}{3!} + \frac{x^5}{5!} - \frac{x^7}{7!} + \cdots + (-1)^k \frac{x^{2k+1}}{(2k+1)!} + \cdots$$

$$\cos x = 1 - \frac{x^2}{2!} + \frac{x^4}{4!} - \frac{x^6}{6!} + \cdots + (-1)^k \frac{x^{2k}}{(2k)!} + \cdots$$

$$\ln(1 + x) = x - \frac{x^2}{2} + \frac{x^3}{3} - \frac{x^4}{4} + \cdots + (-1)^{k+1}\frac{x^k}{k} + \cdots$$

$$\frac{1}{1 - x} = 1 + x + x^2 + x^3 + \cdots + x^n + \cdots .$$

Although the first three series are valid ("converge") for any value of $x$, the logarithmic series is valid only on the interval $(-1, 1)$. The last series, a *geometric series*, converges for $|x| < 1$. In Section C.4, we'll see how Euler used the exponential series to arrive at a formula for the complex exponential function.

A Taylor series is a special type of *power series*. We can differentiate or integrate a power series term by term for values of $x$ within its interval of convergence. If the power series $\sum_{n=0}^{\infty} a_n x^n$ converges to $S(x)$ for $x$ in some interval $I$, then

$$S'(x) = \sum_{n=0}^{\infty} n a_n x^{n-1} = a_1 + 2a_2 x + 3a_3 x^2 + \cdots + n a_n x^{n-1} + \cdots$$

and

$$\int_0^x S(t)dt = \int_0^x \left( \sum_{n=0}^\infty a_n t^n \right) dt = \sum_{n=0}^\infty \int_0^x a_n t^n dt = \sum_{n=0}^\infty \frac{a_n}{n+1} x^{n+1}$$

$$= a_0 x + \frac{a_1}{2}x^2 + \frac{a_2}{3}x^3 + \cdots + \frac{a_n}{n+1}x^{n+1} + \cdots ,$$

where both the differentiated and the integrated series converge for $x$ in $I$.

In Appendix D, we will show how to solve certain differential equations using power series methods.

## A.4 THE FUNDAMENTAL THEOREM OF CALCULUS (FTC)

A function $F$ is an *antiderivative* of the function $f$ if $F'(x) = f(x)$. A very important connection between derivatives and integrals is expressed by the **Fundamental Theorem of Calculus** (**FTC**). This result comes in two flavors:

**A.** If $f(x)$ is continuous on the closed interval $[a, b]$ and if $F(x)$ is any antiderivative of $f(x)$ on this interval, then

$$\int_a^b f(x)dx = F(b) - F(a).$$

**B.** Let $f(x)$ be defined and continuous on a closed interval $[a, b]$ and define the function $G(x)$ on this interval:

$$G(x) = \int_a^x f(t)dt.$$

Then $G(x)$ is differentiable there with derivative $f(x)$: $G'(x) = f(x)$.

Version A simplifies the whole business of finding the value of a definite integral: Just find an antiderivative of the integrand. Of course, as you know, this isn't always as simple as it sounds. At least half of your calculus course was probably devoted to techniques of *substitution* and *integration by parts*, trying to recognize integrands as derivatives resulting from the Product Rule, the Chain Rule, and so forth.

A slight twist on version A tells us that if we integrate $f'$, the *rate* function over $[a, b]$, we get the *total change* in $f$, the *amount* function over the same interval. For example, if $s(t)$, $v(t)$, and $a(t)$ denote the position, velocity, and acceleration, respectively, of a moving object at time $t$, then we have

$$\int v(t)dt = s(t) + C \quad \text{and} \quad \int a(t)dt = v(t) + K,$$

where $C$ and $K$ denote arbitrary constants. Consequently, we can write

$$\int_a^b v(t)dt = \int_a^b s'(t)dt = s(b) - s(a),$$

which says that if we integrate the *velocity* function, we get the *total change in position* of a moving object as $t$ changes from $a$ to $b$. If we integrate the *speed* function—the absolute value of the velocity—we get the *total distance* traveled by the object. (See Example 1.3.4 in the text.)

As useful as version A is in solving differential equations, version B extends the notion of differentiation (and therefore of integration) to functions defined by integrals. (See Problems A9 and A10 in Exercises 1.2.)

### ■ Example

Suppose that $Q(x) = \int_{-2}^x \cos(u^2)du$. Then $Q'(x) = \cos(x^2)$, $Q''(x) = -2x\sin(x^2)$, and so on. ■

## A.5 PARTIAL FRACTIONS

An important and useful result from algebra says that every rational function (quotient of polynomial functions), no matter how complicated, comes from adding simpler fractions. For example, the function

$$\frac{8x + 1}{x^2 - x - 6}$$

comes from the following addition of simpler pieces:

$$\frac{3}{x + 2} + \frac{5}{x - 3} = \frac{3(x - 3) + 5(x + 2)}{(x + 2)(x - 3)} = \frac{8x + 1}{x^2 - x - 6}.$$

In calculus, when we have an integrand that is a rational function, we can reverse this addition process to find the simpler fractions, fractions that we can integrate easily. Thus, for example,

$$\int \frac{8x + 1}{x^2 - x - 6}dx = \int \left(\frac{3}{x + 2} + \frac{5}{x - 3}\right)dx = \int \frac{3}{x + 2}dx + \int \frac{5}{x - 3}dx$$

$$= 3\ln|x + 2| + 5\ln|x - 3| + C.$$

In this example, the algebraic challenge is to find constants $A$ and $B$ such that

$$\frac{8x + 1}{x^2 - x - 6} = \frac{A}{x + 2} + \frac{B}{x - 3}. \tag{*}$$

The fractions $A/(x + 2)$ and $B/(x - 3)$ are called **partial fractions** because each contributes a piece of the whole. In particular, the denominators $x + 2$ and $x - 3$ are parts (factors) of the original denominator $x^2 - x - 6$. The numbers $A$ and $B$ are called **undetermined coefficients** (see Section 4.3, Section 5.6, and Appendix D). To find $A$ and $B$, we clear Equation (*) of fractions by multiplying both sides by $x^2 - x - 6$. The result is

$$8x + 1 = A(x - 3) + B(x + 2).$$

This is supposed to be an identity in $x$. If we let $x = 3$, we find that $8(3) + 1 = 0 + 5B$, or $B = 5$. Similarly, letting $x = -2$, we get $8(-2) + 1 = -5A + 0$, so $A = 3$.

This technique works for a rational function in lowest terms whose denominator can be factored into distinct linear factors. More complicated denominators can also be handled by this kind of algebraic method, and you can find a more detailed discussion in your calculus text. Most computer algebra systems can produce such "partial-fraction decompositions" and can evaluate integrals with integrands that are rational functions. Partial fractions are particularly useful in Section 2.1 and in Chapter 6.

## A.6 IMPROPER INTEGRALS

In dealing with the definite (Riemann) integral $\int_a^b f(x)dx$, we make two basic assumptions: (1) The interval $[a, b]$ is finite, and (2) the integrand $f$ is bounded (that is, does not become infinite) on the closed interval $[a, b]$. If we violate one or both of these assumptions, we encounter a type of **improper integral**.

First, let us assume that we want to consider the interval $[a, \infty)$ or $(-\infty, b]$, where $a$ and $b$ are real numbers. We can define

$$\int_a^\infty f(x)dx = \lim_{B \to \infty} \int_a^B f(x)dx \quad \text{or} \quad \int_{-\infty}^b f(x)dx = \lim_{A \to \infty} \int_{-A}^b f(x)dx$$

provided that each limit exists. If the limit exists, we say that the improper integral **converges**. Otherwise, we say that the improper integral **diverges**. Finally,

$$\int_{-\infty}^\infty f(x)dx = \lim_{A \to \infty} \int_{-A}^c f(x)dx + \lim_{B \to \infty} \int_c^B f(x)dx$$

*provided that each limit on the right-hand side exists individually.* Here, $c$ is an arbitrary real number. It is *not* correct to define $\int_{-\infty}^\infty f(x)dx$ as $\lim_{C \to \infty} \int_{-C}^C f(x)dx$.

### ■ Example

$$\int_1^\infty \frac{dx}{1 + x^2} = \lim_{B \to \infty} \arctan x \big|_1^B = \lim_{B \to \infty} (\arctan B - \arctan 1)$$

$$= \lim_{B \to \infty} \left( \arctan B - \frac{\pi}{4} \right) = \lim_{B \to \infty} \arctan B - \frac{\pi}{4}$$

$$= \frac{\pi}{2} - \frac{\pi}{4} = \frac{\pi}{4}.$$  ■

## ■ Example

Consider $\int_0^\infty \sin x \, dx$. The limit

$$\lim_{B \to \infty} \int_0^B \sin x \, dx = \lim_{B \to \infty} \left( -\cos(B) + \cos(0) \right) = -\lim_{B \to \infty} \cos(B) + 1$$

doesn't exist because $\cos(B)$ oscillates from $-1$ to $1$ as $B$ tends to infinity.

When we are dealing with this first type of improper integral, for which the interval is not finite, sometimes a form of *L'Hôpital's Rule* comes in handy: Suppose that as $x \to a$, where $a$ is $\pm\infty$, $f(x) \to \pm\infty$, and $g(x) \to \pm\infty$. If $\lim_{x \to a} \frac{f'(x)}{g'(x)} = L$, where $L$ is either a real number or $\pm\infty$, then $\lim_{x \to a} \frac{f(x)}{g(x)} = L$.  ■

## ■ Example

Consider Euler's *gamma function*, defined by $\Gamma(x) = \int_0^\infty t^{x-1} e^{-t} dt$. Integration by parts tells us that

$$\Gamma(x) = -t^{x-1} e^{-t} \Big]_0^\infty - \int_0^\infty (x-1) t^{x-2} (-e^{-t}) dt$$

$$= \lim_{c \to \infty} \frac{-t^{c-1}}{e^c} + (x-1) \int_0^\infty t^{x-2} e^{-t} dt$$

$$= (x-1) \cdot \Gamma(x-1),$$

where we have used L'Hôpital's Rule several times in evaluating the limit. (Successive differentiations of the numerator and denominator of $\frac{-t^{c-1}}{e^c}$ eventually give us $-(c-1)!$ in the numerator, whereas the denominator remains $e^c$, so the limit of the quotient as $c$ tends to infinity is 0.) Note that because $\Gamma(1) = \Gamma(2) = 1$, we can conclude that $\Gamma(x+1) = x \cdot (x-1) \cdot (x-2) \cdot (x-3) \cdots 3 \cdot 2 \cdot 1 = x!$ when $x$ is an integer, so the gamma function provides a generalization of $n!$ to the case in which $n$ is not an integer. (See Section D.3, especially footnote 4.)

Now let's suppose that $f$ is defined and finite on the interval $[a, b]$ except at the endpoint $b$. Then the integral $\int_a^b f(x)dx$ is improper, and we define it as

$$\int_a^b f(x)dx = \lim_{B \to b^-} \int_a^B f(x)dx$$

provided that this *left-hand limit* (or *limit from the left*) exists. Similarly, if $f$ is unbounded at the endpoint $a$, then we define

$$\int_a^b f(x)dx = \lim_{A \to a^+} \int_A^b f(x)dx$$

provided that this right-hand limit (or limit from the right) exists. ■

## ■ Example

The function $1/\sqrt{1 - x^2}$ is unbounded at $x = 1$ (and at $x = -1$). The improper integral of this function on the interval $[0, 1]$ converges:

$$\int_0^1 \frac{dx}{\sqrt{1 - x^2}} = \lim_{B \to 1^-} \int_0^B \frac{dx}{\sqrt{1 - x^2}} = \lim_{B \to 1^-} \arcsin(B) - \arcsin(0) = \frac{\pi}{2}.$$

Another possibility is that the function $f$ is defined and finite on $[a, b]$ except at a point $\xi$ *inside* the interval. The improper integral is then defined as

$$\int_a^b f(x)dx = \lim_{c \to \xi^-} \int_a^c f(x)dx + \lim_{d \to \xi^+} \int_d^b f(x)dx$$

provided that both one-sided limits exist. ■

## ■ Example

$$\int_0^2 \frac{dx}{(x - 1)^{2/3}} = \lim_{c \to 1^-} \int_0^c \frac{dx}{(x - 1)^{2/3}} + \lim_{d \to 1^+} \int_d^2 \frac{dx}{(x - 1)^{2/3}}$$

$$= \lim_{c \to 1^-} 3(x - 1)^{1/3} \Big|_0^c + \lim_{d \to 1^+} 3(x - 1)^{1/3} \Big|_d^2$$

$$= \lim_{c \to 1^-} \left[ 3(c-1)^{1/3} - 3(-1)^{1/3} \right] + \lim_{d \to 1^+} \left[ 3(1)^{1/3} - 3(d-1)^{1/3} \right]$$

$$= 3 + 3 = 6. \qquad \blacksquare$$

## A.7 FUNCTIONS OF SEVERAL VARIABLES/PARTIAL DERIVATIVES

Sometimes we encounter functions that depend on more than one independent variable. For example, the area of a rectangle depends on both its length and its width. We can express this relationship as $A = f(l, w) = l \cdot w$. In general, if there are two independent variables ($x$ and $y$) and one dependent variable ($z$), we can express this situation as $z = f(x, y)$. In words, the variable $z$ depends on (is a function of) the variables $x$ and $y$. This means that changes in the value of either $x$ or $y$ (or both) will lead to changes in $z$. The *instantaneous rate of change of z with respect to x* is given by the **partial derivative of z with respect to x**, which is defined by the formula

$$\frac{\partial z}{\partial x} = \lim_{h \to 0} \frac{f(x+h, y) - f(x, y)}{h}.$$

Similarly, the *instantaneous rate of change of z with respect to y* is given by the **partial derivative of z with respect to y**, which is defined by the formula

$$\frac{\partial z}{\partial y} = \lim_{h \to 0} \frac{f(x, y+h) - f(x, y)}{h}.$$

What this means in terms of practical calculation is that to find $\frac{\partial z}{\partial x}$, you just treat $y$ as a constant and differentiate with respect to $x$ as usual. For $\frac{\partial z}{\partial y}$, you treat $x$ as a constant and regard $y$ as the "live" variable.

The Chain Rule takes different forms in this multivariable environment, among them this version: If $x = f(t)$ and $y = g(t)$, where $f$ and $g$ are differentiable and $F$ is a function of $x$ and $y$, then $\frac{dF}{dt} = \frac{\partial F}{\partial x}\frac{dx}{dt} + \frac{\partial F}{\partial y}\frac{dy}{dt}$.

### ■ Example

If $z = f(x, y) = x^2 y^2 - 3xy^3 + 5x^4 y^2$, then

$$\frac{\partial z}{\partial x} = 2xy^2 - 3y^3 + 20x^3 y^2 \quad \text{and} \quad \frac{\partial z}{\partial y} = 2x^2 y - 9xy^2 + 10x^4 y. \qquad \blacksquare$$

### ■ Example

Suppose $w = e^{2x+3y} \sin(xy)$. Then, using the Product Rule and the Chain Rule, we find that

$$\frac{\partial w}{\partial x} = e^{2x+3y} \cos(xy)y + 2e^{2x+3y} \sin(xy)$$

and

$$\frac{\partial w}{\partial y} = e^{2x+3y} \cos(xy)x + 3e^{2x+3y} \sin(xy).$$

■

### ■ Example

Let $F(x, y) = x^2 + y^2$, with $x = x(t) = \cos t$ and $y = y(t) = \sin t$. Then the multivariable Chain Rule gives us $\frac{dF}{dt} = \frac{\partial F}{\partial x}\frac{dx}{dt} + \frac{\partial F}{\partial y}\frac{dy}{dt} = (2x)(-\sin t) + (2y)(\cos t) = -2\cos t \sin t + 2 \sin t \cos t = 0$, which is not surprising since $F(x, y) = x^2 + y^2 = \cos^2 t + \sin^2 t \equiv 1$.

In general, if you have a function of $n$ variables, $z = f(x_1, x_2, x_3, \ldots, x_n)$, then you can define the partial derivative of $z$ with respect to $x_k$ and calculate it by treating $x_k$ as the only true variable, the other $x_i (i \neq k)$ being treated as constants. You can define higher derivatives and *mixed* derivatives in the obvious way: $\frac{\partial^2 z}{\partial x_i \partial x_k}, \frac{\partial^n z}{\partial x_k^n}$, and so on.

■

### ■ Example

Using $z = f(x, y) = x^2 y^2 - 3xy^3 + 5x^4 y^2$ and the results of the first example, we have

$$\frac{\partial^2 z}{\partial x^2} = \frac{\partial}{\partial x}\left(\frac{\partial z}{\partial x}\right) = \frac{\partial}{\partial x}\left(2xy^2 - 3y^3 + 20x^3 y^2\right) = 2y^2 + 60x^2 y^2$$

$$\frac{\partial^2 z}{\partial y^2} = \frac{\partial}{\partial y}\left(\frac{\partial z}{\partial y}\right) = \frac{\partial}{\partial y}\left(2x^2 y - 9xy^2 + 10x^4 y\right) = 2x^2 - 18xy + 10x^4$$

$$\frac{\partial^2 z}{\partial x \partial y} = \frac{\partial}{\partial x}\left(\frac{\partial z}{\partial y}\right) = \frac{\partial}{\partial x}\left(2x^2 y - 9xy^2 + 10x^4 y\right) = 4xy - 9y^2 + 40x^3 y,$$

and so forth.

■

## A.8  THE TANGENT PLANE: THE TAYLOR EXPANSION OF $F(x,y)$

In Section A.1, we saw that the tangent line $y = f(a) + f'(a)(x - a)$ gives the best linear approximation of a single-variable function $f$ near $x = a$. For $F(x, y)$, a function of two variables, the best approximation near a point $(a, b)$ is provided by the tangent plane given by the approximation formula

$$F(x, y) \approx F(a, b) + \frac{\partial F}{\partial x}(a, b)(x - a) + \frac{\partial F}{\partial y}(a, b)(y - b),$$

where $\frac{\partial F}{\partial x}(a, b)$ and $\frac{\partial F}{\partial y}(a, b)$ denote the partial derivatives evaluated at the point $(a, b)$.

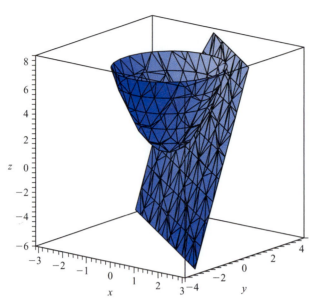

**FIGURE A.2**

*Tangent plane to the surface $z = x^3 - x^2 y^2 + y^3$ at $(1, 2)$*

## ■ Example

Let's calculate the tangent plane approximation of the function $F(x, y) = x^3 - x^2 y^2 + y^3$ near the point $(a, b) = (1, 2)$. We have $\frac{\partial F}{\partial x} = 3x^2 - 2xy^2$ and $\frac{\partial F}{\partial y} = -2x^2 y + 3y^2$, so $F(1, 2) = 1^3 - 1^2 2^2 + 2^3 = 5$, $\frac{\partial F}{\partial x}(1, 2) = 3(1)^2 - 2(1)(2)^2 = -5$, and $\frac{\partial F}{\partial y}(1, 2) = -2(1)^2(2) + 3(2)^2 = 8$. Putting these results into the tangent plane formula, we get

$$F(x, y) \approx F(1, 2) + \frac{\partial F}{\partial x}(1, 2)(x - 1) + \frac{\partial F}{\partial y}(1, 2)(y - 2)$$

$$= 5 - 5(x - 1) + 8(y - 2).$$   ■

Figure A.2 shows the three-dimensional picture of the surface and its tangent plane.

For points $(x, y)$ close to $(1, 2)$, the values of $z$ on the tangent plane are close to the values of $z$ on the surface defined by $z = F(x, y)$.

We can define a full Taylor series expansion of a function of several variables, but for this text we need only the idea of the tangent plane (linear) approximation.

# Vectors and Matrices

Appendix B is intended to provide an expanded view of the vector and matrix algebra needed in this text.

## B.1  VECTORS AND VECTOR ALGEBRA; POLAR COORDINATES

In more abstract courses, a **vector** is an object in a set whose elements obey certain algebraic rules. For physicists, engineers, and other scientists, a **vector**—more properly, a **geometric vector**—is a quantity that has both magnitude (size) and direction. In two-dimensional physical situations, there are two usual ways to represent a vector: (1) as an ordered pair of real numbers, written $(x, y)$ or $\begin{bmatrix} x \\ y \end{bmatrix}$; and (2) as an *arrow* from the origin (usually) of the $x$-$y$ plane to a point $(x, y)$ or $\begin{bmatrix} x \\ y \end{bmatrix}$. The numbers $x$ and $y$ are called the **components** or **coordinates** of the vector. As indicated in Chapter 5, we can also consider vectors with complex-number coordinates and vectors whose components are functions. For the sake of simplicity in this appendix, we'll work with vectors whose components are real numbers.

Both ways of looking at a vector are shown in Figure B.1. In the second ("arrow") view, the vector $\mathbf{v} = (x, y)$ is always the hypotenuse of a right triangle, so by the Pythagorean Theorem its **length**—denoted $|\mathbf{v}|$—is given by the expression $\sqrt{x^2 + y^2}$. The direction of a vector is indicated by the direction of the arrow.

Vectors—often representing forces of various kinds—can be combined to indicate interactions. For example, you can add two vectors as follows: If $\mathbf{v}_1 = (x_1, y_1)$ and $\mathbf{v}_2 = (x_2, y_2)$, then $\mathbf{v}_1 + \mathbf{v}_2 = (x_1, y_1) + (x_2, y_2) = (x_1 + x_2, y_1 + y_2)$. Subtraction is similar. You can also multiply a vector by a real number (or a complex number or a function), which is called a **scalar** in this situation. To do this, just multiply each component of the vector by the scalar: If $\mathbf{v} = (x, y)$ and $r$ is any real number, then $r\mathbf{v} = (rx, ry)$. Because the components of vectors are real numbers, we should expect the usual rules of algebra to apply. If $\mathbf{v}_1$, $\mathbf{v}_2$, and $\mathbf{v}_3$ are vectors and $r$ is a scalar, then $\mathbf{v}_1 + \mathbf{v}_2 = \mathbf{v}_2 + \mathbf{v}_1$ [commutative property], $\mathbf{v}_1 + (\mathbf{v}_2 + \mathbf{v}_3) = (\mathbf{v}_1 + \mathbf{v}_2) + \mathbf{v}_3$

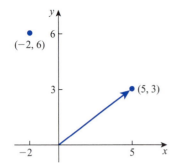

**FIGURE B.1**

*Ways to represent a vector*

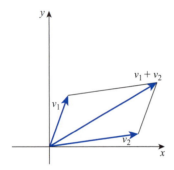

**FIGURE B.2**

*The Parallelogram Law*

[associative property], and $r\,(\mathbf{v}_1 + \mathbf{v}_2) = r\,\mathbf{v}_1 + r\,\mathbf{v}_2$ [distributive property]. There is a **zero vector**, denoted by $\mathbf{0} = (0, 0)$, such that $\mathbf{v} + \mathbf{0} = \mathbf{v} = \mathbf{0} + \mathbf{v}$ for every vector $\mathbf{v}$ [additive identity].

Geometrically, the addition or subtraction of vectors is captured by the **Parallelogram Law** (see Figure B.2 for the two-dimensional version).

Another way of representing a vector in two-dimensional space is by using **polar coordinates** (Figure B.3). If we have a vector corresponding to the point $(x, y)$, then we can describe it using its *length* $r$ (its radial distance from the origin) and the angle $\theta$ that the arrow makes with the positive $x$-axis, measured in a counterclockwise direction. As we saw previously, the length is given by the formula $r = \sqrt{x^2 + y^2}$.

As we look at Figure B.3, simple trigonometry tells us that $x = r \cos\theta$, $y = r \sin\theta$, and $\theta = \arctan\left(\frac{y}{x}\right)$, $x \neq 0$. As indicated in some of the examples in Chapter 7, the polar representation of vectors may be more natural in problems involving expressions that look like $x^2$, $y^2$, $x^2 + y^2$, and so on.

There is no reason to restrict our definition of vectors to two dimensions. In three-dimensional space, a vector is an ordered triplet, $(x, y, z)$ or $\begin{bmatrix} x \\ y \\ z \end{bmatrix}$, of real numbers or an arrow joining the

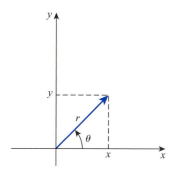

**FIGURE B.3**
*Polar representation of a vector*

origin $(0, 0, 0)$ to the point $(x, y, z)$. In general, an ***n*-dimensional vector** is an ordered *n*-tuple,

$(x_1, x_2, x_3, \ldots, x_n)$ or $\begin{bmatrix} x_1 \\ x_2 \\ x_3 \\ \vdots \\ x_n \end{bmatrix}$, of real numbers. The coordinate-by-coordinate arithmetic/algebra

of vectors generalizes to any dimension in the obvious way.

Given a set of vectors $\{\mathbf{v}_1, \mathbf{v}_2, \ldots, \mathbf{v}_m\}$, any vector of the form $c_1\mathbf{v}_1 + c_2\mathbf{v}_2 + \cdots + c_m\mathbf{v}_m$, where $c_1, c_2, \ldots, c_m$ are scalars, is called a **linear combination** of the set of vectors. The set of vectors $\{\mathbf{v}_1, \mathbf{v}_2, \ldots, \mathbf{v}_m\}$ is called **linearly independent** if the only way we can have $c_1\mathbf{v}_1 + c_2\mathbf{v}_2 + \cdots + c_m\mathbf{v}_m = \mathbf{0}$ (the zero vector) is if $c_1 = c_2 = \cdots = c_m = 0$. Otherwise, the set of vectors is **linearly dependent**. Linear *dependence* implies that at least one vector in the set can be expressed as a linear combination of the others.

## ■ Example

We will determine whether the following vectors are linearly independent:

$$\mathbf{v}_1 = \begin{bmatrix} 1 \\ 1 \\ 0 \\ 0 \end{bmatrix}, \quad \mathbf{v}_2 = \begin{bmatrix} 1 \\ 0 \\ 1 \\ 0 \end{bmatrix}, \quad \mathbf{v}_3 = \begin{bmatrix} 0 \\ 0 \\ 1 \\ 1 \end{bmatrix}, \quad \mathbf{v}_4 = \begin{bmatrix} 0 \\ 1 \\ 0 \\ 1 \end{bmatrix}.$$

The statement $c_1\mathbf{v}_1 + c_2\mathbf{v}_2 + c_3\mathbf{v}_3 + c_4\mathbf{v}_4 = \mathbf{0}$ is equivalent to the system of algebraic equations

$$c_1 + c_2 \qquad\qquad = 0$$
$$c_1 \qquad\quad + c_4 = 0$$
$$c_2 + c_3 \qquad = 0$$
$$c_3 + c_4 = 0.$$

This system is not difficult to solve by hand using substitution or elimination, but we can also use the capability of a graphing calculator or CAS to solve such systems of equations. In any case, we find that $c_1 = 1, c_2 = -1, c_3 = 1$, and $c_4 = -1$ is a solution. Because the scalars are not all zero, we conclude that the four vectors are *linearly dependent*. Note, for example, that we can write the first vector as a linear combination of the remaining vectors: $\mathbf{v}_1 = \mathbf{v}_2 - \mathbf{v}_3 + \mathbf{v}_4$. ∎

## B.2　MATRICES AND BASIC MATRIX ALGEBRA

A **matrix** (the plural is **matrices**) is simply a rectangular arrangement (array) of numbers or other mathematical objects (such as functions) and is usually denoted by a capital letter. It can be considered a generalization of a vector. For example, we can have the matrix

$$A = \begin{bmatrix} 0 & -4 & 1/2 & 9 \\ \pi & 14/5 & -0.15 & 2 \\ 7 & \sqrt{3} & 0 & -3 \end{bmatrix}.$$

The numbers or objects making up a matrix are called its **elements** or **entries**. Most of the time we'll use real numbers, although complex numbers and even functions can appear as entries of matrices (as they can for components of vectors).

One way to describe a matrix is by indicating how many rows and columns it has. Matrix $A$ in the preceding equation has 3 rows and 4 columns and is called a *3 by 4 matrix*, or a $3 \times 4$ *matrix*. A matrix with $m$ rows and $n$ columns is called an *$m$ by $n$ matrix* ($m \times n$ *matrix*). Note that each row or column of a matrix can be considered a vector. An $n \times 1$ matrix is called a **column vector**, whereas a $1 \times n$ matrix is called a **row vector**. Two matrices are called **equal** if they have the same number of rows and columns and their corresponding elements are equal. For example, we can write

$$\begin{bmatrix} 1 & 0 & -5/3 \\ 1/\sqrt{2} & 3 & 0.25 \end{bmatrix} = \begin{bmatrix} 7/7 & 0 & -15/9 \\ \sqrt{2}/2 & 15/5 & 1/4 \end{bmatrix}.$$

You can add and subtract matrices of the same shape by adding or subtracting their corresponding elements, but (for example) you can't add a 3 by 4 matrix and a 4 by 3 matrix or subtract one of these from the other. If we take the matrix $A$ that we have already defined and introduce the matrix $B$, so that we have

$$A = \begin{bmatrix} 0 & -4 & 1/2 & 9 \\ \pi & 14/5 & -0.15 & 2 \\ 7 & \sqrt{3} & 0 & -3 \end{bmatrix} \quad \text{and} \quad B = \begin{bmatrix} 23 & 5 & 21/2 & 4 \\ 3/4 & 6/5 & 0.65 & 8 \\ 29 & \sqrt{2} & 8 & 3 \end{bmatrix},$$

then

$$A + B = \begin{bmatrix} 0 + (-3) & -4 + 5 & 1/2 + (-1/2) & 9 + 4 \\ \pi + 3/4 & 14/5 + 6/5 & -0.15 + 0.65 & 2 + 8 \\ 7 + (-9) & \sqrt{3} + \sqrt{2} & & 0 + 8 & -3 + 3 \end{bmatrix}$$

$$= \begin{bmatrix} -3 & 1 & 0 & 13 \\ \pi + 3/4 & 4 & 0.5 & 10 \\ -2 & \sqrt{3} + \sqrt{2} & 8 & 0 \end{bmatrix}$$

and

$$A - B = \begin{bmatrix} 0 - (-3) & -4 - 5 & 1/2 - (-1/2) & 9 - 4 \\ \pi - 3/4 & 14/5 - 6/5 & -0.15 - 0.65 & 2 - 8 \\ 7 - (-9) & \sqrt{3} - \sqrt{2} & & 0 - 8 & -3 - 3 \end{bmatrix}$$

$$= \begin{bmatrix} 3 & -9 & 1 & 5 \\ \pi - 3/4 & 8/5 & -0.8 & -6 \\ 16 & \sqrt{3} - \sqrt{2} & -8 & -6 \end{bmatrix}.$$

The role of the number zero in matrix algebra is played by the **zero matrix** of the appropriate size—the matrix all of whose entries are zero.

We can also multiply a matrix by a number (or even a function) called a **scalar**, as in the case for vectors. We just multiply every element of the matrix by that scalar:

$$-5 \cdot \begin{bmatrix} 3 & -2 & 0 \\ -7 & 4 & 1/3 \\ 5 & -6 & \sqrt{2} \end{bmatrix} = \begin{bmatrix} -5\,(3) & -5\,(-2) & -5\,(0) \\ -5\,(-7) & -5\,(4) & -5\,(1/3) \\ -5\,(5) & -5\,(-6) & -5\left(\sqrt{2}\right) \end{bmatrix}$$

$$= \begin{bmatrix} -15 & 10 & 0 \\ 35 & -20 & -5/3 \\ -25 & 30 & -5\sqrt{2} \end{bmatrix}.$$

We've just multiplied a 3 by 3 matrix—one type of **square matrix**—by the scalar $(-5)$.

## B.3 LINEAR TRANSFORMATIONS AND MATRIX MULTIPLICATION

The really interesting thing about matrix arithmetic and algebra is how we *multiply* matrices. The natural thing to do—take two matrices with the same shape and multiply their corresponding elements—is not what is meant by matrix multiplication in the theory of linear algebra. Instead, there is a *row-by-column* process that looks strange at first but becomes more natural when you see its applications.

To motivate the multiplication of matrices, let's return to elementary algebra for a moment and look at a system of two equations in two unknowns:

$$-2x + 3y = 5$$
$$x - 4y = -2.$$

In connection with this system, we can think of a point $(x, y)$ in the plane transformed into another point as follows:

$$T(x, y) = (-2x + 3y, x - 4y).$$

For example,

$$T(1, 0) = (-2(1) + 3(0), 1 - 4(0)) = (-2, 1)$$
$$T(-4, 5) = (-2(-4) + 3(5), -4 - 4(5)) = (23, -24)$$

and

$$T(-2.8, -0.2) = (-2(-2.8) + 3(-0.2), -2.8 - 4(-0.2)) = (5, -2).$$

Note that this last calculation says that the ordered pair $(x, y) = (-2.8, -0.2)$ is a solution of our system of linear equations.

Geometrically, the point $(1, 0)$ has been moved to the location $(-2, 1)$, the point $(-4, 5)$ has been changed to $(23, -24)$, and the point $(-2.8, -0.2)$ has been transformed into $(5, -2)$. If we think of a point $(x, y)$ as defining a vector, then the transformation stretches (or shrinks) the vector and rotates it through some angle $\theta$ until it becomes another vector. Figure B.4 shows this interpretation of the effect of $T$ on the vector $(1, 0)$.

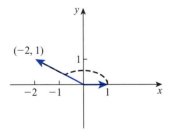

**FIGURE B.4**
*The effect of* $T(x, y) = (-2x + 3y, x - 4y)$ *on the vector* $(1, 0)$

More abstractly, we should be able to see that $T$ is a **linear transformation** of points $(x, y)$ in the plane to other points $(\hat{x}, \hat{y})$ in the plane: If $\mathbf{u} = (x_1, y_1)$ and $\mathbf{v} = (x_2, y_2)$, then $T(c_1\mathbf{u}+c_2\mathbf{v}) = T(c_1\mathbf{u}) + T(c_2\mathbf{v}) = c_1 T(\mathbf{u}) + c_2 T(\mathbf{v})$ for any constants $c_1$ and $c_2$.

Matrix notation was invented by the English mathematician Arthur Cayley precisely to describe linear transformations. If $T(x, y) = (\hat{x}, \hat{y})$, where

$$ax + by = \hat{x}$$
$$cx + dy = \hat{y},$$

then we can pick out the coefficients $a$, $b$, $c$, and $d$ and write them in a square array $A = \begin{bmatrix} a & b \\ c & d \end{bmatrix}$ called a **matrix**. If we know what variables $x$, $y$, $\hat{x}$, and $\hat{y}$ we're using, then knowing this **matrix of coefficients** enables us to understand what $T$ is doing to points in the plane. We can focus on these variables by introducing the vectors $\mathbf{X} = \begin{bmatrix} x \\ y \end{bmatrix}$ and $\hat{\mathbf{X}} = \begin{bmatrix} \hat{x} \\ \hat{y} \end{bmatrix}$. Now we can write our system of equations compactly as

$$\begin{bmatrix} a & b \\ c & d \end{bmatrix} \begin{bmatrix} x \\ y \end{bmatrix} = \begin{bmatrix} \hat{x} \\ \hat{y} \end{bmatrix},$$

or $A\mathbf{X} = \hat{\mathbf{X}}$. To make sense, the "product" of $A$ and $\mathbf{X}$ must be the column matrix $\begin{bmatrix} ax + by \\ cx + dy \end{bmatrix}$, which leads to a *row-by-column multiplication*:

$$\begin{bmatrix} a & b \end{bmatrix} \begin{bmatrix} x \\ y \end{bmatrix} = ax + by \quad \text{and} \quad \begin{bmatrix} c & d \end{bmatrix} \begin{bmatrix} x \\ y \end{bmatrix} = cx + dy.$$

Furthermore, the multiplication of two matrices of the appropriate sizes can be interpreted as a *composition of transformations*—one transformation followed by another.

### ■ Example
Suppose that we have two linear transformations defined by

$$M(x, y) = (x + 2y, 3x + 4y) \quad \text{and} \quad P(x, y) = (-2x, x + 3y).$$

Then

$$(M \circ P)(x, y) = M(P(x, y)) = M(-2x, x + 3y)$$

$$= (-2x + 2(x + 3y), 3(-2x) + 4(x + 3y))$$

$$= (6y, -2x + 12y).$$

In particular, $(M \circ P)(1, 1) = (6, 10)$. ■

The matrices of coefficients for the transformations $M$ and $P$ look like $M = \begin{bmatrix} 1 & 2 \\ 3 & 4 \end{bmatrix}$ and $p = \begin{bmatrix} -2 & 0 \\ 1 & 3 \end{bmatrix}$, so the composition $M \circ P$ takes the form of a *product of $2 \times 2$ matrices*:

$$\begin{bmatrix} 1 & 2 \\ 3 & 4 \end{bmatrix} \begin{bmatrix} -2 & 0 \\ 1 & 3 \end{bmatrix} \begin{bmatrix} x \\ y \end{bmatrix} = \begin{bmatrix} 1 & 2 \\ 3 & 4 \end{bmatrix} \begin{bmatrix} -2x \\ x + 3y \end{bmatrix}$$

$$= \begin{bmatrix} -2x + 2(x + 3y) \\ 3(-2x) + 4(x + 3y) \end{bmatrix} = \begin{bmatrix} 6y \\ -2x + 12y \end{bmatrix}$$

and when $\begin{bmatrix} x \\ y \end{bmatrix} = \begin{bmatrix} 1 \\ 1 \end{bmatrix}$,

$$\begin{bmatrix} 1 & 2 \\ 3 & 4 \end{bmatrix} \begin{bmatrix} -2 & 0 \\ 1 & 3 \end{bmatrix} \begin{bmatrix} 1 \\ 1 \end{bmatrix} = \begin{bmatrix} 6 \\ 10 \end{bmatrix}.$$

You should check to see that $(M \circ P)(x, y) \neq (P \circ M)(x, y)$ or, equivalently, that

$$\begin{bmatrix} 1 & 2 \\ 3 & 4 \end{bmatrix} \begin{bmatrix} -2 & 0 \\ 1 & 3 \end{bmatrix} \neq \begin{bmatrix} -2 & 0 \\ 1 & 3 \end{bmatrix} \begin{bmatrix} 1 & 2 \\ 3 & 4 \end{bmatrix}.$$

Looking at the preceding example, we see that transforming the vector $(x, y)$ by $P$ and then by $M$ is equivalent to transforming the vector by the single transformation $T(x, y) = (6y, -2x+12y)$. In matrix terms, we can express the effect of the composition $M \circ P$ as $\begin{bmatrix} 0 & 6 \\ -2 & 12 \end{bmatrix} \begin{bmatrix} x \\ y \end{bmatrix} = \begin{bmatrix} 6y \\ -2x + 12y \end{bmatrix}$. Note what we get when we add the results of multiplying each element of the first *row* of the matrix associated with $M$, $\begin{bmatrix} 1 & 2 \end{bmatrix}$, by the corresponding element of the first *column* of the matrix associated with $P$, $\begin{bmatrix} -2 \\ 1 \end{bmatrix}$: $(1)(-2) + (2)(1) = 0$, which happens to be the first row, first column element of the matrix corresponding to $M \circ P$. Similarly, for example, combining the second row of the matrix associated with $M$, $(3 \quad 4)$, and the first column of the matrix associated with $P$, $\begin{bmatrix} -2 \\ 1 \end{bmatrix}$, we get the element in the second row, first column of $M \circ P$ : $(3)(-2) + (4)(1) = -2$. In this way we can describe the matrix for $M \circ P$ as the **product** of the matrix representing $M$ and the matrix representing $P$.

In general, if $A$ and $B$ are both $2 \times 2$ matrices, we find the element in row $i$ and column $j$ of the product matrix $C = AB$ by adding the products of each element of row $i$ of matrix $A$ and the corresponding element in column $j$ of matrix $B$. For example, here's what the matrix product corresponding to $M \circ P$ in the last example looks like in full:

$$\begin{bmatrix} 1 & 2 \\ 3 & 4 \end{bmatrix} \begin{bmatrix} -2 & 0 \\ 1 & 3 \end{bmatrix} = \begin{bmatrix} (1)(-2)+(2)(1) & (1)(0)+(2)(3) \\ (3)(-2)+(4)(1) & (3)(0)+(4)(3) \end{bmatrix} = \begin{bmatrix} 0 & 6 \\ -2 & 12 \end{bmatrix}.$$

You should be able to calculate the matrix product corresponding to $P \circ M$. You'll notice that the *order* of composition/multiplication counts: The matrix corresponding to $M \circ$ is not necessarily the matrix corresponding to $P \circ M$. In general, **matrix multiplication is not commutative**: *If A and B are two matrices that can be multiplied (see the next paragraph), then $AB \neq BA$ in general.*

This situation of one function or transformation followed by another is the motivation for matrix multiplication. The general multiplication of matrices remains the row-by-column procedure described for $2 \times 2$ matrices. In order for us to calculate the matrix product $C = AB$, the number of columns of $A$ must be the same as the number of rows of $B$. Let $C = AB$, where $A$ is $m \times r$ and $B$ is $r \times n$. Then the product is a matrix with $m$ rows and $n$ columns:

$$A \cdot B = C$$

$$(m \times r) \cdot (r \times n) = m \times n$$

Thus, if $A$ is a 3 by 5 matrix and $B$ is a 5 by 7 matrix, you can find the product $AB$, which will be a 3 by 7 matrix. However, the product $BA$ does *not* make sense because the number of columns of $B$ (7) does not equal the number of rows of $A$ (3).

If the sizes of $A$ and $B$ are compatible, as described previously, then $c_{ij}$, the element in row $i$ and column $j$ of the product matrix $C$, is just the sum of the products of each element of row $i$ of matrix $A$ and the corresponding element in column $j$ of matrix $B$. Letting $a_{ik}$ denote the entry in row $i$ and column $k$ of matrix $A$, and letting $b_{kj}$ denote the element in row $k$ and column $j$ of matrix $B$, we can write the last sentence more concisely as

$$c_{ij} = \sum_{k=1}^{r} a_{ik} b_{kj} = a_{i1} b_{1j} + a_{i2} b_{2j} + \cdots + a_{ir} b_{rj}. \tag{B.3.1}$$

Schematically, we can represent this matrix multiplication as follows:

$$\begin{bmatrix} c_{11} & c_{12} & \cdots & c_{1j} & \cdots & c_{1n} \\ c_{21} & c_{22} & \cdots & c_{2j} & \cdots & c_{2n} \\ \vdots & \vdots & \vdots & \vdots & \vdots & \vdots \\ c_{i1} & c_{i2} & \cdots & \boxed{c_{ij}} & \cdots & c_{in} \\ \vdots & \vdots & \vdots & \vdots & \vdots & \vdots \\ c_{m1} & c_{m2} & \cdots & c_{mj} & \cdots & c_{mn} \end{bmatrix} = \begin{bmatrix} a_{11} & a_{12} & \cdots & a_{1r} \\ a_{21} & a_{22} & \cdots & a_{2r} \\ \vdots & \vdots & \vdots & \vdots \\ \boxed{a_{i1} \quad a_{i2} \quad \cdots \quad a_{ir}} \\ \vdots & \vdots & \vdots & \vdots \\ a_{m1} & a_{m2} & \cdots & a_{mr} \end{bmatrix} \begin{bmatrix} b_{11} & b_{12} & \cdots & \boxed{b_{1j}} & \cdots & b_{1n} \\ b_{21} & b_{22} & \cdots & \boxed{b_{2j}} & \cdots & b_{2n} \\ \vdots & \vdots & \vdots & \vdots & \vdots & \vdots \\ b_{r1} & b_{r2} & \cdots & \boxed{b_{rj}} & \cdots & b_{rn} \end{bmatrix}.$$

Here are some more examples of matrix multiplication.

## ■ Example

$$\begin{bmatrix} 2 & -3 & 0 \\ 4 & 0 & 1 \end{bmatrix} \cdot \begin{bmatrix} 1 & 2 \\ 3 & 4 \\ 5 & 6 \end{bmatrix} = \begin{bmatrix} 2(1) - 3(3) + 0(5) & 2(2) - 3(4) + 0(6) \\ 4(1) + 0(3) + 1(5) & 4(2) + 0(4) + 1(6) \end{bmatrix}$$

$$= \begin{bmatrix} -7 & -8 \\ 9 & 14 \end{bmatrix},$$

$$\begin{bmatrix} \pi & -2 & 6 \\ 0 & 4 & 1 \\ -3 & 5 & 7 \end{bmatrix} \cdot \begin{bmatrix} 2 & -3 & 0 \\ 9 & 2 & -6 \\ 2 & 1 & 4 \end{bmatrix}$$

$$= \begin{bmatrix} \pi(2) - 2(9) + 6(2) & \pi(-3) - 2(2) + 6(1) & \pi(0) - 2(-6) + 6(4) \\ 0(2) + 4(9) + 1(2) & 0(-3) + 4(2) + 1(1) & 0(0) + 4(-6) + 1(4) \\ -3(2) + 5(9) + 7(2) & -3(-3) + 5(2) + 7(1) & -3(0) + 5(-6) + 7(4) \end{bmatrix}$$

$$= \begin{bmatrix} 2\pi - 6 & -3\pi + 2 & 36 \\ 38 & 9 & -20 \\ 53 & 26 & -2 \end{bmatrix}.$$

■

A particularly important and useful $2 \times 2$ matrix is the **identity matrix** $I = \begin{bmatrix} 1 & 0 \\ 0 & 1 \end{bmatrix}$. You should check to see that this matrix plays the same role in matrix algebra that the number 1 plays in arithmetic—that is, $I \cdot A = A \cdot I$ for any $2 \times 2$ matrix $A$. If matrix $A$ is $2 \times n$, then $I \cdot A = A$, but $A \cdot I$ is not defined unless $n = 2$. Similarly, if $A$ is an $n \times 2$ matrix, then $A \cdot I = A$, but $I \cdot A$ is not defined unless $n = 2$. In general, for any positive integer $n$, the $n \times n$ matrix with ones on the main diagonal (upper left corner to lower right corner) and zeros elsewhere serves as the identity matrix $I$ for $n \times n$ matrix multiplication.

Given an $n \times n$ matrix $A$, the $n \times n$ matrix $B$ is called the (*multiplicative*) *inverse* of $A$ if $AB = I = BA$. If an inverse of $A$ exists, then it is unique and is denoted by $A^{-1}$.

With the definitions we have seen, matrix addition and multiplication satisfy all the familiar basic rules of algebra—except for commutativity. For example, we have the *associative law for multiplication*: If $A$ is an $m \times r$ matrix, $B$ is an $r \times s$ matrix, and $C$ is an $s \times n$ matrix, then $A(BC) = (AB)C$, an $m \times n$ matrix. We also have the *distributive law*: If $A$ is an $m \times r$ matrix and $B$ and $C$ are $r \times n$ matrices, then $A(B + C) = AB + AC$ (which is an $m \times n$ matrix).

Let's prove the distributive law in the situation in which $A$ is an $m \times r$ matrix and $B$ and $C$ are $r \times 1$ matrices (vectors). We expect the product $A(B + C)$ to be a vector having $m$ rows. Now suppose that $a_{ik}$ denotes the element in row $i$, column $k$ of $A$, that $b_k$ and $c_k$ are the elements in row $k(k = 1, 2, \ldots, r)$, and that $p_i$ is the element in row $i$ of the product $A(B + C)$. Then by Equation (B.3.1), we have (for $i = 1, 2, \ldots, m$)

$$p_i = \sum_{k=1}^{r} a_{ik} \left( b_k + c_k \right) = \sum_{k=1}^{r} \left( a_{ik} b_k + a_{ik} c_k \right) = \sum_{k=1}^{r} a_{ik} b_k + \sum_{k=1}^{r} a_{ik} c_k$$

$$= \text{(the entry in row } i \text{ of } AB) + \text{(the entry in row } i \text{ of } AC),$$

so we have shown that $A(B + C) = AB + AC$.

## B.4 EIGENVALUES AND EIGENVECTORS

As we saw in the previous section, if $A$ is an $n \times n$ matrix and $\mathbf{X}$ is a nonzero $n \times 1$ vector $\begin{bmatrix} x_1 \\ x_2 \\ \vdots \\ x_n \end{bmatrix}$,

we can consider the multiplication of $\mathbf{X}$ by matrix $A$ in the form $A\mathbf{X}$ as somehow transforming or changing the vector $\mathbf{X}$. If there is a scalar $\lambda$ such that $A\mathbf{X} = \lambda\mathbf{X}$, then $\lambda$ is called an **eigenvalue** of $A$, and the vector $\mathbf{X}$ is called an **eigenvector** corresponding to $\lambda$. Geometrically, we're saying that *an eigenvector is a nonzero vector that gets changed into a constant multiple of itself.* For example, identifying a vector $\mathbf{X} = \begin{bmatrix} x \\ y \end{bmatrix}$ with the point $(x, y)$ in the familiar Cartesian coordinate system, we can see that an eigenvector is a point (not the origin) such that it and its transformed self lie on the same straight line through the origin. The direction of an eigenvector is either unchanged (if $\lambda > 0$) or reversed (if $\lambda < 0$) when the vector is multiplied by $A$. The matrix equation $A\mathbf{X} = \lambda\mathbf{X}$ is like the functional equation $f(x) = \lambda x$, which represents a straight line through the origin with slope $\lambda$.

If we start with the assumption that $A\mathbf{X} = \lambda\mathbf{X}$, then $A\mathbf{X} - \lambda\mathbf{X} = \mathbf{0}$ (the zero vector), and the distributive property of matrix multiplication allows us to write $(A - \lambda I)\mathbf{X} = \mathbf{0}$. (We must write $A - \lambda I$ instead of $A - \lambda$ because it doesn't make sense to subtract a number from a matrix.) If we can find an inverse for $A - \lambda I$—that is, an $n \times n$ matrix $B$ such that $(A - \lambda I)B = I = B(A - \lambda I)$— then we can divide the factor $A - \lambda I$ out of the matrix equation $(A - \lambda I)\mathbf{X} = \mathbf{0}$ to get $\mathbf{X} = \mathbf{0}$, the $n \times 1$ vector all of whose elements are 0. Therefore, remembering that an eigenvector was defined as a *nonzero* vector, we see that the only interesting situation occurs when the matrix $A - \lambda I$ does *not* have an inverse. (*Do you follow the logic?*)

The equation $(A - \lambda I)\mathbf{X} = \mathbf{0}$ represents a homogeneous system of $n$ algebraic linear equations in $n$ unknowns, and the theory of linear algebra indicates that there is a number $\Delta$, depending on the matrix $A - \lambda I$, with the following important property: If $\Delta \neq 0$, then the system $(A - \lambda I)\mathbf{X} = \mathbf{0}$ has only the zero solution $x_1 = x_2 = \ldots = x_n = 0$. However, if $\Delta = 0$, then there is a solution $\mathbf{X} = \begin{bmatrix} x_1 \\ x_2 \\ \vdots \\ x_n \end{bmatrix}$ with at least one of the $x_i$ different from zero. This number $\Delta$ is the **determinant** of the matrix $A - \lambda I$, denoted by $\det(A - \lambda I)$. Therefore $(A - \lambda I)\mathbf{X} = \mathbf{0}$ has a nonzero solution $\mathbf{X}$ only if $\det(A - \lambda I) = 0$, For any $n \times n$ linear system (homogeneous or not), the nature of the solutions depends on (that is, is *determined* by) whether the determinant is zero. The determinant is often calculated by means of successive operations on the rows and/or columns of the matrix. Rather than spend time learning tedious algorithms for finding determinants, you should learn how to get these numbers from your CAS. Even a graphing calculator will evaluate a determinant if the matrix is not too large. From a more abstract point of view, a determinant is just a special kind of function from a set of square matrices to the real numbers.

For now, let's see how a determinant arises in solving a simple system of algebraic equations.

### ■ Example

Suppose we want to solve the following system of two equations in two unknowns:

$$2x - 3y = 2$$

$$x + 4y = -5.$$

We can use the method of *elimination* to solve this system. For example, we can subtract twice the second equation from the first equation to eliminate the variable $x$ and get $-11y = 12$, or $y = -\frac{12}{11}$. Then we can substitute this value of $y$ in the second equation and solve for $x$. We get $x = -\frac{7}{11}$. Note that when we solve this particular system by elimination, the denominator of each component of the solution is 11.    ■

Now write the system in matrix form:

$$\begin{bmatrix} 2 & -3 \\ 1 & 4 \end{bmatrix} \begin{bmatrix} x \\ y \end{bmatrix} = \begin{bmatrix} 2 \\ -5 \end{bmatrix}.$$

What do we get if we take the matrix of coefficients, multiply the *main-diagonal* (upper left, lower right) elements 2 and 4, and then subtract the product of the other diagonal elements $-3$ and 1? We get $(2)(4) - (-3)(1) = 11$. *Surprise!* The number calculated this way is the determinant of the coefficient matrix. In solving any system of linear equations in two unknowns, you always wind up dividing by the determinant—if it's not zero. *Cramer's Rule*, which you may have seen in a college algebra course, is a general $n \times n$ linear system solution formula that uses determinants.

For a larger system, a CAS or graphing calculator provides important information about a system easily. Let's use technology in the next example to calculate the determinant, eigenvalues, and eigenvectors for a three-dimensional system.

## ■ Example

Suppose we have a system with the matrix of coefficients

$$A = \begin{bmatrix} 2 & 2 & -6 \\ 2 & -1 & -3 \\ -2 & -1 & 1 \end{bmatrix}.$$

A CAS (*Maple* in this case) tells us that $\det(A) = 24$ and that the eigenvalues are $\lambda_1 = 6, \lambda_2 = -2 = \lambda_3$. The corresponding (linearly independent) eigenvectors are

$$\begin{bmatrix} -2 \\ -1 \\ 1 \end{bmatrix}, \begin{bmatrix} 1 \\ -2 \\ 0 \end{bmatrix}, \quad \text{and} \quad \begin{bmatrix} 0 \\ 3 \\ 1 \end{bmatrix}.$$

If you try this example using your own CAS, the eigenvectors may not look like those here, but each should be a constant multiple of one of those given previously. ■

# Complex Numbers

## C.1 COMPLEX NUMBERS: THE ALGEBRAIC VIEW

Historically, the need for complex numbers arose when people tried to solve equations such as $x^2 + 1 = 0$ and realized that there was no real (ordinary) number that satisfied this equation. The basic element in the expansion of the number system is the **imaginary unit**, $i = \sqrt{-1}$. There is an interesting pattern to the powers of $i$: $i^1 = i, i^2 = -1, i^3 = -i, i^4 = 1, i^5 = i, i^6 = -1, i^7 = -i, i^8 = 1, \ldots$. You can use this repetition in groups of four, for example, to calculate a high power of $i$ : $i^{338} = (i^2)^{169} = (-1)^{169} = -1$. A **complex number** is any expression of the form $x + yi$, where $x$ and $y$ are real numbers. If you have a complex number $z = x + yi$, then $x$ is called the **real part**—denoted $\text{Re}(z)$—and $y$ is called the **imaginary part**—denoted $\text{Im}(z)$—of the complex number. (Note that despite its name, $y$ is a real number.) In particular, any real number $x$ is a member of the family of complex numbers because it can be written as $x + 0 \cdot i$. Any complex number of the form $yi(= 0 + yi)$ is called a **pure imaginary number**.

Complex numbers can be added and subtracted in a reasonable way by combining real parts and imaginary parts as follows:

$$(a + bi) + (c + di) = (a + c) + (b + d)i$$

and

$$(a + bi) - (c + di) = (a - c) + (b - d)i.$$

You can also multiply complex numbers as you would multiply any binomials in algebra, remembering to replace $i^2$ whenever it occurs by $-1$:

$$(a + bi) \cdot (c + di) = ac + adi + bci + bdi^2 = (ac - bd) + (ad + bc)i.$$

Division of complex numbers is a bit trickier. If $z = x + yi$ is a complex number, then its **complex conjugate**, $\bar{z}$, is defined as follows: $\bar{z} = x - yi$. (You just reverse the sign of the imaginary part.) The complex conjugate is important in division because $z \cdot \bar{z} = x^2 + y^2$, a real number. (*Check this out.*) In division of complex numbers, the conjugate plays much the same role as the conjugate you learned to use in algebra to simplify fractions. For example, in algebra, if you were asked to simplify the fraction $\frac{3}{\sqrt{5}}$, you would "rationalize the denominator" as follows:

$$\frac{3}{\sqrt{5}} = \frac{3}{\sqrt{5}} \cdot \frac{\sqrt{5}}{\sqrt{5}} = \frac{3\sqrt{5}}{5}.$$

Another example from algebra makes the similarity between conjugates more obvious:

$$\frac{2 + \sqrt{3}}{3 - \sqrt{2}} = \frac{2 + \sqrt{3}}{3 - \sqrt{2}} \cdot \frac{3 + \sqrt{2}}{3 + \sqrt{2}} = \frac{6 + 2\sqrt{2} + 3\sqrt{3} + \sqrt{6}}{9 - 2}$$

$$= \frac{6 + 2\sqrt{2} + 3\sqrt{3} + \sqrt{6}}{7}.$$

In the preceding example, $3 + \sqrt{2}$ is the conjugate of $3 - \sqrt{2}$; when you multiply these conjugates, the radical sign disappears, leaving you with the integer 7. Now if we have to divide two complex numbers, we use the complex conjugate to get the answer, the quotient, to look like a complex number. For example,

$$\frac{2 + 3i}{3 + 5i} = \frac{2 + 3i}{3 + 5i} \cdot \frac{3 - 5i}{3 - 5i} = \frac{21 - i}{9 + 25} = \frac{21}{34} - \frac{1}{34}i.$$

In general, if $z = a + bi$ and $w = c + di$, then

$$\frac{z}{w} = \frac{a + bi}{c + di} = \frac{a + bi}{c + di} \cdot \frac{c - di}{c - di} = \frac{ac + bd}{c^2 + d^2} + \frac{bc - ad}{c^2 + d^2}i.$$

If $z$ and $w$ are complex numbers, you should be able to see that $\bar{\bar{z}} = z$, $\overline{(z + w)} = \bar{z} + \bar{w}$, $\overline{z \cdot w} = \bar{z} \cdot \bar{w}$, and $\overline{\left(\frac{z}{w}\right)} = \frac{\bar{z}}{\bar{w}}$ for $w \neq 0$. Also, $\text{Re}(z) = \frac{z + \bar{z}}{2}$ and $\text{Im}(z) = \frac{z - \bar{z}}{2i}$.

The important algebraic rules of commutativity, associativity, and distributivity work for complex numbers. Furthermore, all the properties in this section extend to vectors and matrices (Appendix B) with complex-number entries. For example, if $\mathbf{V} = \begin{bmatrix} c_1 \\ c_2 \\ \vdots \\ c_n \end{bmatrix}$ is a vector with complex components, then $\overline{\mathbf{V}} = \overline{\begin{bmatrix} c_1 \\ c_2 \\ \vdots \\ c_n \end{bmatrix}} = \begin{bmatrix} \bar{c}_1 \\ \bar{c}_2 \\ \vdots \\ \bar{c}_n \end{bmatrix}$. If $A = [a_{ij}]$ represents a matrix with entry $a_{ij}$ in row $i$ and column $j$, then $\overline{A} = \overline{[a_{ij}]} = [\bar{a}_{ij}]$.

## C.2  COMPLEX NUMBERS: THE GEOMETRIC VIEW

The geometric interpretation of complex numbers occurred at roughly the same time to three people: the Norwegian surveyor and map maker Caspar Wessel (1745–1818), the French-Swiss mathematician Jean Robert Argand (1768–1822), and Karl Friedrich Gauss (1777–1855), the German mathematician-astronomer-physicist.

The idea here is to represent a complex number using the familiar Cartesian coordinate system, making the horizontal axis the **real axis** and the vertical axis the **imaginary axis**. Such a system is called the **complex plane**. For example, Figure C.1 shows how the complex number $3 + 2i$ would be represented as a point in this way.

If we join this point to the origin with a straight line, we get a vector. (See Section B.1.) The sum of $z = a + bi$ and $w = c + di$ corresponds to the point (or vector) $(a + c, b + d)$. This implies that the addition/subtraction of complex numbers corresponds to the Parallelogram Law of vector algebra (Figure C.2).

The **modulus**, or **absolute value**, of the complex number $z = x + yi$, denoted by $|z|$, is the nonnegative real number defined by the equation $|z| = \sqrt{x^2 + y^2}$. The number $|z|$ represents

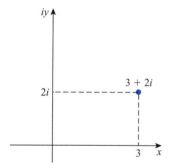

**FIGURE C.1**

*Representation of a complex number*

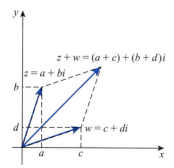

**FIGURE C.2**

*The Parallelogram Law*

the distance between the origin and the point $(x, y)$ in the complex plane, the length of the vector representing the complex number $z = x + yi$. Note that $|z|^2 = z \cdot \bar{z}$.

## C.3  THE QUADRATIC FORMULA

Given the quadratic equation $ax^2 + bx + c = 0$, where $a$, $b$, and $c$ are real numbers with $a \neq 0$, the solutions are given by the **quadratic formula**:

$$x = \frac{-b \pm \sqrt{b^2 - 4ac}}{2a}.$$

The expression inside the radical sign, $b^2 - 4ac$, is called the **discriminant** and enables you to discriminate among the possibilities for solutions. If $b^2 - 4ac > 0$, the quadratic formula yields two real solutions. If $b^2 - 4ac = 0$, you get a single repeated solution—a solution of *multiplicity two*. Finally, if $b^2 - 4ac < 0$, the quadratic formula produces two complex numbers as solutions, a **complex conjugate pair**. To see this last situation, suppose that $b^2 - 4ac = -q$, where $q$ is a positive real number. Then the solution formula looks like

$$x = \frac{-b \pm \sqrt{-q}}{2a} = \frac{-b \pm \sqrt{q(-1)}}{2a} = \frac{-b \pm \sqrt{q}i}{2a},$$

so the two solutions are $x_1 = -\frac{b}{2a} + \frac{\sqrt{q}}{2a}i$ and $x_2 = -\frac{b}{2a} - \frac{\sqrt{q}}{2a}i$, which are complex conjugates of each other.

## C.4  EULER'S FORMULA

Around 1740, while studying differential equations of the form $y'' + y = 0$, Euler discovered his famous formula for complex exponentials:

$$e^{iy} = \cos y + i \sin y.$$

If $z = x + iy$, then we have

$$e^z = e^{x+iy} = e^x e^{iy} = e^x(\cos y + i \sin y).$$

Without fully understanding the way infinite series work, Euler just substituted the complex number $iy$ in the series for $e^x$ (see Section A.3) and then separated real and imaginary parts:

$$e^{iy} = 1 + iy + \frac{(iy)^2}{2!} + \frac{(iy)^3}{3!} + \frac{(iy)^4}{4!} + \frac{(iy)^5}{5!} + \cdots$$

$$= 1 + iy - \frac{y^2}{2!} - i\frac{y^3}{3!} + \frac{y^4}{4!} + i\frac{y^5}{5!} - \cdots$$

$$= \underbrace{\left(1 - \frac{y^2}{2!} + \frac{y^4}{4!} - \cdots\right)}_{\cos y} + i\underbrace{\left(y - \frac{y^3}{3!} + \frac{y^5}{5!} - \cdots\right)}_{\sin y} = \cos y + i \sin y.$$

# Series Solutions of Differential Equations

Appendix D supplements the treatment of linear equations in Chapters 5 and 6.

## D.1 POWER SERIES SOLUTIONS OF FIRST-ORDER EQUATIONS

In Chapters 5 and 6 we discussed solutions for second- and higher-order linear equations with constant coefficients. The methods we discuss in this appendix can be applied to equations— not necessarily linear—with *variable* coefficients, equations that in general do not yield closed-form solutions. Among these are equations important in many areas of applied mathematics.

As an illustration of the key idea, we'll solve a simple first-order equation.

### ■ Example

Consider the equation $y' = 1 - xy$. We make the fundamental assumption that a solution $y$ can be expanded in a power series (Taylor series, Maclaurin series)

$$y(x) = a_0 + a_1 x + a_2 x^2 + a_3 x^3 + \cdots + a_n x^n + \cdots$$

that converges in some interval. (See Section A.3 for the basics.) We have chosen an interval around the origin.

Then, because a convergent power series can be differentiated term by term within its interval of convergence (see Section A.3),

$$y'(x) = a_1 + 2a_2 x + 3a_3 x^2 + \cdots + na_n x^{n-1} + \cdots.$$

Substituting these last two series in the differential equation, we have

$$a_1 + 2a_2 x + 3a_3 x^2 + \cdots + na_n x^{n-1} + \cdots$$
$$= 1 - x\left\{a_0 + a_1 x + a_2 x^2 + a_3 x^3 + \cdots + a_n x^n + \cdots\right\}$$
$$= 1 - a_0 x - a_1 x^2 - a_2 x^3 - a_3 x^4 - \cdots - a_n x^{n+1} - \cdots.$$

Because these power series are equal, coefficients of equal powers of $x$ on both sides must be equal. (This is really the *method of undetermined coefficients* that we first saw in Section 4.3 of the text.) Therefore, we have

$$a_1 = 1, \, 2a_2 = -a_0, \, 3a_3 = -a_1, \, 4a_4 = -a_2, \, 5a_5 = -a_3, \ldots, \, na_n = -a_{n-2}, \ldots,$$

so

$$a_1 = 1, \; a_2 = -\frac{a_0}{2}, \; a_3 = -\frac{a_1}{3} = -\frac{1}{3}, \; a_4 = -\frac{a_2}{4} = -\frac{-\frac{a_0}{2}}{4}$$

$$= \frac{a_0}{2 \cdot 4}, \; a_5 = -\frac{a_3}{5} = \frac{1}{3 \cdot 5}, \ldots, \; a_n = -\frac{a_{n-2}}{n}, \ldots.$$

These formulas, in which we define later coefficients by relating them to earlier coefficients, are called **recurrence** (or **recursion**) **relations**. If we look carefully, we see that for odd indices (subscripts), the pattern is

$$a_1 = 1, \; a_3 = -\frac{1}{3}, \; a_5 = \frac{1}{3 \cdot 5}, \; a_7 = -\frac{1}{3 \cdot 5 \cdot 7}, \ldots.$$

Similarly, for even indices we find the pattern

$$a_0 = \text{arbitrary}, \; a_2 = -\frac{a_0}{2}, \; a_4 = \frac{a_0}{2 \cdot 4}, \; a_6 = -\frac{a_0}{2 \cdot 4 \cdot 6}, \ldots.$$

In general, the pattern is

$$a_{2k} = \frac{(-1)^k a_0}{2 \cdot 4 \cdot 6 \cdots (2k)} \quad \text{for} \quad k = 1, 2, 3, \ldots;$$

$$a_{2k+1} = \frac{(-1)^k}{1 \cdot 3 \cdot 5 \cdot 7 \cdots (2k+1)} \quad \text{for} \quad k = 0, 1, 2, \ldots.$$

Therefore, we can write the power series form of the solution as

$$y(x) = a_0 + x - \frac{a_0}{2} x^2 - \frac{1}{1 \cdot 3} x^3 + \left(\frac{a_0}{2 \cdot 4}\right) x^4 + \frac{1}{1 \cdot 3 \cdot 5} x^5 + \cdots$$

$$= \left(x - \frac{x^3}{1 \cdot 3} + \frac{x^5}{1 \cdot 3 \cdot 5} - \cdots\right) + a_0 \left(1 - \frac{x^2}{2} + \frac{x^4}{2 \cdot 4} - \cdots\right),$$

where $a_0 = y(0)$ is the arbitrary constant that we expect in the general solution of a first-order equation.

To approximate $y(x)$ for a value of $x$ close to zero, we just substitute the value in the series, taking as many terms of this series as are needed to guarantee the accuracy we wish.

If we solve the linear equation in the preceding example using the technique of integrating factors (see Section 2.2), we get the answer

$$y = e^{x^2/2} \int e^{-x^2/2} dx + C e^{x^2/2},$$

which can't be expressed in a more elementary way. If we integrate the power series representation of $e^{-x^2/2}$ term by term, multiply by the series form of $e^{x^2/2}$, and then add the series for $C e^{x^2/2}$, we get the same series solution we found (after collecting terms).

In using this power series method, sometimes we can recognize the series in our solution as a representation of an elementary function. Try using the method on the equation $y' = ay$, where $a$ is a constant, for example. You should recognize the series solution as the Taylor series representation of $C e^{ax}$ about the origin.

All computer algebra systems have the ability to work with series expansions, usually truncating the series after a fixed number of terms that the user can control. However, not all systems can give you a power series solution of an ODE directly. For example, *Maple* has a very useful power series package, *powseries*, and the *dsolve* command (in the package *DEtools*) has a *series* option; but *Mathematica* and MATLAB require the user to do much more work in finding a series solution.

## D.2 SERIES SOLUTIONS OF SECOND-ORDER LINEAR EQUATIONS: ORDINARY POINTS

In this section we'll examine second-order linear equations of the form

$$a(t)y'' + b(t)y' + c(t)y = 0, \tag{D.2.1}$$

where $a(t), b(t)$, and $c(t)$ are polynomial functions. We divide through by $a(t)$ and write Equation (D.2.1) in the *standard form*

$$y'' + P(t)y' + Q(t)y = 0, \tag{D.2.2}$$

where $P(t) = \frac{b(t)}{a(t)}$ and $Q(t) = \frac{c(t)}{a(t)}$.

A point $t_0$ is called an **ordinary point** of Equation (D.2.2) if both $P$ and $Q$ can be expanded in power series centered at $t_0$ that converge for every $t$ in an open interval containing $t_0$. Functions that have such power series representations are called **analytic** at the point $t_0$. If $t_0$ is not an ordinary point, it is called a **singular point** of the equation.

### ■ Example

The point $t = 0$ is an ordinary point of the equation $(t + 2)y'' + t^2 y' + y = 0$ because each of the functions $P(t) = \frac{t^2}{t+2}$ and $Q(t) = \frac{1}{t+2}$ has its own power series expansion that converges near $t = 0$:

$$Q(t) = \frac{1}{2} - \frac{t}{4} + \frac{t^2}{8} - \frac{t^3}{16} + \cdots \quad \text{and} \quad P(t) = \frac{t^2}{2} - \frac{t^3}{4} + \frac{t^4}{8} - \frac{t^5}{16} + \cdots.$$

(See the *geometric series* in Section A.3.) However, $t = -2$ is a singular point because the denominators of $P(t)$ and $Q(t)$ are zero at $t = -2$. ∎

Let's apply the undetermined coefficient method of the last section to a famous second-order linear equation near an ordinary point. The equation is named for the English mathematician Sir George Bidell Airy (1801–1892), who did pioneering work in elasticity and in partial differential equations.

### ■ Example

**Airy's equation**, $y'' + xy = 0$, which occurs in the study of optics and quantum physics, cannot be solved in terms of elementary functions. We can think of the equation as describing a spring-mass system in which the stiffness of the spring is increasing with time. (Maybe the room containing the system is getting colder.)

Noting that $x = 0$ is an ordinary point of this equation, we assume that we can write a solution as

$$y(x) = a_0 + a_1 x + a_2 x^2 + a_3 x^3 + a_4 x^4 + \cdots + a_n x^n + \cdots.$$

Then

$$y'(x) = a_1 + 2a_2 x + 3a_3 x^2 + 4a_4 x^3 + \cdots + na_n x^{n-1} + \cdots$$

and

$$y''(x) = 2a_2 + 6a_3 x + 12a_4 x^2 + \cdots + n(n-1)a_n x^{n-2} + \cdots.$$

Substituting in the differential equation, we get

$$\left(2a_2 + 6a_3 x + 12a_4 x^2 + \cdots + n(n-1)a_n x^{n-2} + \cdots\right)$$
$$+ x\left(a_0 + a_1 x + a_2 x^2 + a_3 x^3 + a_4 x^4 + \cdots + a_n x^n + \cdots\right) = 0.$$

Collecting terms, we can write this last equation as

$$2a_2 + (6a_3 + a_0)x + (12a_4 + a_1)x^2 + \cdots$$
$$+ (n(n-1)a_n + a_{n-3})x^{n-2} + \cdots = 0.$$

Equating coefficients of equal powers of $x$, we see that the preceding equation implies that

$$2a_2 = 0, \text{ or } a_2 = 0; \quad 6a_3 + a_0, \text{ or } a_3 = -\frac{a_0}{2 \cdot 3};$$

$$12a_4 + a_1 = 0, \text{ or } a_4 = -\frac{a_1}{3 \cdot 4}; \quad 20a_5 + a_2 = 0, \text{ or } a_5 = -\frac{a_2}{4 \cdot 5},$$

and so forth, so we can see the recurrence relation as $a_n = -\frac{a_{n-3}}{(n-1)\cdot n}$ for $n = 3, 4, 5, \ldots$ Note that $a_0$ and $a_1$ are arbitrary and that the coefficients are connected by jumps of three in the subscripts. In particular we have $0 = a_2 = a_5 = a_8 = \cdots = a_{2+3k} = \cdots$. Also, we can see the pattern when the subscript is a multiple of 3:

$$a_3 = -\frac{a_0}{2 \cdot 3} \qquad a_6 = -\frac{a_3}{5 \cdot 6} = \frac{a_0}{2 \cdot 3 \cdot 5 \cdot 6} \qquad a_9 = -\frac{a_6}{8 \cdot 9} = -\frac{a_0}{2 \cdot 3 \cdot 5 \cdot 6 \cdot 8 \cdot 9}$$

$$a_{12} = -\frac{a_9}{11 \cdot 12} = \frac{a_0}{2 \cdot 3 \cdot 5 \cdot 6 \cdot 8 \cdot 9 \cdot 11 \cdot 12}$$

and so forth, so the formula is

$$a_{3k} = \frac{(-1)^k a_0}{2 \cdot 3 \cdot 5 \cdot 6 \cdot 8 \cdot 9 \cdots (3k-1) \cdot 3k}.$$

Similarly, we can see that

$$a_4 = -\frac{a_1}{3 \cdot 4} \qquad a_7 = -\frac{a_4}{6 \cdot 7} = \frac{a_1}{3 \cdot 4 \cdot 6 \cdot 7}$$

$$a_{10} = -\frac{a_7}{9 \cdot 10} = -\frac{a_1}{3 \cdot 4 \cdot 6 \cdot 7 \cdot 9 \cdot 10}$$

$$a_{13} = -\frac{a_{10}}{12 \cdot 13} = \frac{a_1}{3 \cdot 4 \cdot 6 \cdot 7 \cdot 9 \cdot 10 \cdot 12 \cdot 13},$$

and so forth, so the recurrence relation is

$$a_{3k+1} = \frac{(-1)^k a_1}{3 \cdot 4 \cdot 6 \cdot 7 \cdot 9 \cdot 10 \cdots (3k) \cdot (3k+1)}.$$

Putting all the pieces together, we get

$$y(x) = a_0 \left[ 1 - \frac{x^3}{2 \cdot 3} + \frac{x^6}{2 \cdot 3 \cdot 5 \cdot 6} - \frac{x^9}{2 \cdot 3 \cdot 5 \cdot 6 \cdot 8 \cdot 9} + \cdots \right]$$

$$+ a_1 \left[ x - \frac{x^4}{3 \cdot 4} + \frac{x^7}{3 \cdot 4 \cdot 6 \cdot 7} - \frac{x^{10}}{3 \cdot 4 \cdot 6 \cdot 7 \cdot 9 \cdot 10} + \cdots \right]$$

$$= y(0) \left[ 1 - \frac{x^3}{2 \cdot 3} + \frac{x^6}{2 \cdot 3 \cdot 5 \cdot 6} - \frac{x^9}{2 \cdot 3 \cdot 5 \cdot 6 \cdot 8 \cdot 9} + \cdots \right]$$

$$+ y'(0)x \left[ 1 - \frac{x^3}{3 \cdot 4} + \frac{x^6}{3 \cdot 4 \cdot 6 \cdot 7} - \frac{x^9}{3 \cdot 4 \cdot 6 \cdot 7 \cdot 9 \cdot 10} + \cdots \right]$$

$$= y(0) \cdot Ai(x) + y'(0)x \cdot Bi(x),$$

where the two series (convergent for all values of $x$) define $Ai(x)$ and $Bi(x)$, the **Airy functions of the first and second kind**, respectively, up to constant multiplicative factors.

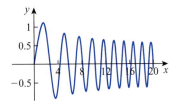

**FIGURE D.1**
*Solution of $y'' + xy = 0$; $y(0) = 0, y'(0) = 1$*

With the aid of technology, we can look at the graph of the solution of Airy's equation with initial conditions $y(0) = 0, y'(0) = 1$ (Figure D.1), which is just the graph of $xBi(x)$.

Both *Maple* and *Mathematica*, for example, have built-in capabilities to deal with Airy functions numerically and graphically—see the commands *AiryAi(x)* and *AiryBi(x)* [in *Maple*] or *AiryAi[x]* and *AiryBi[x]* [in *Mathematica*].    ■

If we want to find a solution near an ordinary point $t_0$ other than zero, we can use the substitution $u = t - t_0$. This substitution transforms the equation in $t$ to one in the variable $u$, which we can solve near the ordinary point $u = 0$. When we have solved the equation in $u$, we can just replace $u$ by $t - t_0$ to return to the original variable.

The method of undetermined coefficients also applies to nonhomogeneous equations and to equations whose coefficients are not polynomials, provided that the function on the right-hand side and the coefficient functions can be expanded in powers of $t$. When we are trying to solve a nonhomogeneous equation, equating coefficients becomes a little more difficult because some of the coefficients of the solution series $y(t) = \sum_{n=0}^{\infty} a_n t^n$ will include numerical values independent of the two arbitrary constants $a_0$ and $a_1$. This part of the general solution $y_{GNH}$ constitutes $y_{PNH}$. Check this out for yourself by using series to solve the equation $y'' - y = e^x$. (You should recognize your solution as $y = c_1 e^x + c_2 e^{-x} + \frac{1}{2} x e^x$.)

## D.3  REGULAR SINGULAR POINTS: THE METHOD OF FROBENIUS

Some singular points are such that special series methods have been developed to handle situations in which they occur. The point $t_0$ is a **regular singular point** of $y'' + P(t)y' + Q(t)y = 0$ if $t_0$ is a singular point and the functions $(t - t_0)P(t)$ and $(t - t_0)^2 Q(t)$ are both analytic at $t_0$. If $t_0$ is a singular point that is not regular, it is called an **irregular singular point**.

For example, $t = 1$ is a singular point of the equation $(t^2 - 1)^2 y'' + (t - 1)y' + y = 0$ because $P(t) = \frac{t-1}{(t^2-1)^2} = \frac{t-1}{(t+1)^2(t-1)^2}$ and $Q(t) = \frac{1}{(t+1)^2(t-1)^2}$ have zero denominators at $t = 1$, so neither $P(t)$ nor $Q(t)$ has a convergent power series expansion in a neighborhood of 1. But if we look at $(t - 1)P(t) = \frac{(t-1)^2}{(t+1)^2(t-1)^2} = \frac{1}{(t+1)^2}$ and $(t - 1)^2 Q(t) = \frac{(t-1)^2}{(t+1)^2(t-1)^2} = \frac{1}{(t+1)^2}$, we see that both $(t-1)P(t)$ and $(t-1)^2Q(t)$ are analytic at $t = 1$, so $t = 1$ is a regular singular point.

Near a regular singular point—say $t = 0$ for convenience—we write Equation (D.2.2) as

$$t^2y'' + tp(t)y' + q(t)y = 0, \qquad (D.3.1)$$

where $p(t) = tP(t)$ and $q(t) = t^2Q(t)$. Because $t = 0$ is a regular singular point, $p$ and $q$ are analytic at $t = 0$. The usual power series method will not work, and we use the **method of Frobenius**,[1] which produces at least one solution of the form

$$y(t) = t^r \sum_{n=0}^{\infty} a_n t^n = \sum_{n=0}^{\infty} a_n t^{n+r}, \qquad (D.3.2)$$

where we assume that $a_0 \neq 0$.

It is important to note that three of the most popular computer algebra systems (*Maple, Mathematica*, and *MATLAB*) cannot apply the method of Frobenius directly to get power series solutions near regular singular points. You must develop a solution in a step-by-step fashion, using the capabilities of your system to handle power series and recursion relations.

We'll illustrate the method of Frobenius using a famous equation in applied mathematics, one that first arose in an investigation of the motion of a hanging chain and has since appeared in such problems as the analysis of vibrations of a circular membrane and planetary motion.

### ■ Example

**Bessel's equation of order $p$** is $x^2y'' + xy' + (x^2 - p^2)y = 0$, which is of the form (D.3.1) and has $x = 0$ as a regular singular point.[2] We'll take the parameter $p$ to be an arbitrary nonnegative real number.

Substituting the type of series given in (D.3.2) for $y$, we find that

$$y' = \sum_{n=0}^{\infty} a_n(n+r)x^{n+r-1}$$

and

$$y'' = \sum_{n=0}^{\infty} a_n(n+r)(n+r-1)x^{n+r-2},$$

---

[1] The German mathematician Ferdinand Georg Frobenius (1849–1917) published his method in 1878. It was based on a technique that originated with Euler (*who else?*). Frobenius made many contributions to analysis and especially to algebra.
[2] Among other achievements, the German astronomer Friedrich Wilhelm Bessel (1784–1846) was the first to measure accurately the distance to a fixed star.

so that we have

$$x^2 y'' + xy' + (x^2 - p^2) y = \sum_{n=0}^{\infty} a_n(n+r)(n+r-1)x^{n+r} + \sum_{n=0}^{\infty} a_n(n+r)x^{n+r}$$

$$+ \sum_{n=0}^{\infty} a_n x^{n+r+2} - \sum_{n=0}^{\infty} a_n p^2 x^{n+r}$$

$$= \sum_{n=0}^{\infty} \left\{ a_n(n+r)(n+r-1) + a_n(n+r) - a_n p^2 \right\} x^{n+r}$$

$$+ \sum_{n=0}^{\infty} a_n x^{n+r+2}$$

$$= \sum_{n=0}^{\infty} \left\{ (n+r)^2 - p^2 \right\} a_n x^{n+r} + \sum_{n=0}^{\infty} a_n x^{n+r+2} = 0.$$

Transposing series and making the substitution (actually a shift of subscripts) $n + 2 = n$ on the right-hand side, we get

$$\sum_{n=0}^{\infty} \left\{ (n+r)^2 - p^2 \right\} a_n x^{n+r} = - \sum_{n=0}^{\infty} a_n x^{n+r+2} = - \sum_{n=2}^{\infty} a_{n-2} x^{n+r}.$$

Now we equate coefficients of equal powers. To start, we have

$$n = 0 : (r^2 - p^2) a_0 = 0$$

$$n = 1 : \left[ (1+r)^2 - p^2 \right] a_1 = 0.$$

Because we have assumed that $a_0 \neq 0$, we must have $r^2 - p^2 = 0$. This last equation is called the **indicial equation**[3] and implies that $r = \pm p$.

Let's assume that $r = p \geq 0$. Then when $n = 1$, the equation $\left[ (1+r)^2 - p^2 \right] a_1 = 0$ reduces to $(2r + 1)a_1 = 0$, so we can conclude $a_1 = 0$.

For $n \geq 2$, equating coefficients of equal powers of $x$ gives us the recurrence relation

$$\left\{ (n+r)^2 - p^2 \right\} a_n = -a_{n-2}$$

or

$$a_n = - \frac{a_{n-2}}{\left\{ (n+r)^2 - p^2 \right\}} = - \frac{a_{n-2}}{n(n+2r)}.$$

---

[3] In general, for the method of Frobenius, the indicial equation has the form $r(r-1) + r p_0 + q_0 = 0$, where $p_0$ and $q_0$ are the constant terms of the series expansions of $p(t)$ and $q(t)$ in Equation (D.3.1).

because $r^2 - p^2 = 0$. We can look at a few terms to see the pattern:

$$a_2 = -\frac{a_0}{2(2 + 2r)} = -\frac{a_0}{2^2(1 + r)}$$

$$a_3 = -\frac{a_1}{3(3 + 2r)} = 0 \quad [\text{because } a_1 = 0]$$

$$a_4 = -\frac{a_2}{4(4 + 2r)} = -\frac{\left(\frac{-a_0}{2^2(1+r)}\right)}{2 \cdot 2^2(2 + r)} = \frac{a_0}{2^4 2!(1 + r)(2 + r)}$$

$$a_5 = -\frac{a_3}{5(5 + 2r)} = 0$$

$$a_6 = -\frac{a_4}{6(6 + 2r)} = -\frac{\left(\frac{a_0}{2^4 2!(1+r)(2+r)}\right)}{6(6 + 2r)} = -\frac{a_0}{2^6 3!(1 + r)(2 + r)(3 + r)}.$$

We can see, for example, that $a_k = 0$ for $k$ odd.

Letting $n = 2k$ and remembering that we're assuming $r = p$, we can express the even coefficients in the form

$$a_{2k} = \frac{(-1)^k a_0}{2^{2k} k!(r + 1)(r + 2) \cdots (r + k)}$$

$$= \frac{(-1)^k a_0}{2^{2k} k!(p + 1)(p + 2) \cdots (p + k)}.$$

In working with Bessel's equation, it is common practice to make things neater by taking $a_0 = \frac{1}{2^p p!}$,[4] so that

$$a_{2k} = \frac{(-1)^k}{2^{2k+p} k!(p + k)!}.$$

The final result is the **Bessel function of order $p$ of the first kind**, $J_p(x)$:

$$y(x) = J_p(x) = \sum_{n=0}^{\infty} \frac{(-1)^n x^{2n+p}}{2^{2n+p} n!(p + n)!} = \sum_{n=0}^{\infty} \frac{(-1)^n \left(\frac{x}{2}\right)^{2n+p}}{n!(p + n)!}$$

$$= \left(\frac{x}{2}\right)^p \sum_{n=0}^{\infty} \frac{(-1)^n \left(\frac{x}{2}\right)^{2n}}{n!(p + n)!}.$$

It can be shown that this series converges for all real values of $x$.

---

[4] Actually, $a_0 = \frac{1}{2^p \Gamma(p+1)}$, where $\Gamma$ denotes Euler's *gamma function* (see Section A.6).

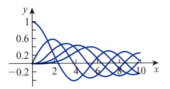

**FIGURE D.2**
$J_p(x)$ for $p = 0, 1, 2, 3, 4; 0 \le x \le 10$

Using technology, we can produce a graph of Bessel functions of order $p$ for $p = 0, 1, 2, 3$, and 4 (Figure D.2).

Both *Maple* and *Mathematica*, for example, can deal with Bessel functions of the first kind numerically and graphically via the command *BesselJ* $(mu, x)$ [in *Maple*] or *BesselJ* $[m, x]$ [in *Mathematica*]. The parameter *mu* or *m* represents the order that we have called $p$. It is interesting to note that a CAS could express the solution of the IVP $y'' + xy = 0; y(0) = 0, y'(0) = 1$ that we considered in Section D.2 as

$$y(x) = \frac{2\,3^{5/6}\pi}{9\,\Gamma\left(\frac{2}{3}\right)}\sqrt{x}\ BesselJ\left(\frac{1}{3}, \frac{2x^{3/2}}{3}\right).$$

■

We should make several comments about the preceding example:

1. In our analysis, we have actually assumed that $x > 0$ to avoid the possibility of fractional powers of negative numbers.

2. In the indicial equation $r^2 - p^2 = 0$, we have assumed that $r = p$, a nonnegative number. If $r$ is in fact a nonnegative *integer*, then the Frobenius series is an ordinary power series with first term $a_0 x^n$. For applications, the choices $p = 0$ and $p = 1$ occur most often.

3. All our efforts have produced just one solution of Bessel's equation for a fixed value of $p$. It can be shown that when $2p$ is not a positive integer,

$$J_{-p}(x) = \left(\frac{2}{x}\right)^p \sum_{n=0}^{\infty} \frac{(-1)^n \left(\frac{x}{2}\right)^{2n}}{n!(-p+n)!}$$

defines a second, linearly independent solution of Bessel's equation. When $p$ is an integer, it can be shown that $J_p(x) = (-1)^p J_{-p}(x)$, so the two solutions are dependent.

4. If $p$ is not an integer, the general solution of Bessel's equation has the form $y(x) = k_1 J_p(x) + k_2 J_{-p}(x)$ for arbitrary constants $k_1$ and $k_2$.

5. The function

$$Y_p(x) = \frac{(\cos p\pi)J_p(x) - J_{-p}(x)}{\sin p\pi}$$

is the standard Bessel function of the second kind. Then

$$y(x) = c_1 J_p(x) + c_2 Y_p(x)$$

is the general solution of Bessel's equation in all cases, whether or not $p$ is an integer. Both *Maple* and *Mathematica* have commands—*BesselY* $(mu, x)$ and *BesselY* $[m, x]$, respectively—that enable users to explore Bessel functions of the second kind numerically and graphically.

There are many treatments of the properties and applications of Bessel functions.

Accessible sources of information include the books *Differential Equations: Theory, Technique, and Practice* by George F. Simmons and Steven G. Krantz (New York: McGraw-Hill, 2007) and *Handbook of Mathematical Formulas and Integrals* (*Fourth Edition*) by Alan Jeffrey and Hui Hui Dai (San Diego: Academic Press, 2008).

## D.4 THE POINT AT INFINITY

In some situations we want to determine the behavior of solutions of the equation

$$y'' + P(t)y' + Q(t)y = 0$$

for large values of the independent variable $t$—the behavior "in the neighborhood of infinity." The way to deal with this problem is to use the substitution $t = \frac{1}{u}$ and investigate the resulting equation near $u = 0$. This substitution converts a problem in large values of $t$ to one in small values of $u$. Once the "$u$-problem" is solved near $u = 0$, we make the substitution $t = \frac{1}{u}$ in the $u$-solution to get the solution near the $t$ point of infinity.

Let $u = \frac{1}{t}$. Then, by the Chain Rule,

$$y' = \frac{dy}{dt} = \frac{dy}{du} \cdot \frac{du}{dt} = \frac{dy}{du} \left( -\frac{1}{t^2} \right) = -u^2 \cdot \frac{dy}{du}$$

and

$$y'' = \frac{d}{dt} \left( \frac{dy}{dt} \right) = \frac{d}{du} \left( \frac{dy}{dt} \right) \cdot \frac{du}{dt} = \left( -u^2 \frac{d^2 y}{du^2} - 2u \frac{dy}{du} \right) (-u^2).$$

Let's use this transformation method to solve an equation for large values of the independent variable.

## ■ Example

Find the general solution of the equation

$$4t^3 \frac{d^2 y}{dt^2} + 6t^2 \frac{dy}{dt} + y = 0$$

for large values of $t$.

First, we write the equation in the standard form

$$\frac{d^2y}{dt^2} + \frac{3}{2t}\frac{dy}{dt} + \frac{1}{4t^3}y = 0.$$

Making the substitution $u = \frac{1}{t}$ and using the calculations for $y'$ and $y''$ given previously, we transform our equation into

$$\left(-u^2\frac{d^2y}{du^2} - 2u\frac{dy}{du}\right)(-u^2) + \frac{3u}{2}\left(-u^2\frac{dy}{du}\right) + \frac{u^3}{4}y = 0$$

or

$$4u\frac{d^2y}{du^2} + 2\frac{dy}{du} + y = 0,$$

which has $u = 0$ as a regular singular point.

If we use the Frobenius method, we find the general solution

$$\Psi(u) = c_1 \sum_{n=0}^{\infty} \frac{(-1)^n u^n}{(2n)!} + c_2 \sum_{n=0}^{\infty} \frac{(-1)^n u^{n+\frac{1}{2}}}{(2n+1)!}.$$

Substituting $u = \frac{1}{t}$, we get the solution

$$y(t) = c_1 \sum_{n=0}^{\infty} \frac{(-1)^n}{(2n)!}\left(\frac{1}{t}\right)^n + c_2 \sum_{n=0}^{\infty} \frac{(-1)^n}{(2n+1)!}\left(\frac{1}{t}\right)^{n+\frac{1}{2}}$$

$$= c_1 \cos\left(\frac{1}{\sqrt{t}}\right) + c_2 \sin\left(\frac{1}{\sqrt{t}}\right).$$

(See Section A.3.) ■

## D.5 SOME ADDITIONAL SPECIAL DIFFERENTIAL EQUATIONS

Many famous functions, such as the Airy and Bessel functions, arise as power series solutions of second-order differential equations. These functions form a particular class of what are usually called *special functions*. [A classic reference is *Special Functions* by Earl D. Rainville (New York: Chelsea, 1971). A more recent book is *Special Functions* by George E. Andrews, Richard Askey, and Ranjan Roy (New York: Cambridge University Press, 1999).]

Among these important second-order equations that have been significant in solving problems in applied mathematics, science, and engineering are the following, which you are invited to investigate using the methods of this appendix.

**Chebyshev's equation**: $(1 - x^2)y'' - xy' + p^2 y = 0$, where $p$ is a constant. (When $p$ is a nonnegative integer, the solution is an $n$th-degree polynomial.)

**Gauss's hypergeometric equation**: $x(1 - x)y'' + [c - (a + b + 1)x]y' - aby = 0$, where $a$, $b$, and $c$ are constants.

**Hermite's equation**: $y'' - 2xy' + 2py = 0$, where $p$ is a constant.

**Laguerre's equation**: $xy'' + (1 - x)y' + py = 0$, where $p$ is a constant.

**Legendre's equation**: $(1 - x^2)y'' - 2xy' + \left[ k(k + 1) - \frac{m^2}{1-x^2} \right] y = 0$, where $m$ and $k$ are constants, $k > 0$.

# Answers/Hints to Odd-Numbered Exercises

## Exercises 1.1

**A1. a.** The independent variable is $x$ and the dependent variable is $y$.
    **b.** First-order.
    **c.** Linear.

**A3. a.** The independent variable is not indicated, but the dependent variable is $x$.
    **b.** Second-order.
    **c.** Nonlinear because of the term $\exp(-x)$—the equation cannot be written in the form (1.1.1), where $y$ is replaced by $x$ and $x$ is replaced by the independent variable.

**A5. a.** The independent variable is $x$ and the dependent variable is $y$.
    **b.** First-order.
    **c.** Nonlinear because you get the terms $x^2(y')^2$ and $x\,y'\,y$ when you remove the parentheses.

**A7. a.** The independent variable is $x$ and the dependent variable is $y$.
    **b.** Fourth-order.
    **c.** Linear.

**A9. a.** The independent variable is $t$ and the dependent variable is $x$.
    **b.** Third-order.
    **c.** Linear.

**A11. a.** The independent variable is $x$, the dependent variable is $y$; first-order, nonlinear because of the exponent $y'$.

**A13. a.** Nonlinear; the first equation is nonlinear because of the term $4xy = 4x(t)y(t)$.
    **b.** Linear.
    **c.** Nonlinear; the first and second equations are nonlinear because each contains a product of dependent variables.
    **d.** Linear.

**B1.** $a = 1$.

## Exercises 1.2

**A11. a.** For example, $c\,y' = 1$ is a possible differential equation satisfied by $y$.
    **b.** For example, $y' - ay = be^{ax}\cos bx$.

**c.** For example, $y' - y = Be^t$. Other possibilities are the equations $y'' - y = 2Be^t$ and $y'' - y' = Be^t$.

**d.** One answer is $y' = -3e^{-3t} + ty$, or $y' - ty = -3e^{-3t}$.

**A13.** $y' = \left(\dfrac{y^2+1}{y^2+2}\right)\left(\dfrac{x^2+2}{x^2+1}\right)$, a first-order nonlinear equation.

**A15.** One solution is $y' = (-2xy)/(x^2+4)$.

**B3. a.** The given equation is equivalent to $(y')^2 = -1$. Since there is no real-valued function $y'$ whose square is negative, there can be no real-valued function $y$ satisfying the equation.

**b.** The only way that two absolute values can have a sum equal to zero is if each absolute value is itself zero. This says that $y$ is identically equal to zero, so that the zero function is the only solution. The graph of this solution is the $x$-axis (if the independent variable is $x$).

**B5.** If $x > c$ or $x < -c$, then $c^2 - x^2 < 0$ and then the functions $y = \pm\sqrt{c^2 - x^2}$ do not exist as real-valued functions. If $x = \pm c$, then each function is the zero function, which is not a solution of the differential equation.

**B7.** One solution is $y(x) = (1/2)(\sin x - \cos x)$.

# Exercises 1.3

**A1.** $R(t) = -\pi(1 + \cos t)$.

**A3.** $r(t) = (a/b)(e^{bt}/b - t - 1/b)$.

**A5.** $A = -1/4, B = 1/37, C = -6/37$.

**B3.** Recall the Product Rule and the Fundamental Theorem of Calculus (Section A.4).

**B5.** No. An equation of order $n$ requires an $n$-parameter family of solutions.

**B7.** The length of the runway must be $5/6$ mile (five-sixths of a mile).

**B11. a.** Deriving inspiration from Example 1.2.1, we get $u(t) = u(0)e^{kat} = Ae^{kat}$.

**b.** We have $w(t) = \dfrac{(k-1)A}{k}(1 - e^{kat})$ for $0 < k \le 1$; $w(t) = aAt$ for $k = 0$.

**C1. b.** As $t \to \infty$, $e^{-kEt} \to 0$, so that $W(t) \to \dfrac{C}{E}$. Note that $C/E$ is in pounds per day.

**c.** About 79 days for 20 lb, 138 days for 30 lb, and about 177 days for 35 lb. This says that the weight is a concave up decreasing function of time—that is, the rate of weight loss slows down with time.

**C3. b.** Here's the graph of $y(t) = \dfrac{387.9802}{1 + 54.0812\,e^{-0.02270347\,t}}$:

**c.**

| | Actual Population | Logistic Pop. Value |
|---|---|---|
| 1790 | 3,929,214 | 7,043,786 |
| 1980 | 226,545,805 | 225,066,248 |
| 1990 | 248,709,873 | 246,050,716 |

**d.** $\displaystyle\lim_{t\to\infty} y(t) = \lim_{t\to\infty} \dfrac{387.9802}{1 + 54.0812\,e^{-0.02270347\,t}} = 387.9802$ million people.

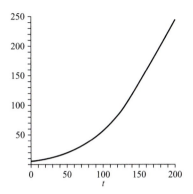

## Exercises 2.1

**A1.** $y = \dfrac{A}{2} + \dfrac{C}{x^2}$.

**A3.** $y = (t - 2)^3 = t^3 - 6t^2 + 12t - 8$. The solution $y \equiv 0$ is a **singular solution** of the basic ODE and satisfies the initial condition.

**A5.** $y = 2 - 3\cos x$. The only possible singular solution is $y \equiv 2$, but this can be obtained by letting $C = 0$.

**A7.** $\dfrac{y^2}{2} + y + \ln|y - 1| = -\dfrac{1}{x} + C$. The constant function $y \equiv 1$ is a **singular solution**. Notice that the implicit solution formula is not defined for $y = 1$.

**A9.** $z = \dfrac{\ln(C - 10^x)}{\ln 10}$. Note that for each particular value of the parameter $C$, the solution is defined only for $10^x < C$—that is, for $x < \ln C / \ln 10$ (or $x < \log_{10} C$).

**A11.** $y = -\dfrac{x^2}{2} + C$ or $y = Ce^{-x}$.

**A13.** $x + 2y - 2\ln|x + 2y + 2| = x + C$; $y = -(x + 2)/2$ is a singular solution.

**A15.** $\arctan\left(\dfrac{y}{x}\right) - \dfrac{1}{2}\ln\left(\dfrac{x^2 + y^2}{x^2}\right) - \ln|x| - C = 0$.

**A17.** We have two one-parameter families of solutions: $y = x\sqrt{2\ln|x| + C}$ and $y = -x\sqrt{2\ln|x| + C}$.

**B3. a.** $x(t) = \dfrac{1}{2 - t}$.

  **b.** The interval $I$ can be as large as $(-\infty, 2)$ or $(2, \infty)$. Any such interval $I$ cannot include the point $t = 2$, at which $x(t)$ is not defined.

  **c.**

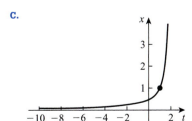

  **d.** The only solution is $x \equiv 0$, a **singular solution** that satisfies the initial condition $x(0) = 0$.

**B5.** $t = 60$.

**B7.** $V(4) = 32\pi$ cubic units.

**C1.** $V = 300$ ft/sec.

**C3. a.** If we let $p(x) = dy/dx$, then the original equation becomes $\frac{dp}{dx} = k\left[1 + p^2\right]^{1/2}$.

   **b.** $y = \frac{C_1}{2k}e^{kx} + \frac{1}{2kC_1}e^{-kx} + C_2 = \frac{1}{2k}\left(C_2 e^{kx} + \frac{1}{C_2}e^{-kx}\right) + C_3$.

**C5. a.** The equilibrium solution occurs when $C = D/\mu$.

   **b.** The formula for the concentration is $C = \frac{D}{\mu} - \left(\frac{D - \mu C_0}{\mu}\right)e^{-\mu t}$. As $t \to \infty$, $e^{-\mu t} \to 0$, so $C(t) \to \frac{D}{\mu} - 0 = \frac{D}{\mu}$, the equilibrium solution found in part (a).

   **c.**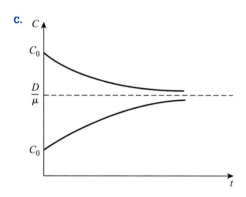

# Exercises 2.2

**A1.** $y = 2x - 1 + Ce^{-2x}$.

**A3.** $x = \dfrac{t^2}{2} - \dfrac{1}{2} + Ce^{-t^2}$.

**A5.** $y = \dfrac{t^3}{6} - \dfrac{t^2}{5} + \dfrac{C}{t^3}$.

**A7.** $y = x\sin x + Cx$.

**A9.** $x = e^t\left(\ln|t| + t^2/2 + C\right)$.

**A11.** $y(x) = \dfrac{e^x + ab - e^a}{x}$.

**A13.** For $m \neq -a$, we have $y = \dfrac{e^{mx}}{a + m} + Ce^{-ax}$. If $m = -a$, then

$$e^{ax}y = \int e^{(a-a)x}\,dx = \int 1\,dx = x + C, \text{ so that } y = xe^{-ax} + Ce^{-ax} = (x + C)e^{-ax}.$$

*Note:* A CAS that can solve ODEs may miss the need for an analysis of two cases.

**A15.** $x(t) = \left(\dfrac{t}{t+1}\right)(t + \ln|t| - 1).$

**A17.** $y = -\ln(x + Cx^2).$

**B1.** $y = \dfrac{t^4}{t^6 + C};$ $y \equiv 0$ is a **singular solution.** ✓

**B3.** $y = \dfrac{\pm 1}{\sqrt{x + \dfrac{1}{2} + Ce^{2x}}};$ $y \equiv 0$ is a **singular solution.**

**B5.** $y = 2e^{t^2}\big/\left(2C - e^{t^2}\right);$ $y \equiv 0$ is a **singular solution.**

**B7.** **a.** $W(t) = \left(\dfrac{\alpha}{\beta} + Ce^{-\beta t/3}\right)^3.$

**b.** $W_\infty = \left(\dfrac{\alpha}{\beta}\right)^3.$

**c.** $W(t) = W_\infty\left(1 - e^{-\beta t/3}\right)^3.$

**d.**

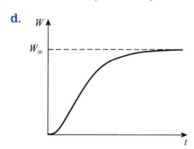

**B9.** **a.** $I(t) = \dfrac{E}{R} - \dfrac{E}{R}e^{-(R/L)t} = \dfrac{E}{R}\left(1 - e^{-(R/L)t}\right).$

**b.** $\lim\limits_{t\to\infty} I(t) = \dfrac{E}{R}.$

**c.** $t = \dfrac{L}{R}\ln 2.$

**d.** $I(t) \equiv \dfrac{E}{R}.$

**B11.** $Q(t) = \dfrac{E_0 C\,[\sin(\omega t) - \omega RC\cos(\omega t)]}{1 + (RC\omega)^2} + \dfrac{\omega E_0 RC^2}{1 + (RC\omega)^2}e^{-t/RC}$

$= \dfrac{E_0 C}{1 + (RC\omega)^2}\left\{\sin(\omega t) - \omega RC\cos(\omega t) + \omega RCe^{-t/RC}\right\}.$

**B13.** $p(t) = \dfrac{v}{\mu + v} + \left(p_0 - \dfrac{v}{\mu + v}\right)e^{-(\mu+v)t} = \dfrac{v}{\mu + v}\left[1 - e^{-(\mu+v)t}\right] + p_0 e^{-(\mu+v)t}$

$$q(t) = 1 - \left\{ \frac{v}{\mu + v} \left[ 1 - e^{-(\mu+v)t} \right] + p_0 e^{-(\mu+v)t} \right\}$$

$$= \frac{\mu}{\mu + v} + \left( q_0 - \frac{\mu}{\mu + v} \right) e^{-(\mu+v)t}$$

$$= \frac{\mu}{\mu + v} \left[ 1 - e^{-(\mu+v)t} \right] + q_0 e^{-(\mu+v)t}.$$

**C1.** Note that $(1 - n)y' = y^n u'$.

## Exercises 2.3

**A1.** $p' = bp - dp = (b - d)p.$

**A3.** $p' = kp^2 - dp = (kp - d)p.$

**B1. a.** $P = \frac{\alpha t}{k} + \frac{\alpha}{k^2} + \left( 1.285 - \frac{\alpha}{k^2} \right) e^{kt} \approx (0.0452\,t + 1.275) + 0.01045\,e^{0.0355\,t}.$

**b.** $P(20) \approx 2{,}200{,}736$ people.

**B3.** $A(t) = \frac{25}{2} - \frac{25}{2} e^{-t/50} = \frac{25}{2} \left( 1 - e^{-t/50} \right).$

**B5. a.** 100 gallons.

**b.** 175 pounds.

**B7.** 165.12 minutes.

## Exercises 2.4

**A1.**

**A3.**

**A5.**

**A7.**

**A9.**

**A11.**

**A13.**

**A15.**

**A17.**

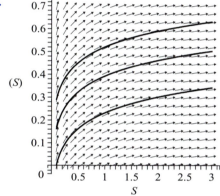

**A19.** The equations in Problems A4, A5, A8, and A14 are autonomous.

**B1. a.**

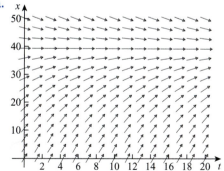

**b.** The amount of substance $X$ approaches 40.

**B3.** Circles centered at the origin with radii $\sqrt{1/C^2 - 1}$, where $C \in [-1, 0) \cup (0, 1]$ is the slope.

**B5.** The horizontal lines $x = \alpha$ and $x = \beta$.

**B7. a.** Field 3.
   **b.** Field 1.
   **c.** Field 2.

**B9.**

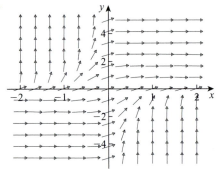

The slope field indicates that any solution must be an increasing function. Some solutions in the second quadrant seem to have vertical asymptotes, so that they "blow up in finite time," whereas other solutions starting out in this area flatten out (approach some finite value asymptotically) as they pass through the first quadrant. Solutions with initial points in the third quadrant are almost flat until they pass into the first or fourth quadrants. Starting out in the fourth quadrant, a solution will start out having a very large slope, but will move into the first quadrant and approach a positive finite value asymptotically. Overall then, we see that as $x \to \infty$, we have both $y \to \infty$ and $y \to a$, where $a$ is a positive real number. As $x \to -\infty$ (i.e., as we look at the slope field from right to left), we see that $y \to \infty$ or $y \to 0$.

## Exercises 2.5

**A1.**

**A3.**

**A5.**

**A7.**

**A9.**

**A11.**  ⟶  • 0  ⟵  • 1  ⟶  • 3  ⟵  — x

**A13.**  ⟶  ⟵  • 0  ⟵  — x

**A15. a.**  ⟶  | 0  ⟶  — x

 **b.** There are no critical points. Any solution must be an increasing function.

**B1. a.**  ⟶  • 1.2  ⟵  — I

 **b.** If the initial current, $I(0)$, is 3 amps, it is to the right of the critical point, so that the current tends to *decrease* toward 1.2 amps as $t$ gets larger.

**B3. a.**  ⟶  | 0  ⟶  • $\sqrt{9.8/c}$  ⟵  — v

 **b.** The positive equilibrium solution is $\sqrt{9.8/c}$, which is a sink representing "terminal velocity."

**B5. a.**

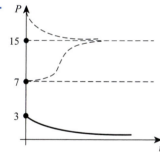

 **b.** $P(t) \to 0$ as $t \to \infty$

**B7.**

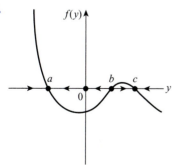

**C1.** If $\alpha < 1/2$, the equilibrium solution $(1 - 2\alpha)^{-1}$ is a sink. If $\alpha > 1/2$, the equilibrium solution $(1 - 2\alpha)^{-1}$ is a source.

**C3. a.**

**b.** $x(t) \to \sqrt{a/b}$.

**c.** $x(t)$ stays at zero.

**d.** $x(t) \to \sqrt{a/b}$.

## Exercises 2.6

**A1.** The equilibrium points are $y = 0$ and $y = 1$. Both of these are nodes.

**A3.** The only equilibrium point is $y = 0$, a source.

**A5.** The equilibrium points are $x = -a/b$ and $0$. We find that $x = -a/b$ is a sink and $x = 0$ is a source.

**A7.** The equilibrium points are $x = 0$ and $1$. The solution $x = 1$ is a sink. But $x = 0$ is a node.

**A9.** The equilibrium points are $x = 0, 2$, and $4$. We see that $x = 0$ is a source, $x = 2$ is a sink, and $x = 4$ is a source.

**A11.** The only equilibrium point is $x = 0$. A careful examination reveals that $x = 0$ is a source.

**A13.** There is only one equilibrium point, $x \approx 0.74$, which is a source.

**A15.** We see that $x = -1$ is a sink, $x = 0$ is a source, and $x = 0.5$ is a node.

**B1. a.** $u = \sqrt[3]{\dfrac{8P}{bS}} = 2\sqrt[3]{\dfrac{P}{bS}}$.

**b.** The equilibrium speed is a sink.

**c.** A rower may start from rest with maximum acceleration but then tire a bit so that his or her speed would level off at the equilibrium speed. If the rower's speed is *greater* than the equilibrium speed, we can reasonably believe that he or she may tire or the "drag force" $bSu^2$ may exceed the "tractive force" $\frac{8P}{u}$ and so slow the boat down.

**B3. a.**

**b.** With $a = 10$ and $b = 2$, the graph is

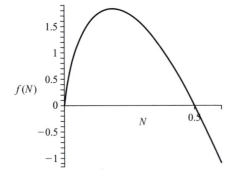

**c.** The only equilibrium point is $N = 1/b$. The phase portrait given in (a) indicates that this is a sink.

## Exercises 2.7

**A1. 1.**

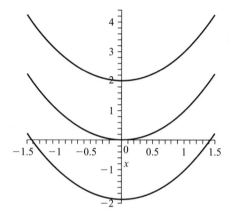

**2.** The only bifurcation point is $c = 0$.
**3.** The bifurcation diagram is

**A3.** **1.**

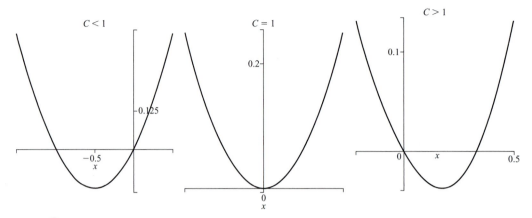

**2.** The only bifurcation point is $c = 1$.

**3.**

**A5.** **1.**

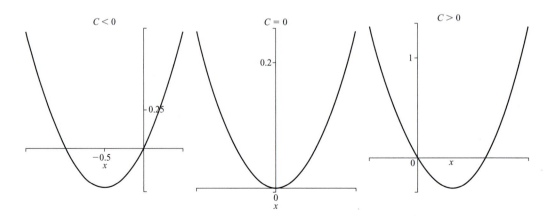

**2.** There is one bifurcation point, $c = 0$.

**3.**

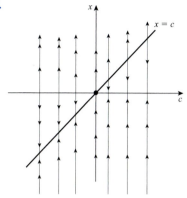

**B1.** We find that $h^* = 25/4$ is the maximum harvest rate beyond which any population will become extinct.

**B3.** The value $c = 0$ is a *pitchfork bifurcation*.

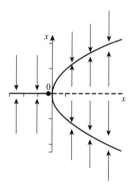

**B5.** The values $\alpha = -2$ and $\alpha = 2$ are the only bifurcation points:

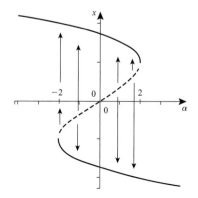

**C1. c.** The bifurcation point $R = R_c$ is a *sink*.

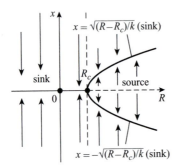

## Exercises 2.8

**A1.** For example, take any rectangle centered at $(0, 3)$ that avoids the $t$-axis ($x = 0$).

**A3.** There is no rectangle $R$ containing the origin that does not also include points of the $x$-axis, where $t = 0$.

**A5.** There is no rectangle $R$ satisfying the requirements of the Existence and Uniqueness Theorem.

**A7.** For example, take any rectangle in the $t$-$y$ plane that does not include part of the line $t = -1$.

**A9.** The solution's domain is $I = (-\pi/2, \pi/2)$, an interval of length $\pi$.

**A11.** $x(t) = \left(\frac{t}{3} + \sqrt[3]{x_0}\right)^3$; the initial condition of Example 2.8.2 is $x(0) = 0$, so that we don't expect uniqueness in that case. In the current exercise, both $f$ and $\frac{\partial f}{\partial x}$ are continuous at $(0, x_0)$ if $x_0 < 0$, so that we are guaranteed existence and uniqueness on some $t$-interval $I$.

**B1. a.** $\dfrac{\partial f}{\partial Q}$ is not defined at $Q = 1$.
　　**b.** The constant function $Q \equiv 1$ is a solution because $Q' = 0 = |Q - 1|$ and $Q(0) = 1$. This solution is in fact unique.

**B3.** The conditions of the Existence and Uniqueness Theorem are not satisfied, so uniqueness is not guaranteed.

**C1.** The conditions of the Existence and Uniqueness Theorem are satisfied, and so we expect to find an interval $I = (2 - h, 2 + h)$ centered at $x = 2$ such that the IVP has a unique solution on $I$.

**C3.** No. If a solution near $P \equiv b$ were to equal the equilibrium solution—that is, if another solution curve intersects the horizontal line $P \equiv b$ at the point $(t^*, b)$—then we would have *two* solutions of the IVP $\frac{dP}{dt} = kP(b - P), P(t^*) = b$.

**C5. a.** Show that $[y(t)\, y(-t)]' = 0$.
　　**b.** What does part (a) imply about the signs of $y(t)$ and $y(-t)$?

## Exercises 3.1

**A1.**

| $t_k$ | $y_k$ |
|-------|-------|
| 0 | 1.000000 |
| 0.25 | 0.750000 |
| 0.50 | 0.625000 |
| 0.75 | 0.589844 |
| 1.00 | 0.643490 |

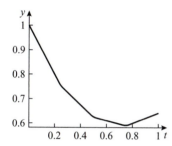

**A3.**

| $t_k$ | $y_k$ |
|-------|-------|
| 1.0 | 2.000000 |
| 1.5 | 3.359141 |
| 2.0 | 4.266010 |
| 2.5 | 5.065065 |
| 3.0 | 5.807155 |

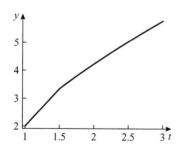

**A5.** $y(\pi/2) \approx 1.14884140143$. The absolute error of the approximation is $|1 - 1.14884140143| = 0.14884140143$.

**A7.** $y(1) \approx 1.38556107091$.

**A9.** $y(1) \approx 0.80998149723$; since the solution of the equation is $y(x) = \arctan x$, we have $y(1) = \arctan 1 = \pi/4$. Multiplying our approximation for $y(1)$ by 4, we get an approximation for $\pi$.

**A11. a.**

$$y_1 = y_0 + 0.2y_0^2 = 1 + 0.2(1)^2 = 1.2$$

$$y_2 = y_1 + 0.2y_1^2 = 1.2 + 0.2(1.2)^2 = 1.488$$

$$y_3 = y_2 + 0.2y_2^2 = 1.488 + 0.2(1.488)^2 = 1.9308288$$

$$y_4 = y_3 + 0.2y_3^2 = 1.9308288 + 0.2(1.9308288)^2 = 2.67644877098$$

$$y_5 = y_4 + 0.2y_4^2 = 2.67644877098 + 0.2(2.67644877098)^2 = 4.10912437572$$

$$y_6 = y_5 + 0.2y_5^2 = 4.10912437572 + 0.2(4.10912437572)^2 = 7.48610500275.$$

**b.** The equation is separable.

**c.** The following table compares approximate and actual values:

| $t_k$ | $y_k$ | Actual $y(t_k)$ |
|---|---|---|
| 0 | 1 | 1 |
| 0.2 | 1.2 | 1.25 |
| 0.4 | 1.488 | 1.6667 |
| 0.6 | 1.9308 | 2.5000 |
| 0.8 | 2.6764 | 5.0000 |
| 1.0 | 4.1091 | UNDEFINED |
| 1.2 | 7.4861 | −5.0000 |

The solution graph indicates the difficulty:

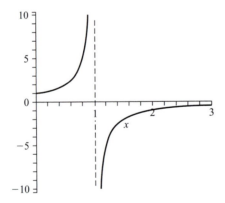

**B1. a.** $P(1) \approx 1.330624$ million people $= 1{,}330{,}624$ people.

**b.** $P(0) = 1.284999\ldots$ people $\approx 1{,}285{,}000$ people.

**B3.** $V(0) = 166.390541\ldots \approx 166.39$ meters per second.

**B5.** With $h = 0.5$, we get $x(2) \approx 2.746746$, with absolute error about $0.253254$. With $h = 0.25$, we get $x(2) \approx 2.870814$, with absolute error about $0.129186$.

**B7.**  **a.** $x'' = 3x^5$.

    **b.** If $x(0) = 1$, then $x'' = 3x^5 > 0$ for all $x > 0$, implying that the solution curve is concave up.

    **c.** Euler's method *underestimates* the true value of the solution at $t = 0.1$.

**B9.**  **a.** Note that the direction field is not meaningful for $y < -1$ or $y > 1$.

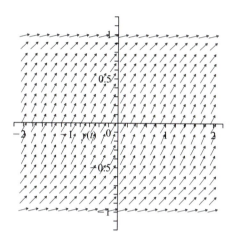

    **c.** We see that $y(1) \approx 0.8950$ and $y(1.2) \approx 1.0235$. However $y(1.3)$ and $y(1.3)$, for example, can't be calculated because the values would involve the square roots of negative numbers. With a relatively large step size of 0.4, once you get a little past $t = 1.2$, Euler's method produces values of $y$ that are greater than 1.

**C1.**  The only way a solution curve can coincide with its tangent line segments is if the solution curve is a straight line—that is, if $y(x) = Cx + D$, so that the differential equation is $\frac{dy}{dx} = c$.

**C3.**  **a.** $\dfrac{2500}{2501} \cos x + \dfrac{50}{2501} \sin x - \dfrac{2500}{2501} e^{-50x}$; $y(0.2) = 0.9836011240\ldots$.

    **b.** $y(0.2) \approx 1.7466146068$. The absolute error is 0.7630134828.

    **c.** $y(0.2) \approx 1.1761983279$. The absolute error is 0.1925972039.

    **d.** $y(0.2) \approx 1.8623800769$. The absolute error is 0.8787789529.

    **e.** No.

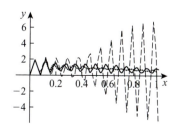

## Exercises 3.2

**A1.** The table of approximations is

|  | TRUE VALUE | Euler's Method | Absolute Error | Improved Euler Method | Absolute Error |
|---|---|---|---|---|---|
| $h = 0.1$ | 5.93977 | 5.69513 | 0.24464 | 5.93266 | 0.00711 |
| $h = 0.05$ | 5.93977 | 5.81260 | 0.12717 | 5.93791 | 0.00186 |
| $h = 0.025$ | 5.93977 | 5.87490 | 0.06487 | 5.93930 | 0.00047 |

**A3. a.** $x(t) = -t - 1 + 2\,e^t$.

    **b.** $x(1) \approx 3.42816$.

    **c.** The following table shows the absolute error at each step of part (b):

| $t_k$ | $x_k$ | TRUE VALUE | Absolute Error |
|---|---|---|---|
| 0 | 1 | 1 | 0 |
| 0.1 | 1.11000 | 1.11034 | 0.00034 |
| 0.2 | 1.24205 | 1.24281 | 0.00076 |
| 0.3 | 1.39847 | 1.39972 | 0.00125 |
| 0.4 | 1.58180 | 1.58365 | 0.00185 |
| 0.5 | 1.79489 | 1.79744 | 0.00255 |
| 0.6 | 2.04086 | 2.04424 | 0.00338 |
| 0.7 | 2.32315 | 2.32751 | 0.00436 |
| 0.8 | 2.64558 | 2.65108 | 0.00550 |
| 0.9 | 3.01236 | 3.01921 | 0.00685 |
| 1.0 | 3.42816 | 3.43656 | 0.00840 |

**B1. a.** $P(1) \approx 1.330624$ million people $= 1,330,624$ people.

    **b.** $P(0) \approx 1.285363$ million people $= 1,285,363$ people.

**B3. a.** $y(t) = [(1 - \alpha)\,t]^{1/(1-\alpha)}$.

    **b.** Take $f(t, y) = y^\alpha$, $t_0 = 0$, and $y_0 = 0$ in the improved Euler method. Then we have, for any $h$, $y_1 = y_0 + \frac{h}{2}\{f(t_0, y_0) + f(t_1, y_0 + hf(t_0, y_0))\} = \frac{h}{2}f(t_1, 0) = 0 = y_2 = \cdots = y_k = y_{k+1}$ for all positive integer values of $k$.

    **c.** If $y_0 = y(0) = 0.01$, then for any $h$, the improved Euler's method produces a nonzero sequence of approximate values.

## Exercises 3.3

**A1.**

|  | TRUE VALUE | Euler's Method | Improved Euler Method | RK4 Method |
|---|---|---|---|---|
| $h = 0.1$ | 2.7182818 | 2.5937425 | 2.7140808 | 2.7182797 |
| $h = 0.05$ | 2.7182818 | 2.6532977 | 2.7171911 | 2.7182817 |
| $h = 0.025$ | 2.7182818 | 2.6850638 | 2.7180039 | 2.7182818 |

**A3.** $y(1) = e \approx 2.71828181139414093$.

**A5. a.** $x = \dfrac{2}{t^2 + C}$.

    **b.** The *rkf45* method yields $x(1) \approx 0.99999999727228860$.

**B1. a.** The following table provides the required data:

| $t$ | $V(t)$ |
|---|---|
| 5 | 100.163 |
| 10 | 104.984 |
| 15 | 105.045 |
| 16 | 105.046 |
| 17 | 105.046 |
| 18 | 105.046 |
| 19 | 105.046 |
| 20 | 105.046 |

    We guess that the terminal velocity is $105.046$ ft/sec.

**b.**

**C1. a.**

| $t$ | $(0.5, 1)$ | $(0.5, 2)$ | $(1.5, 1)$ | $(1.5, 2)$ | $(2, 2)$ |
|---|---|---|---|---|---|
| 1 | 1.845 | 1.707 | 2.320 | 1.918 | 2.051 |
| 2 | 2.663 | 2.069 | 4.207 | 2.215 | 2.233 |
| 3 | 3.349 | 2.190 | 4.893 | 2.235 | 2.236 |
| 4 | 3.872 | 2.224 | 4.988 | 2.236 | 2.236 |
| 5 | 4.246 | 2.233 | 4.999 | 2.236 | 2.236 |
| 6 | 4.504 | 2.235 | 5.000 | 2.236 | 2.236 |
| 7 | 4.677 | 2.236 | 5.000 | 2.236 | 2.236 |
| 8 | 4.791 | 2.236 | 5.000 | 2.236 | 2.236 |
| 9 | 4.865 | 2.236 | 5.000 | 2.236 | 2.236 |
| 10 | 4.913 | 2.236 | 5.000 | 2.236 | 2.236 |

    **b.** $(r, q) \approx (0.5, 1.4)$.

## Exercises 4.1

**A1.** $y = (c_1 + c_2 t)e^{2t}$.

**A3.** $x = e^t(c_1 \cos t + c_2 \sin t)$.

**A5.** $x = c_1 + c_2 e^{-2t}$.

**A7.** $y = c_1 \cos 2t + c_2 \sin 2t$.

**A9.** $r(t) = e^{2t}(c_1 \cos 4t + c_2 \sin 4t)$.

**A11.** $x(t) = -e^{2t} + 2e^t$.

**A13.** $y(t) = \frac{1}{4}e^{(2t-\pi)} \sin 4t$.

**B3.** **a.** $I(t) = \frac{3}{2}\left(e^{-10t} - e^{-50t}\right)$.

**b.**

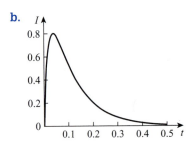

**c.** The maximum value of $I$ is approximately 0.8.

**d.** The maximum value of $I$ is achieved when $t \approx 0.04$ seconds.

**B5.** $x(t) = 3 \cos 12t + \frac{5}{6} \sin 12t$.

**B7.** **a.** $x(t) = -\frac{1}{30}e^{-2t}\left(11\sqrt{3} \sin(2\sqrt{3}t) + 3 \cos(2\sqrt{3}t)\right)$.

**b.** The graphs are

**c.** The greatest distance is approximately 33 cm.

**C3. c.** $u(t) = C_1 t + C_2$.

## Exercises 4.2

**A7.** $y_p = 3x^2$.

**B1.** $y(t) = \frac{3}{4}x - \frac{1}{16} + e^{-3t/2}\left(C_1 \cos\left(\frac{1}{2}\sqrt{7}x\right) + C_2 \sin\left(\frac{1}{2}\sqrt{7}x\right)\right)$.

**B3.** $y(x) = c_1 e^{-x} + c_2 e^{2x/3} - \frac{5}{13}\cos x + \frac{1}{13}\sin x$.

**B5.** $y(x) = c_1 \cos x + c_2 \sin x - \frac{1}{2}x\cos x$.

**B7.** $x(t) = c_1 e^{-t} + c_2 + \frac{1}{2}e^t - te^{-t}$.

**C1.** $x(t) = \begin{cases} 1 & \text{if} \quad 0 \le t < 1 \\ -\cos \pi t & \text{if} \quad t \ge 1. \end{cases}$

## Exercises 4.3

**A1.** $y_{GH} = c_1 + c_2 e^{-3t}$; $\quad y_{PNH} = Kt$.

**A3.** $y_{GH} = c_1 + c_2 e^{-7x}$; $\quad y_{PNH} = K x e^{-7x}$.

**A5.** $y_{GH} = c_1 \cos 5x + c_2 \sin 5x$; $\quad y_{PNH} = Ax(\cos 5x + \sin 5x)$.

**A7.** $y_{GH} = e^{-3x}(c_1 \cos 2x + c_2 \sin 2x)$; $\quad y_{PNH} = xe^{-3x}(c_1 \cos 2x + c_2 \sin 2x)$.

**A9.** $y_{GH} = c_1 \cos kt + c_2 \sin kt$; $\quad y_{PNH} = C$.

**A11.** $y_{GNH} = C_1 e^{3t} + C_2 e^{-t} + \frac{1}{5}e^{4t}$.

**A13.** $x_{GNH} = C_1 e^t \cos t + C_2 e^t \sin t + \left(-\frac{2}{5}t - \frac{14}{25}\right)\sin t + \left(\frac{1}{5}t + \frac{2}{25}\right)\cos t + e^t$.

**A15.** $x_{GNH} = C_1 + C_2 e^{-t} - 2\cos t - 2\sin t$.

**A17.** $y_{GNH} = (C_1 + C_2 x)e^{-5x} + 2x^2 e^{-5x}$.

**A19.** $x_{GNH} = C_1 e^{-t} + C_2 e^{-2t} + \left(-\frac{3}{10}t + \frac{17}{50}\right)\cos t + \left(\frac{1}{10}t + \frac{3}{25}\right)\sin t$.

**B1.** $y(x) = \frac{3}{5}xe^{4x} - \frac{3}{25}e^{4x} + \frac{3}{25}e^{-x}$.

**B3.** $y(t) = e^{-t/2}\left(7\cos\left(\frac{\sqrt{3}}{2}t\right) + \frac{11\sqrt{3}}{3}\sin\left(\frac{\sqrt{3}}{2}t\right)\right)$
$\qquad + \left(-t^2 e^{-t} - 6e^{-t} - 4te^{-t}\right)\sin t + \left(-2te^{-t} - 6e^{-t}\right)\cos t$.

**B7. a.**

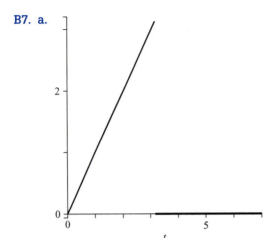

**b.** $y(t) = t - \sin t.$

**c.** $y(t) = -\pi \cos t - 2 \sin t.$

**d.** $y(t) = \begin{cases} t - \sin t, & 0 \le t \le \pi \\ -\pi \cos t - 2 \sin t, & t \ge \pi. \end{cases}$

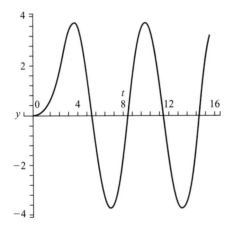

**C1.** $y(x) = c_1 e^x + c_2 e^{2x} + \frac{3}{130}(9 \cos 3x - 7 \sin 3x) - \frac{1}{6970}(27 \cos 9x - 79 \sin 9x).$

**C3.** $y(x) = c_1 \cos 2x + c_2 \sin 2x + \frac{1}{12} - \frac{1}{24}\cos^2 2x - \frac{1}{32}\cos^3 2x.$

**C5. b.** No, any solution is unbounded unless the constant is zero.

   **c.** If $c = 0$, the solution is unbounded unless the constant is zero. If $b = 0$ also, the solution is a polynomial and thus unbounded as $t \to \infty$.

**C7.** $y(x) = c_1 x^{(\sqrt{10}-2)} + c_2 x^{(-\sqrt{10}-2)} + (10x^2 - 2)/5x^2.$

## Exercises 4.4

**A1.** $x_{GNH} = C_1 t e^t + C_2 e^t + t \ln t e^t - t e^t = K_1 t e^t + K_2 e^t + t e^t \ln t$.

**A3.** $r_{GNH} = C_1 t e^t + C_2 e^t + t \ln t e^t - t e^t = K_1 t e^t + K_2 e^t + t e^t \ln t$.

**A5.** $y_{GNH} = c_1 x e^{-2x} + c_2 e^{-2x} + \frac{1}{2} x^3 e^{-2x}$.

**A7.** $y_{GNH} = c_1 x e^{-x} + c_2 e^{-x} + \frac{1}{4} x^2 e^{-x} (2 \ln x - 3)$.

**A9.** $y_{GNH} = e^x \left( c_1 e^x + c_2 - e^x \cos(e^{-x}) \right)$.

**B1.** $y_{GNH} = c_1 x \ln x + c_2 x + \frac{1}{2} x \ln^2 x$.

**B3.** $y_{GNH} = c_1 x^2 + c_2 x + \frac{1}{4} x^3 (2 \ln x - 3)$.

**B5.** $y_{GNH} = c_1 \tan x + c_2 + \frac{1}{2} x \tan x$.

**B7.** $y_{GNH} = c_1 x^3 + c_2 x^2 + \frac{1}{2} x^4$.

## Exercises 4.5

**A1.** $y(x) = C_1 e^{3x} + C_2 e^{-x} + C_3$.

**A3.** $y(x) = C_1 x e^{-x} + C_2 e^{-x} + C_3$.

**A5.** $y(t) = C_1 e^{10t} + e^t (C_2 \cos t + C_3 \sin t)$.

**A7.** $y(t) = C_1 e^{-2t} + C_2 e^{2t} + C_3 e^{-3t} + C_4 e^{3t}$.

**A9.** $y(t) = C_1 + (C_2 t + C_3) e^t + C_4 e^{-2t}$.

**B1.** $y(t) = (C_1 t^2 + C_2 t + C_3) e^t + (C_4 t + C_5) e^{2t} + C_6 e^{3t} + C_7 e^{4t}$.

**B3.** $y(x) = e^{-0.7289x}(C_1 \cos (0.6186t) + C_2 \sin (0.6186t))$
$\qquad + e^{0.4765x}(C_3 \cos (0.7591t) + C_2 \sin (0.7591t))$.

**B5. a.** $y_{GH} = C_1 x^3 + C^2 x^2 + C_3 x + C_4$.
   **b.** An intelligent guess would be $y_{PNH} = (R/24) x^4$.

**B7.** $y(t) = (C_1 t^{n-3} + C_2 t^{n-4} + \cdots + C_{n-3} t + C_{n-2}) + C_{n-1} e^{-t} + C_n e^t$.

**B9.** $y(t) = \frac{1}{6} e^{3t} + C_1 + C_2 e^{-6t} + C_3 e^t$.

**B11.** $y(x) = C_1 e^{4x} + C_2 e^{3x} + C_3 e^{-2x} - \frac{x e^{3x}}{375} (25x^2 + 60x + 126)$.

**C1.** $y(x) = C_1 \cos x + C_2 \sin x + C_3 x + C_4 + x \sin x + 2x \cos x + \frac{1}{4} x^4 - 3x^2$.

**C3. a.** $y_{PNH} = C_1 y_1 + C_2 y_2 + C_3 y_3$.

**b.** $\begin{bmatrix} y_1 & y_2 & y_3 \\ y_1' & y_2' & y_3' \\ y_1'' & y_2'' & y_3'' \end{bmatrix} \begin{bmatrix} C_1' \\ C_2' \\ C_3' \end{bmatrix} = \begin{bmatrix} 0 \\ 0 \\ g \end{bmatrix}$.

**c.** $y(x) = c_1 e^x + c_2 e^{-x} + c_3 e^{2x} - \frac{1}{2} x e^x$.

## Exercises 4.6

**A1.** $\left\{ \frac{dx_1}{dt} = x_2, \frac{dx_2}{dt} = 1 + x_1 \right\}.$

**A3.** The nonautonomous system is $\left\{ y_1' = y_2, y_2' = (5 \ln x + 3xy_2 - 4y_1)/x^2 \right\}$. Replacing $x$ by $y_3$ and adding the equation $y_3' = 1$ yields an autonomous system.

**A5.** $\left\{ x_1' = x_2, x_2' = x_3, x_3' = t x_3 - x_2 + 5x_1 - t^2 \right\}.$

**A7.** The nonautonomous system is
$\left\{ w_1' = w_2, w_2' = w_3, w_3' = w_4, w_4' = 6 \sin(4t) + 2w_4 - 5w_3 - 3w_2 + 8w_1 \right\}.$ To get an
*autonomous* system, replace $t$ by $w_5$ and add the equation $w_5' = 1$.

**A9.** $\left\{ x_1' = x_2, x_2' = 1 - 3x_2 - 2x_1; \quad x_1(0) = 1, x_2(0) = 0 \right\}.$

**A11.** $\frac{dx_1}{dt} = x_2, \frac{dx_2}{dt} = \frac{1}{2} \left\{ x_1 - y_2^2 \right\}, \frac{dy_1}{dt} = y_2,$ and $\frac{dy_2}{dt} = \frac{4t + y_1}{x_1}.$

**A13.** $\frac{d^2 u}{dx^2} - 4u - 2 = 0$ or $\frac{d^2 v}{dx^2} - 4v + 2 = 0.$

**A15.** $\frac{d^2 x}{dt^2} + 9 \frac{dx}{dt} + 6x - 12 = 0$ or $\frac{d^2 y}{dt^2} + 9 \frac{dy}{dt} + 6y - 18 = 0.$

**B1.** $\left\{ \frac{dx_1}{dt} = x_2, \frac{dx_2}{dt} = -4x_1 - 4x_2; \quad x_1(0) = 2, x_2(0) = -2 \right\}$; the first equation represents the
*velocity* of the mass at time $t$, whereas the second equation represents the *acceleration* of the
mass.

**B3.** $\left\{ \frac{dy_1}{dt} = y_2, \frac{dy_2}{dt} = -\left( \frac{g}{s_0} \right) y_1 \right\}.$

**B5.** $\left\{ w_1' = w_2, w_2' = w_3, w_3' = w_4, w_4' = 6 \sin(4t) + 2w_4 - 5w_3 - 3w_2 + 8w_1 \right\}$; to get an
*autonomous* system, replace $t$ by $w_5$ and add the equation $w_5' = 1$.

**B7.** $\left\{ u_1' = u_2, u_2' = u_3 u_2 + u_3^2 u_1, u_3' = 1; \quad u_1(0) = 1, u_2(0) = 2, u_3(0) = 0 \right\}.$

**C1.** $\frac{d^4 x}{dt^4} - 16x = 120e^{-t} - 64t; \quad x(0) = 6, x'(0) = 8, x''(0) = -48,$ and $x'''(0) = -8$ or
$\frac{d^4 y}{dt^4} - 16y = 540e^{-t} - 96, y(0) = -24, y'(0) = 0, y''(0) = -12,$ and $y'''(0) = 84.$

## Exercises 4.7

**A1. a.** $\{ x_1' = x_2, x_2' = -x_2; \quad x_1(0) = 1, x_2(0) = 2 \}.$

**b.**

c.

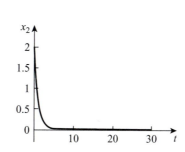

**A3. a.** $\{\dot{y}_1 = y_2, \dot{y}_2 = -y_1; \quad y_1(0) = 2, y_2(0) = 0\}$.

b.

c.

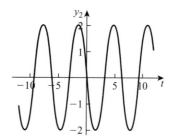

**A5. a.** $\{\dot{x}_1 = x_2, \dot{x}_2 = x_2; \quad x_1(0) = 1 = x_2(0)\}$.

**b.**

**c.**

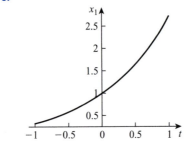

**A7. a.** $\{x_1' = x_2, x_2' = x_1 - x_1^3; \quad x_1(0) = 0, x_2(0) = 1\}$.

**b.**

**c.**

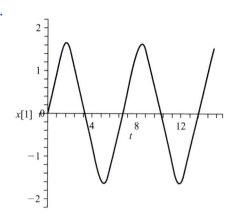

**B1.** $2 \arctan \left( \frac{y}{x} \right) + \ln \left( x^2 + y^2 \right) - C = 0.$

**B3. a.** $\{ \dot{Q}_1 = Q_2, \dot{Q}_2 = \frac{1}{2} \sin t - 14Q_1 - 9Q_2; \quad Q_1(0) = 0, Q_2(0) = 0.1 \}.$

**b.**

**c.**

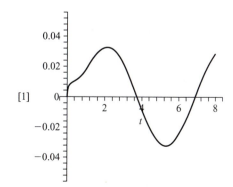

**B5.** Every point on the $x$- and $y$-axes is an equilibrium point. The phase portrait looks like

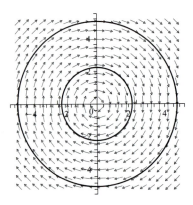

**C1.** The equilibrium solutions are $(0, 0)$, $(-1, 0)$, and $(1, 0)$.

**C3. a.** With $\lambda = 1$, the phase portrait is

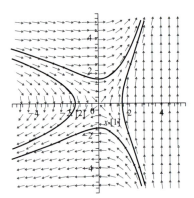

**b.** With $\lambda = -1$, the phase portrait is

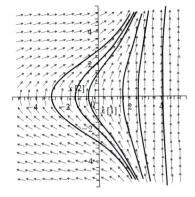

**c.** The value $\lambda = 0$ is a *bifurcation point*, a point at which the phase portrait changes drastically.

## Exercises 4.8

**A1. a.** $\{\dot{x}_1 = x_2, \dot{x}_2 = -64x_1 - 20x_2; \quad x_1(0) = 1/3, x_2(0) = 0\}$.

**b.**

**c.**

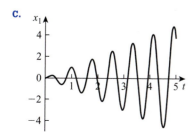

**d.** The mass approaches its equilibrium position but doesn't quite reach it because of the large damping force. In particular, the mass doesn't overshoot its equilibrium position.

**A3. a.** $\{\dot{x}_1 = x_2, \dot{x}_2 = 16\cos 8t - 64x_1; x_1(0) = 0, x_2(0) = 0\}$.

**b.**

**c.**

    **d.** The half lines $x = t$ and $x = -t$ are asymptotes for the graph in (c) for $t \geq 0$:

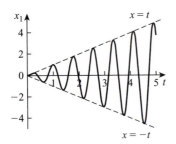

**B1. a.** $(M + 0.5)\ddot{x} + 10\,\dot{x} + kx = 0$.

    **b.** $n = \sqrt{1000\,M + 400}/(2\,M + 1)$.

    **c.** $t = \dfrac{2M + 1}{10}\ln 4$.

    **d.** Removing the damping force means that the scales would continue to oscillate and would not settle down to allow a reading to be taken.

**C1. a.** $Q'' + 8Q' + 15Q = 0$: This equation *can* represent a spring-mass system with spring constant 15 and damping constant 8. (We could also have derived a single second-order equation in $R$.)

    **b.** $\ddot{x} - 6\dot{x} + 10x = 0$: This second-order equation *cannot* represent a spring-mass system because the equation implies that any damping force works in the same direction as the mass's motion. (We could also have derived a single second-order equation in $y$.)

## Exercises 4.9

**A3. a.** $(-\infty, \infty)$.

    **b.** $(-\infty, 0)$ or $(0, \infty)$.

    **c.** $(-\infty, 0)$ or $(0, 1)$ or $(1, \infty)$.

    **d.** $(0, \infty)$.

**B1.** There is no contradiction.

**C1.** Extend the Existence and Uniqueness Theorem to six dimensions.

**C3.** (c) If the solution with initial condition $x(0) = 1/2, y(0) = 0$ satisfies $x^2(t) + y^2(t) \geq 1$ for any finite value of $t$, this means that the solution intersects the solution given in part (b) at some point $(x(t^*), y(t^*))$ on the unit circle. Thus, two solutions of the system pass through the same point in the open disk $x^2 + y^2 < 4$, contradicting the uniqueness established in part (a).

## Exercises 4.10

**A1. a.** $x_{k+1} = x_k + \dfrac{h}{2}\left\{f(t_k, x_k, y_k) + f(t_{k+1}, x_k + hf(t_k, x_k, y_k)), y_k + hg(t_k, x_k, y_k))\right\}$

             $y_{k+1} = y_k + \dfrac{h}{2}\left\{g(t_k, x_k, y_k) + g(t_{k+1}, x_k + hf(t_k, x_k, y_k)), y_k + hg(t_k, x_k, y_k))\right\}$.

    **b.** $x(0.5) \approx 1.1273, y(0.5) \approx 0.5202$.

   **c.** For $x(0.5)$, the absolute error is approximately 0.0003; while for $y(0.5)$, the absolute error is approximately 0.0009.

**A3. a.** $\{u_1' = u_2, u_2' = 2x + 2u_1 - u_2; u_1(0) = 1, u_2(0) = 1\}$.

   **b.** Using Euler's method with $h = 0.1$, we find that $u_1(0.5) \approx 1.8774$ and $u_2(0.5) \approx 4.1711$; $u_1(1.0) \approx 5.5515$ and $u_2(1.0) \approx 13.3031$.

   **c.** Using a fourth-order Runge-Kutta method with $h = 0.1$, we get $u_1(0.5) \approx 2.1784$ and $u_2(0.5) \approx 4.7536$; $u_1(1.0) \approx 6.7731$ and $u_2(1.0) \approx 14.7205$.

**B1. a.** $(x(t), y(t), z(t)) = (0, 5, 0)$ for all the values of $t$ specified. The particle doesn't seem to be moving.

   **b.** The values of $x$, $y$, and $z$ seem to be increasing without bound as $t$ grows larger, with the values of $x$, $y$, and $z$ approaching each other.

**B3.** $t^* = 3.72$, to two decimal places.

**B5. a.** $\left\{ \frac{du}{dx} = v, \frac{dv}{dx} = -\frac{2}{x}v - u^3; u(0) = 1, v(0) = 0 \right\}$.

   **b.** $x \approx 6.9$.

**C1. a.** $\frac{dS}{dt} + \frac{dI}{dt} + \frac{dR}{du} = \frac{d}{dt}(S + I + R) = 0$. This means that the total population does not change.

   **b.**

   **c.**

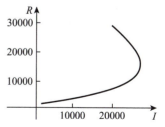

   **d.** We have used the *rkf45* method and all values are rounded to the nearest whole number. The values show the steady increase in the number of people who have recovered, the decreasing number of susceptible people, and the fact that the number of infected people probably peaks between days 10 and 15.

| $t$ | $S$ | $I$ | $R$ |
|---|---|---|---|
| 1 | 44,255 | 3062 | 2682 |
| 2 | 42,649 | 4405 | 2947 |
| 3 | 40,460 | 6217 | 3323 |
| 10 | 13,044 | 25,547 | 11,408 |
| 15 | 3447 | 25,638 | 20,915 |
| 16 | 2681 | 24,609 | 22,710 |
| 17 | 2108 | 23,464 | 24,428 |

    **e.** We conclude that $t \approx 161$ if we round down; but $t \approx 171$ if we round $I$ to the nearest integer.

**C3. a.**

| $t$ | $x(t)$ | $y(t)$ |
|---|---|---|
| 0.01 | 0.4492 | −0.0158 |
| 0.02 | 0.4468 | −0.0113 |
| 0.03 | 0.4432 | −0.0068 |
| 0.04 | 0.4385 | −0.0024 |
| 0.05 | 0.4330 | 0.0019 |
| 0.06 | 0.4266 | 0.0062 |
| 0.07 | 0.4196 | 0.0105 |
| 0.08 | 0.4120 | 0.0146 |
| 0.09 | 0.4039 | 0.0187 |
| 0.10 | 0.3952 | 0.0227 |

**C5.** The direction of the solution curve is *counterclockwise*.

    **b.**

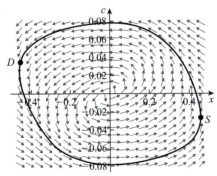

    **c.** $t \approx 1.1$.

    **d.** Diastole: $(x, c) \approx (-0.46, 0.02)$ when $t \approx 0.52$; systole: $(x, c) \approx (0.46, -0.02)$ when $t \approx 1.05$.

## Exercises 5.1

**A1. a.** $\begin{bmatrix} 3 & 4 \\ -1 & -2 \end{bmatrix} \begin{bmatrix} x \\ y \end{bmatrix} = \begin{bmatrix} -7 \\ 5 \end{bmatrix}$;

**b.** $\begin{bmatrix} \pi & -3 \\ 5 & 2 \end{bmatrix}\begin{bmatrix} a \\ b \end{bmatrix} = \begin{bmatrix} 4 \\ -3 \end{bmatrix}.$

**c.** $\begin{bmatrix} 1 & -1 & 1 \\ -1 & 2 & -3 \\ 2 & -3 & 5 \end{bmatrix}\begin{bmatrix} x \\ y \\ z \end{bmatrix} = \begin{bmatrix} 7 \\ 9 \\ 11 \end{bmatrix}.$

**A3.** $V = \begin{bmatrix} 1 \\ 0 \end{bmatrix}.$

**A5.** $\dot{X} = \begin{bmatrix} \dot{x} \\ \dot{y} \end{bmatrix} = \begin{bmatrix} 2 & 1 \\ 3 & 4 \end{bmatrix}\begin{bmatrix} x \\ y \end{bmatrix}.$

**A7.** $\dot{X} = \begin{bmatrix} \dot{x} \\ \dot{y} \end{bmatrix} = \begin{bmatrix} 2 & 1 \\ -1 & 4 \end{bmatrix}\begin{bmatrix} x \\ y \end{bmatrix}.$

**A9.** $\dot{X} = \begin{bmatrix} \dot{x} \\ \dot{y} \end{bmatrix} = \begin{bmatrix} -2 & 1 \\ 0 & -2 \end{bmatrix}\begin{bmatrix} x \\ y \end{bmatrix}.$

**B1. a.** The system is $\{y_1' = y_2, y_2' = 3y_2 - 2y_1\}$, which can be written as

$$\begin{bmatrix} y_1' \\ y_2' \end{bmatrix} = \begin{bmatrix} 0 & 1 \\ -2 & 3 \end{bmatrix}\begin{bmatrix} y_1 \\ y_2 \end{bmatrix}.$$

**b.** The system is $\left\{y_1' = y_2, y_2' = \frac{1}{5}y_1 - \frac{3}{5}y_2\right\}$, which can be written as

$$\begin{bmatrix} y_1' \\ y_2' \end{bmatrix} = \begin{bmatrix} 0 & 1 \\ \frac{1}{5} & -\frac{3}{5} \end{bmatrix}\begin{bmatrix} y_1 \\ y_2 \end{bmatrix}.$$

**c.** The system is $\{y_1' = y_2, y_2' = -\omega^2 y_1\}$, which can be written as

$$\begin{bmatrix} y_1' \\ y_2' \end{bmatrix} = \begin{bmatrix} 0 & 1 \\ -\omega^2 & 0 \end{bmatrix}\begin{bmatrix} y_1 \\ y_2 \end{bmatrix}.$$

# Exercises 5.2

**A1. a.** 17;

**b.** 0;

**c.** $6t^4 + 4\sin t$;

**d.** 1.

**A3.** $\begin{bmatrix} -1 & 2 \\ -4 & 5 \end{bmatrix}$, for example.

**A5. a.** $\begin{bmatrix} \dot{x} \\ \dot{y} \end{bmatrix} = \begin{bmatrix} 1 & -1 \\ -4 & 1 \end{bmatrix}\begin{bmatrix} x \\ y \end{bmatrix}.$

**b.** The characteristic equation is $\lambda^2 - (1+1)\lambda + (1(1) - (-1)(-4)) = \lambda^2 - 2\lambda - 3 = 0$.

    **c.** $\lambda_1 = 3$ and $\lambda_2 = -1$.

    **d.** Any nonzero vector of the form $\begin{bmatrix} x \\ -2x \end{bmatrix} = x \begin{bmatrix} 1 \\ -2 \end{bmatrix}$ is an eigenvector corresponding to the

    eigenvalue $\lambda = 3$. Any nonzero vector of the form $\begin{bmatrix} x \\ 2x \end{bmatrix} = x \begin{bmatrix} 1 \\ 2 \end{bmatrix}$ is an eigenvector

    corresponding to the eigenvalue $\lambda = -1$.

**A7. a.** $\begin{bmatrix} \dot{x} \\ \dot{y} \end{bmatrix} = \begin{bmatrix} 1 & 0 \\ 0 & 1 \end{bmatrix} \begin{bmatrix} x \\ y \end{bmatrix}$.

    **b.** The characteristic equation is $\lambda^2 - (1+1)\lambda + (1-0) = \lambda^2 - 2\lambda + 1 = 0$.

    **c.** $\lambda_1 = 1 = \lambda_2$.

    **d.** Clearly, any nonzero vector $\begin{bmatrix} x \\ y \end{bmatrix}$ is an eigenvector corresponding to the eigenvalue $\lambda = -1$.

    As we see in Section 5.4, a $2 \times 2$ system that has a repeated eigenvalue (an eigenvalue of "multiplicity two") must be handled carefully. In this problem, we can find two eigenvectors that do not lie on the same straight line—$V_1 = \begin{bmatrix} 1 \\ 0 \end{bmatrix}$ and $V_2 = \begin{bmatrix} 0 \\ 1 \end{bmatrix}$, for example.

**A9. a.** $\begin{bmatrix} \dot{x} \\ \dot{y} \end{bmatrix} = \begin{bmatrix} 5 & -1 \\ 2 & 1 \end{bmatrix} \begin{bmatrix} x \\ y \end{bmatrix}$.

    **b.** $\lambda^2 - 6\lambda + 7 = 0$.

    **c.** $\lambda_1 = 3 + \sqrt{2}, \lambda_2 = 3 - \sqrt{2}$.

    **d.** $V_1 = \begin{bmatrix} 1 \\ 2 - \sqrt{2} \end{bmatrix}, V_2 = \begin{bmatrix} 1 \\ 2 + \sqrt{2} \end{bmatrix}$.

**B1. b.** $x = \begin{vmatrix} e & b \\ f & d \end{vmatrix} \Big/ \begin{vmatrix} a & b \\ c & d \end{vmatrix}$ and $y = \begin{vmatrix} a & e \\ c & f \end{vmatrix} \Big/ \begin{vmatrix} a & b \\ c & d \end{vmatrix}$.

**B5.** The phase portrait corresponding to the system in Problem B4 is

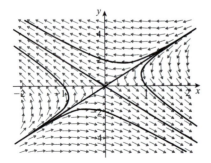

The trajectories are moving away from the origin as $t$ increases. Algebraically, this is a consequence of the fact that one eigenvalue, 3, is positive. Furthermore the trajectories approach the line determined by the representative eigenvector $\begin{bmatrix} 1 \\ -2 \end{bmatrix}$ associated with the larger eigenvalue 3 as $t \to \infty$. The trajectories approach the line determined by $\begin{bmatrix} 1 \\ 2 \end{bmatrix}$, the representative eigenvector associated with the smaller eigenvalue $-1$ as $t \to -\infty$.

**B7. a.** $c_1(x) = \alpha_1 \left( \dfrac{C_0 - c_0}{\alpha_2 - \alpha_1} \right) e^{(\alpha_2 - \alpha_1)x} + \dfrac{\alpha_2 c_0 - \alpha_1 C_0}{\alpha_2 - \alpha_1}$ and

$c_2(x) = \alpha_2 \left( \dfrac{C_0 - c_0}{\alpha_2 - \alpha_1} \right) e^{(\alpha_2 - \alpha_1)x} + \dfrac{\alpha_2 c_0 - \alpha_1 C_0}{\alpha_2 - \alpha_1}.$

**B9. a.** The system is $\{ \dot{x}_1 = x_2, \ \dot{x}_2 = -kx_2 - \frac{g}{L}x_1 \}$.

  **b.** $\lambda^2 + k\lambda + \frac{g}{L} = 0.$

  **c.** $\lambda_1 = \left( -kL + \sqrt{k^2 L^2 - 4gL} \right) \big/ 2L, \lambda_2 = \left( -kL - \sqrt{k^2 L^2 - 4gL} \right) \big/ 2L.$

  **d.** $V_1 = \begin{bmatrix} 1 \\ \left( -kL + \sqrt{k^2 L^2 - 4gL} \right) \big/ 2L \end{bmatrix}, V_2 = \begin{bmatrix} 1 \\ \left( -kL - \sqrt{k^2 L^2 - 4gL} \right) \big/ 2L \end{bmatrix}.$

  **e.** As $t \to \infty$, the pendulum tends to its equilibrium position $(0, 0)$.

**C1. d.** Part (c) shows that the sign of $d\theta/dt$ equals the sign of $x\dot{y} - y\dot{x}$, which has just been shown to equal $c\left[ x + \left( \frac{d-a}{2c} \right)y \right]^2 + \frac{y^2}{4c} \left[ 4(ad - bc) - (a + d)^2 \right]$. The first term of this last expression is $c$ times a perfect square, while in the second term the bracketed expression is positive by part (b). Given the presence of another perfect square, $y^2/4$, it is clear that $x\dot{y} - y\dot{x}$ (and so $d\theta/dt$) must have the same sign as $c$.

## Exercises 5.3

**A1. a.** $\lambda_1 = 3$ and $\lambda_2$: Any nonzero vector of the form $\begin{bmatrix} x \\ 0 \end{bmatrix} = x\begin{bmatrix} 1 \\ 0 \end{bmatrix}$ is an eigenvector corresponding to the eigenvalue $\lambda = 3$. Any nonzero vector of the form $\begin{bmatrix} 0 \\ y \end{bmatrix} = y\begin{bmatrix} 0 \\ 1 \end{bmatrix}$ is an eigenvector corresponding to the eigenvalue $\lambda = 2$.

  **b.** Here is a plot of several trajectories with the eigenvectors (essentially, the $x$- and $y$-axes) shown:

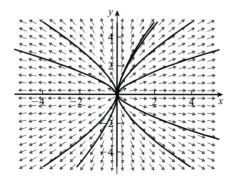

**A3.** **a.** $\lambda_1 = 1$ and $\lambda_2 = -2$: Any nonzero vector of the form $\begin{bmatrix} x \\ -4x \end{bmatrix} = x \begin{bmatrix} 1 \\ -4 \end{bmatrix}$ is an eigenvector

corresponding to the eigenvalue $\lambda = 1$. Any nonzero vector of the form $\begin{bmatrix} x \\ -x \end{bmatrix} = x \begin{bmatrix} 1 \\ -1 \end{bmatrix}$ is an eigenvector corresponding to the eigenvalue $\lambda = -2$.

**b.** Here's the plot of several trajectories and the eigenvectors:

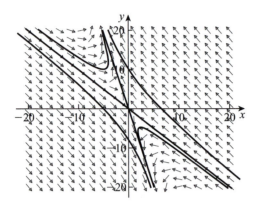

**A5.** **a.** $\lambda_1 = 2$ and $\lambda_2 = -4$: Any nonzero vector of the form $\begin{bmatrix} x \\ \frac{1}{5}x \end{bmatrix} = x \begin{bmatrix} 1 \\ \frac{1}{5} \end{bmatrix} = x \begin{bmatrix} 5 \\ 1 \end{bmatrix}$ is an eigenvector corresponding to the eigenvalue $\lambda = 2$. Any nonzero vector of the form $\begin{bmatrix} x \\ -x \end{bmatrix} = x \begin{bmatrix} 1 \\ -1 \end{bmatrix}$ is an eigenvector corresponding to the eigenvalue $\lambda = -4$.

**b.** Trajectories and eigenvectors are shown here:

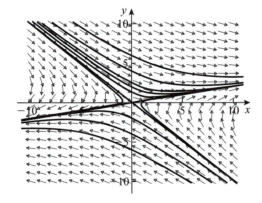

**A7.** **a.** $\lambda_1 = \frac{-5+\sqrt{17}}{2}$ and $\lambda_2 = \frac{-5-\sqrt{17}}{2}$, both irrational numbers; using a CAS, we find corresponding representative eigenvectors $V_1 = \begin{bmatrix} \frac{-1+\sqrt{17}}{8} \\ 1 \end{bmatrix}$ and $V_2 = \begin{bmatrix} \frac{-1-\sqrt{17}}{8} \\ 1 \end{bmatrix}$.

**b.** Here are some trajectories and the eigenvectors (which are difficult to pick out):

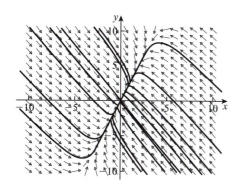

**A9. a.** $\lambda_1 = \sqrt{5}$ and $\lambda_2 = -\sqrt{5}$, irrational numbers: Any nonzero vector of the form

$$\begin{bmatrix} x \\ -\left(2 + \sqrt{5}\right) \end{bmatrix} = x \begin{bmatrix} 1 \\ -\left(2 + \sqrt{5}\right) \end{bmatrix} = x \begin{bmatrix} 2 - \sqrt{5} \\ 1 \end{bmatrix} \text{ is an eigenvector corresponding to the}$$

eigenvalue $\lambda = \sqrt{5}$. Any nonzero vector of the form $\begin{bmatrix} x \\ -\left(2 - \sqrt{5}\right)x \end{bmatrix} = x \begin{bmatrix} 1 \\ -\left(2 - \sqrt{5}\right) \end{bmatrix} =$

$x \begin{bmatrix} 2 + \sqrt{5} \\ 1 \end{bmatrix}$ is an eigenvector corresponding to the eigenvalue $\lambda = -\sqrt{5}$.

**b.** Here are some trajectories and the eigenvectors:

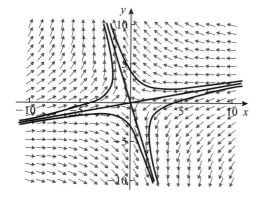

**A11. a.** $\lambda_1 = 0, V_1 = \begin{bmatrix} -2 \\ 1 \end{bmatrix}; \lambda_2 = 1, V_1 = \begin{bmatrix} -3 \\ 1 \end{bmatrix}.$

**b.** Here are the eigenvectors and some trajectories:

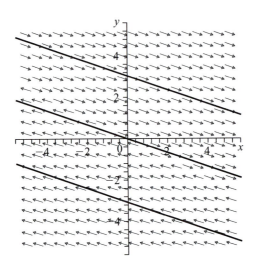

The two eigenvectors are very close together and may be difficult to see as separate vectors.

**B1. a.** $\lambda = 0$ and $\lambda = -2$.

    **b.** Representative eigenvectors corresponding to the eigenvalues $\lambda_1 = 0$ and $\lambda_2 = -2$ are

$$V_1 = \begin{bmatrix} 3 \\ 4 \end{bmatrix} \text{ and } V_2 = \begin{bmatrix} 1 \\ 2 \end{bmatrix}, \text{ respectively.}$$

    **c.** Here's a plot of some trajectories:

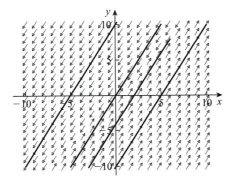

Every point of the line $y = \frac{4}{3}x$ is an equilibrium point. The origin is a *sink*, while every other point on the line is a *node*. All other trajectories (straight lines) seem to be parallel to the trajectory determined by the eigenvector $V_1$.

    **d.** $X(t) = c_1 e^{0 \cdot t} \begin{bmatrix} 3 \\ 4 \end{bmatrix} + c_2 e^{-2t} \begin{bmatrix} 1 \\ 2 \end{bmatrix} = c_1 \begin{bmatrix} 3 \\ 4 \end{bmatrix} + c_2 e^{-2t} \begin{bmatrix} 1 \\ 2 \end{bmatrix}.$

**B3.** $X(t) = \frac{1}{2}e^{3t} \begin{bmatrix} 1 \\ 5 \end{bmatrix} + \frac{1}{2}e^{-t} \begin{bmatrix} 1 \\ 1 \end{bmatrix}$; the solution approaches $\frac{1}{2}e^{3t} \begin{bmatrix} 1 \\ 5 \end{bmatrix}$ (the line $y = 5x$) asymptotically.

**B5. b.** Here are some trajectories for different values of $\alpha$:

$\alpha = 0$

$\alpha = 5$

$\alpha = -5$

When $\alpha = 0$, the trajectory is a straight line from $(0, 0.5)$ to the origin. For $\alpha > 0$, the trajectory swirls down from $(0, 0.5)$ toward the origin in a clockwise direction, flattening as $\alpha$ increases. For $\alpha < 0$, the trajectory swirls from $(0, 0.5)$ toward the origin in a counterclockwise direction, flattening as $\alpha$ increases in the negative direction.

**B7.** $\left(\frac{1}{CR_2} - \frac{R_1}{L}\right)^2 - \frac{4}{CL} > 0$ [Also: $L + CR_1R_2 > 2R_2\sqrt{LC}$, or $L > R_2\left(2\sqrt{LC} - CR_1\right)$.

**C1. a.** $X(t) = \begin{bmatrix} r(t) \\ s(t) \end{bmatrix} = c_1 e^{\left(\frac{-2+\sqrt{2}}{2}\right)t}\begin{bmatrix} -\sqrt{2} \\ 1 \end{bmatrix} + c_2 e^{\left(\frac{-2-\sqrt{2}}{2}\right)t}\begin{bmatrix} \sqrt{2} \\ 1 \end{bmatrix}$

$$= \begin{bmatrix} -\sqrt{2}c_1 e^{\left(\frac{-2+\sqrt{2}}{2}\right)t} + \sqrt{2}c_2 e^{\left(\frac{-2-\sqrt{2}}{2}\right)t} \\ c_1 e^{\left(\frac{-2+\sqrt{2}}{2}\right)t} + c_2 e^{\left(\frac{-2-\sqrt{2}}{2}\right)t} \end{bmatrix} : \text{The origin is a } sink.$$

**b.** $X(t) = \begin{bmatrix} r(t) \\ s(t) \end{bmatrix} = c_1 e^{\left(\sqrt{2}-1\right)t}\begin{bmatrix} 1 \\ -\sqrt{2} \end{bmatrix} + c_2 e^{\left(-1-\sqrt{2}\right)t}\begin{bmatrix} 1 \\ \sqrt{2} \end{bmatrix}$

$$= \begin{bmatrix} c_1 e^{\left(\sqrt{2}-1\right)t} + c_2 e^{\left(-1-\sqrt{2}\right)t} \\ -\sqrt{2}c_1 e^{\left(\sqrt{2}-1\right)t} + \sqrt{2}c_2 e^{\left(-1-\sqrt{2}\right)t} \end{bmatrix} : \text{The origin is a } saddle\ point.$$

**c.** $\lambda_1 = -1 + \sqrt{\beta}$ and $\lambda_2 = -1 - \sqrt{\beta}$; the bifurcation point occurs at $\beta = 1$.

## Exercises 5.4

**A1. a.** $\lambda_1 = 3 = \lambda_2$; $V_1 = \begin{bmatrix} 1 \\ 0 \end{bmatrix}$ and $V_2 = \begin{bmatrix} 0 \\ 1 \end{bmatrix}$.

**b.** Here's a plot of the eigenvectors and some trajectories:

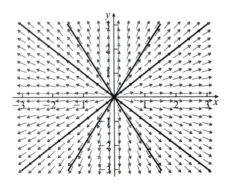

**A3.** **a.** $\lambda_1 = 3 = \lambda_2$: Any nonzero vector of the form $\begin{bmatrix} x \\ x \end{bmatrix} = x \begin{bmatrix} 1 \\ 1 \end{bmatrix}$ is an eigenvector corresponding to the repeated eigenvalue $\lambda = 3$. All eigenvectors lie on the straight line determined by $\begin{bmatrix} 1 \\ 1 \end{bmatrix}$ and there is only one linearly independent eigenvector.

**b.** Here's a plot of the eigenvector and some trajectories:

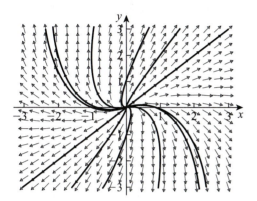

**A5.** **a.** $\lambda_1 = -1 = \lambda_2$: Any nonzero vector of the form $\begin{bmatrix} x \\ x \end{bmatrix} = x \begin{bmatrix} 1 \\ 1 \end{bmatrix}$ is an eigenvector corresponding to the repeated eigenvalue $\lambda = -1$. All eigenvectors lie on the straight line determined by $\begin{bmatrix} 1 \\ 1 \end{bmatrix}$ and there is only one linearly independent eigenvector.

**b.** Here's a plot of the eigenvector and some trajectories:

**A7. a.** $\lambda_1 = -2 = \lambda_2$: Any nonzero vector of the form $\begin{bmatrix} -x \\ x \end{bmatrix} = x \begin{bmatrix} -1 \\ 1 \end{bmatrix}$ is an eigenvector corresponding to the repeated eigenvalue $\lambda = -2$. All eigenvectors lie on the straight line determined by $\begin{bmatrix} -1 \\ 1 \end{bmatrix}$ and there is only one linearly independent eigenvector.

**b.** Here's a plot of the eigenvector and some trajectories:

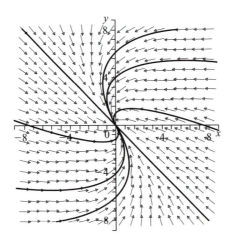

**B1.** $\alpha^2 = 4\beta$.

**B3.** Two such systems are $\{\dot{x} = -2x, \dot{y} = -2y\}$ and $\{\dot{x} = x + 3y, \dot{y} = -3x - 5y\}$.

**B7. b.** The sole linearly independent eigenvector is $V = \begin{bmatrix} 1 \\ 0 \end{bmatrix}$.

## Exercises 5.5

**A1. a.** $\lambda_1 = -1 + 2i$ and $\lambda_2 = -1 - 2i$; $V_1 = \begin{bmatrix} i \\ 1 \end{bmatrix}$ and $V_2 = \overline{V}_1 = \begin{bmatrix} -i \\ 1 \end{bmatrix}$.

**b.** Here's a plot of some trajectories, spirals swirling into the origin (a sink):

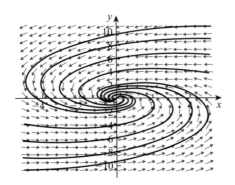

**A3. a.** $\lambda_1 = -0.5 + i$ and $\lambda_2 = -0.5 - i$; $V_1 = \begin{bmatrix} -1 \\ i \end{bmatrix}$ and $V_2 = \begin{bmatrix} -1 \\ -i \end{bmatrix}$.

**b.** Some trajectories, spirals swirling toward the origin (a sink), follow:

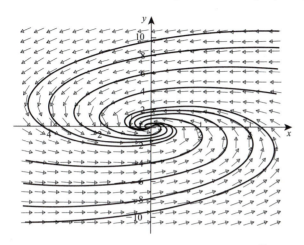

**A5. a.** $\lambda_1 = \left(1 + \sqrt{3}i\right)/2$ and $\lambda_2 = \left(1 - \sqrt{3}i\right)/2$; $V_1 = \begin{bmatrix} 1 \\ \left(-3 + \sqrt{3}i\right)/2 \end{bmatrix}$ and

$V_2 = \begin{bmatrix} 1 \\ \left(-3 - \sqrt{3}\,i\right)/2 \end{bmatrix}$.

**b.** Here are some trajectories, spirals swirling away from the origin (a source):

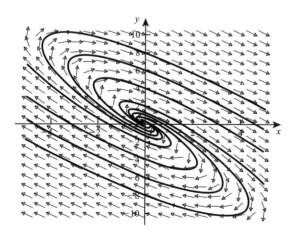

**A7. a.** $\lambda_1 = -6 + i$ and $\lambda_2 = -6 - i$; $V_1 = \begin{bmatrix} 1 - i \\ 2 \end{bmatrix}$ and $V_2 = \begin{bmatrix} 1 + i \\ 2 \end{bmatrix}$.

**b.** Some trajectories, spirals swirling toward the origin (a sink), follow:

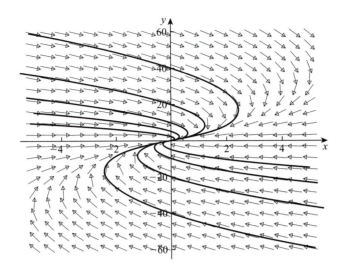

**A9. a.** $\lambda_1 = 5 + 2i$ and $\lambda_2 = 5 - 2i$; $V_1 = \begin{bmatrix} 1 \\ 1 - 2i \end{bmatrix}$ and $V_2 = \begin{bmatrix} 1 \\ 1 + 2i \end{bmatrix}$.

    **b.** Some trajectories, spirals swirling away from the origin (a source), follow:

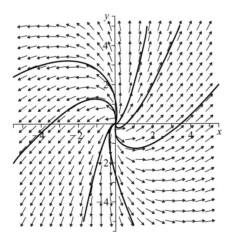

**B1. a.** One such system has the matrix of coefficients $\begin{pmatrix} 2 & -4 \\ 1 & 2 \end{pmatrix}$, which yields the system

    $\{\dot{x} = 2x - 4y, \ \dot{y} = x + 2y\}$.

    **b.** One such system has the matrix of coefficients $\begin{pmatrix} 1 & -2 \\ 5 & -1 \end{pmatrix}$, which yields the system

    $\{\dot{x} = x - 2y, \ \dot{y} = 5x - y\}$.

**c.** One such system has the matrix of coefficients $\begin{pmatrix} -1 & 2 \\ -2 & -1 \end{pmatrix}$, which yields the system $\{\dot{x} = -x + 2y, \dot{y} = -2x - y\}$.

**B5.** $X(t) = e^{-\frac{2}{3}t} \begin{pmatrix} \cos\left(\frac{\sqrt{2}}{3}t\right) - 2\sqrt{2}\sin\left(\frac{\sqrt{2}}{3}t\right) \\ -3\sqrt{2}\sin\left(\frac{\sqrt{2}}{3}t\right) \end{pmatrix}$.

## Exercises 5.6

**A1.** $\left\{ x(t) = \frac{1}{2}e^t - \frac{1}{2}(\sin t + \cos t), \; y(t) = t - 1 + 2e^{-t} \right\}$.

**A3.** $X_{\text{GNH}} = \begin{bmatrix} c_1 e^t - c_2 e^{-t} + te^t - \frac{1}{2}e^t - t^2 - 2 \\ c_1 e^t + c_2 e^{-t} + te^t - \frac{3}{2}e^t - 2t \end{bmatrix} = \begin{bmatrix} \left(c_1 - \frac{1}{2}\right)e^t - c_2 e^{-t} + te^t - t^2 - 2 \\ \left(c_1 - \frac{3}{2}\right)e^t + c_2 e^{-t} + te^t - 2t \end{bmatrix}$.

**A5.** $X_{\text{GNH}} = \begin{bmatrix} 2c_1 e^{4t} - c_2 e^t + 3e^{5t} \\ c_1 e^{4t} + c_2 e^t + e^{5t} \end{bmatrix}$.

**A7.** $X_{\text{GNH}} = \begin{bmatrix} -c_1 e^{3t} + c_2 e^{2t} + te^{2t} \\ c_1 e^{3t} - 2c_2 e^{2t} + 2e^{2t} - 2te^{2t} \end{bmatrix} = \begin{bmatrix} -c_1 e^{3t} + (t + c_2)e^{2t} \\ c_1 e^{3t} - (2t - 2 + 2c_2)e^{2t} \end{bmatrix}$.

**A9.** $X_{\text{GNH}} = \begin{bmatrix} 3c_1 e^{4t} + c_2 e^{2t} - 4e^{3t} - e^{-t} \\ c_1 e^{4t} + c_2 e^{2t} - 2e^{3t} - 2e^{-t} \end{bmatrix}$.

**A11.** $X_{\text{GNH}} = \begin{bmatrix} -c_1 e^{3t} + c_2 e^t + 2e^t \cos t - e^t \sin t \\ c_1 e^{3t} + c_2 e^t + 3e^t \cos t + e^t \sin t \end{bmatrix}$.

**A13.** $X_{\text{GNH}} = \begin{bmatrix} -\frac{1}{2}c_1 e^{-2t} - c_2 e^{-t} + \frac{1}{10}\sin t - \frac{3}{10}\cos t + \frac{1}{6}e^t \\ -c_1 e^{-2t} + c_2 e^{-t} + \frac{1}{10}\cos t + \frac{3}{10}\sin t + \frac{1}{6}e^t \end{bmatrix}$.

**A15.** $X_{\text{GNH}} = \begin{bmatrix} (c_1 + (c_2 + 1)t)e^t + \frac{1}{4}e^{-t} \\ c_2 e^t - \frac{1}{2}e^{-t} \end{bmatrix}$.

**B7. a.** $X_{\text{GNH}} = \begin{bmatrix} x(t) \\ y(t) \end{bmatrix} = \begin{bmatrix} -\frac{I}{k_1}e^{-k_1 t} + I/k_1 \\ \frac{I}{k_1 - k_2}e^{-k_1 t} - \frac{k_1 I}{k_2(k_1 - k_2)}e^{-k_2 t} + I/k_2 \end{bmatrix}$

$= \begin{bmatrix} \frac{I}{k_1}\left(1 - e^{-k_1 t}\right) \\ \frac{I}{k_2}\left(1 + \frac{k_2}{k_1 - k_2}e^{-k_1 t} - \frac{k_1}{k_1 - k_2}e^{-k_2 t}\right) \end{bmatrix}$.

**b.** $\lim\limits_{t \to \infty} x(t) = \frac{I}{k_1}$ and $\lim\limits_{t \to \infty} y(t) = \frac{I}{k_2}$.

**c.** The graphs of $x(t)$ and $y(t)$ for the decongestant follow:

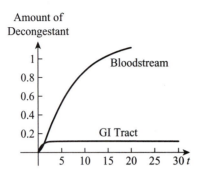

**d.** The graphs of $x(t)$ and $y(t)$ for the antihistamine follow:

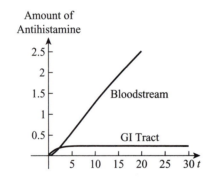

**B9. a.** $ax - by + e > 0$.

**b.** $-cx + dy - e > 0$.

**c.** We have the equilibrium point $(x^*, y^*) = \left( \frac{e(b-d)}{ad-bc}, \frac{e(a-c)}{ad-bc} \right)$, provided that $b \geq d, a \geq c$, and $ad - bc \neq 0$. (We can't have a negative number of supporters.)

**C1. a.** $X_{\text{GNH}} = \begin{bmatrix} x(t) \\ y(t) \end{bmatrix} =$

$$\begin{bmatrix} \left[ \frac{1}{2}\left(\alpha - \frac{l-d}{b}\right) - \frac{\sqrt{ab}}{2b}\left(\beta - \frac{k-c}{a}\right) \right]e^{\sqrt{ab}t} + \left[ \frac{1}{2}\left(\alpha - \frac{l-d}{b}\right) + \frac{\sqrt{ab}}{2b}\left(\beta - \frac{k-c}{a}\right) \right]e^{-\sqrt{ab}t} + \frac{l-d}{b} \\ \left[ -\frac{\sqrt{ab}}{2a}\left(\alpha - \frac{l-d}{b}\right) + \frac{1}{2}\left(\beta - \frac{k-c}{a}\right) \right]e^{\sqrt{ab}t} + \left[ \frac{\sqrt{ab}}{2a}\left(\alpha - \frac{l-d}{b}\right) + \frac{1}{2}\left(\beta - \frac{k-c}{a}\right) \right]e^{-\sqrt{ab}t} + \frac{k-c}{a} \end{bmatrix}.$$

**b.** $\beta - \frac{k-c}{a} < \frac{\sqrt{ab}}{a}\left(\alpha - \frac{l-d}{b}\right)$.

**c.** Here are the graphs of $x(t)$ and $y(t)$ for $0 \le t \le 50$:

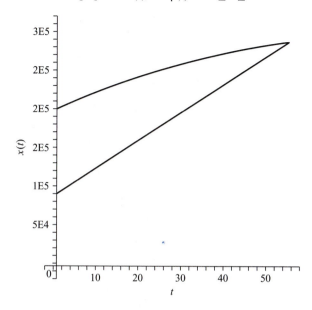

Extending the time axis a bit and using the "trace" or "zoom" capabilities of a CAS or graphing calculator, we find that $x(t^*) = y(t^*)$ when $t^* \approx$ **54 days**. From the graph we see that side "$y$" is winning after 50 days.

## Exercises 5.7

**A1. a.** $\begin{bmatrix} \dot{x} \\ \dot{y} \\ \dot{z} \end{bmatrix} = \begin{bmatrix} 1 & -1 & 1 \\ 1 & 1 & -1 \\ 2 & -1 & 0 \end{bmatrix} \begin{bmatrix} x \\ y \\ z \end{bmatrix}$.

**b.** $\lambda_1 = -1, V_1 = \begin{bmatrix} 1 \\ -3 \\ -5 \end{bmatrix}; \lambda_2 = 1, V_2 = \begin{bmatrix} 1 \\ 1 \\ 1 \end{bmatrix}; \lambda_3 = 2, V_3 = \begin{bmatrix} 1 \\ 0 \\ 1 \end{bmatrix}$.

**c.** $X(t) = \begin{bmatrix} c_1 e^{-t} + c_2 e^{t} + c_3 e^{2t} \\ -3c_1 e^{-t} + c_2 e^{t} \\ -5c_1 e^{-t} + c_2 e^{t} + c_3 e^{2t} \end{bmatrix}$.

**A3. a.** $\begin{bmatrix} \dot{x} \\ \dot{y} \\ \dot{z} \end{bmatrix} = \begin{bmatrix} 3 & -1 & 1 \\ 1 & 1 & 1 \\ 4 & -1 & 4 \end{bmatrix} \begin{bmatrix} x \\ y \\ z \end{bmatrix}$.

**b.** $\lambda_1 = 1, V_1 = \begin{bmatrix} 1 \\ 1 \\ -1 \end{bmatrix}; \lambda_2 = 2, V_2 = \begin{bmatrix} 1 \\ -2 \\ -3 \end{bmatrix}; \lambda_3 = 5, V_3 = \begin{bmatrix} 1 \\ 1 \\ 3 \end{bmatrix}$.

**c.** $X(t) = \begin{bmatrix} c_1 e^t + c_2 e^{2t} + c_3 e^{5t} \\ c_1 e^t - 2c_2 e^{2t} + c_3 e^{5t} \\ -c_1 e^t - 3c_2 e^{2t} + 3c_3 e^{5t} \end{bmatrix}$.

**A5. a.** $\begin{bmatrix} \dot{x} \\ \dot{y} \\ \dot{z} \end{bmatrix} = \begin{bmatrix} 2 & -1 & 1 \\ 1 & 2 & -1 \\ 1 & -1 & 2 \end{bmatrix} \begin{bmatrix} x \\ y \\ z \end{bmatrix}$.

**b.** $\lambda_1 = 2, V_1 = \begin{bmatrix} 1 \\ 1 \\ 1 \end{bmatrix}; \lambda_2 = 3, V_2 = \begin{bmatrix} 1 \\ 0 \\ 1 \end{bmatrix}; \lambda_3 = 1, V_3 = \begin{bmatrix} 0 \\ 1 \\ 1 \end{bmatrix}$.

**c.** $X(t) = \begin{bmatrix} c_1 e^{3t} + c_3 e^{2t} \\ c_2 e^t + c_3 e^{2t} \\ c_1 e^{3t} + c_2 e^t + c_3 e^{2t} \end{bmatrix}$.

**A7.** The space trajectory through $(0, 1, 0)$ when $t = 0$ is

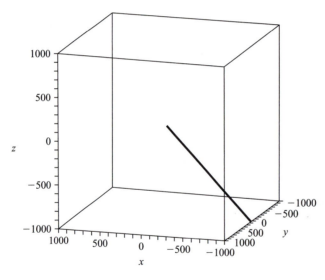

**B3.** $X(t) = \begin{bmatrix} x(t) \\ y(t) \\ z(t) \end{bmatrix} = \begin{bmatrix} -e^{-t} \\ e^{-t} \\ 0 \end{bmatrix}$.

**B5. a.** $\{y_1' = y_2, \; y_2' = -2y_1 + y_3, \; y_3' = y_4, \; y_4' = y_1 - 2y_3\}$.

**b.** $\lambda_1 = i, \; \lambda_2 = -i, \; \lambda_3 = \sqrt{3}i, \; \lambda_4 = -\sqrt{3}i$.

**c.** $Y(t) = \begin{bmatrix} y_1(t) \\ y_2(t) \\ y_3(t) \\ y_4(t) \end{bmatrix} = \begin{bmatrix} x_1(t) \\ x_1'(t) \\ x_2(t) \\ x_2'(t) \end{bmatrix} \begin{bmatrix} 2\cos t + \sin t \\ -2\sin t + \cos t \\ 2\cos t + \sin t \\ -2\sin t + \cos t \end{bmatrix}$.

The masses oscillate in sync because $x_1(t) = x_2(t)$ and $x_1'(t) = x_2'(t)$ for all positive values of $t$.

**d.** $Y(t) = \begin{bmatrix} y_1(t) \\ y_2(t) \\ y_3(t) \\ y_4(t) \end{bmatrix} = \begin{bmatrix} x_1(t) \\ x_1'(t) \\ x_2(t) \\ x_2'(t) \end{bmatrix} = \begin{bmatrix} 2\cos\left(\sqrt{3}t\right) + \sin\left(\sqrt{3}t\right) \\ -2\sqrt{3}\sin\left(\sqrt{3}t\right) + \sqrt{3}\cos\left(\sqrt{3}t\right) \\ -2\cos\left(\sqrt{3}t\right) - \sin\left(\sqrt{3}t\right) \\ 2\sqrt{3}\sin\left(\sqrt{3}t\right) - \sqrt{3}\cos\left(\sqrt{3}t\right) \end{bmatrix}.$

The masses are now out of sync because at any time $t$ the masses are located at opposite sides of their respective equilibrium positions and are moving either toward each other or away from each other.

**e.** There are two natural frequencies, $1/2\pi$ for the system in part (c), and $\sqrt{3}/2\pi$. The "1" and the "$\sqrt{3}$" in the numerators of the frequencies are the imaginary parts of the eigenvalues. A third mode of oscillation is possible, combining the two natural frequencies already found.

**C1.** $X(t) = \begin{bmatrix} A(t) \\ B(t) \\ C(t) \end{bmatrix} = \begin{bmatrix} 11{,}000e^{-0.02t} + (11{,}000/3)e^{-0.06t} + 25{,}000/3 \\ -(22{,}000/3)e^{-0.06t} + 25{,}000/3 \\ -11{,}000e^{-0.02t} + (11{,}000/3)\,e^{-0.06t} + 25{,}000/3 \end{bmatrix}.$

**C3. a.** $\dot{X}(t) = \begin{bmatrix} -0.1 & 0.02 & 0 & 0 & 0 & 0 \\ 0.1 & -0.14 & 0 & 0 & 0 & 0 \\ 0 & 0 & -0.1 & 0.02 & 0 & 0 \\ 0 & 0 & 0.1 & -0.14 & 0 & 0 \\ 0 & 0 & 0 & 0 & -0.1 & 0.02 \\ 0 & 0 & 0 & 0 & 0.1 & -0.14 \end{bmatrix} \begin{bmatrix} x_1(t) \\ x_2(t) \\ y_1(t) \\ y_2(t) \\ z_1(t) \\ z_2(t) \end{bmatrix} + \begin{bmatrix} 4 \\ 2 \\ 0 \\ 0 \\ 0 \\ 0 \end{bmatrix}.$

**b.** $x_1(t) = 50 - 4.59e^{-0.169t} - 45.41e^{-0.071t}$, $x_2(t) = 50 + 15.82e^{-0.169t} - 65.82e^{-0.071t}$, $y_1(t) = 14.79e^{-0.169t} + 35.21e^{-0.071t}$, $y_2(t) = -51.03e^{-0.169t} + 51.03e^{-0.071t}$, $z_1(t) = -10.21e^{-0.169t} + 10.21e^{-0.071t}$, $z_2(t) = 35.21e^{-0.169t} + 14.79e^{-0.071t}$.

**c.** Here are $x_1(t)$, $y_1(t)$, and $z_1(t)$ on the same set of axes:

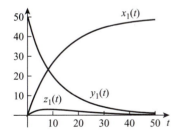

**d.** Here are $x_2(t)$, $y_2(t)$, and $z_2(t)$ on the same set of axes:

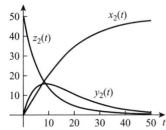

**C5.** $X_{\text{GNH}} = \begin{bmatrix} x(t) \\ y(t) \\ z(t) \end{bmatrix} = \begin{bmatrix} t^2 + \frac{c_3}{12} \\ t^3 + t^2 + \frac{c_3}{4}t + \frac{c_2}{4} - \frac{c_3}{48} \\ t^4 + \frac{5}{3}t^3 + \frac{1}{2}(c_3 + 1)t^2 + c_2 t + c_1 \end{bmatrix}.$

## Exercises 6.1

**A1.** $2/s^3, s > 0.$

**A3.** $3!/s^4, s > 0.$

**A5.** $1/(s-a)^2, s > a.$

**A7.** $10/s + 100/(s-2), s > 2.$

**A9.** $\left(12 - 14s + 5s^2 - 17s^3\right)/s^4,\ s > 0.$

**A11.** $2/(s-1) - 3/(s+1) + 8/s^3,\ s > 1.$

**A13.** $\left(1 - e^{-4s} - 4se^{-4s}\right)/s^2,\ s > 0.$

**A15.** $\left(s + e^{-2s} - e^{-s}\right)/s^2,\ s > 0.$

**A17.** $\mathcal{L}[y(t)] = 1/(s-1).$

**A19.** $\mathcal{L}[y(t)] = (2s+5)/\left((s+1)(s+2)\right).$

**A21.** $\mathcal{L}[y(t)] = s/\left(s^2+1\right).$

**A23.** $\mathcal{L}[y(x)] = \left(s^3 - 2s^2 + s + 3\right)/\left((s^2+1)(s^2 - s - 2)\right).$

**A25.** $\mathcal{L}[y(x)] = 2\left(s^2 + s - 1\right)/\left(s^3\,(s-1)^3\right).$

**B1.** $\mathcal{L}[\sinh(at)] = 1/[2(s-a)] - 1/[2(s+a)] = a/\left(s^2 - a^2\right)\,;\ \mathcal{L}[\cosh(at)] = s/\left(s^2 - a^2\right).$

**B3.** $\frac{1}{s}\left(\frac{e^{as}-1}{e^{as}+1}\right) = \frac{1}{s}\tanh\left(\frac{as}{2}\right).$

**B9.** For example, $h(t) = \pi$ for $t = 1$, $h(t) = 1$ for $t \ne 1$. Functions that differ at only a finite number of points have equal integrals.

## Exercises 6.2

**A1.** $\frac{1}{3}\sin 3t.$

**A3.** $\cos\sqrt{2}t.$

**A5.** $\frac{1}{2}\left\{1 - e^{-t}(\cos t + \sin t)\right\}.$

**A7.** $4e^{2t} - 3\cos 4t + \frac{5}{2}\sin 2t.$

**A9.** $-\frac{1}{6}e^{-t} - \frac{4}{3}e^{2t} + \frac{7}{2}e^{3t}.$

**B7.** $s/\left(s^2+1\right)^2.$

**B9.** $y(t) = 2t^2 - 6t + 7 + e^{-2t} - 8e^{-t}$.

**B11.** $y(x) = x^2 + 4x + 4 + x^2 e^x - 4e^x$.

**B13.** $Q(t) = \begin{cases} e^{-t} \sin t & \text{for } 0 \le t < \pi \\ \frac{2}{5} \cos t - \frac{1}{5} \sin t - \frac{1}{5} e^{(-t+\pi)} \{2 \cos t + \sin t\} + e^{-t} \sin t & \text{for } t \ge \pi. \end{cases}$

**B15.** $f(t) = 4t + \frac{2}{3}t^3$.

**C1.** $x(t) = 2 - e^{-t}$.

**C3.** $x(t) = \frac{1}{10} t^5 e^{2t} + \frac{1}{4} t^4 e^{2t} = \frac{1}{20} t^4 e^{2t} (2t + 5)$.

# Exercises 6.3

**A1. a.**

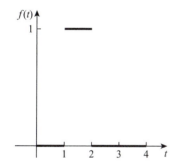

**b.** $f(t) = 1 \cdot U(t - 1) + U(t - 2)[0 - 1] = U(t - 1) - U(t - 2)$.

**A3. a.**

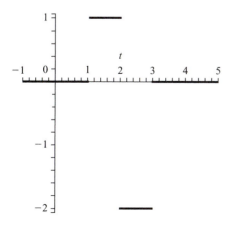

**b.** $f(t) = U(t - 1) - 3U(t - 2) + 2U(t - 3)$.

**A5. a.**

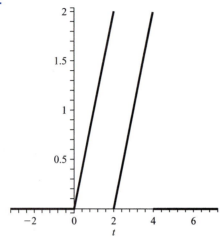

**b.** $f(t) = t \cdot U(t) + U(t-2)[(t-2)-t] + U(t-4)[0-(t-2)] = tU(t) - 2U(t-2) + (2-t)U(t-4)$.

**A11.** $\mathcal{L}[f(t)] = \left(2 - 2e^{-2s} + 4s^2 e^{-2s}\right)/s^3$.

**A13.** $\mathcal{L}[f(t)] = \left(1 - 2e^{-2s} + e^{-4s}\right)/s^2$.

**B1.** $\left[\frac{1}{2}U(t-\pi) - \frac{1}{2}U(t-3\pi)\right]\sin^2 t = \frac{1}{2}\sin^2 t \quad \text{for } \pi < t < 3\pi \quad \text{and } 0 \text{ elsewhere.}$

**B3.** $y(t) = \begin{cases} -\frac{14}{5}e^{5t/4} + 6t + \frac{24}{5} & \text{for } 0 \le t < 1 \\ -\frac{14}{5}e^{5t/4} + \frac{54}{5}e^{\frac{5}{4}(t-1)} & \text{for } t \ge 1. \end{cases}$

**B5.** $y(t) = 1 - \cos t + \sin t - U\left(t - \frac{\pi}{2}\right)(1 - \sin t) = \begin{cases} 1 - \cos t + \sin t & \text{for } t < \pi/2 \\ -\cos t + 2\sin t & \text{for } t \ge \pi/2. \end{cases}$

**B7.** $y(t) =$
$$\begin{cases} 4 + \left(\frac{10\sqrt{17}}{17} - 2\right)e^{-(\sqrt{17}+5)t/2} - \left(2 + \frac{10\sqrt{17}}{17}\right)e^{(\sqrt{17}-5)t/2} & \text{for } t \le 1 \\ 4 - e^{t-1} + \left(\frac{1}{2} - \frac{7\sqrt{17}}{34} - 2 + \frac{10\sqrt{17}}{17}\right)e^{-\left((\sqrt{17}+5)(t-1)/2\right)} + \left(\frac{1}{2} + \frac{7\sqrt{17}}{34}\right)e^{(\sqrt{17}-5)(t-1)/2} \\ + \left(\frac{10\sqrt{17}}{17} - 2\right)e^{-\left((\sqrt{17}+5)t/2\right)} - \left(\frac{10\sqrt{17}}{17} + 2\right)e^{-(\sqrt{17}+5)t/2} & \text{for } t > 1. \end{cases}$$

**B9.** $y(t) = \begin{cases} \frac{1}{9}e^{3t} - \frac{1}{3}t - \frac{1}{9} & \text{for } 0 \le t \le 1 \\ \frac{1}{9}e^{3t} - \frac{2}{9}e^{3(t-1)} + \frac{1}{3}t - \frac{5}{9} & \text{for } 1 \le t \le 2 \\ \frac{1}{9}e^{3t} - \frac{2}{9}e^{3(t-1)} + \frac{1}{9}e^{3(t-2)} & \text{for } t > 2. \end{cases}$

**B11. a.** $P(t) = \begin{cases} Ae^{kt} + \frac{h}{k}\left(1 - e^{kt}\right) & \text{for } 0 \le t \le 30 \\ Ae^{kt} + \frac{h}{k}\left(e^{-k(30-t)} - e^{kt}\right) & \text{for } t > 30. \end{cases}$

**b.** $A = \frac{h}{k}\left(e^{330k} - e^{360k}\right)\Big/\left(1 - e^{360k}\right)$.

**C1.** **a.** The graph of $W(t)$ is

**b.** $y(t) \equiv 0$.

**c.** $y(t) = \begin{cases} 0 & \text{for} \quad 0 \le t < 1 \\ \frac{1}{2}e^{2(1-t)} + \frac{1}{2} - e^{1-t} & \text{for} \quad 1 \le t \le 2 \\ \frac{1}{2}e^{2(1-t)} - e^{1-t} + e^{2-t} - \frac{1}{2}e^{2(2-t)} & \text{for} \quad t > 2. \end{cases}$

## Exercises 6.4

**A1.** $-1$.

**A3.** $e^{-\pi s} \cos \pi^3$.

**A5.** $y(t) = \begin{cases} 0 & \text{for} \quad t < 1 \\ e^{-8(t-1)} & \text{for} \quad 1 \le t < 2 \\ e^{-8(t-1)} + e^{-8(t-2)} & \text{for} \quad t \ge 2. \end{cases}$

**A7.** $y(t) = \frac{1}{6}\left(e^{2t} - e^{-4t}\right)$.

**A9.** $y(t) = e^{-t} + 2te^{-t} + \left[2(t-1)e^{-(t-1)}\right] U(t-1)$.

**A11.** $y(t) = \begin{cases} 1 & \text{for} \quad t < 2\pi \\ 1 + \sin t & \text{for} \quad t \ge 2\pi. \end{cases}$

**A13.** $y(t) = -\frac{1}{4}\left(2\sin(t-1) - e^{t-1} + e^{-(t-1)}\right) U(t-1)$.

**B1.** $y(x) = $ [after expanding and simplifying] $\begin{cases} \frac{W}{6EI}x^2\left(\frac{3}{2}L - x\right) & \text{for} \quad 0 \le x < \frac{L}{2} \\ \frac{WL^2}{24EI}\left(3x - \frac{L}{2}\right) & \text{for} \quad x \ge \frac{L}{2}. \end{cases}$

**C1. a.** $x(t) = \begin{cases} 0 & \text{for } t < a \\ \frac{H\sqrt{mk}}{m}\sin\left(\sqrt{\frac{k}{m}}(t-a)\right) & \text{for } t \geq a. \end{cases}$

  **b.** Because $m$ and $k$ are constants directly related to the spring-mass system, varying the value of $H$ affects the *amplitude* of the oscillations, the maximum distance of the mass from its equilibrium position. Viewed another way, in the original equation, the quantity $kH$ represents the *magnitude* of the jerk upward.

  **c.** If we want $A$ to be the maximum displacement of the mass from equilibrium, we must have $\frac{H\sqrt{mk}}{m} = A$, or $H = \frac{A\sqrt{mk}}{k}$.

**C3. a.** $\mathcal{L}[y(t)] = \left(1/(s^2+2)\right)\sum\limits_{n=1}^{\infty} e^{-ns}.$

  **b.** $y(t) = \left(1/\sqrt{2}\right)\sum\limits_{n=1}^{\infty} U(t-n)\sin\sqrt{2}(t-n).$

  **c.** The solution oscillates indefinitely.

## Exercises 6.5

**A1.** $y(t) = \frac{5}{2}e^{-2t} + \frac{1}{2}e^{-4t}.$

**A3.** $x(t) = 3e^{4t} + 5e^{-t}, y(t) = -2e^{4t} + 5e^{-t}.$

**A5.** $x(t) = 5e^t - 18te^t, y(t) = -3e^t - 54te^t.$

**A7.** $x(t) = e^t\cos 2t + \frac{1}{2}e^t\sin 2t,\ y(t) = x' - x = e^t\cos 2t - 2e^t\sin 2t.$

**A9.** $x(t) = -e^{-3t}(-2\sin 4t + \cos 4t), y(t) = 2e^{-3t}\cos 4t.$

**A11.** $x(t) = \frac{3}{10}e^{-t} + \frac{7}{10}e^{2t}\cos t - \frac{11}{10}e^{2t}\sin t, y(t) = -\frac{2}{5}e^{-t} + \frac{2}{5}e^t\cos t + \frac{9}{5}e^{2t}\sin t.$

**A13.** $x(t) = 3e^t - 9t^2 + 6t + 2,\ y(t) = -e^t - 6t.$

**B1.** $x(t) = \frac{4}{15}\cos^2\sqrt{3}t - \frac{2}{15} + \frac{4}{15}\sqrt{3}\sin\sqrt{3}t\cos\sqrt{3}t - \frac{2}{15}\cosh\sqrt{3}t + \frac{1}{15}\sqrt{3}\sinh\sqrt{3}t$

  $y(t) = -\frac{32}{15}\cos^2\sqrt{3}t + \frac{16}{15} + \frac{32}{45}\sqrt{3}\sin\sqrt{3}t\cos\sqrt{3}t + \frac{1}{15}\cosh\sqrt{3}t - \frac{2}{45}\sqrt{3}\sinh\sqrt{3}t.$

**B3.** $x(t) = \frac{11}{20}\cos 2t + \frac{9}{20}\cos\left(\sqrt{2}t\right), \theta(t) = \frac{11}{20}\cos 2t - \frac{9}{20}\cos\left(\sqrt{2}t\right).$

**B5.** $I_1(t) = \frac{11}{4} - \frac{1}{20}e^{-6t} - \frac{27}{10}e^{-t}, I_2(t) = \frac{3}{4} + \frac{3}{20}e^{-6t} - \frac{9}{10}e^{-t}.$

**C1.** $x(t) = \frac{1}{4}e^t - \frac{3}{4}e^{-t} - \frac{1}{2}\cos t + \sin t + e^{-2t}$

  $y(t) = \frac{1}{2}\cos t - \sin t + \frac{1}{4}e^t - \frac{3}{4}e^{-t}.$

## Exercises 6.6

**A1.** The solution $x(t) \to 0$ as $t \to \infty.$

**A3.** The solution $x(t)$ oscillates as $t \to \infty.$

**A5.** The solution $x(t)$ becomes unbounded as $t \to \infty$.

**A7.** There are oscillations with decreasing amplitudes: $x(t) \to 0$ as $t \to \infty$.

**A9.** There are transient terms in $x(t)$, but one pole is $s = 1$, implying that $x(t)$ becomes unbounded as $t \to \infty$.

**A11. a.** $\frac{1}{s^2-1}$.

    **b.** The poles are $s = \pm 1$.

    **c.** $x(t)$ becomes unbounded as $t$ grows large.

**A13. a.** $X = \frac{2s^3+10s^2+48s+44}{(s^2+2s+2)(s^2+4s+20)}$.

    **b.** $s = -1 \pm i, -2 \pm 4i$.

    **c.** $x(t) \to 0$ as $t \to \infty$.

**B1. a.** $\ddot{x} + \dot{x} + x = f(t)$.

    **b.** $x(t) = \frac{2}{13}\left[e^{-(t-1)/2}\cos\frac{\sqrt{3}}{2}(t-1)\right]U(t-1) + \frac{2}{13}\left[\frac{7\sqrt{3}}{3}e^{-(t-1)/2}\sin\frac{\sqrt{3}}{2}(t-1)\right]U(t-1)$
           $- \frac{2}{13}\left[\cos 2(t-1) + \frac{3}{2}\sin 2(t-1)\right]U(t-1)$.

**B5. a.** $\frac{1}{s^2-s-6}$.

    **b.** $\frac{1}{5}\left\{e^{3t} - e^{-2t}\right\}$.

    **c.** $y(t) = \frac{1}{5}e^{3t}\int_0^t e^{-3u}g(u)\,du - \frac{1}{5}e^{-2t}\int_0^t e^{2u}g(u)\,du + 2e^{3t} - e^{-2t}$.

**B7. a.** $\frac{1}{a_1 s + a_0}$.

# Exercises 7.1

**A1.** $(0,0)$ and $\left(\frac{1}{2}, 1\right)$.

**A3.** $(0,0)$, $(1, 1)$, and $(-1, 1)$.

**A5.** $(0,0)$ and $(-1, -1)$.

**A7.** $(0,0)$, $\left(0, \frac{3}{2}\right)$, $(1, 0)$, and $(-1, 2)$.

**A9.** $(0, 2n\pi)$ and $(2, (2n+1)\pi)$, $n = 0, \pm 1, \pm 2, \dots$.

**A11.** $(0,0)$, $(1, -1)$, and $(1, 1)$.

**A13.** The entire $x$-axis except the origin plus the point $(1, 1)$.

**B1.** $(0,0)$, $(1, 0)$, $(-1, 0)$, and $(-4, 0)$.

**B3.** $(0, a_2)$ and $\left(\frac{a_1(-1+a_1 a_2 - a_2)}{a_1 - 1}, \frac{1}{a_1 - 1}\right)$, provided that $a_1 \neq 1$.

**C1.** $(-1.016, 0.166)$, $(-0.798, -1.450)$, $(-0.259, -1.208)$, $(0.355, 1.551)$, $(0.634, 1.900)$, and $(1.085, -0.956)$.

## Exercises 7.2

**B1.  a.** With $u = 4$ and $v = 2$, the phase portrait is

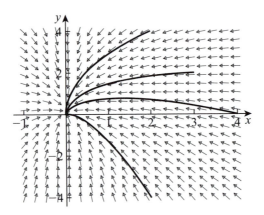

The boat eventually winds up at $(0, 0)$.

**B3.  a.** $\{\dot{x} = y, \dot{y} = 0.25x^2 - x\}$.

  **b.** Yes.

  **c.** $\{\dot{x} = y, \dot{y} = -x\}$.

**C1.  d.** If $a < 0$, then $\dot{r} < 0$, implying that a trajectory will spiral into $(0, 0)$, so that the origin is a *stable spiral point* (a sink). If $a = 0$, then $\dot{r} = 0$, so that $r$ is a constant and the origin is a *stable center.* If $a > 0$, the origin is an *unstable spiral point* (a source). In the language of Section 2.6, the parameter value $a = 0$ is a *bifurcation point.*

$a < 0$

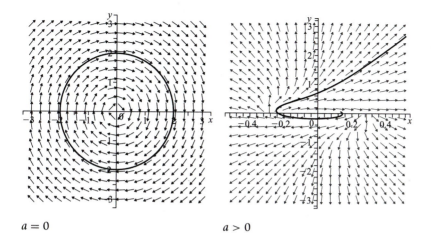

$a = 0$                                         $a > 0$

# Exercises 7.3

**A1.** The origin is a *spiral source*.

**A3.** The origin is a *stable node*, a *sink*.

**A5.** The origin is a *saddle point*.

**A7.** The origin is a *center*.

**A9.** The origin is a *spiral sink*.

**B1. a.** $(0, 0)$ and $(4, 1)$.

   **b.** $(0, 0)$ is a *saddle point*, whereas $(4, 1)$ is a *sink*.

**B3. a.** $(0, 0)$, $(2, 0)$, and $(9, 3)$.

   **b.** $(0, 0)$ is a *source*, $(2, 0)$ is a *sink*, and $(0, 3)$ is a *saddle point*.

**C1. a.** The only equilibrium point is $\left(1, \frac{a}{b}\right)$.

   **b.** $\dot{x} = (a-1)(x-1) + b\left(y - \frac{a}{b}\right), \dot{y} = -a(x-1) - b\left(y - \frac{a}{b}\right)$. Letting $u = x - 1$ and $v = y - \frac{a}{b}$, this becomes the system $\dot{u} = (a-1)u + bv, \dot{v} = -au - bv$.

   **c.** $\lambda = -\frac{1}{2} - \frac{b}{2} + \frac{a}{2} \pm \frac{1}{2}\sqrt{(a-b)^2 - 2(a+b) + 1}$.

   **d.** (1) $(1, 3)$ is a *spiral source*;   (2) $(1, 2/7)$ is a *sink*;   (3) $(1, 1/4)$ is a *sink*.

# Exercises 7.4

**A1.** The only nontrivial equilibrium point is $\left(\frac{1}{4}, \frac{3}{2}\right)$.

**A3.** The only nontrivial equilibrium point is $\left(\frac{3}{4}, \frac{1}{4}\right)$.

**A5.** The only nontrivial equilibrium point is $(3, 3)$.

**A7.** $[t = .1, x(t) = .905130981942487424, y(t) = 1.99036351756532892]$

   $[t = .2, x(t) = .820792722038854339, y(t) = 1.96310057595484810]$

$[t = .3, x(t) = .746918149815353428, y(t) = 1.92096012363624435]$

$[t = .4, x(t) = .682988624646712062, y(t) = 1.8668278660690446]$

$[t = .5, x(t) = .628223724693027674, y(t) = 1.80349863147592626]$

$[t = .6, x(t) = .581725902364229608, y(t) = 1.73353191257205697]$

$[t = .8, x(t) = .509916830817235156, y(t) = 1.58232127656877064]$

$[t = .9, x(t) = .482944734671241827, y(t) = 1.504546664756024171]$

$[t = 1, x(t) = .460967197688796904, y(t) = 1.42710548153511119]$.

**C1.** **c.** $y = \pm\sqrt{2\cos x + C}$.

**C3.** **a.** $\theta(t) = \sin 2t$.

   **b.** The period of $\sin 2t$ is $2\pi 2 = \pi$.

   **c.** 19 ticks.

   **d.** Halving the length of the pendulum reduces the amplitude and the period by a factor of $\frac{\sqrt{2}}{2} = \sqrt{\frac{1}{2}}$. In other words, shortening the pendulum makes it run faster, yielding more ticks per minute. (In this case, the clock will tick 27 times per minute.)

## Exercises 7.5

**A3.** **a.**

   **b.**

  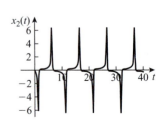

**c.** Each trajectory indicates the existence of a stable limit cycle. However, the shapes of the trajectories and the limit cycles change as $\varepsilon$ changes. Similarly, $x_1(t)$ and $x_2(t)$ are periodic but not trigonometric; and when $\varepsilon$ changes from $1/4$ to $4$, $x_1(t)$ changes to a flatter shape, while $x_2(t)$ develops spikes.

**B3. a.** $r(t) = 1 + \sqrt{\frac{1}{Ce^{2t}-1}}$.

**b.** $\theta(t) = t + C$.

**c.** $(x(t), y(t)) = (r(t)\cos\theta(t), r(t)\sin\theta(t))$.

**B5.**

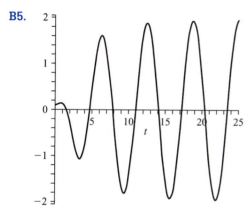

**B11. a.** $\{\dot{x}_1 = x_2, \dot{x}_2 = -a(x_1)x_2 - b(x_1)\}$.

**C1. a.** The only equilibrium points are $(0,0)$ and $(-2,-1)$.

**b.** The phase portrait near the origin and near the point $(-2,-1)$ looks like

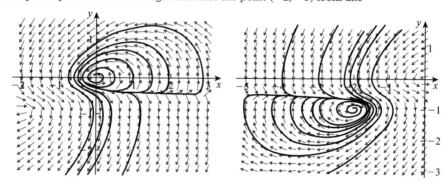

**c.** There seems to be an unstable limit cycle around $(0,0)$ (a source) and a stable limit cycle (a sink) around $(-2,-1)$.

# Index

## A

Above, 110
Absolute error, 99, 106, 399
Absolute value, 427
Adaptive methods, 124, 129
Air pollution, 51–53
   rates, 52
Airy functions
   of first kind, 433
   of second kind, 433
Airy, George Bidell, 432
Airy's equation, 432
Algebra, 304
   complex numbers and,
     425–426
   linear, 221
   matrix, 224–225, 414–415
   systems, 320
   vector, 411–414
Algorithm, Runge-Kutta-Fehlberg,
   123, 128–129, 210–213
Allen's Speculative Model, 96
Almost linear systems, 360–363,
   395
Amounts, 165, 403
Amplitude, 137
   of oscillations, 200, 392
   time-varying, 195
Amplitude-phase angle form,
   136–138
Analytic, 431
Analytic solutions, 27
Andrews, George E., 440
Angles, phase, 137
Angular velocity, 379, 389f

Antiderivatives, 403
Approximation methods, 160
   at equilibrium points,
     358–367
   Euler's method and, 358
   five-step, 106f
   for integrals, 103f
   local linearity, 101
   misleading, 359–360
   numerical, 99–129
   slope fields and, 364
   for solutions, 127
   tangent line, 399–400
   Taylor polynomial, 361, 362,
     400–403
   three-step, 101, 102f
   trajectories and, 364, 365
Arbitrary constants, 290
Argand, Robert, 427
Arms race model, 180–181
Askey, Richard, 440
Associated homogeneous
   equations, 141–142
Associated linear system, 363,
   395
Associative law, 420
Associative rule, 225
Asymptotes, 236
Asymptotically stable
   equilibrium points, 372
Asymptotically stable solutions,
   74, 96
Attractors, 74
Autonomous equations, 60, 95
   slope field for, 63–64

Autonomous nonhomogeneous
   systems, stability of,
     276–277
Autonomous systems, 172
   converting nonautonomous
     equations into, 168–169
   nonlinear, 361
   qualitative analysis, 173–185
   three-dimensional, 169
Auxiliary equation, 132–133
Averages, weighted, 122–123

## B

Backward difference quotients,
   129
Backward Euler formula, 129
Balance equation, 49
Balance law, 49
Basic matrix algebra, 414–415
Basic notions, 8
Bending movement, 327
Bendixson's criterion, 393
Bernoulli equation, 46
Bessel, Friedrich Wilhelm,
   435n2
Bessel functions, 111
   of first kind, 437
   *Maple* and, 438
   *Mathematica* and, 438
Bessel's equation, 435
Bifurcations, 81–88
   diagram, 81
   differential equations and,
     81–82
   pitchfork, 85